WIRELESS COMMUNICATIONS
Theory and Techniques

Related Recent Titles

WIRELESS COMMUNICATIONS
Theory and Techniques

by

Asrar U. H. Sheikh

KLUWER ACADEMIC PUBLISHERS
Boston / Dordrecht / New York / London

Distributors for North, Central and South America:
Kluwer Academic Publishers
101 Philip Drive
Assinippi Park
Norwell, Massachusetts 02061 USA
Telephone (781) 871-6600
Fax (781) 871-6528
E-Mail: <kluwer@wkap.com>

Distributors for all other countries:
Kluwer Academic Publishers Group
Post Office Box 322
3300 AH Dordrecht, THE NETHERLANDS
Telephone 31 78 6576 000
Fax 31 78 6576 254
E-Mail: <services@wkap.nl>

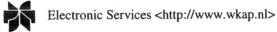 Electronic Services <http://www.wkap.nl>

Library of Congress Cataloging-in-Publication Data

Wireless Communications: Theory and Techniques
Asrar U. H. Sheikh
ISBN 1-4020-7621-5

Printed on acid-free paper.

Printed in the United States of America

DEDICATION

This book is dedicated to my wife Parveen, my children Farhana, Fahim and Samia. It would have been impossible to write this book unless my family had not forgone their claim on the time they deserved.

Contents

Chapter 3   83

MICRO-CHARACTERIZATION OF WIRELESS CHANNELS

Chapter 4

149

SIGNAL TRANSMISSION OVER MOBILE RADIO CHANNELS

Chapter 5

INTERFERENCE, DISTORTION AND NOISE

Chapter 6

ANTENNAS FOR MOBILE RADIO SYSTEMS

Chapter 7

SIGNAL PROCESSING IN WIRELESS COMMUNICATIONS

Chapter 8   387

MULTIPLE ACCESS COMMUNICATIONS

Chapter 9

443

MOBILE WIRELESS SYSTEMS AND SERVICES

Chapter 10   **457**

WIRELESS DATA SYSTEMS

Chapter 11   **493**

WIRELESS CELLULAR SYSTEM DESIGN PRINCIPLES

Chapter 12

MOBILE WIRELESS CELLULAR SYSTEMS

Chapter 14

705

WIRELESS SYSTEMS BEYOND 3G

Preface

Wireless communication systems, since their inception in the form of cellular communications, have spread rapidly throughout the western world and the trend is catching on in the developing countries as well. These systems have caused revolutionary changes in the way we live. Cellular Communications have become important both as means of communication and as a new domain of commercial enterprise. Hand held telephones are now rapidly replacing the fixed telephone and in less than twenty years, the number of subscribers has reached nearly three quarters of a billion. In a short span of twenty years, the cellular communications progressed from the first generation to the third generation systems, which started operations in Japan on October 1, 2001. The first generation wireless technology, which was thought to be obsolete is now being used for fixed wired telephony in several countries of Asia, Africa and Latin America. As some commentator said in 1983, the cellular system is the best thing that has happened in telecommunications since the introduction of computers to the masses.

This book is written to provide readers with the fundamental concepts of wireless communications. It is intended for a graduate course on wireless communications but it could be easily adopted at the senior level by skipping material involving difficult mathematical manipulations. The text does not go through the rigorous material on mathematical treatment of electromagnetic waves and propagation, rather it emphasizes more on the practical aspects of this. Besides providing some fundamental concepts, the text goes into discussions on implications of these principles on the wireless system design.

The book is divided into three parts. Following the introductory chapter on the overview of wireless communications, Chapters 2 and 3 examine the arduous radio signal environment in which mobile communications has to provide the subscriber with reliable telecommunication services. The second chapter reviews the basic radio wave propagation theory with a view to predict signal strength at any point within and beyond the radio coverage area of a transmitter. This is termed macro (large scale) characterization of mobile radio channels. The starting point is a theoretical model accurate only under ideal conditions, e.g., a planar perfectly conducting surface surrounding the transmitter. Modifications are then introduced to account for the effects of terrain variation on signal strength. Methods are presented for the prediction of mean signal strength for a given transmitter-receiver geometry. Chapter 3 proceeds with a study of the micro (small scale) structure of signal variations caused primarily by scatterers in the immediate vicinity of the mobile unit. The starting point is the channel impulse response. The length and variability of the impulse response characterizes channel delay spread, fading and path strengths. In practice, vehicle movement causes rapid variations about a mean, and deep fades will be encountered, which preclude reliable communications over the channel unless sophisticated signal processing methods are used.

The second part of this book consists of five chapters that consider methods of transmission, reception, and processing of signals, which had been affected by the harsh propagation conditions. Chapter 4 examines signal transmission and reception techniques. The main emphasis in this chapter is on performance analysis under fading and frequency selective conditions. The performance of several digital modulation systems are examined under the above mentioned conditions. Interference in wireless communications has a very significant impact on the performance, quality of service and system capacity. It is a major cause of inefficiencies in spectrum utilization. Chapter 5 goes into considerable depth of interference, distortion and noise. Interference, modeling, signal outage because of channel impairments are the main topics of discussion. The impact of interference and channel fading on frequency reuse and receiver performance is analyzed.

Antennas are essential part of any wireless communications. Chapter 6 describes the various types of antennas and their characteristics. Again rigorous mathematical treatment is avoided. The chapter considers methods whereby a suitable antenna system can be employed to combat fading and substantially enhance performance. It is seen that fading can severely degrade the performance of various modulation schemes. This theme is followed up in Chapter 7 in the context of signal processing. Multiple signals distinguished, for example, by frequency, space, time or carrier polarization can be combined to minimize the effect of fading. The impact of diversity on the signal restoration is emphasized. The role of equalization, coding and interleaving in improving the performance is discussed in this chapter. Multiple access communications is introduced in Chapter 8. It discusses the methods of using spectrum resource in the multi-user environment. In this chapter, analyses on the impact of access schemes on the system capacity are presented. This chapter sets the scene for the requirements on multiuser communications introduced in the third part of the book. Conventional and multichannel trunked systems in addition to modern cellular systems are introduced.

The third part of the book consists of six chapters mostly on system aspects. This part of the book does not have solved examples or problems.

Chapter 9 introduces the principles of several wireless systems. It describes, paging, private land mobile systems and gives a brief account of wireless systems. The chapter takes a broader perspective, dealing with details on the implementation of systems operating in the mobile wireless environment. A closely related topic, treated in Chapter 10, is that of Private Mobile Data Communications Systems (PMDCS) such as packet switched radio, and voice-data integrated systems. An important application of packet switched radio is the mobile data distribution network, which has both civilian and military applications. Other applications include transportation, public safety, ambulance, and wireless local area networks (WLANS) and high speed wireless networks using ATM technology. Chapter 11 outlines the principles of designing cellular systems. It makes distinction between designing a TDMA and CDMA systems. Principles of mobile wireless systems; frequency assignment, control architecture, vehicle location and handoff are included in the discussion.

The details on several mobile communications systems in operation are explored in Chapter 12. In particular, the first two generations of cellular systems are described. These include AMPS, GSM, and CDMA. The new emerging applications in telecommunications is steering wireless communications towards multimedia wireless communications. In Chapter 13, five new standards for 3G are described. However, WCDMA (IMT-DS) is described in some detail. Chapter 14 examines a logical extension of the cellular network concept, that is systems beyond 3G. The chapter presents a window into the future by speculating on the shape of the future networks. This chapter also looks at technologies suitable for the wideband 4G systems.

I would like to acknowledge the contributions made by several of my present and former colleagues who encouraged me into writing the book. In particular, I would like to thank Professor J.S. Riordon who agreed to co-author this book. Unfortunately, he could not continue because of his heavy commitments to the administration and later due to my departure from Carleton University, Ottawa, Canada. However, his contributions at the start of this project are much appreciated. My thanks also go to Mohammad Abdullah Bugshan who persuaded Lucent Technologies in creating the Chair in Telecommunications that I currently hold. I would like to acknowledge the support of King Fahd University of Petroleum and Minerals in providing research facilities and reducing my teaching load to facilitate completion of this project. I should not forget to thank my students who kept me on toes by asking interesting questions and insisting on seeing the book in print.
Asrar U. H. Sheikh
Dhahran, Saudi Arabia
June 30, 2003

Chapter 1

OVERVIEW OF WIRELESS COMMUNICATIONS

The communication of ideas is essential for a society to function and develop. Since the beginning of time, humans have been laboring to find efficient methods to communicate. The exchange of information between humans is believed to have originated through the use of incomprehensible audio sounds accompanied by some sign language. Written communication began with the use of images/pictures depicting objects as these were seen by the people. From there on pictorial languages evolved into words which were simpler to write and reproduce†. The written form has since become the dominant mode by which knowledge has been passed down from generation to generation.

Communications may be categorized into three classes: interactive, stored, and broadcast [1]. Face to face communication is the most natural form of interactive communication. Interactive communication defines the transmission of messages, which either have been influenced by the previously received messages, or influences the nature of the future messages. Today, interactive communication is demonstrated through the use of the telephones and remote computer access terminals.

The stored message, written or otherwise, may be copied, disseminated, and stored. Beginning with etching of stones and making of tablets, the stored information became the most important mode of communications upon which humans built civilizations. Over the years, several different modes of information storage have been added to this category. These include the letter, the book, the gramophone record, the cine film, the magnetic tape, the disc, and the compact disc. The exponential growth in information, requires development of high density small in size storage media.

Remote broadcasting, the third category began with a shout to propagate messages and progressed through the use of smoke signals to the development of the radio and television. The latter two forms are the modern means of broadcasting information. The communication in this case is instant, that is it is received effectively at the same instance at which it is transmitted. It is interesting to note that on December 12, 2001, the hundredth birthday of

---

†. *For example, Chinese written language is developed from pictorial representation of objects. We could image the names of the objects as code words.*

radio communications - a technology which was thought to be useless in 1900 was celebrated [2]. During the past sixty years, radio and television broadcasting developed as dominant source of information. The television broadcast on a single local channel broadcasting a few hours a day nearly fifty years ago has grown to become a truly global network operating in excess of 100 channels 24 hours a day. By the year 2010, a network of five hundred channels operating around the clock is a distinct possibility. The instant dissemination of information is significantly influencing practically every aspect of human life; in particular the business scene has become so dynamic that significant changes could be expected every minute. In order to manage business and personal affairs more effectively, a continuous watch on the political and social developments around the globe has become necessary. Some television broadcast stations have been broadcasting these developments on a continuous basis.

International Switched Telephone Networks (STN) dominate in providing interactive personal and business communications. The convergence of computer and communication technologies in the modern digital telecommunication networks has created many possibilities whereby integration of voice, data, and image information is taking place. Modern switched telephone networks acquire greater intelligence to provide a larger assortment of services by simply upgrading the network software. These networks now offer a greater variety of services than ever before; call identification, call forwarding, conference calls, call waiting, call blocking, group calls and voice mail are now being offered as a matter of routine. Depending on the length of the local loop, ISDN carry data at rates from 64 kbits/sec to 10 Mbits/sec. Asynchronous Transfer mode (ATM) technologies are now implemented for rates in hundreds of megabits/sec to provide high speed data and image services.

These rapid advances in telecommunications have been brought on by a revolution in VLSI electronics, computers, and software technologies. The revolution continues and already we are observing integration of the three categories of communications mentioned earlier in the form of multimedia communications. The introduction of intelligent networks in the Switched Telephone Network (STN) allows the users to demand bandwidth that matches their application. Certainly, these developments will have a major influence on the way people interact, behave, and live; noticeable changes are already beginning to occur due to the availability of Internet access, electronic super highways, and worldwide web (WWW). Internet is now being used for banking, shopping, and commerce. Internet is also said to be responsible for the current information explosion. It is estimated that by year 2005, the information volume available on the Internet will double every 67 days.

In parallel with these advances in information technologies, varied modes of transportation have also seen rapid development in the twentieth century. These technologies have spurred a rise in international borderless businesses spanning the world with the result that society has become more mobile with the concomitant need of individuals to maintain links with the office and the family from varying, remote, and often mobile locations. Two important trends are worth noting at this point. One is the increase in the use of digitally encoded text and video messages in interactive communications. The second is the availability of greatly enhanced capability of signal processing using very large scale integration (VLSI) of circuitry because of impressive advances in microelectronics. Taken together, these trends are leading the way towards the development of widespread, low cost, reliable mobile wireless communications networks.

The convergence of computers, telecommunications and radio that took place in the past decade has resulted in proliferation of new systems that target new applications in telecommunications, law enforcement, emergency medical services, and transportation. The telecommunication sector has seen major advances, but cellular mobile systems, personal communication systems, and in particular wireless local area networks have gone through a phenomenal growth. The first generation of cellular networks matured rapidly, and was quickly replaced by the second generation. The second generation networks have became congested in some areas with unforeseen penetration rates of around 80%. The third generation (3G) systems became operational in Japan on October 1, 2001. The 3G mobile communication networks are designed to offer multimedia services. Such networks are beginning to form one of the major integrating factors referred to earlier. In summary, the wireless mobile networks will continue to play an increasingly important role in everyday life.

This book introduces the subject of mobile wireless communication systems and networks with an emphasis on theory, design principles, and applications. In particular, fundamentals of transmission of radio signals to and from mobile terminals over mobile radio channels are considered in much detail. The terminal mobility in an environment consisting of randomly located and oriented obstacles give rise to channel impairments of fading, shadowing, multipath propagation, interference and noise. These impairments sometimes are so severe that reliable communication becomes very difficult. It becomes necessary that countermeasures to these impairments be used to recover intelligible messages. Furthermore, details on how these strategies are used in the development of practical systems will provide guidelines for the design of newly emerging systems. The advancements in signal processing during the past two decades have contributed enormously to make communications over radio channels reliable. A chapter is included to familiarize the reader various signal processing techniques used in mobile communications. A large portion of the book describes cellular networks and their development in the future. At this stage, however, it may be of some interest to the reader to become familiar with the historical perspective of mobile wireless communications.

1.1 Historical Perspective

It has long been recognized that radio is the only practical method of establishing communication between mobile users[†]. As early as 1910, a steam bus was fitted experimentally with a large antenna and wireless equipment [3]. The frequency of operation was about 2 MHz and amplitude modulation was used for transmission. With the primitive technology of the day, the resulting radio equipment was bulky and impractical; however, further attempts continued to reduce the size of the equipment and improve portability.

In 1920, the Detroit Police Force commissioned the development of a mobile radio system. It is interesting to note that mobile communications was invented to fight crime and radio communications and it still forms the backbone of law enforcement agencies. The first mobile receiver was installed in 1928, and one-way communication was achieved. By 1930, police radio existed in 29 American cities and eight one way paging frequencies were allocated for the exclusive use of the police. The first mobile communication system that allowed two-way communications was licensed to the Bayonne, New Jersey Police Department in 1932.

The operation of these systems revealed several important technical problems. The major difficulty was the lack of portability due to the large size equipment. Reduction in antenna size required the use of higher frequencies, which in turn awaited advances in high frequency electronic circuit components and design techniques. Another problem was interference. In part this was again a reflection of the state of equipment design; in addition the use of amplitude modulation rendered systems susceptible to noise, channel variability and fading.

Technical advances made inroads into each of these areas. In 1933 the frequency of operation was raised from 2 MHz to 35 MHz. In the late 1930's Professor D. E. Noble of the University of Connecticut used frequency modu-

---

[†]. *We obviously exclude interactive face to face verbal communications. This may be regarded, in strict sense, a wireless communication system. Underwater communication is also excluded. In this book, we shall confine ourselves to mobile communications using radio waves.*

lation for the first time in the Connecticut State Police Mobile Communication System. The trial was a resounding success. The value of FM modulation in reducing noise and interference by a significant margin was immediately recognized, since this has been the dominant modulation format used in mobile radio systems. By the close of the 1930's, mobile dispatch systems had spread to fire departments and taxicabs; the major emphasis, however, remained on police communications. For instance, in 1934 the FCC reported that on eleven frequencies there were 194 municipal radio stations, 58 state police stations, and 5260 police cars equipped with receivers [4]. However, with the increased use of mobile communications, a difficult problem of spectrum scarcity emerged. With the exception of terminal size, the other two problems of interference and spectrum scarcity are still of major concern.

In parallel to police communications, the demand for public mobile telephone systems started to emerge. To meet the growing demand of mobile telephony, 11 channels were carved out of the spectrum allocated in 150 MHz band by reducing the channel bandwidth from 60 kHz to 30 kHz. Up until 1949, mobile telephone services were offered by telephone companies; that year the FCC also authorized Radio Common Carriers (RCCs) to offer mobile services to the public. In 1956, 12 additional channels were added in the 450 MHz band. The push-to-talk mode of mobile telephony was replaced by automatic system in 1964. This allowed the users to dial directly[†].

1.2 Services

Until recently mobile wireless systems were defined on the basis of applications. These were (a) one-way paging; (b) conventional dispatch; (c) trunked; and (d) cellular. More recently, the systems have been classified taking a broader view of offered services. The services are now divided into three distinct classes: voice oriented, data-oriented and multimedia services. In the voice oriented services, we may include two way simplex push to talk conventional mobile radio systems (CMRS), two way full duplex mobile telephony in the form of high power Vehicular Mobile Cellular Communications Systems (VMCCS), and low power cordless and Personal Communication Systems (PCS). In the second category of data oriented services, we include one way paging, mobile radio data services (MRDS) including cellular data services, and high speed wireless local area networks. In the third category of multimedia, the emphasis has been to integrate voice, data and image services, thus blurring the distinction between voice and data services. The application of multimedia services are found in situations where interactive

---

† Nearly all the developments described here pertain to U.S.A.

communications using voice, data, and image becomes necessary. Remote medical diagnostic service is an example of multimedia service, which include other emerging applications like Internet web browsing, mobile computing, and mobile commerce (MCom).

In conventional mobile radio systems (CMRS), the base station and mobile units typically operate on the same frequency. Simplex technology allows the use of only one voice path; those using it take turns, giving rise to the well known 'push-to-talk' protocol. Conventional broadcast techniques are normally used for this purpose. Broadcasting is characterized by low efficiency and lack of privacy. These services sometimes known as dispatch services, involve brief two-way communications, typically used for deploying and monitoring vehicles such as taxis and trucking fleets. More advanced dispatch systems use terrestrial as well as satellite systems and cover very wide areas. Monitoring systems for commercial trucks operating in Continental Europe are better examples of present day dispatch systems.

Interconnection with the wired telephone and the ability to call or be called by any other telephone, fixed or mobile characterizes the mobile telephone service. Trunked systems use broadcast-like conventional systems but have access to tens of channels controlled by a central computer. The mode of operation, however, is full duplex. Users request permission to transmit and a computer assigns the first available unused channel. This, alleviating the two major drawbacks in CMRS, increases both privacy and spectral efficiency. A higher spectrum efficiency is achieved by using larger trunking groups (number of channels available in a group). This will be discussed further in Chapter 8.

Cellular systems achieve high spectral efficiency through channel trunking and the reuse of each channel many times over within a service area. The service area is divided up into "cells", each having a diameter anywhere between one to twenty kilometers. The smaller cell sizes, 1 to 5 km diameter are called microcells, the larger, macrocells. Each individual, cell operates like a trunked system with a given (trunking) set of frequencies. Ideally, the transmitter power of each base station is adequate only to cover the entire cell area. Adjacent cells use distinct sets of channels, but the same frequency may be used again in a non-adjacent cell. The continuation of calls while users roam through the service area is an essential feature of the cellular system. A mobile user leaving a cell and entering another is handed over to the new cell without any interruption to the call. This feature requires the establishment of a new duplex radio link with the new cell site; to do this a complex network management procedure is followed.

In data oriented services, the simplest is the one-way paging system. One-way paging employs a central base station, which transmits a coded signal to a designated mobile unit. Usually this takes the form of the familiar beeper that alerts its user to contact the base station for a message using another communication facility. The critical characteristic is that the user can not respond through the mobile unit. Short messages allow for high spectral efficiency and low cost per unit; however, very short messages are also undesirable as they rob the system of its efficiency due to relatively high overhead. The paging overhead consists of preamble, addressing and protection bits. The enhanced paging systems allow users to display a message or to store it for future use.

Several new systems providing data services have been developed more recently. These include Private Mobile Radio Systems (PMRS) where data rates up to 19.2 kbits/sec are used in many fixed point to point and mobile applications like credit card verifications, point of sale transactions, vehicle monitoring and stolen vehicle recovery systems, rapid transit systems and railways, and after sale maintenance support. Several commercial systems (APCO, TETRA, etc.), though incompatible with each other, have been installed.

All of the above mentioned services, can be offered effectively within a cellular environment. Many new applications are beginning to emerge; cellular digital packet data (CDPD) where data services are offered over the first generation analog system is an important example of some newly emerging services. The distinction, between voice oriented and data oriented systems is now beginning to disappear. The analog cellular system primarily designed for voice has been superseded by second generation digital systems - narrowband TDMA systems in North America (DAMPS) and Pacific Digital Cellular (PDC) in Japan, and Global System for Mobile communications (GSM) in Europe and around the world. These systems can be used to provide both voice and data services. New services such as Short Message Service (SMS) and *i*-Mode are experiencing spectacular growths.

The entry of new direct sequence code division multiplex (DS-CDMA) IS-95 system fueled the debate on the relative merits of the two competing TDMA and CDMA systems. We may classify IS-95 belonging to two and a half generation system. Licences to operate Personal Communications Systems (PCS) have been awarded in the United States, Canada and other countries around the world. These systems belonging to the second generation systems, are low power systems, and operate in the 1.8 to 1.9 GHz band. The third generation systems have been under study for several years and in June 1998 eleven alternative systems were proposed as IMT-2000 candidates. Since the third generation systems are expected to be truly global, four mobile satellite systems were also proposed. The proposed systems were thoroughly evaluated by simulations and field trials. Ultimately, a family of five air interface standards consisting of IMT-DS, IMT-SC, IMT-TF, IMT-MC, IMT-TC were selected. The dream of approving a single universally acceptable standard, however, remains unfulfilled.

Wireless LANs (802.11b standard for 11 Mbits/sec LAN) have been developed for easier implementation and to save the cost of permanently wiring buildings. For wireless LAN applications, the Industrial Scientific and Medical (ISM) bands in 900 MHz, 2.4 GHz, 5.7 GHz and other frequencies in the 20-60 GHz band are considered to be the choice bands. Currently, high speed wireless networks based on ATM technology are being developed. Wireless TV distribution networks, local multipoint communication systems (LMCS) or wireless local loop (WLL), are now operational in several areas in the United States. These systems with the potential of offering on demand low rate data to high quality video services, are likely to replace wire or cable to home and the last kilometer will become wireless.

1.3 Spectrum and Management Issues

The scarcity of spectrum assigned to mobile radio, and indeed allocations to other communication systems such as commercial broadcast, had remained a significant hurdle in problem. Consequently spectrum use continues to be tightly controlled by governmental regulatory agencies. As technology developed, new and higher frequencies became feasible for commercial use, and new spectrum allocations were therefore made from time to time. In 1974, in the United States, the FCC allocated 40 MHz in the 800-900 MHz band on a temporary basis for experiments to demonstrate the feasibility of a spectrum efficient concept known as cellular mobile radio. This allocation consisted of a total of 666 thirty kHz wide duplex channels assigned between 825 and 890 MHz. Each channel consisted of a 25 kHz FM signal with a five kHz guard band. Under this mandate AT&T and American Radio Telephone Service (ARTS) installed experimental cellular mobile systems in Chicago and the Washington-Baltimore-Philadelphia area respectively.

Following the decision in 1979 by the World Administrative Radio Conference (WARC), to allocate much of the 800-900 MHz band to mobile communications, the FCC confirmed this temporary spectral allocation for commercial cellular system use in 1981. In 1983, first commercial mobile cellular system was installed in the United States. The next year saw an additional allocation of 12 MHz in two blocks, 845-851 MHz and 890-896 MHz. By July 1986, the demand of cellular radio reached to a level that additional spectrum of 10 MHz had to be allocated in the United States in the areas of high traffic congestion. Table 1.1 shows the allocated frequencies in the United States since 1970 for domestic public mobile radio use on land [5].

Table 1.1   ***Frequency Assignments for Mobile Communications***

System	Base transmit (MHz)	Mobile Transmit (MHz)	Number of Channels	Bandwidth
IMTS	35.26-35.66	43.26-43.66	10	40 kHz
IMTS	152.51-152.81	157.77-158.07	11	30 kHz
IMTS	454.375-454.654	59.375-459.65	12	25 kHz
Cellular	869-894	824-849	832	30 kHz
PCS	1850-1910	1930-1990	Operator's choice	Operator's choice
902-928 MHz, 2.4-2.5 GHz, 4.2 to 4.28, and 5.7 GHz bands are ISM bands				

In a more recent WARC'92 meeting, further allocations in the 900 MHz and 1.9 GHz were made for licensed and unlicensed systems. The allocation (1850-1910 and 1930-1990 MHz) is primarily licensed for newly emerging Personal Communications systems; 2.11-2.15 and 2.16-2.20 GHz bands were added to this allocation. In WRC'2000 meeting in Istanbul, Turkey, additional 175 MHz were added to the spectrum for mobile communications. Within the 2 GHz band, allocations to various services are summarized in Table 1.2. The spectrum allocation in the 1800-2250 MHz band in different regions of the world are shown in Figure 1.1. It is seen that allocation to PCS systems in the US conflicts with the allocation to IMT-2000/UMT in the rest of the world. This may create some problem in realizing the dream of totally transparent global roaming.

The unlicensed ISM (Industrial, Scientific and Medical) bands (902-928 MHz, 2.4 to 2.48 GHz, 4.2 to 4.28 GHz, and 5.7 GHz) are intended for uncoordinated indoor systems. A maximum transmit power is limited to 1 W for systems operating in ISM band. Bluetooth is such a system that uses 2.45 GHz band to provide interconnection with other Bluetooth devices and with home and business phones and computers. In addition to data, up to three voice channels are available. Each device has a unique 48-bit address from the IEEE 802 standard. Connections over a limited range of 10 meters can be point to point or multipoint. Attempts are continuing to explore higher frequency bands, and the possibility of some form of mobile systems operating in 20-60 GHz band is real.

| Table 1.2 | *Allocations in 2 GHz Band* |

Frequency, MHz	Use and Status
1710 -1850	Federal Government's fixed-services band
1850 - 1990	Allocated to Emerging Technologies (1850-1865/1930-1945;1865-1880/1945-1960;1880-1895/1960-1975)
1880 - 1900	Allocated in much of Europe to Digital European Cordless Telecommunications. (DECT)
1885 - 2025	Designated by ITU for Future Public Land Mobile Telecommunications Services (FLPMTS), a globally compatible PCS network. Japan has allocated 1900 - 1911 to Personal Handi Phone (PHP).
1910 - 2200	The European Community has proposed PCS allocations in this band. 1910-1930 allocated by the FCC to Emerging Technologies, e.g. for unlicensed PCS devices for cordless telephony and data communications. 1990-2110 allocated for used in broadcast auxiliary systems. 2110-2130 previously allocated to point to point microwave service e.g. for short haul interconnection by cellular telephone companies, now reallocated by the FCC to emerging technologies 2110-2200 designated by the ITU for the FPLMTS 2130-2150 a private operational fixed services band reallocated by the FCC to Emerging Technologies 2150-2160 Used by microwave distribution service (Wireless Cable) for entertainment and educational programming 2160-2180 Point to Point Microwave Service, used for short haul interconnection by cellular companies (reallocated to Emerging Technologies) 2180-2200 A Private Operational Fixed Service band reallocated to Emerging Technologies.

In addition to terrestrial mobile communications, satellite based mobile communication systems have been developed. Mobile services are provided by INMARSAT and MSAT systems [2]. The mobile satellite program (M-SAT) in Canada provides voice and data services to remote areas in L-band (1.5 GHz). Low orbit satellites are considered to be ideal platforms for providing personal communications services. The Iridium[†] system consisting of 66 low earth orbit satellites for universal access is in operation. Globalstar,

---

† *At the time of writing, Iridium system went out of business because of lack of subscribers, high development cost, expensive handset, and keen competition from terrestrial cellular systems. Another company has acquired the system and it is operational under different name. Global star has started offering satellite service in 2001.*

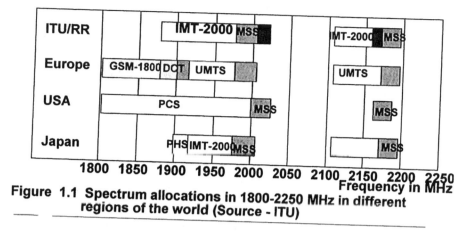

Figure 1.1 Spectrum allocations in 1800-2250 MHz in different regions of the world (Source - ITU)

with 48 satellites is now operational and ICO another system is currently in the planning stage.

1.4 Technologies

The phenomenal rapid growth in mobile wireless communications in recent years has spurred the development of new modulation methods, forward error control coding, voice and video compression techniques, and signal processing technologies. The goal is to make the use of the limited amount of available spectrum more efficient and to minimize the impact of interference and fading. These objectives have been the principal reasons behind recent advances in wireless technologies.

Low bit rate voice encoders and efficient modulation techniques are desirable in transmission of maximum number of bits/Hz with a minimum signal power penalties. To this end, efforts are underway to develop 4.8 kbits/sec or lower high quality voice encoder. Also, advances have been made in video compression techniques.

Interference is a natural consequence of extensive frequency reuse, which is essential in increasing spectrum utilization. An efficient control of inter-channel interference could improve spectrum efficiency. Several technologies, including equalization, adaptive antennas, and optimum detection receivers are being examined. Other interference rejecting signal processing techniques include smart antennas to create nulls in the direction of arrival of interference and diversity. The receiver technology is changing from specifically designed radios to a generic receiver structure which could be programmed into a receiver for a specific system. This technology, known as software radio technology, opens doors for the development of multi-standard terminals and may facilitate intercontinental roaming. New signal transmission schemes like Multicarrier CDMA are being implemented to increase frequency reuse factor.

In the network area, high speed integrated networks are the ultimate goal. Intelligent networks will play significant role in optimizing network resources and mobility management.

1.5 Prognosis and Challenges

Since its inception, the demand for cellular communications has exceeded every prediction, and has caused frequency congestion so severe that, with the present allocations, future subscriber demand cannot be met, unless new spectrum conservation methods are invented. To meet the demand, second generation cellular systems have been designed for Europe, North America, and Japan. In Europe Global System for Mobile Communications (GSM), based on 200 kHz bandwidth TDMA, has been operational for more than ten years; and is emerging as a defacto world standard for cellular communications. By the end of 2000, more than 170 countries had adopted GSM as the standard system for mobile communications. The other second generation systems, the North American NBTDMA IS-54 and DS-CDMA IS-95 are also has been operational in several areas since July 1993. The IS-54 is mostly confined to the United States, Canada and Mexico. South Korea has the largest DS-CDMA network in the world.

By December 2000, there were more than 550 million cellular mobile radio customers worldwide. The cellular mobile radio customers are projected to exceed 900 million by the end of 2003. The distribution of users in different regions of the world are shown in Figure 1.2.

The demand for mobile telephones is outstripping supply; as the latter can not cope with the annual growth of mobile communications touching 20%. In Scandinavian countries, land mobile communications has a penetration rate (the number of subscribers per 100 eligible subscribers) between 75 and 81%. The penetration rate in Germany is expected to surpass 80% by 2002. Japan and Hong Kong are not far behind with current penetration rates of 54% and 51%. Many first and second generation systems are currently in operation, the third generation systems became operational on October 1, 2001.

Future applications of mobile communications systems are numerous, including portable telephones (personal communications), more sophisticated

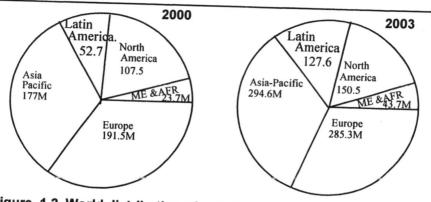

Figure 1.2 World distribution of cellular system users and forecast

pagers, portable data terminals and information retrieval systems for sales, law enforcement and medical emergency systems. Other applications such as the provision of communication facilities to remote areas through satellites are being examined [6] [7]. Personal communications system (PCS) is heralded to be a truly anywhere to everywhere communication that will integrate paging, indoor systems, cellular systems, and large constellation low earth orbit satellites. The PCS systems will offer multimedia services including voice, data and image. Several small cell high capacity systems have been designed and many have been deployed. European efforts in this direction have resulted in DCS1800 (DCS1900 in USA), Digital European Cordless Telephone (DECT), CT3, and CT2. On this side of the Atlantic Canadian CT2 Plus, Omnipoint, Wireless Access Communications Systems (WACS) and Qualcomm's Direct Sequence Spread Spectrum Systems (IS-95) are significant new developments. Other new system such as self-organizing wireless ad-hoc networks are being built. Transport congestions are looking forward to mobile traffic control systems, collision avoidance systems, and navigational systems. In short, many diverse types of systems are being developed and installed.

The prognosis of wireless communications is stunning. Besides rapid acceptance of truly mobile systems by the mobile users, fixed point to point systems to complement switched telephone network are being installed. However, for a lustrous future of wireless communications several technical, commercial and policy challenges have to be met.

Some of these technical challenges were indicated in Section 1.4. Other significant technical challenges towards realization of a truly universal communications system are: telephone numbering scheme, and development of network interfaces and interoperability standards so that personal communication networks using different protocols within a country or across the world may be interconnected.

The recent inroads made by the Internet in wireless telecommunications, have necessitated new developments in the core networks. The current circuit switched network optimized for voice has to give way to high speed packet switched network. In this respect IP based networks are expected to provide backbone networks.

The spectrum shortage is expected to remain a problem. There are two alternatives to alleviate this problem. The first alternative is to devise systems that are highly efficient in spectrum usage. Multiuser detection, multiple input multiple output (MIMO) systems, interference cancellation are some of the techniques that are likely to considerably improve the spectrum efficiency. The second alternative is to rationalize the spectrum usage by assigning it where it is needed the most. For example, large chunks of frequency spectra are currently being used in TV broadcasting. Is such use absolutely necessary or other alternatives are available to deliver TV signals? In the near future reclamation of the spectrum from fixed services to mobile services is expected to be a dominant as well as a contentious issue. A proper spectrum rationalization will ensure continual rapid growth of wireless communications.

1.6 Summary

An overview on wireless communications has been presented in this chapter. A historical perspective was presented and this followed by a discussion on issues concerning assignment of spectrum and its management and future trends in using higher and higher frequencies. Prognosis and future challenges were spelled. It was stressed that either new spectrum is found or new methods are discovered to create bandwidth or increase efficiency.

Problems

Problem 1.1

The cellular concept was proposed in 1947, but the first commercial cellular system became operational in 1981. What impeding factors stalled the development of mobile cellular communication systems?

Problem 1.2

Enlist the most important differences between the fixed telecommunication networks and wireless communication networks.

Problem 1.3

What are the salient differences between Public and Private mobile communication systems.?

Problem 1.4

Give at least four reasons, why the first generation analog cellular systems were replaced very quickly by digital standards.

Problem 1.5

Give reasons why efforts to define a single worldwide standard failed even 3G systems.

Problem 1.6

Bluetooth is a new wireless technology developed for point to point or point to multipoint communications. What are the salient features of Bluetooth system and what are the major applications of Bluetooth?

Problem 1.7

(a) Give three reasons for Iridium ended in failure.
(b) From a custormer's point of view, list two advantages and two disadvantages of using terrestrial cellular systems over LEO satellite based systems.

Problem 1.8

What services do you think will be provided by the 3G systems? Give some examples in target markets of transportation, business, personal communications, and emergency services.

Problem 1.9

What is the most important technical challenge the wireless communication will be facing in the future?

References

[1] C.P. Sandbank, "Communications in the 21st century," *Proceedings IEE*, vol. 127 Pt.A, No. 1, January 1980, pp. 12-20.
[2] T. Logsdon, *Mobile Communication Satellites*, McGraw-Hill, 1995.
[3] J. Oetting "Cellular mobile radio- An emerging technology," *IEEE Comm. Mag.*, vol. 21, pp 10-15, 1983.

[4] R.Bowers, A. M. Lee and C. Hershey (Edit.), *Communications for a Mobile Society*, Sage Publications, Beverly Hills, 1978.

[5] H. Ware "The competitive potential of cellular mobile telecommunications," *IEEE Comm. Mag.* vol. 21, No.8, pp 16-23, 1983.

[6] P.A. Castruccio, C.S. Marantz, J. Freibaum, "Need for and financial feasibility of satellite aided land mobile communications," *ICC Proceedings*, 1983.

[7] A. Ghais, G. Berzins, and D. Wright, "INMARSAT and the future of mobile satellite services," *IEEE Journal of Selected Areas in Communications*, vol. SAC-5, no. 4, pp. 592-600, May 1987.

Chapter 2

MACRO-CHARACTERIZATION OF WIRELESS CHANNELS

The golden rule - *no signal no communications* applies to all communication systems, more so in wireless communications where a transmitted signal disperses in almost every direction. Thus, provision of an adequate signal level at the receiver is the first step towards a reliable wireless communication. A typical user of a mobile wireless system wants consistently good quality communication link and in general not interested in knowing why sometimes the reception quality degrades. It is therefore the responsibility of the system designer to ensure that adequate signal strength is available at any selected location within the designated service area. In other words, each terminal irrespective of its location within the service area must be guaranteed to receive reliable communications regardless of the presence of distance related signal attenuation, mobility related fading, interference from the other users, and noise.

It is imperative that the designers of mobile radio systems have good insight into the mechanics of signal propagation in complex and time variant environments. To this end, a comprehensive understanding of the signal environment on the part of system designers opens many doors to innovate methods to make communications reliable. In this chapter, we discuss methods to estimate signal strength with a reasonable accuracy at any chosen location and intervening environmental geometry between the transmitter and the receiver. It should be pointed out that because of wide variability of terrain features from one location to another, it is practically impossible to predict signal strength precisely. If we are able to estimate the signal strength within a few dBs of actual value, we are doing very well indeed.

2.1 Signal Transmission Over Mobile Radio Channels

Let us consider a typical wireless signal environment elucidated by a transmitter-receiver geometry and the intervening environmental features shown in Figure 2.1. In terms of specific locations of the transmitter and receiver, the received signal strength may be written as a function of:

(i) transmitter and receiver parameters i.e. transmit power, transmitter, receiver antenna gain and noise figure;

(ii) the distance between the transmitter and the receiver;

(iii) terrain irregularities between the transmitter and the receiver;

(iv) the movement of terminal through the intervening terrain irregularities.

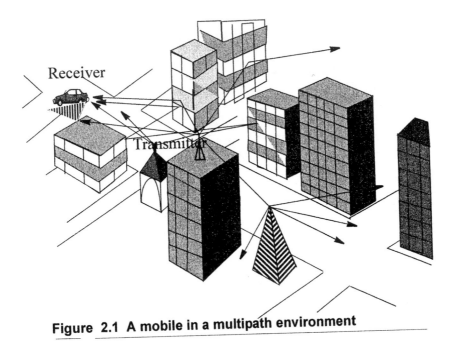

Figure 2.1 A mobile in a multipath environment

The first function is obvious, the received signal increases with an increase in the transmit power and antenna gains. The second effect is one of simple signal attenuation; the received signal continues to go weaker as the distance between the transmitter and the receiver increases. Reflections, diffractions, and refractions of radio waves by the terrain irregularities in the propagation path (e.g., hills, buildings, trees, humidity, atmospheric pressure etc.) produce the third effect. The signal rays after experiencing reflection or diffraction arrive at the receiver from different directions and having different strengths. In doing so, both the received signal amplitude and phase undergo random changes. In the absence of obstacles in the immediate vicinity of the receiver, the signal amplitude does not vary significantly for small movement of the terminal.

The fourth effect arises due to the movement of the mobile through standing waves, which are created by numerous waves produced by scattering of the signal by obstacles in the receiver locale. These rays arrive at the receiver with infinitesimally small and random differential time delays. Thus, the rays have independent carrier phases. As the mobile unit moves through the local frequency sensitive standing wave pattern, the mean signal strength gradually diminishes with an increase in the distance from the transmitter. In addition, the large obstacles interceding the transmitter-receiver path cast a shadow, which extends over a relatively large area. As the mobile unit enters the shadow zone and progresses through it, the mean signal strength first decreases and then increases as the vehicle emerges from it. The shadow causes slower signal variations because of their relatively lower rate of angular change as perceived by the receiver. At the same time, the movement of the terminal in the spatial standing waves setup by multipath propagation results in a faster signal variations. These faster variations are superimposed on relatively slower changes in the signal strength due to shadow effect. Figure 2.2 shows a typical received signal pattern as the mobile terminal travels a distance.

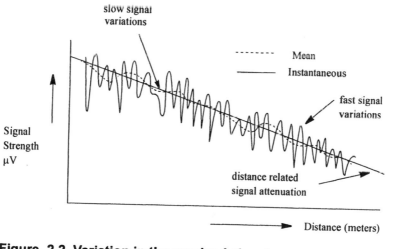

Figure 2.2 Variation in the received signal

In this figure, it is seen that the received signal fluctuates rapidly around a slow varying mean. It is also noticed that the rate of rapid fluctuations changes as the mobile progresses through the distance. This is due to an increase in the mobile speed.

The signal received by a moving terminal exhibits each of the four types of variations mentioned above. In this chapter, however, we shall consider only the second and the third, those caused by the distance and environment related signal loss and shadowing. The rapid fluctuations are removed by taking short term average of the received composite signal leaving the large scale (or slower) signal variations[†]. The 'short term' average is obtained by averaging the received signal over a distance (equivalently a short period of time, if the vehicle speed is held constant), which is long enough to smooth out rapidly multipath fluctuations, but short enough to retain slow variations caused by distant terrain irregularities. Lee and Yeh [1] have shown that slow variations or the mean signal with a standard deviation of 1 dB could be obtained by averaging the received signal over a distance of approximately 40λ. However, several other averaging intervals have also been used [2][3]. It is important to note that in the absence of a well defined boundary between the fast and slow variations, it is very difficult to completely separate the two types of variations.

A comment is worthwhile at this point on the measures of the signal strength and the methodology used to obtain them. The mean of a random variable indicates the central tendency or mass centre of the data. As a single measure, however, it has the disadvantage of being biased appreciably by a small number of extreme values. In mobile communication environments extreme values of signal strength do occur and the mean signal strength is found to be somewhat misleading. A more appropriate measure is the median signal strength, which has the advantage of being relatively insensitive to extremes. On the other hand, mean values are much more easily obtained in the field. Once the statistical description becomes known median and mean levels can be related. The measurement procedure followed in the field measurement is shown in Figure 2.3.

An area where measurements are to take place is divided by means of a grid into a set of elementary areas in such a way that the characteristic of each elementary area is approximately uniform in its terrain or topology. A typical elementary area might be about 500 meters square in an urban environment. In rural areas, a larger elementary area may be selected, while in hilly areas smaller elementary areas are chosen. Field strength or signal level measurements within the elementary area are made over a route that follows a winding pattern so that the receiver takes as many orientations relative to the transmitter (refer to Figure 2.3) as possible. The measured data is then averaged over intervals of approximately 40λ, to obtain the average signal strength. Since the measurement is recorded in real time, the mobile speed during the measure-

---

†. *These variations are also called fast and slow fading in the mobile communication literature. We will use fading and signal variations interchangeably.*

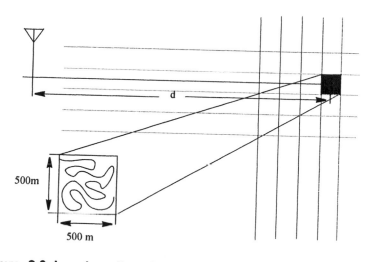

Figure 2.3 Local median signal measurement procedure

ments is also recorded so that the inter-sample time could be adjusted according to the vehicular speed so as to obtain uniform sampling in distance. The median value of these average signal strengths is then designated to the center of the elementary area as the measure of the median signal strength.

It has been observed that in addition to the factors previously mentioned, street orientation relative to the line between transmitter and receiver could affect the results significantly and so do tree foliages. It can thus be imagined that the signal strength at a given point is a complex function of the precise geometry of all the terrain irregularities surrounding the receiver. With so many factors influencing the propagation loss, the designer's task of predicting signal strength by pure analytical means is rather impossible. The alternative is to use field measured data to define empirical models, which can be used to estimate signal strength at desired locations. The total signal attenuation loss is then given by adding basic attenuation (free space loss) and losses due to environmental factors. These additional losses are estimated from correction factors developed from measurements and are given as a function of distance, frequency, antenna heights, and terrain features. Of the numerous publications detailing these factors [4]-[24], the most important and widely cited work is that of Okumura *et al* [22]. This has been adopted by ITU-R with some modifications. In the remainder of this chapter, we begin with a simple, ideal abstraction and work towards a propagation loss prediction model by successive refinements. The final model thus obtained will allow estimation of the signal strength in any environment.

We begin with the simplest of models, that of free space propagation presented in section 2.2. Since any land mobile radio communication system has at least one antenna situated close to ground level, free space propagation is rarely an effective model. However, in certain environment free space propagation loss may be used as a basis for calculating overall propagation loss. It must therefore be modified to account for reflecting and/or diffracting surfaces between transmitter and receiver. In this respect it is convenient, initially, to consider that the earth is flat as well as perfectly conducting. This first modification is derived in Section 2.3. After the discussion of the plane earth propagation model, terrain features that reduce conduction are considered and their effects on signal strength are described in Section 2.4. Section 2.5 examines effects related to the Earth's curvature, while Section 2.6 relaxes the requirement of Section 2.3 that the surface between the antennas be smooth. Signal strength prediction becomes to some degree an art when the intervening terrain between the transmitter and receiver is not uniform. This situation is introduced in Section 2.6, and definitions of terrain classification are presented in Section 2.7. Sections 2.8 to 2.10 examine some of the results available in the literature for statistical signal prediction over certain types of terrain irregularity, including sloping land, hills, bodies of water, mountain ridges and city streets.

Another important area of application of radio mobile systems is that of indoors wireless or portable communications for use in enclosed spaces. The effects on propagation loss of building characteristics such as materials, structures and area configurations are considered in Section 2.11. Also included is a brief summary of radio propagation in tunnels. On a more global basis, Section 2.12 comments on published results relating to signal effects caused when satellites are used for mobile communications.

It should be emphasized that the nature of signal strength prediction is inherently statistical. Thus, the models that result from available measurements and analyses based on these can do no better than to predict the service area in terms of a percentage of locations where adequate signal strength is statistically possible. Further discussion of this point is presented in Section 2.13.

2.2 Free Space Field

A radio signal transmission path that is free of all objects that might absorb, diffract or reflect radio energy is termed free space propagation path. The free space field intensity, E_d in volts/meter, at a distance d meters from the transmitting antenna, is given by

$$\frac{E_d^2}{R} = \frac{P_b g_b}{4\pi d^2}$$

$$E_d = \frac{\sqrt{30 g_b P_b}}{d}$$

(2.1)

where P_b = radiated power in watts, g_b = power gain ratio of transmitting antenna, and R is the real part of the free space impedance, which equals 120 π. From now onwards, the subscript b will be used to refer to the base station and m will refer to the mobile unit. The value for g_b is unity for an isotropic antenna; for a dipole of size smaller than $\lambda/2$, its value is 1.5, and it increases to 1.64 in the direction of maximum radiation for a half wave dipole [25]. The maximum useful power P_m in watts that can be delivered to a matched mobile receiver is given by

$$P_m = \frac{E_d^2}{120\pi}\left(\frac{\lambda^2}{4\pi}\right) g_m$$

(2.2)

where E_d = received field intensity (volts per meter) at a distance d from the transmitter, λ = wavelength (meters), f = frequency (Hz), and g_m = power gain ratio of the receiving mobile antenna. The factor $\lambda^2/4\pi$ is the aperture or the capture area of the antenna. The ratio of the radiated power to the received power is called transmission loss and this ratio may, in general, be written as

$$\frac{P_m}{P_b} = \left(\frac{\lambda}{4\pi d}\right)^2 g_b g_m$$

(2.3)

which can be expressed conveniently as the free space path loss of L dB

$$L_{dB} = 32.44 + 20\log(f) + 20\log(d) - 20\log(g_b) - 20\log(g_m)$$

(2.4)

where frequency f is expressed in MHz and distance d is in km. The path loss for various frequencies has been plotted against distance in Figure 2.4. Note that the path loss is proportional to d^2 or the received signal power decreases by 20 dB for every decade increase in the distance or for every decade increase in frequency.

2.3 Plane Earth Propagation

The free space propagation assumes the absence of any obstacles in the propagation path. This can happen, though rarely in mobile communica-

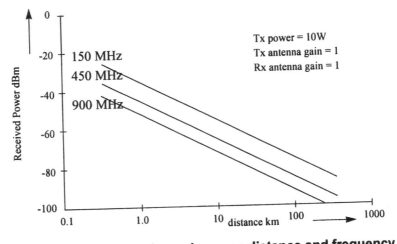

Figure 2.4 Free space loss dependence on distance and frequency

tions; mobile terminal on an elevated bridge is an example of such environ-ment. Nevertheless, the free space propagation loss is used as the basic propagation loss in all signal strength prediction problems.

In mobile communications, it is customary to install base station at an elevated location. The mobile terminal height is usually low and lies closer to the ground. The presence of ground influences the signal picked up by the mobile terminal. Therefore, the effect of the ground on the propagation loss should be included. The ground by acting as a partial reflector and a partial absorber disturbs the distribution pattern of the signal energy, and therefore affects the propagation loss. Neglecting induction fields and other secondary affects due to ground, field strength at the receiver is determined by the sum-mation of free space, reflected, and surface wave components; it is given by [4][5],

$$E_r = E_d\left[1 + Re^{j\Delta_1} + (1 - R)Ae^{j\Delta_2}\right]$$
(2.5)

where R the reflection coefficient of the ground, generally complex, is approx-imately equal to -1 when the angle of reflection is small. Δ_1 and Δ_2 are the phase angles of the reflected and the ground wave components. We are partic-ularly interested in small grazing angles since one of the antennas in the link is almost always located near the ground; therefore for finite conductivity of the ground surface, the reflection coefficient approaches -1 for both vertical and horizontal polarization. The magnitude and the phase of the reflection coefficient may be calculated from following equations [6][8][14][18]:

$$R = \frac{\sin\theta - z}{\sin\theta + z} \qquad (2.6)$$

where,

$$z_v = \frac{\sqrt{\varepsilon_o - \cos^2\theta}}{\varepsilon_o} \qquad (2.7)$$

$$z_h = \sqrt{\varepsilon_o - \cos^2\theta} \qquad (2.8)$$

$$\varepsilon_o = \varepsilon - j60\sigma\lambda \qquad (2.9)$$

ε = relative permittivity of the ground; σ = conductivity of the ground in mhos per meter; λ = wavelength in meters; θ = angle of incidence; z_h and z_v are for horizontal and vertical polarization respectively.

The surface wave attenuation A depends on signal frequency, ground conductivity and type of polarization. It is never greater than unity, and decreases with increasing frequency and distance. It is approximately given by

$$A \cong \frac{-1}{1 + \frac{j2\pi d}{\lambda(\sin\theta + z)}} \qquad (2.10)$$

The surface wave component is due to energy absorption by imperfect ground conduction. Its effect is significant up to only a few wavelengths above the earth's surface. At the relatively high frequencies (above 300 MHz) used for mobile communications, this effect is limited to few inches from the ground, and is therefore generally ignored. Under the assumption that $d > 5(h_b + h_m)$, the phase difference, Δ_1, between the reflected and the incident waves, is approximately given by

$$\Delta_1 \cong \frac{4\pi h_b h_m}{\lambda d} \qquad (2.11)$$

Equation (2.5) describes the physical mechanism of the propagation, but an equivalent expression, which is more convenient for a discussion on surface effects is given by

$$\frac{E_r}{E_d} = 2\sin\frac{\Delta_1}{2} + j[1 + R + (1-R)A]e^{\frac{j\Delta_1}{2}} \tag{2.12}$$

In this case the principal effect of the surface wave is to produce inter-ference fringes or lobes so that the field intensity, at a given distance and for a given frequency, varies about the free space field as either of the antenna heights is varied. When the angle Δ_1 is greater than about 0.5 radian, the terms inside the brackets, which include the surface wave, are usually negligible; a sufficiently accurate approximation is then given by

$$\frac{E_r}{E_d} \approx 2\sin\left(\frac{2\pi h_m h_b}{\lambda d}\right) \tag{2.13}$$

When the angle Δ_1 is smaller than 0.5 radian and the receiving antenna is physically placed below the maximum of the first lobe, the surface wave may become important. An approximation for this condition is given by

$$\frac{E_r}{E_d} = \frac{4\pi h'_m h'_b}{\lambda d} \tag{2.14}$$

In this equation $h' = h + h_o$, where h is the actual antenna height and h_o is the minimum effective antenna height, given by

$$h_o = \frac{\lambda}{2\pi z} \tag{2.15}$$

The surface wave is the dominant factor for receiver antenna heights less than h_o. In this region $(0, h_o)$ the received signal is not appreciably affected by a change in antenna height. However, for frequencies of interest in mobile communications, the antenna height is typically greater than λ (e.g., λ is 0.3m at a frequency of 900 MHz.), and the surface wave contribution is insignifi-cant. If the antenna height exceeds h_o, then the received signal power is increased by 6 dB for each doubling of the height until free space transmission conditions are reached. The ratio of the received to the transmitted power for a transmission over a plane earth surface is obtained by substituting (2.14) into (2.3), yielding

$$\frac{P_m}{P_b} = \left(\frac{\lambda}{4\pi d}\right)^2 g_m g_b \left(\frac{4\pi h'_b h'_m}{\lambda d}\right)^2 = \left(\frac{h'_b h'_m}{d^2}\right)^2 g_m g_b \tag{2.16}$$

h'_m and h'_b are the effective antenna height of the mobile and the base station respectively. It is shown in Figure 2.5 for an ideal isotropic antenna ($g_b = g_m = 1$). It is interesting to note that this relation is independent of frequency. However, the received power has been found to be a function of operating frequency, antenna heights, and the link distance. This suggests that the plane earth propagation model is approximate and it ignores the secondary effects, which may be present.

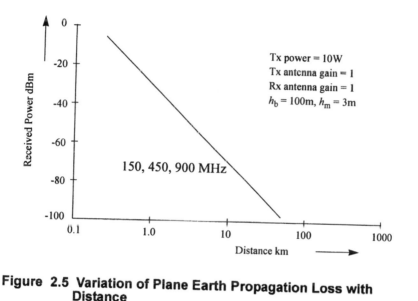

Figure 2.5 Variation of Plane Earth Propagation Loss with Distance

2.4 Egli's Model for Path Loss Estimation

To this point, it has been assumed that the earth's surface is planar and perfectly conducting. The second of these assumptions is relaxed in this section, and an empirical model is presented, which takes into account the loss associated with transmission over a surface that is planar on a large scale, but has a significant number of local obstructions. Egli [11] showed that an increase in the propagation loss with distance over a uniform terrain with local obstructions present is the same as that for a perfectly conducting earth; however, the actual propagation loss is substantially higher than that predicted by the theoretical model. Several other workers [3][9][12][19][20][23] reported similar results. To account for this additional loss, a clutter factor (or correction factor), β, is introduced. Experimental data for the median path loss in urban areas indicates that, despite different locations and different frequencies which have been used, a fourth power law relates path loss to range, particularly when the distances involved are less than 20 km. Thus, it appears that the plane earth equation (2.16) with an additional factor to account for the environmental clutter provides a suitable propagation model for a uniform terrain with some local obstructions. Equation (2.16) can now be rewritten as

$$\frac{P_m}{P_b} = \left(\frac{h_m h_b}{d^2}\right)^2 \frac{g_m g_b}{\beta} \tag{2.17}$$

where β, the clutter factor, tends to be dependent on frequency but insensitive to distance. A suitable choice of β is very much an empirical matter, and the model itself is probably still incomplete. The difficulty may be observed from measurements of signal strength made in major cities. Alloesbrook and Parsons report values of β_{dB} as 16.5, 19 and 34 dB at frequencies of 85.5, 167.2 and 441.5 MHz respectively [19][12] for the city of London. Another measurement gives a value of 29 dB at 168 MHz and 37 dB at 455 MHz also in London but for a range of 10 km [3]. It is of interest to compare these values with the clutter factor obtained by Young [23] in New York. He estimated the clutter factor to be 26 dB and 36 dB at 150 and 450 MHz. Differences in the values of the clutter factor between these two cities are probably due to differences in the environments. Streets in London have more winding patterns, in contrast to New York's grid layout. The latter causes a ducting effect, while the former may lead to extensive shadowing owing to the irregular orientations of the streets and buildings. While this difference in layout tends to cause higher attenuation in London, there is a countervailing condition in the much greater building heights in New York. Measurements indicate that the clutter factor model is useful only to estimate median signal levels in a given region. Variations of ± 6 dB in the median level may be expected according to specific locale. In addition to the clutter factor, other corrections due to antenna heights and frequency should be included while estimating the total propagation loss [26]. These factors will be considered in Section 2.8.

2.5 Propagation Beyond the Line of Sight

A basic characteristic of electromagnetic waves is that their energy is propagated in the direction perpendicular to the equi-phase surface. Thus, the radio waves will travel in a straight line as long as the phase front is plane and infinite in extent. The environmental irregularities affect the shape of the phase front with the result that the signal may propagate in almost any direction. The bending of phase front can result in signal transmission beyond the horizon.

The signal energy can be transmitted beyond the horizon by three principal modes: reflection, refraction and diffraction. Reflection and refraction are associated with either sudden or gradual changes in the direction of the phase front, while diffraction is an edge effect due to the finite extent of the phase front. For each of the three effects the direction of the phase front is modified in different manners. When all three mechanisms act simultaneously, the phase front becomes irregular and unpredictable in amplitude and/or position. Furthermore, it becomes extremely difficult to distinguish the roles played by the above mentioned effects. The result is scattering of energy in all directions.

A gradual decrease in the dielectric constant of the atmosphere with an increase in height results in wave propagation beyond the horizon. The radio wave directed at a certain elevation angle is thus bent towards the ground. As long as the change in the dielectric constant is linear with height, the net effect of refraction is the same as if the radio wave continued to travel in a straight line over an earth whose modified radius is:

$$k_a = \frac{a}{1 + \left(\frac{a}{2}\right)\frac{d\varepsilon}{dh}} \qquad (2.18)$$

where a = true radius of earth = 6378 km and $d\varepsilon/dh$ = rate of change of dielectric constant with height. A value of $k_a = 4/3$ (for a decrease in dielectric constant of 7.839×10^{-8} per meter) is considered to be standard.

A rate of decrease of 3.136×10^{-7}/m in the value of the dielectric constant corresponds to $k_a = \infty$ which means that radio wave will travel along a path 'parallel' to the earth's surface. Under these conditions, the earth may be considered to be flat for the purpose of radio wave propagation. When the dielectric constant decreases more rapidly than 3.3×10^{-7} per meter of height, radio waves that are radiated parallel to or at an angle above the earth's surface bend downward sufficiently to be reflected from the earth's surface. Repeated bending downs and reflections makes the radio energy to appear as trapped within a duct or waveguide between the earth and the maximum height of the radio path. This mode of propagation is known as multiple-hop propagation or ducting.

Under certain atmospheric conditions, the dielectric constant may increase over a reasonable height range, thereby causing the radio waves to bend away from the earth, this happens when $0 < k_a < 1$. In this case, reception of radio signal is not possible at the surface of the earth.

Radio waves also propagate by the phenomenon of diffraction. Diffraction is a fundamental property of wave motion; it is the correction applied to geometrical optics (ray theorem) to obtain more accurate wave optics. In other words, all shadows are somewhat fuzzy on the edges and transitions from light to dark areas are gradual rather than sharp. The diffraction of radio waves makes propagation around the earth possible. The magnitude of the diffraction loss increases with frequency and distance; in addition there is some dependence on antenna height [26].

A mobile link becomes attainable if the receiver is within the radio horizon of the transmitter antenna. The radio horizon is defined as the maximum distance from the transmitter at which the transmitter antenna is visual to the receiving antenna. The distance to the radio horizon for a given antenna height is given by

$$d_t = 2.9\sqrt{\frac{3h'}{2}} \tag{2.19}$$

$$d_a = 2.9\sqrt{2h'} \tag{2.20}$$

where d_a and d_t are the actual and true distances as shown in Figure 2.6. In case, propagation beyond horizon exists, the total distance over which a reliable communication can take place (refer to Figure 2.6) may be written as,

$$d = d_1 + d_2 + d_3 \tag{2.21}$$

where d_1 = distance to the horizon from antenna 1; d_2 = distance to horizon

$$d_{t1} = d_1, d_{t2} = d_2$$

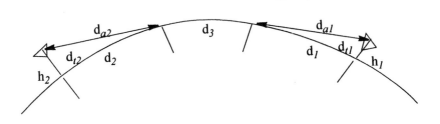

Figure 2.6 Distance to horizon calculations

from antenna 2; d_3 = distance beyond the line of sight. The distance to the horizon over a smooth earth is given by $d_i = \sqrt{2k_a h_i}$, i=1,2 where $h_{1,2}$ = appropriate antenna height (meters); k_a= earth's effective radius (meters). Bullington in [14] produced a nomogram to calculate beyond line of sight over a smooth earth diffraction decibel loss relative to free space loss.

EXAMPLE E-2.5.1 Propagation Beyond the Line of Sight

a) A mobile receiver with an antenna height of 3m is moving away at 100 km/hr from a base station whose antenna height is 60m. In the absence of surface obstructions, how long will it take before the receiver loses line-of-sight contact with the transmitter?

b) For a mobile antenna height of 3m, find the minimum base station antenna height so that a line-of-sight communication is maintained over a radius of 10 km?

Solution:

a) For an antenna height of h meters, the distance d_a to the radio horizon is given by $d_a = 2.9 \, (3h/2)^{1/2}$ km. For antenna heights of 3m and 60m, the maximum distance d_m over which straight line communication is possible is therefore $d_m = 2.9 \, [(3 \times 3/2)^{1/2} + (3 \times 60/2)^{1/2}] = 25.6$ km. At a velocity of 100 km/hr, this corresponds to a time $t = 25.6/100$ hr, or 15.36 minutes.

b) The required antenna height is found as the value of h_b which satisfies the equation $d_m = 2.9 \, [(3h_b/2)^{1/2} + (3h_m/2)^{1/2}]$ where $d_m = 10$ km and $h_m = 3$m to yield a value of h_b of 1.173 m.

Obviously the use of such a low base station antenna is unrealistic. If the terrain is in fact largely free of obstructions, then the region is one of low population density and hence almost certainly of low radio traffic density; in this case coverage of a large area is required, so that a tall base station antenna is necessary. If the region is one of high population density, man-made obstructions are predominant, and a good antenna height is again required. In the latter case, the usual practice is to place the base station antenna on the roof of a tall building. An exception to this may be when coverage is expected to be aided by multiple reflections and diffractions. In this case lower height may have an advantage. The one condition in which a low antenna may be used is for direct mobile-to mobile communication, without the use of base station, in an open country environment. It is easily shown that communication over a distance of 10 km requires two antennas of height at least 1.5 m each.

2.6 Terrain Definitions

Just as the plane perfectly conducting earth model is inadequate, so is the model that assumes a smooth earth surface. Specifically, the latter model fails when ground irregularities become comparable to the wavelength of the transmit frequency. We, therefore, need a criterion to make a distinction between different types of terrain surface. One such criterion is related to the phase difference, ψ, between the rays reflected from the crest and the trough of the surface. The roughness criterion, in general, is based on $\psi = \pi/2$, although different values of ψ such as $\pi/4$, or even $\pi/8$, have been used [11]. When the phase difference exceeds $\pi/2$ radian, the terrain is classified as rough. Figure 2.7 shows that the path difference between the rays reflected from crest and trough is $2h \sin \psi$; thus for an undulation height of Δh, the phase difference is given by $(4\pi\Delta h/\lambda) \sin \psi$. For example, consider a grazing angle of 0.25 degrees at a frequency of 900 MHz, an undulation height of 9.54m may be considered as rough. For lower incident angles, larger undulation heights satisfy the roughness criterion.

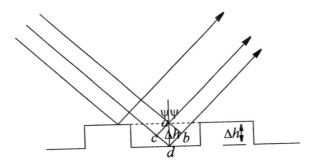

Figure 2.7 Definition of rough surface leading to Rayleigh criterion

Unfortunately this criterion is applicable to a single location. For wide area, roughness can only be defined statistically. The so called Rayleigh criterion gives a statistical measure of terrain classification. The roughness parameter, C, is defined as:

$$C = \frac{4\pi\sigma\theta}{\lambda} \tag{2.22}$$

where σ = standard deviation of the surface irregularities relative to mean surface level; λ = wavelength; and θ = angle of incidence. If $C < 0.1$, the specular reflected component is significant, and the ground may be regarded as smooth, whereas if $C > 10$ the surface is regarded as irregular. In between these two values, $0.1 \leq C \leq 10$, the terrain is regarded as quasi-smooth.

2.6.1 Okumura's Terrain Classification

The 'smooth' terrain, as defined above, is rarely encountered in practice. Even a flat ground is usually interrupted by trees and buildings. Extensive work by Okumura et al [22] suggests that, for the purposes of signal strength prediction, classification of the terrain in two broad categories is sufficient i.e. terrain is considered either quasi-smooth or irregular; the latter class have several subclasses.

a) Quasi-smooth terrain;

b) Irregular terrain: Terrain other than quasi-smooth is classified as irregular; subclassifications are as follows:

      b1) rolling hilly terrain (includes mountainous),

      b2) general sloping terrain,

      b3) mixed land-sea path,

      b4) mountain ridges.

While Okumura's terrain definitions are largely empirical, there is a good correspondence with the definitions given in the previous section. Okumura's method to predict propagation loss will be presented in Section 2.8. Prior to this, however, it is necessary to review the definitions of terrain profiles and link parameters as defined by Okumura.

2.6.2 Effective Antenna Height

Okumura et al[22] defined the effective antenna height as the height of the antenna tip above the mean ground level, the latter averaged over a distance of 3 to 15 km from the base station. The averaging distance equals the link distance if d does not exceed 15 km. The effective antenna height (h_{te}) definition is illustrated in Figure 2.8. For example, if the antenna height above sea level is h_{ts} and the average ground level above sea level is h_{ga}, then effective transmitter antenna height is given by $h_{te} = h_{ts} - h_{ga}$. Okumura in his measurements used 200m as a standard effective antenna height. For other antenna heights, an antenna height correction factor is defined.

EXAMPLE E-2.6.1   Effective Antenna Height

Consider the placement of base station and mobile receiving antennas as shown in Figure E-2.6.1. The base station is situated on a hill, while the

---

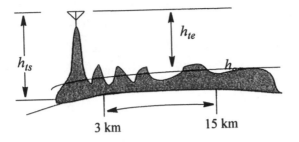

3 km   15 km

Figure 2.8 Okumura's Definition of Effective Antenna Height

mobile receiver is located at a certain distance on another hill. Calculate the
effective base station antenna height.

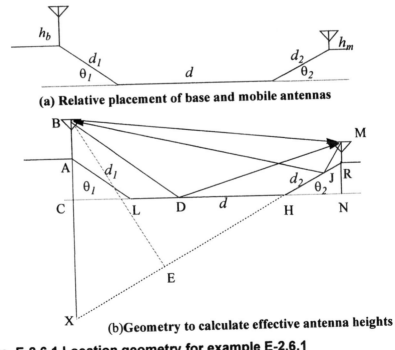

(a) Relative placement of base and mobile antennas

(b)Geometry to calculate effective antenna heights

Figure E-2.6.1 Location geometry for example E-2.6.1

Solution:

Figure E-2.6.1(a) shows the geometry of the antennae positions. First,
it is important to determine the path(s) of reflective rays that reach the

receiver. If the only reflected component that reaches the receiver is from the segment HX, then it is valid to assume the plane earth model, in this case the effective antenna height is given by the length of segment BE.

The problem is then cast as follows: given antenna heights h_m, h_b, Distances d, d_1, d_2, Angles θ_1, θ_2, find effective antenna height. The antenna transmit antenna height could be either BE or BC depending on which plane reflects the signal.

To solve, augment the distances BA and RH (shown in Figure E-2.6.1(b)) by extending these lines to intersect at X. Also, signal could reach the receiver via direct line of sight, or reflected paths BJR and BDR; this construction establishes points J and D as well. Consider first that reflection plane is CN. The transmit antenna height is then given by $h_b+d_1\sin\theta_1$. If the plane of reflection is RX, the transmitter effective antenna height is given by BE. It is seen that BX = $\{d_1 \sin \theta_1 + (d + d_1 \cos \theta_1) \tan \theta_2\} + h_b$ so that the effective antenna height is BE = $\{h_b + d_1 \sin \theta_1 + (d + d_1 \cos \theta_1) \tan\theta_2\} \cos \theta_2$. Further analysis shows that the effective link distance is RX = $[(d + d_1 \cos\theta_1) \tan \theta_2 + h_b + d_1 \sin\theta_1] \cos^{-1}\theta_2 + d_2$.

2.7 Propagation Over Quasi-smooth Terrain

Okumura classified quasi-smooth terrain as the basic terrain. It is defined as the terrain when judged from the propagation path profiles, it is a superimposed smooth undulations having peak-to-peak magnitude of about 20 meters or less around a flat surface. "Flat", in this context, indicates a piece of terrain whose average level is approximately constant; thus, a more or less planar region exhibiting a steady drop or a rise in level of more than about 20 meters could not be considered as "quasi-smooth". Okumura considered quasi-smooth terrain as standard, for all other terrains additional correction factors are determined.

Okumura [22] produced curves representing median propagation loss for several terrain types, and parameters such as base station antenna height h_b, frequency f_c, and vehicular antenna height h_m. The use of these curves is usually cumbersome; a reason why efforts were made to transform them into empirical expressions mainly for use in computer prediction models [27]-[31]. Hata [27] produced empirical formulas for urban centers built on a quasi-smooth terrain. For other types of terrain - rural and sub-urban - correction terms have been suggested.

Okumura's prediction curves give the received field strength from a dipole radiating 1kW ERP (Effective Radiated Power). It is necessary to transform from dipole ERP to EIRP (Effective Isotropic Radiated Power).

This transformation is accomplished by adjusting for the power gain differ-ence between the dipole antenna and the isotropic antenna gain of 0 dB. Since the absolute power gain of the dipole antenna is 2.2 dB [31], therefore

$$P_t(dBEIRP) = P_t(ERP(dipole)) + 2.2(dB) \qquad (2.23)$$

For example, when P_t' is 1 kW (ERP(dipole) = 30dBW), P_t (dB EIRP) is 32.2 dBW. Using (2.2), the predicted propagation loss between two isotropic anten-nas is found as:

$$L_p(dB) = 178 - 10\log\frac{\lambda^2}{4} - E_d(dB\mu V/m) \qquad (2.24)$$

where E_d is the received field strength in dBμV/m. In an urban environment built over a quasi-smooth terrain, the propagation loss over a distance of R km is given by the empirical equation [27]

$$E_d(dB\mu V/m) = \gamma + \beta\log R \qquad (2.25)$$

where β and γ are constants determined by h_b(meters) and f_c(MHz). Note that a correction factor is added due to mobile antenna. The propagation loss of L_p dB can therefore be written in an alternative form as:

$$L_p = A + B\log R \qquad (2.26)$$

$$A = 178 - 10\log\frac{\lambda^2}{4} - \gamma + a(h_m) \qquad (2.27)$$

$$B = \beta$$

In (2.27), we have included the correction factor for the mobile antenna height, $a(h_m)$. In Okumura's measurements, a mobile antenna height h_m of 1.5m has been used and the correction factors for the other heights have also been reported. It is therefore convenient to use 0 dB as a reference value for $a(h_m)$ when the antenna height is 1.5m.

Hata [27] empirically determined the values of A, α, and β as follows:

$$A = \alpha - 13.82\log h_b - a(h_m) \qquad (2.28)$$

$$\alpha = 69.55 + 26.16\log f_c \qquad (2.29)$$

and

$$B = 44.96 - 6.55\log h_b \qquad (2.30)$$

Combining equations (2.23), (2.26), and (2.27) we get a standard formula for median propagation loss L_p in dB as follows:

$$L_p = 69.55 + 26.16\log f_c - 13.82\log h_b - a(h_m) + (44.9 - 6.55\log h_b)\log R \quad (2.31)$$

where $a(h_m)$ is the correction factor for h_m. ($a(h_m) = 0$ dB for $h_m = 1.5$ m). Note that (2.31) gives median propagation loss for an urban area built on quasi-smooth ground.

The correction factor $a(h_m)$ depends on environmental factors. Its value for a small or medium sized city has been found to be considerably different from that for a large city. It is given by

$$a(h_m) = h_m(1.1\log f_c - 0.7) - (1.56\log f_c - 0.8) \quad (2.32)$$

Hata similarly modeled correction factors for suburban and open areas. For suburban areas the correction factor is given by [27]:

$$K_r(dB) = 2\left(\log\frac{f_c}{28}\right)^2 + 5.4 \quad (2.33)$$

where f_c = frequency in MHz. K_r the correction factor is obtained from Okumura's prediction results. A good approximation for the propagation loss $L(dB)$ is then calculated by

$$L(dB) = L_p - K_r \quad (2.34)$$

The open area correction factor $Q_r(dB)$ is given by

$$Q_r(dB) = 4.78(\log f_c)^2 + 18.33\log(f_c) + 40.94 \quad (2.35)$$

and the propagation loss in the open area is then given by

$$L_{po}(dB) = L_p - Q_r \quad (2.36)$$

This empirical expression has been found to be accurate within the following limits:

f_c: 150-1500 MHz, h_b: 30-200 m, h_m = 1-10 m, R: 1-20 km.

Hata's model is applicable up to 1500 MHz. The frequency allocation to GSM (DCS-1800) in 1800 and rest of the world and 1850-1970 MHz to PCS in the United States requires that Hata's model be extended to these frequencies. Under the direction of the European Cooperative for Scientific and Technical research (COST), a committee COST-231 extended the Hata model to 2 GHz. The median path loss in urban area is given by [29]

$$L_p = 46.3 + 33.9 \log f_c - 13.82 \log h_b - a(h_m) + (44.9 - 6.55 \log h_b) \log R + C_m \quad (2.37)$$

where $a(h_m)$ is defined as in (2.32). $C_m = 0$ dB for medium sized city and sub-urban areas and 3 dB for urban heavily built areas. The applicable limits to the system parameters are: frequency, f from 1500 and 2000 MHz, transmitter antenna height, h_b, from 30 to 200m, mobile antenna height from 1 to 10m, and distance R from 1 to 20km.

Several other urban propagation loss models are also available in [31]-[35]. These have been used in prediction of signal strength mainly to select the best location of the base station antenna site as well as coverage areas but with limited results and therefore are not that important. For example, Kozono *et. al.*[36] proposed a model for propagation loss estimation in an urban environment based on occupied area and the volume of the buildings. The reader is referred to the references cited for further details.

EXAMPLE E-2.7.1   Loss Estimation using Hata's Empirical Model

A base station with an antenna height of 200m is transmitting 40 Watts of power at 896 MHz to a mobile unit with an antenna height of 3m. Use Hata's empirical formulations to find the received signal strengths in the three sectors of the coverage area at indicated distances shown in Figure E-2.7.1.

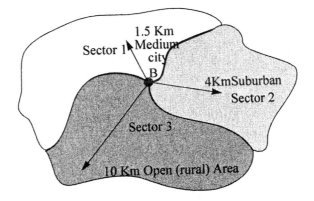

Figure E-2.7.1 Location of a base station with surrounding terrain

Solution:

Sector 1 (Medium city, urban area):

$$L_p = 69.55 + 26.16 \log f_c - 13.82 \log h_b - a(h_m) + (44.9 - 6.55 \log h_b)$$
$\log R = 112.57 - a(h_m) \ dB;$

Correction factor for the antenna applies: $a(hm) = (1.1 \log fc) - 0.7)h_m - (1.56 \log fc - 0.8) = 3.84 \ dB$. Total Loss is 116.4 dB.

Sector 2 (Sub-urban area):

We use (2.33) to calculate correction factor for the sub-urban area. However, we must include adjustment for an increase in the distance from 1.5 km to 4 km. This adjustment is 12.7 dB. Now the L_p is increased by 12.7 dB. Applying correction factor for sub-urban area, we have the propagation loss as,

$$L_{ps} = L_p - 2 \ \{log \ (f_c/28))\}^2 + 5.4 \ dB = 116.4 + 12.7 - 5.5 + 5.4 = 129 \ dB$$

Sector 3 (Open Area):

After making allowance for an increase in the distance from 1.5 km to 10 km in this case, the loss in an urban environment without applying antenna correction factor is found to be 140.96 dB. The open area loss is given by:

$$L_{po} = L_p - 1.78 \ \{log f_c\}^2 + 18.33 \log f_c - 40.94 = 138.6 \ dB.$$

The received field strengths in the three sectors using Equation (2.24), $E(dB\mu V/m) = 178 - 10 \log (\lambda^2/4) - L_p - 10 \log_{10} (40/1000)$, are found to be:

- = 91.09 dBμV/m (Sector 1, medium city urban area at a distance of 1.5 km)

- = 86.0 dBμV/m (Sector 2, suburban area at a distance of 4 km)

- = 72.69 dBμV/m   (Sector 3, open area at a distance of 10 km).

2.8 Propagation Loss Over Irregular Terrain

In this section, we extend the study of the propagation loss to that over irregular terrain. In general, the basic free space loss is calculated, excess loss due to quasi-smooth earth is applied to it before considering correction factors due to other terrain irregularities. The empirical formulation by Hata assumes urban, sub-urban and rural environments on quasi-smooth terrain. For other types of irregular terrain, the propagation loss can be predicted using terrain correction factors obtained for the given transmit power, base and mobile station antenna heights. Correction factors proposed by Thiessen [37] and ITU-R are applicable for mobile antenna heights of 10m; however for lower mobile antenna heights between 1 to 3m, signal strength is reduced considerably, this is typically about 10 dB. In these cases, the correction factors proposed in [37] cannot be used. Okumura *et.al.*[22] provided correction factors for the following terrain types:

(a) rolling hilly terrain;

(b) general sloping terrain;

(c) mixed land-sea path; and

(d) mountain ridges.

2.8.1 Rolling Hilly Terrain and Parameter Δh

Almost all propagation paths have some surface irregularities. In a rolling terrain, the terrain undulation height Δh, indicating the degree of irregularity is defined as the difference between the tenth and ninetieth percentiles of terrain heights within a distance d of 10 km from the receiving point toward the transmitting point as shown in Figure 2.9. Similarly, Δh could also be obtained for mountainous terrain when several mountains affect propagation by multiple diffraction between the transmitter and the receiver. This technique, however, cannot be applied to a simple sloping terrain or to a path with one mountain or a single undulation. When the total link distance d is shorter than 10 km, the following intervals are used to define the inter-decile height [12]:

Interval	Link distance
1 km to d,	for 1 km $< d <$ 6 km
2 km to d,	for 6 km $< d <$ 12km
d -10km to d,	for $d >$ 12 km

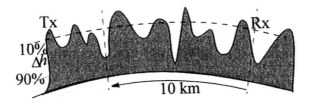

Figure 2.9 Rolling hilly terrain parameter Δh definition

2.8.1.1 Rolling Hilly Terrain Correction Factors

The median signal strength correction factor for rolling hilly terrain is given in terms of interdecile height Δh as defined in Section 2.8.1, under the condition that the mobile unit is on a hillside facing the transmitter, and is therefore illuminated by it. However, if the mobile unit is on the opposite side of the hill i.e. slope in a shadow, signal attenuation is significantly higher, and an additional fine correction factor is applied. These correction factors are shown in Figure 2.10(a) and (b).

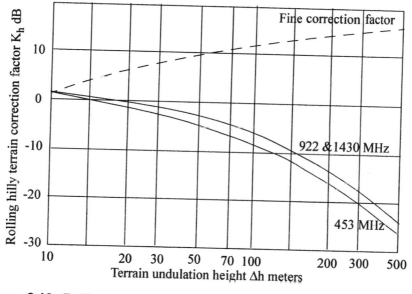

Figure 2.10 Rolling hilly terrain correction factors (after Okumura *et al*)

2.8.2 General Sloping Angle Parameter

When terrain, whether planar or undulating, slopes over a distance of at least 5 to 10 km (along the line connecting the transmitter and the receiver), the terrain parameter θ_m may be used. With reference to Figure 2.11, θ_m is expressed as

$$\theta_m = \frac{(h_n - h_m)}{d_n} \tag{2.38}$$

Note that if $h_n > h_m$ (uphill), the slope angle is positive ($+\theta_m$) and if $h_n < h_m$ (down hill), it is negative ($-\theta_m$). The average ground slope for link distances less than 9 km is defined over the following intervals:

	Interval	Link distance
a)	1 km to d	for $1\,\mathrm{km} < d < 4$ km
b)	2 km to d	for $4\,\mathrm{km} < d < 7$ km

c) d-5km to d for 7km $< d <$ 9 km

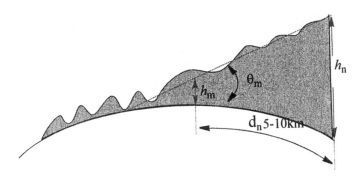

Figure 2.11 Definition of average angle of general slope
(Okumura *et al*)

When $d > 9$ km, the larger of the computed slope values over the following intervals is chosen:

d-5 km to d and d-10km to d for $d >$12 or

2 km to d, for $d <$ 12 km.

2.8.2.1 General Sloping Terrain Correction Factors

When the terrain is sloped for a distance of at least 5 to 10 km, the terrain slope correction factor in relation to the average angle of slope, θ_m in milliradian, at frequencies between 450 and 900 MHz, has been given by Okumura *et al* and is shown in Figure 2.12.

2.8.3 Distance Parameter for Mixed Land-Sea Path

Where there is an expanse of water as well as land in the propagation path, three different cases are possible:

i) An expanse of water lies between the transmitter and the receiver; the receiver is located at the edge of the water or upon it (i.e., at the shoreline, or aboard a boat or ship), while the transmitter is located inland at some distance from the water.

ii) The transmitter is at the shoreline and the receiver is located at a considerable distance from the shoreline.

iii) The body of water is in between the transmitter and receiver. This category also includes a land-water-land-water-land configuration in which an island or

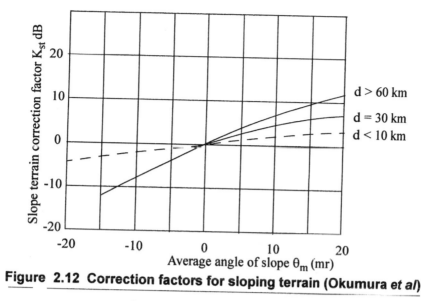

Figure 2.12 Correction factors for sloping terrain (Okumura *et al*)

number of islands may exist within the central expanse of water.

These situations are shown in Figure 2.13a, b, and c, along with the definitions of the distance parameters that are used to find the correction factors. Note that the case in which both the transmitter and the receiver are located at the water's edge or upon ships is distinct from those presented above, as it relates to transmission over water only. The water may be regarded as a reflecting surface of a planar nature, smooth or irregular depending upon the magnitude of waves on its surface relative to the wavelength of the transmitted carrier signal.

2.8.3.1 Mixed Land-Sea Path Correction Factor

When a significant expanse of water, e.g., a lake or the sea, lies along the propagation path, field strength is usually higher than on land alone. Calculation of field strength is based on the land-sea path parameter (C_{ls}), defined in Section 2.7.1 from which a median signal strength correction factor can be determined. This is shown in Figure 2.14. If the mobile or base station is very close to water, a secondary correction factor is needed. This correction factor is larger if the water adjoins the mobile unit than if it is near the base station.

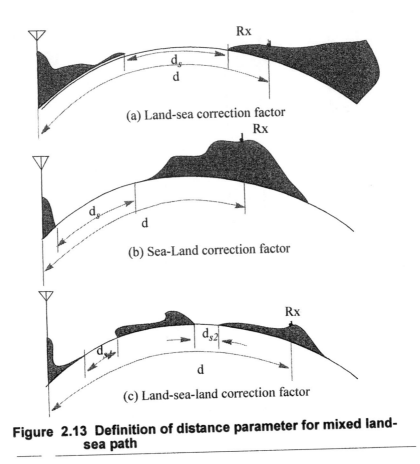

(a) Land-sea correction factor

(b) Sea-Land correction factor

(c) Land-sea-land correction factor

**Figure 2.13 Definition of distance parameter for mixed land-
sea path**

EXAMPLE E-2.8.1 Mixed Land - Sea Propagation Loss

A base station antenna radiating at 900 MHz is located on a high rise building at a height of 150 meters and it is connected via a cable to a 40 W transmitter located at a height of 120 meters. The cable attenuation is known to be 0.12 dB/m. The transmitter antenna gain is 6 dB. The high rise building is located 2 km from a shoreline of a lake.

(a) Estimate propagation loss to a boat which is 15 km away from the shoreline. Assume that the intervening terrain between the high rise and the shoreline is classified as quasi-smooth. The boat antenna height is 4m. Figure E-2.8.1 shows the geometry of the terrain.

(b) If the noise figure of the receiver is 4 dB and its bandwidth is 30 kHz, estimate receiver input carrier-to-noise ratio.

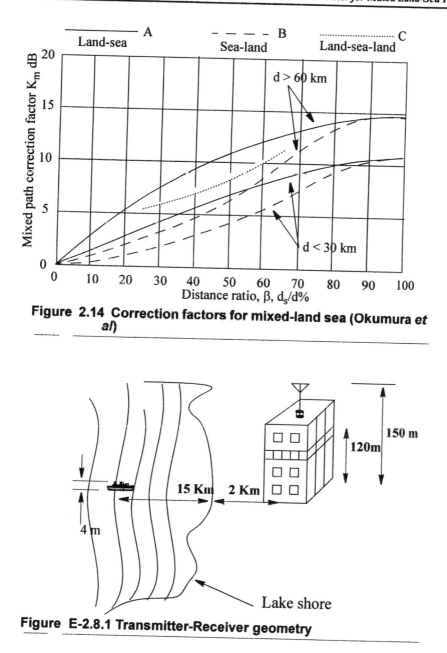

Figure 2.14 Correction factors for mixed-land sea (Okumura *et al*)

Figure E-2.8.1 Transmitter-Receiver geometry

Solution:

(a) The total loss is determined by first calculating the free space loss and adding to it the excess loss due to quasi-smooth ground using Okumura's

measurements, and applying correction factors for base and mobile antenna heights. The total loss is given by:
$$L_T = L_{QS} + K_s + \text{Adj } (h_m) + \text{Adj } (h_b) + \text{free space loss.}$$
From Figure 2.8.2, L_{QS} is determined to be 31 dB. The free space loss is calculated (assuming both g_m and g_b are equal to unity) to be 115 dB. Adj(h_m) = 20 log (4/3) = 2.5 dB; Adj(h_b) = 20 log (150/200) = -2.5 dB For land-sea loss, we have $d/d_1 = 0.866$ and the loss from Figure 2.14 is approximately equal to 10.2 dB. The total loss is then estimated to be 156.2 dB.

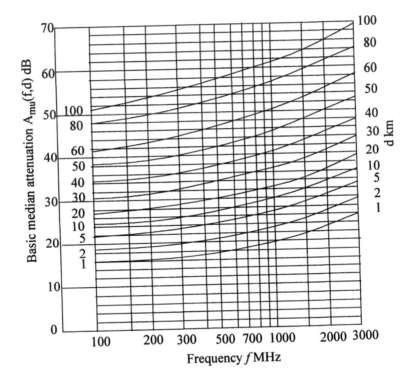

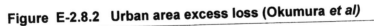

Figure E-2.8.2 Urban area excess loss (Okumura *et al)*

(b) Effective transmitter power = 46 - (0.12 x 30) + 6.0 = 48.4 dBm.
Received Power = 48.4 - 156.2 = -107.8 dBm
Receiver noise power = kT_oBF = -125.2 dBm
where k = Boltzmann's constant = 1.38 x 10^{-23} J/K^o, T_o = Ambient temperature in Kelvin, B is the noise bandwidth in Hz, and F is the noise factor $10^{FdB/10}$. Therefore the carrier-to-noise ratio at the receiver is estimated to be 17.4 dB.

2.8.4 Correction Factors for Mountain Ridges

When large man-made structures or terrain features obstruct the transmission path, the total propagation loss can be much larger than that predicted by the models discussed above. The excess loss is due to diffraction of the radio waves over and around the obstruction. The exact calculation of the diffraction loss is difficult if not impossible, however the order of magnitude estimation of the loss may be obtained by modeling such an obstruction as a perfectly absorbing knife edge. The loss due to a perfectly absorbing knife edge can be calculated in terms of Fresnel-Kirchoff parameter, κ, defined as [40]:

$$\kappa = \theta \sqrt{\frac{2}{\lambda}\left(\frac{d_1 d_2}{d_1 + d_2}\right)} \tag{2.39}$$

where d_1, d_2, λ and θ are defined in Figure 2.15. The buildings adjacent to the receiver but distant from the transmitter act as diffractors, with the diffraction angle given by:

$$\theta = \frac{h_o - h_r}{d_2}; \quad d_1 \gg d_2 \tag{2.40}$$

The usual assumption concerning the height h of the knife edge is [39]

$$\lambda \ll h \ll \frac{d_1}{10} \tag{2.41}$$

It is also assumed that the knife edge is thin across the line joining the transmitter and the receiver. One simple criterion to characterize a typical knife edge obstacle path is that the first Fresnel zones around TM and MR shown in Figure 2.15 are not obstructed. The excess diffraction loss L_m, in dB, can be expressed as a function of h/r where r is the radius of the first Fresnel zone constructed around TR. With distance in kilometers and frequency f in megahertz, the value of r in meters is [39]

$$r = 548 \sqrt{\frac{d_1 d_2}{f(d_1 + d_2)}} \tag{2.42}$$

For $h > r$ the diffraction loss, L_m, is [39]

$$L_m = 20\log\frac{h}{r} + 16 \tag{2.43}$$

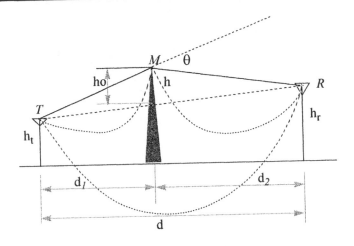

Figure 2.15 Single knife edge diffraction

Usually a parameter, v is introduced and it is defined as:

$$v = -h_o \sqrt{\frac{2}{\lambda}\left(\frac{1}{d_1} + \frac{1}{d_2}\right)} \tag{2.44}$$

This parameter indicates the extent to which the diffractor blocks the line of sight. The diffraction loss is approximately given by simple empirical functions for different ranges of parameter v. The loss is given by [90]:

$$
\begin{aligned}
L_m &= 0; \qquad v \geq 1 \\
L_m &= 20\log(0.5 + 0.62v); \qquad 0 \leq v \leq 1 \\
L_m &= 20\log(0.5\,exp(0.95v)); \qquad -1 \leq v \leq 0 \\
L_m &= 20\log 0.4 - \sqrt{0.1184 - (0.1v + 0.38)^2}; \qquad -2.4 \leq v \leq -1 \\
L_m &= 20\log\frac{0.225}{v}; \qquad v < -2.4
\end{aligned}
\tag{2.45}
$$

EXAMPLE E-2.8.2 Propagation Loss for a Single Knife Edge

Find the total propagation loss for a link of geometry shown in Figure E-2.8.3 for operating frequencies of 450 and 896 MHz.

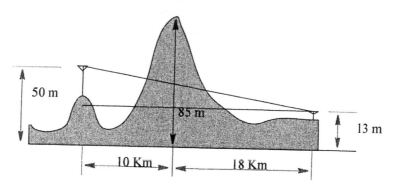

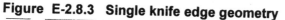

Figure E-2.8.3   Single knife edge geometry

Solution:

At 450 MHz, λ = 0.67m and at 896 MHz, λ = 0.335m h_p = effective
ridge height = 85 - {(50-13)/28000) x 18000 +13}   = 48.21 m

Loss Calculation at 450 MHz (L_{pf1})

v = -48.21 $[(2/\lambda)(1/d_1 + 1/d_2)]^{1/2}$ = -1.039,
The excess diffraction loss = 20log{0.4 - (0.1184 -(0.1 v+ 0.38)^2)$^{1/2}$}
– 14.17 dB.

Free space propagation loss = 32.44 + 20 log$f$+ 20 log d = 114 dB,
The total loss = 128.17 dB.

Loss Calculation at 896 MHz (L_{pf2})

r = 548 (18 x 10/(896 x 28))$^{1/2}$ = 46.41 (using (2.42)) L_m = 20
log(48.21/46.41) + 16 = 16.33 dB Or using empirical relations given Equation
(2.43), we have v = -1.471 L_m =20 log {0.4 - (0.1184 - (0.1v+ 0.38)^2)$^{1/2}$}   =
16.67 dB

Total loss of 16.66+114=146.66 dB.

The total loss of 144.84 dB obtained using diffraction loss formula is 1.82 dB
lower than that calculated using (2.43), which is insignificant considering that
signal strength prediction is a low accuracy estimation. Thus, we can say that
the two approaches give similar results.

2.8.5 Loss Due to Multiple Mountain Ridges

It is common to come across multiple ridges; mountainous terrain and
high-rise buildings in urban environment are two common examples of multiple ridges. Several models have been suggested for estimation of the excess
loss for multiple knife edge diffractors. Bullington [18] assumed that the total

diffraction loss may be estimated by an equivalent single knife edge with a height calculated from the geometric construction shown in Figure 2.16. This model suffers from inaccuracies since many different geometries of multiple diffractors of different heights and locations can result in the same equivalent knife edge height and location. This model, however, gives reasonable accuracy when the knife edges are located close to each other.

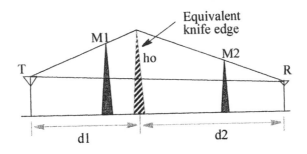

Figure 2.16 Bullington's Model for Multiple Knife Edge Diffractors

Epstein and Peterson [40] proposed that the total excess loss for two diffractors can be considered simply as the sum of the individual losses calculated independently as illustrated in Figure 2.17.This method gives accurate results only when diffractors are far apart; otherwise it is a poor predictor.

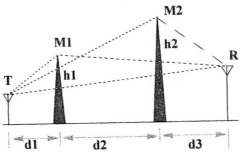

Figure 2.17 Epstein and Peterson Model for multiple knife edge diffraction loss

Picquenard [41] developed an alternative method, shown in Figure 2.18, for the estimation of loss due to two diffractors. This method removes the limitations of Bullington's and Epstein-Peterson's models. The loss due to a double diffractor is obtained by first estimating the loss due to the first diffractor (closer to the transmitter i.e. TM_1R) while ignoring the second diffractor near the receiver. The loss due to the second diffractor (M_1M_2R) is then calculated on the assumption that the transmitter is located on top of the first diffractor. The sum of these losses yields a fairly good estimate of actual diffraction loss.

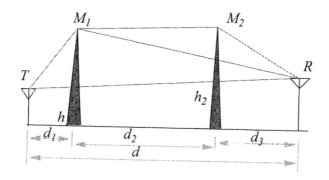

Figure 2.18 Plcquenard's Model for Multiple knife edges

EXAMPLE E-2.8.3 Propagation Loss for Double Knife Edge

Estimate the excess diffraction loss at a transmission frequency of 821 MHz for the double knife edge shown in Figure E-2.8.4 using models due to Bullington's, Epstein-Peterson's and Piquenard.

Solution: $\lambda = 0.365$ m

(a) Bullington's Method:

Let the equivalent knife edge occurs at a distance X from point A, and has a height of H. Then from geometry we get X = 2670m and height of the equivalent knife edge above the line of sight joining the antennas is 70.50 m. The parameter v is calculated as:

$$v = -h_o\sqrt{\frac{2}{\lambda}\left(\frac{1}{d_1}+\frac{1}{d_2}\right)} = -70.50\sqrt{\frac{2}{0.365}\left(\frac{2670 + 1080}{2670 \times 1080}\right)} = -5.95 \qquad \text{(E-2.8.4)}$$

Excess loss = $L_e = L_T = -20\log_{10}(-0.225/v) = 28.44$ dB.

(b) Epstein-Peterson Method:

The total excess loss L_T is given by

$$L_T = L(TM_1R) + L(TM_2R) = L_1 + L_2.$$

$h_{p1} = 19.8$ m and $hp_2 = 65.27$; v_1 and v_2 are calculated to be -1.89 and -6.2869.

Using these values the loss is calculated to be, $L_1 = 18.89$ dB, and $L_2 = 28.85$ dB. The total excess loss is therefore $L_T = 47.74$ dB.

(c) Picquenard's Method:

$L_T = L(TM_1R) + L(M_1M_2R) = L_1 + L_2$. L_1 is identical to the value calculated in part (b) and is equal to 18.89 dB.

For L_2, $h_p = 58.67$, $v = -5.32$, giving the loss $L_2 = 20 \log(-0.225/-5.32) = 27.47$ dB. Therefore the total excess loss = 45.80 dB.

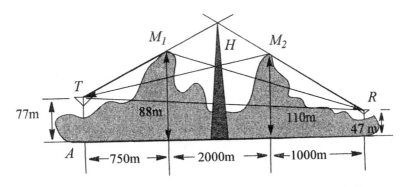

Figure E-2.8.4 Double knife edge geometry example 2.8.3

These calculations show that Bullington's model produces optimistic loss figures, while the Epstein-Peterson's and Picquenard's models produce comparable results.

2.8.6 Street Orientation and Environmental Clutter

2.8.6.1 Environmental Clutter:

Okumura classified environmental clutter and ground obstacles into three distinct groups [22]:

    (i) Open Area: This category comprises open spaces devoid of tall trees or buildings in the propagation path over a span of 300 meters or more. Typical examples include open fields and farmland.

    (ii) Suburban Area: A village, or a highway scattered with trees and houses, falls into this category; such an area has some obstacles near the mobile antenna, but the area is not very congested.

    (iii) Urban area: A downtown centre crowded with large buildings, or a densely packed residential area with tall

---

trees and houses of two or more floors fall into this category.

It was noted in Section 2.4 that "ducting" caused by streets with tall buildings on either side have a significant effect on the signal strength. This phenomenon was investigated in [43] where measurements in Manhatten, New York were reported. When measured along the street where the transmitter is located the signal may be as much as 20 dB stronger than that along the street which is oriented perpendicular to the other street [42].

Trees form an important class of environmental clutter. The presence or absence of leaves affect their signal attenuation characteristics. A tree with heavy foliage causes a shadow zone, whereas a tree without leaves may be regarded as a knife edge diffractor. Measurements at UHF conducted by Reudink and Wazowicz [44] indicate that a tree full of foliage in summer, has approximately 10 dB higher loss mainly because of shadows as compared to the loss due to the same tree without leaves in winter when it acts as a radio-wave diffractor. Seasonal differences in the propagation loss have also been reported in [45]. An attempt was made by LaGrone *et. al.* [46] to predict losses at a receiver within a grove of trees on the assumption that trees can be represented by knife edge diffractors. At frequencies of 500 MHz and above, measurement results agreed well with the theoretical diffracting model over distances greater than five times the average tree height, but at smaller distances diffraction theory does not hold due to the imprecise definition of distance from the tree. At lower frequencies the absorption is found to be 1.6 dB per 30m at 82 MHz, rising to 2.4 dB at 210 MHz. This result agrees well with the diffraction theory.

ITU-R P.833-2 gives terrestrial path attenuation with one terminal in the woodlands as [47]

$$A_{ev} = A_m \left[1 - \exp\left(-\frac{d\gamma}{A_m} \right) \right] \tag{2.46}$$

where d is the length in meters of the path in woodlands, γ is the specific attenuation for very short vegetative paths in dB/m, and A_m is the maximum attenuation for one terminal within a specific type and depth of vegetation in dB. The value of specific attenuation due to vegetation, γdB/m, depends on the species and density of vegetation. Its value is a function of frequency. For horizontal polarization it changes as a decade change in attenuation for a decade in change in frequency. At 100 MHz, its value is 0.02 dB/m. The value of maximum attenuation, A_m, is limited by scattering from the surface wave. Measured A_m in the frequency range of 900-1800 MHz is given by

$$A_m = 0.18 f^{0.752} \qquad\qquad (2.47)$$

where f is the frequency in MHz.

2.9 Propagation Considerations Towards Total Integration

Until recent past, the term "mobile communications" normally presupposed the use of vehicles by subscribers; the term has become more general in its use as it includes all users irrespective of their locations and speeds at which they move. A number of new applications of wireless systems have recently emerged; these include portable telephone facilities, mobile data terminals, data distribution systems, various emergency services, communications to and from a moving train, paging, communications in the office and multimedia communications [48]-[53]. Indeed, many of these applications are already commercially available, many others are still in the advanced stages of development. The frequency range of considerable interest is 800 MHz to 2.5 GHz, though higher frequencies in 5GHz, 20 GHz and even 60 GHz are being considered for indoors wireless LANs and outdoor Local Multipoint Communication Systems (LMCS) services. These frequencies are preferable because sufficiently small portable subscriber equipment can be designed at these frequencies. The WARC (World Administrative Radio Conference) decisions to allocate much of the 825-845/870-890 MHz and 2.5 GHz bands to mobile and personal/portable communications respectively triggered considerable activity in the measurement and characterization of radio wave propagation in these bands. A similar increase in the number of reports published on propagation characterization in the millimeter band is also noticed.

A number of results on measurements made inside and around buildings, and losses due to penetration of radio waves when a building is illuminated by an external transmitter have been reported [54]-[64]. Personal portable communication often takes place within such an environment, so that building materials and its interior configuration are major determinants of the propagation loss. Measurements relating to such losses are presented in the following sections.

2.9.1 Street Level Signal Strength Variations

Portable communication devices operating at street level in an urban setting encounter an environment similar to that of vehicular mobile communications with one exception. The rate of change of signal strength caused by the relatively slowly moving terminal is much slower in the case of portable communications. Also, the received signal can experience rapid changes due to changes in the antenna orientation. As in the vehicular case, propagation losses vary considerably with antenna height and with the nature of the path from transmitter to receiver.

The signal strength received by a hand-held terminal is dependent on the type of structure surrounding it, e.g., multi-floor buildings, residential dwellings, open space etc. While within fixed structures, typically two cases occur: one for the transmitting antenna close to the ground level (mobile transmitting) and the other when it is located at a considerable height (base station transmitting). Moreover, due to proximity of portable terminals to the buildings, the signal strength may be more susceptible to shadows. Researchers from Bell Laboratories and Nippon Laboratories conducted outdoor measurements. Their findings are summarized in Table 2.2.

2.9.2 Propagation Within Buildings: External Signal Source

The first case that we shall consider is that of signal strength measurements within a building when the transmitter is located outside the building. The signal is affected by the loss due to the building shell, and the loss due to propagation of the signal within the building. Table 2.4 provides a summary of work reported on this aspect.

Floor height effects on the signal strength have been measured by a number of researchers [57][58][63]. The general conclusion is that the penetration loss in dB is inversely proportional to floor height. The slope of the line representing the path loss with the floor number has been estimated to be -1.9 dB/floor by Walker [57]; a similar figure has been suggested by Cox et.al. [56], with a condition that this slope does not apply to locations below ground level i.e. basements, where an additional penetration loss of approximately 10 dB should be added. Pillmeir's results indicate that penetration loss reduces by 2-3 dB for each higher floor [63].

Losses are higher than predicted by straight line approximations if the subject building is sandwiched between two adjacent buildings. The standard deviation of the signal strength for floors lower than the 12th has been found to be about 10 dB; for higher floors a standard deviation of 5 dB has been reported. Note that these apply to the cases when the transmitting antenna is installed much higher than the surrounding buildings.

2.9.3 Loss Sensitivity to Transmitter Location and Building Materials

The way in which a building is illuminated also affects the penetration loss. The orientation of the serving transmitter to the building could result in a penetration loss of as much as 5 dB with a standard deviation of about 4 dB

Table 2.3		800-900 MHz outdoor Propagation studies	
Measurements conducted at	**Frequency MHz**	**Propagation loss d^{-n}**	**Statistics**
Nippon telegraph and Telephone [55] [a]	836.675	(i) Close to ground: d^{-4} + excess loss of 18 dB as clutter factor in heavily built-up areas. Propagation loss deviates from free space loss. (ii) High above the ground: d^{-2} + 4 dB clutter factor.	Over a distance of 300m the signal fluctuations follow Rayleigh distribution in urban areas. For residential area the loss tends to follow lognormal.
Bell Labs [b]	815	High above the ground: d^{-2} + 2 dB, closer to ground: $d^{-4.5}$ (average), $d^{-3.1}$ (over distance of 400-800 m: $d^{-4.1}$ (over distance of 800-1600m)	Lognormal

a. : The measurement conditions were:

Transmitter power: 0 dBm

Antenna type: Standard half wave dipole for both transmitter and receiver.

Antenna Height for near ground measurements: Transmitter 1.5 m and the receiver 1.2 m above the ground and for high above ground measurements both the transmitter and receiver antennas were 7-15m above ground.

Antenna polarization: Vertical

b. The measurement conditions were:

Transmitter power: 0.8 watt

Transmitter antenna: $\lambda/2$ coaxial sleeve

Receiver antenna: collinear array (four dipole elements)

Receiver antenna gain: 5.8 dB over dipole

Receiver antenna height: 27 feet above the ground.

[58]. Penetration loss is sensitive to the type of building material. Metal clad buildings are expected to show the highest attenuation, wooden building the least. The difference between the respective penetration losses has been measured to be between 16 and 30 dB [57]. A complete blockage of the signal has been reported to occur by copper sputtered windows [58]. Buildings with windows may have interior signal levels up to 5 dB higher than similar areas without windows. For instance, an enclosed area such as a group of individual offices will typically show an additional loss of 3 dB when compared to one for open spaces (cafeterias, conference halls, open-concept offices, etc.) [57][64].

As in the exterior transmitter case, there are two distinct aspects of propagation:

Table 2.4		Propagation loss due to building penetration and distance inside the buildings
Measurements conducted by	**Frequency MHz**	**Type of Building, penetration loss and signal statistics**
Walker [58]	850	Average penetration loss is 18 dB for downtown high rise buildings with reinforced concrete structures, 13dB for typical suburban buildings. Fluctuation statistics are not reported.
Cox et.al.	815	Penetration loss for all-steel buildings is 24 dB, for metal siding 11-20 dB, for wooden buildings 4-8 dB, (loss depends on position and orientation of the receiver with respect to the transmitter). The signal fluctuations have been reported to follow a Rayleigh distribution with constant median.

(i) spatial distribution of signal amplitude;

(ii) distance dependency of the median signal power.

Signal amplitude distribution depends on relative locations of the transmitter and the receiver. If the transmitting antenna is located outside the building, the specular component of the signal may be completely missing and propagation occurs through the multipath mode. Under these conditions measurements indicate that the signal amplitude is described by a Rayleigh distribution [65]. However when the line of sight between the transmitting and the receiving antenna is present, the signal magnitude statistics become Rician or Nakagami.

The distance dependency of the median signal strength inside the building has been found to be dependent on floor height. Measurements [57][66] indicate that the variation with distance from the transmitter is $d^{-3.9}$ for the first floor, $d^{-3.0}$ for the second and $d^{-3.2}$ for the basement. Inside a building the rate of decrease in signal strength with distance from the transmitter has been found to be less than the rate exterior to the building. Penetration loss is also sensitive to the interior of the building. Rappaport [59] and Violette [60] have studied propagation losses due to common building materials. More research is necessary to determine more precisely the effects of materials, floor height and configuration on signal propagation. Penetration losses of some materials are: concrete wall (8-15 dB), concrete floor (10 dB), sheetrock (2-5 dB), plywood (1-14 dB) etc. These losses increase at higher frequencies.

2.9.4 Propagation Within Buildings: Internal Signal Source

In a typical indoors portable wireless system, a base station antenna is installed at an elevated position most likely at ceiling level to communicates with a number of portables within that building. The indoors radio environment is determined by the internal structure i.e. partitions, ceiling heights and corridors of the building. Two types of measurements are commonly made: narrowband and wideband.

Alexander [61] studied propagation in a building and measured signal strength within rooms at a frequency of 941 MHz; the transmit power was 30 mW, radiated by a 1.2 meters high vertically polarized dipole. The distance-power law,

$$S(dB) = -10\alpha \log D \qquad (2.48)$$

was tested to fit the measured results. The departure from the best fitting straight line was less than 10 dB in most cases. The distance-power law gradient for steel partitioning was found to be 5.7 compared to a theoretical value of 2 for free space (d^{-2}). Whiter and Gibson [62] installed a transmitter on one of the floors of a three floor building and observed that average signal strength remained almost the same on all three floors examined. It was concluded that the presence of steel reinforcing mesh in the floor did not produce significant attenuation. However, in [59] a propagation loss figure of 8-15 dB is given. Inside elevators, the signal was found to be weaker by as much as 20 dB depending on the location and orientation of the elevator. This excess loss is probably due to the metallic cage forming the elevator box.

More recently, a large volume of measurement results on this topic appeared on this topic in the open literature. The significant finding suggests that the model of (2.48) generally applies at most all frequencies of interest but no agreement exists on the value of α. Such variations are not beyond expectations, since path loss factor is strongly dependent on the internal layout of the building and furnishing therein. For example, the reported values of α are 1.5 to 1.8 for LOS, 1.4 to 3.3 for several manufacturing floors [67]; less than 2 for hallways and 3 within a room in an office building [68]; equal to unity at 1700 MHz and ranges from 1 to 3 at 433 and 861 MHz for LOS measurements in the same corridor [69]. For non-LOS links values of α between 3.3 and 4.5 are suggested.

Besides the distance power law, other loss models have been suggested. A distance loss model with a distance dependent breakpoint is considered to be more realistic. For example, in [70], the mean power of received signal is

$$P = \frac{C_p}{d^\alpha (1 + d/g)^\beta} \qquad (2.49)$$

where C_p is a constant, d is the distance from the transmitter, g is the turning point of the path loss curve, α is a basic attenuation rate and β is an additional attenuation rate for distance greater than the breakpoint.

A more refined model, Ericsson's multiple breakpoint model, was proposed after extensive measurements in a multi-floor building [71]. The model presents an upper and lower bound loss and has four breakpoints. The model assumes attenuation of 30 dB at a distance of 1m from the transmitter. The attenuation slopes change from 20 dB/decade to the first breakpoint to an attenuation of D^{-12} for the final section of the model.

At millimeter wave frequencies, loss in dB per unit distance is frequently quoted. Measurements at 850 MHz, 1.9, 4, and 5.8 GHz showed that a free space loss plus a linear loss of 0.3 to 0.6 dB per meter is sufficient to describe the environment with very little dependence on frequency [72].

2.9.5 Propagation in Tunnels

Radio frequencies suffer considerable propagation loss in tunnels, making radio communications difficult to establish. However, radio propagation within tunnels is of considerable interest to designers who are required to provide communications to people riding in trains, vehicles and mine trolleys during their passage through tunnels. Signal reception inside a tunnel when the transmitting antenna is situated outside the tunnel is practically impossible unless the tunnel is very short. Coverage within tunnels is therefore provided either by transmitting antennas located at the tunnel ends to illuminate the tunnel, or by a leaky feeder radiator, which is run along the whole length of the tunnel. The first method is more appropriate for short length tunnels while the second method is typically used for long tunnels. Experiments have been conducted to measure signal attenuation while the tunnel is illuminated [73] at frequencies of 153, 300, 600, 900, 2400, 6000, and 11200 MHz. Frequencies within the range 2400-11,000 MHz suffer an attenuation of 2-4 dB per 300 meters. At frequencies below 2400 MHz, the propagation loss is proportional to $d^{-\alpha}$ and the value of 'α' has been found to be between 2 and 4, the higher figure of α is measured at 900 MHz. At 153 MHz the signal attenuates at a rate 40 dB/300 meters, whereas the attenuation rate drops to 20 dB/300 meters at 300 MHz. Radio propagation within mines and tunnels, in the frequency range of 438 MHz to 24 GHz have been subject of research by Davis et.al. [74]. It was concluded that major source of attenuation are the bends and obstructions in the tunnels. The variation of loss with distance over the considered frequency range is presented in Figure 2.19.

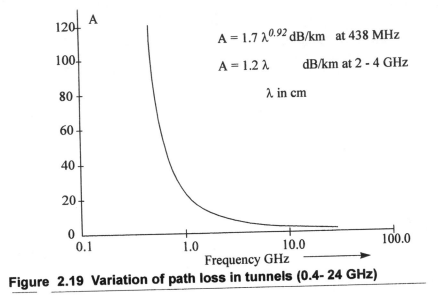

$$A = 1.7\,\lambda^{0.92}\ \text{dB/km}\quad\text{at 438 MHz}$$

$$A = 1.2\,\lambda\qquad\text{dB/km at 2 - 4 GHz}$$

$$\lambda\ \text{in cm}$$

Figure 2.19 Variation of path loss in tunnels (0.4- 24 GHz)

2.10 Shadowing in Wireless Communications

It was noted earlier that when the mobile terminal moves behind a large structure, the mean signal strength suffers a gradual loss. The phenomenon of slow degradation, in the mean signal level as the terminal moves into the shadow zone and gradual improvement as the mobile moves out of it, is known as shadowing. The relative rate at which signal level decays and then improves depends on the position of the terminal relative to the obstruction. When the terminal is nearer an obstacle, the fade rate due to shadow is faster than for the case when the distance between the structure and the mobile is relatively large. The shadowing affects the mean signal at the mobile terminal and produces slower variation relative to faster Rayleigh fading discussed in detail in Chapter 3. Lognormal distribution is an accepted statistical model for shadowing. The pdf of the mean signal (or median) strength is given by:

$$p(r) = \frac{1}{r\sigma_l\sqrt{2\pi}}exp\left(-\frac{1}{2}\frac{(\log r - \mu_l)^2}{\sigma_l^2}\right);\qquad r > 0 \tag{2.50}$$

$$= 0;\quad\text{otherwise}$$

where r the signal magnitude is expressed in linear units, μ_l is its mean and σ_l is its standard deviation. When the signal is expressed in dBs, the resulting pdf is given by the standard normal distribution:

$$p(r_d) = \frac{1}{\sigma_d\sqrt{2\pi}}\,exp\left(-\frac{(r_d-\mu_d)^2}{2\sigma_d^2}\right) \qquad (2.51)$$

where μ_d and σ_d is the mean and the standard deviation expressed in dB respectively.

The environmentally dependent standard deviation represents shadowing depth. Standard deviation for several terrain types have been measured by many researchers [3][9][12][23]. The shadowing depth, in dB, varies from a low of approximately 2 dB in open areas to around 12 dB in heavily built up urban areas; typical downtown areas of a large American city is a good example of urban environment which would give rise to a large shadowing depth. Deep shadowing depths are encountered in the indoor environment because of the presence of walls with changing layout. It will be seen from an example later that shadowing could extend the coverage area beyond that predicted by the distance loss models. Increased interference and larger reuse distance are important consequences of shadowing.

2.11 Statistical Estimation of Area Coverage

Signal strength prediction techniques presented earlier in the foregoing sections are based on local median signal strength. It should be noted that over an irregular terrain, the actual received signal level at a given point may vary by as much as 30 dB from the mean as the receiver relocates from one point to another. Clearly, a statistical measure of the grade of service (GoS) is needed.

The grade of service specifies the mobile radio system service quality. Grade of service is defined in several ways; minimum acceptable signal to noise ratio at certain percentage locations of the service area is a criterion most widely used. Other criteria like minimum bit error rate, outage probability rate, signal to noise plus interference are also used. The system operator is expected to maintain a certain grade of service over the entire service area. We shall first determine signal coverage in a small area.

2.11.1 Signal coverage in small area

In a small area the local mean has a lognormal distribution defined in (2.51). The probability that r_d exceeds the threshold of reception r_o is

$$P(r_d \geq r_o) = \int_{r_o}^{\infty} p(r_d) dr_d = \frac{1}{2} + \frac{1}{2} erf\left(\frac{r_o - r_d}{\sigma_d \sqrt{2}}\right) \tag{2.52}$$

If the threshold of reception, r_o, is defined and the mean signal level, r_d, is measured at a certain distance from the base, then (2.52) calculates the percentage of locations where reception threshold is met. For example, for a threshold of reception of -115 dBm, σ_{rd} of 7 dB, and r_d of -108 dBm, 84% of locations at this distance will satisfy the reception threshold.

2.11.2 Wide Area Coverage Estimation

In addition to finding percentage of a small area that has adequate coverage area, it is of interest to extend this to larger coverage area, that is the percentage of locations within a circle of radius R where the received signal strength exceed a given threshold.

Let service area be defined as

$$A_c = \int p da \tag{2.53}$$

where p is the probability of adequate service for the incremental area da. When the area is divided in elementary areas i.e. a grid, the integral may be approximated by a summation

$$A_c = \sum_i p_i a_i \tag{2.54}$$

In this case, the probability of adequate service in the elementary areas a_i is p_i. The fraction of the total area A with adequate received signal strength is given by

$$Q = \frac{\sum_i p_i a_i}{A} \tag{2.55}$$

The choice of a_i clearly depends on the terrain. For the statistical approach to be valid, the principal requirement is that there exists a reasonable homogeneity within the elementary area a_i. Thus, grid dimensions may be different from terrain to terrain. Thus, incremental areas in flat regions can be considerably larger than those in hilly areas. If the terrain within a service area is sufficiently uniform, large area signal variation may be modeled as a log-normal distribution (shadowing) whose standard deviation depends on the type of terrain. Two approaches, one analytical and the other graphical, can be used. The analytical approach requires the behavior of the mean signal strength with the distance i.e. $r^{-\alpha}$. Thus, r_d at a distance R may be written as

$$r_d = r_r - 10\alpha \log_{10}\left(\frac{r}{R}\right) \tag{2.56}$$

where r_r, expressed in dB, is a constant determined from the transmitter power, antenna height and gains etc. It could be 1 km intercept. Equation (2.52) may be rewritten as

$$P(r_d \ge r_o) = \int_{r_o}^{\infty} p(r_d) dr_d = \frac{1}{2} - \frac{1}{2} erf\left(\frac{r_o - r_r + 10\alpha \log_e\left(\frac{r}{R}\right)}{\sigma_d \sqrt{2}}\right) \tag{2.57}$$

By letting $a = (r_o - r_r)/\sigma_d \sqrt{2}$, $b = 10\alpha \log_{10} e / \sigma_d \sqrt{2}$ and using in (2.53), we get

$$A_c = \frac{R^2}{2}\left\{1 + erf(a) + \exp\left(\frac{2ab+1}{b^2}\right)\left[1 - erf(\frac{ab+1}{b})\right]\right\} \tag{2.58}$$

The above equation can be used to estimate the fraction of the area where the signal strength exceeds the threshold. The graphical method was proposed by Barsis [88], we use the following example to illustrate how to estimate the service area using the graphical technique.

EXAMPLE E-2.11.1   Calculation of Cell Coverage

The base station antenna shown in Figure E-2.11.1 is used for coverage of a single cell of a radio mobile system. As the figure shows, the surrounding areas include a part of the city core to the north, suburban housing to the east and west, and an open area to the south. The 25 meter antenna has a 6 dB gain. It is fed from a 10 watt source through a 12 meter cable with a total attenuation of 4 dB. The mobile receiving antenna is 3 meters high, and has a gain of 0 dB. The receiver noise figure is 12 dB and the noise bandwidth is 30 kHz. It is required to determine the contours of signal strength providing a receiver signal-to-noise ratio of 18 dB, with 90% and with 50% coverage. For this purpose, Egli's model can be used, with clutter factors of 34, 22 and 12 dB, and standard deviations of 8 dB, 4 dB and 2 dB respectively for urban, suburban and open areas. Assume both the transmitting and receiving antennas have unity gain. The Boltzmann's constant is equal to 1.38×10^{-23} K/J.

Solution:

The noise power at the receiver is $N_o = KT_oBF = 1.38 \times 10^{-23} \times 290 \times 30 \times 10^3 \times 15.85 = 1.902 \times 10^{-15}$ watts. For 8 dB carrier-to-noise ratio requirement the carrier power P_r at the receiver must be $P_r = 1.902 \times 10^{-15} \times 63.10 =$

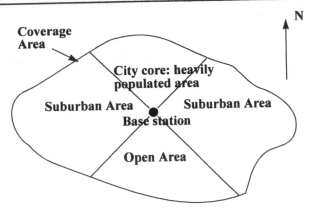

Figure E-2.11.1Geographical makeup of a single cell mobile radio system

1.200 × 10^{-13} watts or -99.2 dBm. The transmitter output amplifier generates a signal level of 10 dBW. Including antenna gain and cable loss, the transmitted power, P_t, is 10 + 6 - 4 = 12 dBW or 42 dBm. The propagation loss P_p in dB is $P_p = P_t - P_r = 141.2$ dB.

Egli's model assumes plane earth propagation with a clutter factor, which increases the theoretically predicted attenuation. Consider the ratio of mobile/base signal strength at a distance of 1 km from the base station. Assuming plane earth propagation, this is given by $P_m/P_b = (h_m\ h_b/d^2)^2 =$ 5.625 × 10^{-9} or -82.50 dB so that with the additional clutter factor, we have, loss at 1 km from the transmitter:

    (i) Loss in urban sector = -82.5 - 34 = -116.5 dB

    (ii) Loss in suburban sectors = -82.5 - 22 = -104.5 dB

    (iii) Loss in open sector = -82.5 - 12 = - 94.5 dB

The transmitted power is 42 dBm. Therefore, the received power levels, P_r at 1 km, corresponding to the above figures are: urban sector = 42 - 116.5 = -74.5 dBm, in suburban sectors = 42 - 104.5 = -62.5 dBm and in open sector = 42 - 94.5 = -52.5 dBm

Since it is known that received power loss decreases with distance by 40 dB per decade, the following Table E:2.11.1 can be constructed from the data available for 2, 5, 10 and 20 km points.

The signal strength data shown above are the median values of normal distributions, and therefore represent signal power corresponding to 50% coverage at the distance indicated. More generally, probability of reception can be plotted against signal strength, on a Gaussian probability scale, with distance as a parameter. Figure E-2.11.2 shows such a plot for the urban area with a standard deviation of 8 dB. It can be seen that, for a received signal power of -99.2 dBm, 90% coverage exists to a radius of 2.4 km in the urban zone, while 50% coverage extends to a radius of 4.4 km.

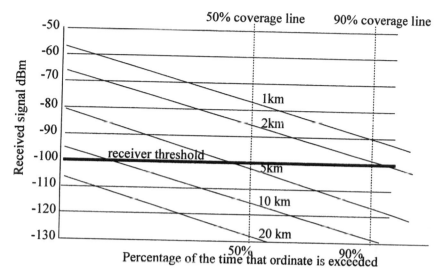

Figure E-2.11.2 Signal distribution, urban environment

Table E:2.11.1   *Median signal Strength as a function of Distance*

Distance	Received signal (dBm)		
(Kms)	urban	Suburban	Open
1	-74.5	-62.5	-52.5
2	-86.5	-74.5	-64.5
5	-102.5	-90.4	-92.5
10	-114.5	-102.5	-92.5
20	-126.5	-114.5	-104.5

Similar plots for suburban and open regions are shown in Figure E-2.11.3 and Figure E-2.11.4 respectively.

Use of this approach yields corresponding results for the suburban and open areas, as shown in Table E:2.11.2. The data in Table E:2.11.1 and Table E:2.11.2 can be used to draw approximate contour lines of coverage, as shown in Figure E-2.11.5.

For purposes of the contour map, it is assumed that there is a smooth transition between zones where they abut. Note that the resulting pattern is

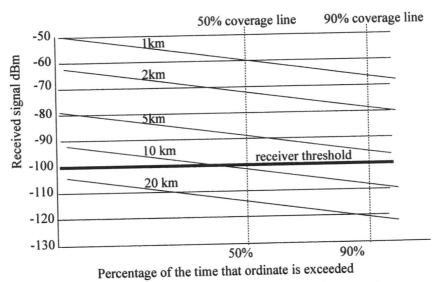

Figure E-2.11.3Signal distribution, suburban environment

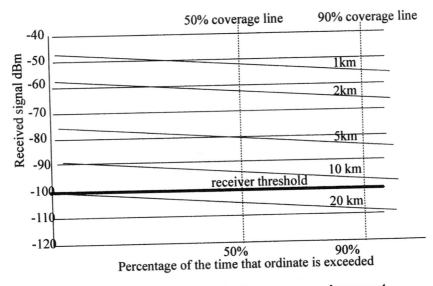

Figure E-2.11.4 Signal distribution, open environment

highly asymmetrical along a north-south axis because of the relatively large attenuation caused by the built-up region to the north of the transmitter.

Table E:2.11.2   *Radius of Coverage by Zone (Covered radius is in Km)*

Zone	50% Coverage	90% Coverage
Urban	4.4	2.4
Suburban	8.2	6.1
Open	15.0	13.0

In the case of non-uniform terrain, estimation of coverage using the method proposed by Barsis will result in considerable error. Under these conditions the designer has to use computer based coverage estimation techniques in which path profiles are stored in a database. Several such methods have been proposed [20][24][30][31][33]-[35][89]. Most of these use variations of the technique advocated by Longley and Rice [31], with modifications to the empirical expressions for propagation loss, representing whether or not the line of sight is blocked.

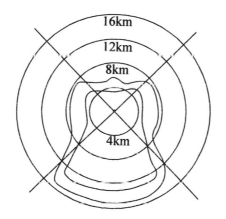

Figure E-2.11.5 Radio propagation coverage

2.12 Land Mobile Satellite Path Loss

A satellite based land mobile communication system was first proposed by the Communications Research Centre of Canada [75][76]. Later, this idea has been taken up by the FCC and NASA in the United States [77]-

[81] and in Sweden, where a similar system was proposed for a truck dispatch system [82]. Two major differences between satellite and terrestrial mobile channels are the lower elevation angle of the former and the larger distances involved. Another difference lies in the small signal characteristics, which will be discussed in section 3.2. Estimates of land mobile satellite excess path loss measurements were obtained by Hess [83] at 860 MHz, using the ATS-6 satellite. The aim was to investigate the effect of the environment on the signal. Measurements indicate that excess path loss due to environmental clutter varies from 25 dB in urban environments to 10 dB in the suburban/rural areas. The higher loss in the urban environment results mainly from shadows created by buildings, accentuated by the low elevation angle of the satellite. The excess path loss difference between 860 MHz and 1550 MHz links was found to be negligible. Huck et.al.[84] used INMARSAT's MARECS-A Satellite to simulate MSAT propagation conditions in both 806-890 MHz and 1.6/1.5 GHz bands. Their results suggest that propagation is significantly affected by both multipath and shadowing. They recommend that a conditional probability model is needed to describe the signal level distribution. Their work was extended by Loo [85].

Loo proposed a statistical model of a land mobile and obtained expressions for probability density function of the envelope and average level crossings and average fade duration, details of which will be discussed in Section 3.5. He also validated his theory experimentally using an airborne helicopter which represented a satellite. Vogel [86] measured the characteristics of the land mobile satellite path by using a hot air balloon for use in simulation studies of mobile satellite link. His results are similar to those obtained by Huck et.al. and Loo. The difficulty with geostationary mobile satellite is of large excess propagation loss due to large distance of the satellite from the earth (38,000 to 42,000 km). The large propagation loss restricts the use of geostationary mobile satellites to large power terminals e.g. in vehicle or briefcase

Motorola proposed the cellular system in the sky, Iridium [87]. The system uses a constellation of 66 low orbit satellites in six orbital planes. The satellite altitude is 780 km. Iridium provides global access to personal communications. Voice, data, fax, paging, messaging and position location services will be offered. The channel data rate is 2.4 kbits/sec and 1100 to 3840 voice circuits are available. The frequency of operation is in L-band (1500-1700 MHz) for the user terminals and Ka-band (27-40 GHz) for intersatellite links, gateways and feeder links. The mobile terminals were promised to be light, low power and small. The launching of satellites was completed in 1997. The system became commercially available in the later part of 1998. The major difficulty faced by Iridium is lack of customers probably due to very expensive rate per unit time or large customer terminal. The future viability of

Iridium remained in question and in year 2000, Iridium system was shut down. More recently, another company is trying to revive the system.

Globalstar services becomes available in 2001. The constellation of the Globalstar system consists of 48 satellites in orbit inclined at 52° with eight planes at an altitude of 1414 km. The system will cover regions of the earth from 70° S to 70° N. Globalstar will use Qualcomm's terrestrial CDMA technology for mobile link. It will offer variable data rates from 1200 bits/sec to 9600 bits/sec.

ICO[†] is another mobile satellite system which is expected to become operational in year 2001. It will consist of ten operational satellites in medium earth orbit (MEO) in two orbital planes at an altitude of 10,390 km. This is a compromise between GEO and LEO. The satellite antenna size and power are relatively modest and the latency is small. ICO system will use TDMA with 4500 voice circuits. The ICO is expected to reuse GSM technology as far as possible.

Teledesic system is far the most ambitious of the proposed satellite based mobile communication systems. The Teledesic will consists of 288 satellites divided in 12 orbital planes, each with 24 satellites. The system will offer a wide ranging bit rates - 16 kbits/sec to 2.048 Mbits/sec. The satellite altitude will be less than 1400 km.

EXAMPLE E-2.12.1 Satellite Link

A mobile satellite downlink communication path uses an SCPC/TDM system transmitting 10 W at a data rate of 2.048 Mbits/sec on a 1.56 GHz channel. The Mobile G/T ratio is -10.5 dB/^{o}K. The satellite antenna gain is 24 dB and duplexer loss is 1.5 dB. Other incidental losses are estimated to be 2 dB. Rain attenuation is estimated to be 0.5 dB. The distance of the satellite from ground is 780 km. Find E_b/N_o at the mobile ground terminal.

Solution

We prepare a link budget to find E_b/N_o. In doing so, we ignore channel fade margin or reduction of signal level because of fading and shadowing.

Description	Gains (dB relative to 1W)	Losses
Satellite Power	10.0	-
Satellite Antenna Gain	24.0	-
Duplexer	-	1.5 dB
Free space propagation	-	154.15 dB
Incidental losses	-	2.0 dB

†. *ICO stands for Intermediate Circular Orbit.*

Description	Gains (dB relative to 1W)	Losses
Rain attenuation	-	0.5 dB
Boltzmann's constant	228.6	-
Mobile G/T	-	10.5 dB
Bandwidth	-	63.0 dB
Net gains	262.6	-
Net Losses	-	231.65 dB
Carrier to Noise Ratio (E_b/N_o)	262.6 - 231.65 ≡ 31 dB	

2.13 Summary

While theoretical models for the prediction of field strength under ideal conditions have been well understood for half a century, it is only in recent years that a significant effort has been put into development of models that can be applied to more realistic conditions. In this chapter we have seen that starting from the plane earth model successive modifications result in models that are more realistic. Many of these are empirical in nature, and yield results that predict a mean signal level within a degree of accuracy. In practice, the instantaneous received signal may vary considerably from this value. It is therefore necessary to consider reception on a statistical basis. Thus, an acceptable signal may be present in a certain part of the coverage area but in other parts the signal may be so poor that communications could become difficult; alternatively, a moving vehicle may receive an acceptable signal during a part of the intended duration of reception. Signal prediction techniques in different environments were presented. These include terrestrial mobile systems for use indoors and outdoors. Mobile satellite channel characteristics were also discussed. A more detailed examination of the instantaneous signal received by a moving terminal reveals an envelope whose magnitude varies rapidly with time. The rate of variation is related to the speed of the terminal. To design reliable mobile communications systems, it is necessary to statistically analyze these variations. Such analyses are the subject of the next chapter.

Problems

Problem 2.1:

Okumura's measurements form the basis of several path loss prediction models, identify the applicable range of frequency, link distance, and base station antenna height in Okumura's model. How Okumura's measurements fit into Hata's empirical model? What are the limitations of Hata's model?

Problem 2.2

The received power at a distance of 1 km is equal to 1μW, find the received powers at a distance of 3km, 7 km, 15 km and 20 km from the same transmigrator the following path loss models:

(a) free space,
(b) $\alpha = 3$,
(c) $\alpha = 4.5$,
(d) two ray plane earth model,
(e) extended Hata Model.

The other link parameters are, $f = 1900$ MHz, $h_t = 40$m, $h_m = 3.5$m, $G_t = G_r = 0$ dB.

Problem 2.3

A 10W, 1800 MHz signal is fed to a 6-dB gain transmit antenna. Determine the received signal power in dB by a 3dB receive antenna at a free space reference distance of 100m. Assume that the system loss factor not related to propagation equals 3dB.

Problem 2.4

(a) An improvement in the propagation coverage can be obtained by raising the base station antenna height. On the assumption that the received signal at a certain location is -112 dBm when the base station antenna height is 80m, to what height the base station antenna must be raised in order to increase the signal strength at that location by 7 dB?

(b) If the propagation path loss varies as d^{-4}, what percentage increase in coverage distance is possible if the receiver is made 5 dB more sensitive?

Problem 2.5

In certain measurements over a quasi-smooth terrain, the 1 km intercept was measured to be -67 dBm. It was determined that over a distance of 10 km, the signal attenuates at a rate of 39.8 dB per decade increase in distance. The measurements were conducted with a base station having the following parameters: Base station Power 10W, Antenna height 50m, Antenna Gain G_t 6 dB, Mobile antenna height 3m, Mobile antenna gain 0 dB. If the base station parameters are modified to allow for greater range, find the distance beyond which the signal power falls below -112 dBm. The new parameters are: Base station Power 40 W, Antenna gain 8 dB, height 60 m, Mobile antenna height 3m, Mobile antenna gain 1.5 dB.

Problem 2.6

A tour company operates a river excursion boat and wants to install a mobile link with a base station located on the roof of a building 60 meter high. The building is located at a distance of 7 km from the north river bank. The initial experiments with the link proved successful at all times except when the boat goes behind a high rise building located 3 km from the north bank. The obstructing building is 93 m tall.

(a) Find total propagation loss between the base and the mobile when:

(i) the mobile is at midstream,

(ii) the mobile is at the south bank if the river is 1 km wide.

(b) The height of boat's mobile antenna is 6m, and the frequency of transmission is 840 MHz. If the intervening land is quasi-smooth and contains urban construction, determine the total propagation loss when the boat goes behind the high rise building.

Problem 2.7

A consultant is hired by a taxi dispatch operator to design a mobile link in a large city. The operator has a licence to operate radio equipment at 463 MHz with a power of 100 Watts. He can install the base station antenna on a building at a height of 80m. The topography of the area surrounding the base is shown in Figure P-2.7. Find the signal power at the airport 40 km from the base station. The intervening terrain is made up of urban terrain ($\alpha = 3.9$) for the first six km and suburban terrain ($\alpha = 3.4$) for the remainder of 34 km. The propagation loss varies as $d^{-\alpha}$.

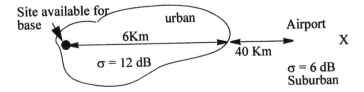

Figure P-2.7 Basestation location relative to the airport

Problem 2.8

For the link shown in Figure P-2.8, it is required to measure the signal strength when the mobile unit is moved towards the transmitter starting from a location 12 km away. The signal received at the mobile at a distance of 15 km is measured to be -98dBm, and the intervening terrain is such that the signal attenuates 12.4 dB for every doubling of the distance. Draw a graph showing variation of signal strength with distance.

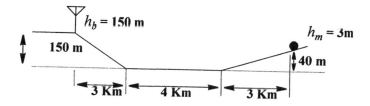

Figure P-2.8 Base station and mobile locations for Problem P-2.8

Problem 2.9

A cellular system antenna is located on top of a tower at a height of 98 m and is connected to 40 Watt transmitter located at a height of 79m via a cable which attenuates the signal by 0.12 dB per meter. The transmitting antenna gain is 6 dB. The tower is located 2.5 km from the shoreline of a lake. Estimate the propagation loss to a boat which is 12 km from the shore.

Problem 2.10

For the link parameters shown in Figure P-2.10, estimate the propagation loss at frequencies of 462 and 928 MHz, based on models of:
(i) Bullington, (ii) Petersen and Epstein, (iii) Picquenard.

Problem 2.11

A receiver in an urban cellular radio system detects a 1mW signal at d $=d_o=1$ meter from the transmitter. In order to mitigate co-channel interference effects, it is required that the signal received at any base station receiver from another base station operating at the same frequency must be below -100 dBm. A measurement has determined that the average path loss exponent in the system is 3. Determine the separation distance between the base stations.

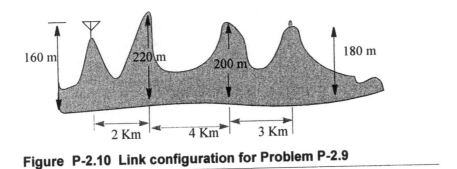

160 m 220 m 200 m 180 m

2 Km   4 Km   3 Km

Figure P-2.10 Link configuration for Problem P-2.9

Problem 2.12

(a) A 150 m high base station serves a downtown core where the mean signal strength is log normally distributed with a standard deviation of 11 dB. The mean signal strength at a distance of 1 km from the base station is measured to be -57.8 dBm, and the signal attenuates at a rate of 41.3 dB per decade increase in distance. Estimate what fraction of locations will receive adequate signal strength (-102 dBm) at a distance of 23 km.

(b) Find the radius of the circle within which 90% of the locations have adequate signal strength.

Problem 2.13

A transmitter provides 10W to an antenna with 6 dB gain. The 200 kHz bandwidth receiver is connected to an antenna having a gain of 2 dB. If the receiver noise figure is 4 dB and the carrier frequency is 920 MHz, determine the maximum transmitter-receiver separation that will ensure a SNR of 25 dB for 95% of the time. Assume propagation loss varies as 37 dB/decade beyond a distance of 1 km. Free space loss may be considered to be applicable up to a distance of 1 km.

Problem 2.14

(a) A medium size ancient city is said to have been built on a quasi-smooth terrain sloping at an angle of $2°$ as shown in Figure P-2.14. A 35m long base station antenna is installed on a 100m high ridge as shown. The mobile antenna is 1.5m high. The transmitter power is measured to be +50 dBm. Use Hata's empirical formula and Okumura's correction factors to find the signal strength a the location marked •. The frequency of operation is 850 MHz. Assume antenna gains to be unity.

(b) If the receiver at the location has a noise figure of 4.5 dB and the receiver is attached to a 3m antenna. Find the signal to noise ratio when the receiver bandwidth is 1.25 MHz.

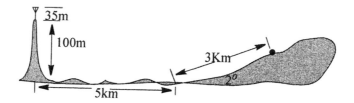

Figure P-2.14 The geometry of medium sized city built on a quasi-smooth sloping terrain

Problem 2.15

In a 1800 MHz cellular system, the base transmitter is transmitting an EIRP of 150W using a 0 dB gain antenna. The receiver having a bandwidth of 30 kHz has a noise figure of 8 dB. A SNR at the receiver is desired to be 24 dB. The terrain intervening the transmitter and the receiver is flat with some built up areas, which causes shadowing of standard deviation of 6 dB. Determine the percentage of the time that the desired signal is available at a distance of 6 km.

Problem 2.16

Explain what is fade margin and why it is required in the link analysis of wireless communication systems. Also, discuss what happens if the fade margin is not set appropriate level.

A transmitter delivers 10W to a 10dB gain antenna. The receiver antenna having a gain of 2.5 dB is connected to a receiver of bandwidth 200 kHz. The receiver has a noise figure of 8 dB, the carrier frequency is 900 MHz and the antenna temperature is 18°C. Determine the maximum T-R separation that will ensure that a SNR of 18 dB is provided for 95% of the time. Assume propagation constant of 4, $\sigma = 8$dB and $d_o = 1$km.

Problem 2.17

Calculate the maximum permissible path loss P_{Lmax} for the following link.

Transmitter power = 30 dBm
Transmit antenna gain = 6 dB

Receiver antenna gain = 1.25 dB
Line losses in the transmitter system = 1 dB
Line losses in the receiver system = 1 dB
Fade Margin = 15 dB
Noise power spectral density = -170 dBm/Hz
Bit rate 384 kbits/sec
Required E_b/N_o = 27 dB.

Problem 2.18

(a) Find the transmitter power from a portable terminal operating at 1678 MHz to deliver E_b/N_o of 6 dB at one of Irridium satellite happens to be at a height of 770 Km above ground. The noise figure of the satellite receiver is 2.5 dB. The transmitted signal occupies a bandwidth of 10 MHz.

(b) If the terrain in which the mobile is operating results in shadows with a standard deviation of 4 dB. Find the probability of signal outage i.e. Prob $(E_b/N_o < 6$ dB).

Problem 2.19

Determine the missing items in the following satellite up-link budget.

Earth station EIRP	110 dBW
Additional up-link atmospheric losses	0.8 dB
Free space path loss	205 dB
Carrier power at satellite	???
Satellite line losses	1.2 dB
Satellite G/T_e	-6dBK-1
Satellite C/T_e	????
Satellite C/N_o	????
Satellite E_b/N_o	????
Satellite C/N	????
Bit rate	128 Mbits/sec
Modulation	8PSK (3 bits/symbol)

References

[1] W.C.Y. Lee and Yeh, "On the estimation of the second order statistics of lognormal fading in mobile radio environment," *IEEE Trans. on Communications*, vol.COM-22, pp. 869-873, June 1974.

[2] D.O. Reudink, "Properties of mobile radio propagation above 400 MHz," *IEEE Trans. on Veh.Tech*, vol. VT-23, no. 4, November 1974.

[3] R.C. French, "Radio propagation in London at 462 MHz," *The Radio and Electronic Engineer*, vol.46, no. 7, pp 333-336, July 1976.

[4] C.M. Burrows and M.C. Grayest, "Effects of earth's curvature on ground wave propagation," *Proc. IRE*, vol. 29, pp.16-24, Jan.1941.

[5] Norton, "The calculation of ground wave field intensities over a finitely conducting spherical Earth," *Proc.IRE*, vol.29, pp.623-639, Dec.1941.

[6] K.Bullington, "Radio propagation for vehicular communications," *IEEE Trans. on Veh. Tech.*, vol.VT-26, pp 295-325, no. 3, August,1980.

[7] N.H.Shepherd, "Radio wave loss deviation and shadow loss at 900 MHz," *IEEE Trans.*, vol. VT-26, no. 4, pp. 309-313, November 1977.

[8] K. Bullington, "Radio propagation at frequencies above 30 Megacycles," *Proc. IRE*, vol. 35, pp 1122-1136, Oct. 1947.

[9] K. Alloesbrook and J.D. Parsons "Mobile radio propagation in British cities at frequencies in the VHF and UHF Bands," *IEEE Trans. on Veh. Tech.*, vol. VT-26, no. 4, pp 313-323, Nov.1977.

[10] G.Y. Delisle, J.-P. Leferere, M. Lecours and J.-Y. Chouinard, "Propagation loss prediction: A comparative study with application to mobile radio channel," *IEEE Trans. Veh. Tech.*, vol. VT-34, pp.86-96, May 1985.

[11] J.J. Egli, "Radio propagation above 40 MC over irregular terrain," *Proc. IRE*, vol. 45, pp 1383-1391, Oct. 1957.

[12] K. Alloesbrook and J.D.Parsons, "Mobile radio propagation in British cities at frequencies in the VHF and UHF Bands," *Proc. IEE*, vol. 124, no. 2, pp 95-102, Feb. 1977.

[13] J.D. Parsons, M.F.Ibrahim and R.J. Holbeche, "Mobile radio propagation in rural areas at 168,455, and 896 MHz," *IEE Proc.*, vol 130, Pt. F, no. 7, pp.701 706, July 1983.

[14] K. Bullington, "Radio propagation for vehicular communications," *IEEE Trans. Veh. Tech.*, vol. VT-26 No. 4., pp 295-308, Nov.1977.

[15] B.R. Davis, R.E. Bogner, "Propagation at 500 MHz for mobile radio," *IEE Proc.*, vol. 132, pt. F, pp. 307-320, August. 1985.

[16] M.F. Ibrahim and J.D.Parsons, "Urban mobile radio propagation at 900 MHz," *Electronic Letters*, vol. 18, No. 3, pp. 113-115, 1982.

[17] K. Watanabe, "800 MHz Band radio propagation characteristics for cellular land mobile telephone system," *Proc. ICC*, pp 57.3.1-57.3.6 1979.

[18] K.Bullington, "Radio propagation fundamentals," *Bell System Technical Journal*, vol.36, pp. 593-626, May 1957

[19] J. Steel and R.A. Court, "Measurement of field strength for mobile services at VHF and UHF for the city of Melbourne," *Elect. Letters*, vol. 13, no.24, pp. 710-712, Oct. 1977.

[20] J.D. Parsons, M.F. Ibrahim and R.J. Samuel, "Median signal strength prediction for mobile radio propagation in London," *Elect. Lett.*, vol. 16, No. 5,pp 172-173, Feb. 1980.

[21] M.M. Peritsky, "Statistical estimation of mean signal strength in rayleigh fading environment," *IEEE Trans. on Comm.*, vol. COM-21, No.11, pp. 1207-1213, November, 1973.

[22] Y.Okumura, E.Ohmori, T.Kawano and K.Fakuda,"Field strength and its variability in VHF and UHF mobile service,", *Rev. Elect.Comm. Lab.*, vol 16, pp. 825-875, 1968.

[23] W.R. Young Jr., "Comparison of mobile radio transmission at 150, 450, 900, and 3700 MHz," *Bell Syst. Tech. Joun.*, vol.31, pp. 1068-1085, 1952.

[24] M.F. Ibrahim and J.D. Parsons, "Signal strength prediction techniques in built-up areas, Part I: median signal strength," *IEE Proc. Part F*, vol. 130, No.5, pp. 377-384. Part II: signal variability, Ibid. pp. 385-391.

[25] W.C.Jakes, Jr.(ed), *Microwave Mobile Communications*, J.Wiley & Sons, New York, pp. 20, 1974.

[26] W. C. Y. Lee, "Studies of base station antenna height effects on mobile radio," *IEEE Trans. on Veh. Tech.*, vol. VT-29, no. 2 pp252-260, May 1980.

[27] M.Hata, "Empirical formula for propagation loss in land mobile radio services," *IEEE Trans. Veh. Tech.*, vol. VT-29, pp. 317-325, August 1980

[28] J. Durkin, "Computer prediction of service areas for VHF and UHF land mobile radio services," *IEEE Trans. on Veh. Tech.*,vol. VT-26, no. 4, pp. 323-327, Nov. 1977.

[29] European Cooperation in the Field of Scientific and Technical Research EURO-COST 231, "Urban Transmission Loss Models for Mobile Radio in the 900 and 1800 MHz bands," Revision 2, The Hague, September 1991.

[30] F.H. Palmer, "The Communications Research Centre VHF/UHF Propagation Prediction Program: An Overview," *Canadian Elec. Eng. J.*, vol. 6, No.4, pp3-9, 1981.

[31] A.G. Longley and P.L.Rice, *Prediction of Tropospheric Transmission Loss Over Irregular Terrain- a Computer Method-1968*, ESSA Tech Rept. ERL 79-ITS 67, U.S. Government Printing Office, Washington, D.C. 1968.

[32] P.L.Rice;et.al, *Transmission Loss Prediction for Tropospheric Communication Circuits vol.I and II*, National Bureau of Standards Technical Note 101,(1966).

[33] R. Edwards and J. Durkins, "computer prediction of service areas for vhf mobile radio networks," *Proc. IEEE*, vol. 66, pp. 1493-1500, 1969.

[34] C.E.Dadson, J.Durkin, and R.E. Martin, "Computer prediction of field strength in planning of radio systems," *IEEE Trans. on Veh. Tech.*, vol.VT-24, pp. 1-8, 1975.

[35] E.A. Neham, "An approach to estimating land mobile radio coverage," *IEEE Trans on Veh. Tech.*, vol. VT-23, no.4, pp. 135-139,1974.

[36] S. Kozono and K. Watanabe, "Influence of environmental buildings on uhf land mobile propagation", *IEEE Trans. on Communications*, vol. COM-25, no.10, pp. 1133-1143, Correction *IEEE Trans.*, vol. COM-26, pp. 128-135, 1977.

[37] Thiessen, *CCIR, Doc. Correction Factors for Frequency Band I-V*, Doc. V/28.E, pp.330, 1965.

[38] L.J.Anderson and L.G. Trolese, "Simplified method for computing knife edge diffraction in the shadow region," *IRE Trans. on Antennas and Propagation*, vol.AP-6, pp.281-286, July 1958.

[39] J.Degout, "Multiple knife edge diffraction of microwaves," *IEEE Trans. on Veh.Tech.* vol. VT-23, pp. 143-159, 1971.

[40] J. Epstein and D.W. Peterson, "an experimental study of wave propagation at 850 Mcs," *Proc. IRE*, vol.AP-1, pp 595-611, May 1953.

[41] A. Picquenard, *Radio Wave Propagation*, J. Wiley, New York, 1974, pp.296.

[42] D.O. Ruedink, "Preliminary investigation of mobile radio transmission at x-band in an urban area," *1967 Fall URSI Meeting* at Ann Arbor, Mich. 1967.

[43] A. J. Goldsmith and L. J. Greenstein, "A measurement-based model for predicting coverage areas of urban microcells," *IEEE Journal on Selected Areas in Communications*, vol. 11, no. 7, pp. 1013-1023, 1993.

[44] D.O. Reudink and M.F. Wazowicz, "Some propagation experiments relating foliage loss and diffraction loss at x-band at uhf frequency," *Joint IEEE Comm.Soc.-VTS Special Trans. On Mobile Radio Communication*, vol.VT-22, pp.1198-1206, November 1973.

[45] W.J. Vogel and E.K. Smith, "Propagation considerations in land mobile satellite transmission," *Microwave Journal*, vol. 28, No.10, pp.111-130, October,1985.

[46] A.H. LaGrone, P.E.Martin and C.W. Chapman, "Height gain measurements at vhf and uhf behind a grove of trees," *IRE Trans. on Ant. and Prop.*, vol. 9, pp.487-491, 1961.

[47] *Attenuation in Vegetation*, ITU-R Recommendation ITU-R P.833-2, 1999.

[48] W.M. Pannell, *Frequency Engineering in Mobile Radio Bands*, Granta Tech. Editions, Cambridge,1979.

[49] D.C. Cox, "Universal portable radio communications," *IEEE Trans. on Veh. Tech.*, vol. VT-34, No.3, pp. 117-121,1985.

[50] A.U. H. Sheikh, and F.Hadziomerovic, "A simulation study of in-building personal communication systems", *Proc. IEEE Veh. Tech. Conf.*, pp.175-178,1986.

[51] D. Holmes, "Cordless telecommunications in the U.K.", *Proc. IEEE Veh. Tech.*, Conf. pp. 434, 1986.

[52] H.L. Lester, S.S. Rappaport, C.M. Puckette, "PRS- A consumer mobile radio communication system with distributed control," *GLOBECOM 83*, pp.1417-1424, Dec.1983.

[53] M. Morinaga, M. Nakagawa, and R. Kohno, "New concepts and technologies for achieving highly reliable and high capacity multimedia wireless communications system," *IEEE Comm. Mag.*, vol. 35, pp. 34-40, January 1997.

[54] A. Akeyama, T. Tsuruhara, and Y. Tanaka, "920 MHz mobile propagation for portable telephone," *Trans. IECE (Japan)*, vol. E65, No.9, pp.542-543, 1982.

[55] T.Ooi, T.Kikuchi and K. Taujimura, "UHF radiowave propagation characteristics for cordless telephone," *IEEE NTC Record*, pp. 25.5.1-25.5.3, 1978.

[56] D.C.Cox, R.R.Murray and A.W. Norris, "Measurements of 800 MHz radio transmission into buildings with metallic walls," *Bell System Technical Journal*, vol 62, No.9, pp. 2695-2717, November 1983.

[57] E.H. Walker, "Penetration of radio signals into buildings in the cellular radio environment," *Bell System Technical Journal*, vol. 26, No.9, pp. 2719-2734, 1983.

[58] S.E. Alexandar, "Characterizing buildings for propagation at 900 MHz," *Electronic Letters*, vol.19, No. 20, pp.860, September 1983.

[59] T. S. Rappaport, "The wireless revolution," *IEEE Communication Magazine*, vol. 29, pp. 52-71, November 1991.

[60] E. J. Violette, R. H. Espeland, and K. C. Allen, *Millimeter-Wave Propagation Characteristics and Channel Performance for Urban-Suburban environments*, National Telecommunications and Information Administration, NTIA report 88-239, December 1988.

[61] S.E. Alexandar, "900 MHz propagation within buildings," *Proc. Second Conference on Spectrum Conservation*, IEE Conf. Pub. No. 224, 1983.

[62] T.W. Whiter and R.W. Gibson, "Radio field strength measurements at 900 MHz within a building," *IEE Colloquium on Terrestrial Mobile Communication Above 500 MHz* (Digest), pp 4/1-4/4, 1982.

[63] R.J. Pillmeier, "In-building signal correlation for an urban environment," *IEEE Veh. Tech. Conf.*, pp.156-161,1984.

[64] J.Shefer, "Propagation statistics of 900 MHz and 450 MHz signals inside buildings," *Microwave Mobile Radio Symposium*, March 7-9, Boulder, Colorado, 1985.

[65] S.E.Alexander and G.P.Pugliese, "Cordless communication within buildings: Results of measurements at 900 MHz and 60 GHz," *Br. Telecom. Journal*, vol.1,no.1, pp. 99-105, July, 1983.

[66] T. Tsuruhara, A. Akeyama, and K. Satori, "Mobile radio propagation characteristics in urban areas in the UHF Band," *Trans. IECE (Japan)*, vol. E-66, no. 12, pp. 724-726, Dec.1983.

[67] T.S. Rappaport, "Indoor radio communications for factories of the future," *IEEE Commun. Mag.*,pp. 15-24, May 1989.

[68] A.A.M. Saleh and R.A. Valenzuela, "A statistical model for indoor multipath propagation," *IEEE J. Selected Areas in Comm.*, vol., SAC-5, no.2, pp. 128-137, Feb. 1987.

[69] R.J.C. Bultitude, "Measurement of wideband propagation characteristics for indoor radio with predictions for digital system performance," *Proc. Wireless '91 Conf.*, July 8-10, 1991, Calgary, Alberta, Canada.

[70] P. Haley, "Short distance attenuation measurements at 900 mhz and 1.8 ghz using low antenna heights for microcells," *IEEE J.Select., Areas Comm.*, vol. SAC-7, no.1, pp. 5-11, Jan. 1989.

[71] D. Akerberg, "Properties of a TDMA picocellular office communication system," *IEEE Globecom*, pp. 1343-1349, 1988.

[72] D.M.J. Devasirvatham, R.R. Murray, and C. Banerjee, and D.A. Rappaport, "Multi-frequency propagation measurements and models in a large metropolitan commercial building for personal communication," *Proc. Second IEEE Int. Symp. Personal, Indoor and Mobile Radio Commun.*, London, pp. 98-103, 1991.

[73] D.O. Reudink, "Mobile radio propagation in tunnels," *IEEE Veh. Tech. Con.*, Dec. 1968.

[74] Q.V. Davis, D.J.R. Martin, R.W. Haining, "Microwave radio in mines and tunnels," *IEEE Veh. Tech. Cof. Record*, pp. 31-36, 1984.

[75] *Mobile Communication Via Satellite*, DOC Report to MOSST, June 1981.

[76] *MSAT Commercial Viability Study*, ADGA/Touch-Ross and Partners, Ottawa, Ontario, DOC-CR-SP-82-028,June 1982.

[77] P.A. Castruccio, C.S.Marantz, and J. Freibaum, "Need for, and feasibility of satellite-aided land mobile communications," *ICC Conf. Record*, pp.7H.1.1-7H.1.7,1982.

[78] R.E. Anderson,R.L.Frey, and J.R. Lewis, "Technical feasibility of satellite aided land mobile radio," *ICC Conf. Record*, pp. 7H.2.1-7H.2.5, 1982.

[79] B.E. LeRoy, "satellite-aided land mobile communications system implementation considerations," *ICC Conf. Record*, pp. 7H.3.1-7H.3.5, 1982.

[80] T.M.Abbot, "Requirements of a mobile satellite service," *International Conference on Mobile Radio Systems and Techniques*, IEE Pub. No. 238, pp. 212-215, 1984.

[81] R.J. Cocks, P.S.Hansell, R. Krawec, and D.Lewin, "Personal mobile communications by satellite," *International Conference on Mobile Radio System and Techniques*, IEE Pub. No. 238, pp.135-139, 1984.

[82] T.Freygard, "Feasibility of international transport communication system," *ICC, Record*, pp. 7H.4.1-7H.4.6, 1982.

[83] G.C. Hess, "Land mobile satellite excess path loss measurements," *IEEE Trans. on Veh. Tech.*, vol. VT-29, no.2, pp. 290-297, 1980.

[84] R.W.Huck, J.S. Butterworth and E.E. Matt, "Propagation measurements for land mobile satellite services," *IEEE Veh. Tech. Conference*, pp. 265-268, 1983.

[85] C.Loo, "A statistical method for a land mobile satellite link," *IEEE Trans. on Veh. Tech.* vol. VT-34, no. 3,pp. 121-127, August, 1985.

[86] W.J. Vogel and E.K. Smith, "Theory and measurements of propagation for satellite to land mobile communication at UHF,"*IEEE Veh. Tech. Conf.*, pp. 218-226, 1985.

[87] R. Frieden, "Satellite-based personal communication services," *Telecommunications*, vol. 27, no. 12, pp. 25-28, December 1993.

[88] A.P. Barsis, "Determination of service area for VHF/UHF land mobile and broadcast operations over Irregular Terrain," *IEEE Trans. on Veh. Tech.*,vol. VT-22, pp. 21-29, 1973.

[89] R.T. Forestall, "Practical approach to radio propagation measurements- As used in the design of mobile radio communications systems", *IEEE Trans. on Veh. Tech.*, vol.VT-24, no.4, 1975.

[90] W.C.Y. Lee, *Mobile Communications Engineering*, McGraw Hill, 1983.

Chapter 3

MICRO-CHARACTERIZATION OF WIRELESS CHANNELS

The previous chapter concentrated on estimation of the mean or median signal strength for a given transmitter and receiver geometry characterized by a certain terrain definition in a coverage area. The concept of statistical signal coverage was also presented. It was considered sufficient to assume transmission of a pure tone at the frequency of interest. It was emphasized that rapid fluctuations of the envelope had already been removed by either selecting an appropriate averaging interval or a filter with a suitable cutoff frequency to yield signal mean or median. It was also stressed that due to the absence of definitive boundary between the slow and fast fluctuations, the selection of an appropriate cutoff frequency is difficult. Furthermore, the median signal level in a small area within a given cell was taken as a primary indication of the signal strength. Though necessary as a starting point, such a rough approximation is quite adequate for signal coverage prediction, but a single sinusoid is hardly used for transmission of intelligent information. The information bearing signals always occupy a finite bandwidth. Therefore, it is important that impairments induced to the intelligent signals during propagation through the channel be characterized. A knowledge on the channel character is essential in the design of suitable signal transmission formats. In other words given a certain channel characteristic, decisions on appropriateness of modulation and detection processes can be taken so that even in the presence of distortion of the signal amplitude and phase, the signal is recovered with minimum of distortion.

Even if it is known that adequate average signal strength at the receiver exists at most of locations within the coverage area, as system designers we need to know the character of the signal fluctuations and how these impact the reception reliability. The system performance may be measured in terms of either the average bit error rate (BER) or the probability that the received signal falls below the minimum acceptable level i.e. the outage probability. In short, the understanding of the impact of the channel vagaries on the transmitted signal microstructure is an important step towards a reliable communication system design. Once we understand how the channel induces impairments to the signal, we can investigate methods to unravel these impairments.

Micro-characterization of mobile wireless channel relates to the identification of those channel parameters that are responsible for degrading the reception quality and it is the channel characterization that leads to a reliable communication system.

3.1 Channel Impulse Response

Impulse response of a channel completely characterizes it. Thus, the channel characterization starts with exciting the channel with an impulse or a wideband probing signal and analyzing the response. Consider the response of a channel to a RF signal pulse of a short duration. If the response resembles the transmitted pulse but is modified in amplitude and arrives with a certain delay of τ seconds, we conclude that the channel is distortion free. The probing narrow pulse may be replaced by a unit impulse, $\delta(t)$, to yield an output called impulse response, $h(t)$. It is given by,

$$h(t) = A\delta(t - \tau).\tag{3.1}$$

where A is the strength of the received impulse and τ is the delay. The channel transfer function, $H(f)$, obtained by the Fourier transform of $h(t)$, is given by,

$$H(f) = Ae^{-j\omega\tau}.\tag{3.2}$$

The transfer function of the channel is therefore $A\ exp(-j\omega\tau)$, that is the magnitude response is constant over the signal bandwidth while the phase response, $-\omega t$ is linear, where $\omega = 2\pi f$. These are the conditions for distortionless transmission and any deviation from these conditions will result in distortion of the received signal. For example, consider a channel transfer function that has linear phase response but does not meet the requirement of constant magnitude response,

$$H(f) = \begin{cases} (1 + k\cos 2\pi fT)e^{-j2\pi f\tau}; & |f| < B \\ 0; & \text{elsewhere} \end{cases}\tag{3.3}$$

The output, $Y(f)$, due to a bandlimited signal, $S(f)$, is given by,

$$Y(f) = S(f)H(f) = S(f)e^{-j2\pi f\tau} + kS(f)\cos 2\pi fTe^{-j2\pi f\tau}.\tag{3.4}$$

Taking the inverse Fourier transform of (3.4), the time domain output is given by,

$$y(t) = s(t-\tau) + \frac{k}{2}[s(t-\tau-T) + s(t-\tau+T)] \tag{3.5}$$

where $s(t)$ is the inverse Fourier transform of $S(f)$. Equation (3.5) shows that the received signal consists of three replicas of the transmitted signal modified in amplitude and arrive at times τ, $\tau+T$ and $\tau-T$ seconds. It is easy to show that nonlinearity in the channel phase also produces multitude of responses. The number of the signal replicas and their arrival times depend on the shape of the channel magnitude and the phase responses. Thus, if the channel bandwidth[†] is narrower than the signal bandwidth, the channel response to $s(t)$ will be long and it is possible that several responses merge into one. On the other hand, for a channel bandwidth wider than the signal bandwidth, the channel output may consist of a single response resembling $s(t)$.

A similar situation occurs in mobile radio. The signal after being reflected off a multitude of buildings and other obstacles reaches the receiver in the form of several signal reproductions. It is easy to imagine that with the movements of either the receiver or the transmitter or both, the amplitudes of the signal replicas and their arrival times will change continuously. For a time variant channel, the impulse response is written as $h(\tau;t)$.

3.1.1 Channel Impulse Response Measurements

As mentioned earlier, the channel impulse response is obtained by exciting the channel with an impulse or another suitable wideband signal. It customary to replace the pulse by a carrier modulated by a sufficiently long pseudonoise (PN) sequence of short duration pulses called chips. The inverse of chip duration, τ_c, determines the baseband bandwidth of the probing signal. However, wideband data thus obtained generally results in processing difficulties. To alleviate these difficulties, channel sounders have been designed. A channel sounder sends out a carrier binary phase modulated by a pseudonoise sequence with good correlation properties. At the receiver, the demodulated signal is correlated using a sliding correlator. The sliding correlator correlates the recovered sequence with a locally generated sequence clocked at a rate slightly lower than that used at the transmitter. The output of the sliding correlator is integrated over a time ranging from a fraction to the full period of the PN sequence. To explain the advantage of using a sliding correlator, we take an example.

Suppose, a 1023 chip long sequence clocked at 50 MHz is used at the transmitter. The chip duration is then 20 nsec and the sequence period is 20.46 μsec. The width of the sequence autocorrelation function is therefore 40

[†]. *The channel bandwidth is defined as the bandwidth in which frequency components of the signal remain significantly correlated. More discussion on channel bandwidth is given in the latter sections.*

nsec. The receiver correlator is clocked at a slightly lower rate; the clock off-set is called slip rate. Considering a clock offset of 5 kHz, the scaling factor is given by the ratio of the chip rate and the slip rate; in this example the scaling factor is 10,000[†]. For a sequence length of 1023, the delay profile of 20.46 μsecond is recorded. With this time scaling, the measurement length will be 204.6 msec. Thus, a sampling rate of 10 kHz for each impulse response measurement will be sufficient to provide resolution of 1 chip or 20 nanoseconds. The impulse response measurements using sliding correlator or direct pulse excitation typically results in a set of impulse responses in time-space plane. For more detail on sliding correlators, the reader is referred to [1]. Figure 3.1 reproduces a profile of measurement results taken in the Hong Kong Polytechnic University while the receiver was moved at a speed of approximately 1m/sec. It is seen that as the receiver is moved, the impulse response profile changes with some new paths appearing while some others that existed previously disappeared [2]. The channel impulse response profile i.e. number of paths, their locations and magnitudes determine the severity with which the channel distorts the signal.

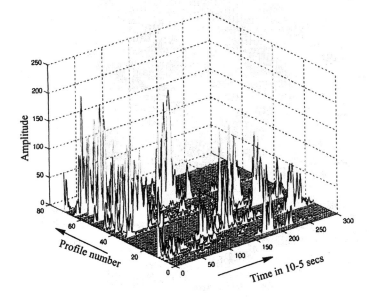

Figure 3.1 A typical set of channel impulse responses

---

† *The clock offset will induce some inaccuracies in the impulse response, but these being insignificant can be ignored for practical purposes.*

The Fourier transform of the time variant impulse response with respect to τ gives time variant transfer function of the channel, $T(f:t)$. Taking Fourier transform of $T(f:t)$ with respect to t results in Doppler variant transfer function $H(f:v)$ and inverse Fourier transform of $H(v;f)$ results in Doppler variant impulse response $s(\tau;v)$. Furthermore, an ensemble of impulse responses taken at different times or at locations in space contains information on temporal and spatial variation of the channel. By appropriately processing the four channel functions we can obtain several correlation functions and spectral densities which are interrelated by Fourier transforms. These relations are also shown in Figure 3.2. In general, four correlation functions dependent on four channel parameters are used to determine other channel parameters of practical importance.

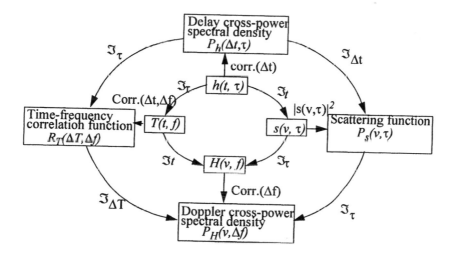

Figure 3.2 Relations between functions defining wireless channels

$$\Re_s(v, v';\tau, \tau') = E[s(v, \tau)s^*(v', \tau')]$$

$$\Re_h(t, t';\tau, \tau') = E[h(t, \tau)h^*(t', \tau')]$$

$$\Re_T(t, t';f,f) = E[T(t,f)T^*(t',f)]$$

$$\Re_H(v, v';f,f) = E[H(v,f)h^*(v',f)]$$

(3.6)

Under the assumption that the channel is Wide Sense Stationary Uncorrelated Scattering (WSSUS), the four correlation functions defined in (3.6) may be rewritten as

$$\Re_s(v, v';\tau, \tau') = \delta(v' - v)\delta(\tau' - \tau)P_s(v, \tau)$$

$$\Re_h(t, t + \Delta t;\tau, \tau') = \delta(\tau' - \tau)P_h(\Delta t, \tau)$$

$$\Re_T(t, t + \Delta t; f, f + \Delta f) = \Re_T(\Delta t, \Delta f)$$

$$\Re_H(v, v'; f, f + \Delta f) = \delta(v' - v)P_H(v, \Delta f)$$

(3.7)

where $P_h(\Delta t, \tau)$, $R_T(\Delta t, \Delta f)$, $P_s(v, \tau)$ and $P_H(v, \Delta f)$ are called Delay Cross-Power Spectral density, Time-Frequency Correlation Function, Power Density, and Doppler Cross-Power Spectral Density respectively. The most important of these are the Delay Power Density Spectrum, $P_h(\tau)$ and Doppler Power Density Spectrum $P_H(v)$ which are $P_h(\Delta t, \tau)$ and $P_H(v, \Delta f)$ with $\Delta t = 0$ and $\Delta f = 0$ respectively. From $P_h(\tau)$ we obtain rms delay spread and average delay and from $P_H(v)$ average Doppler and rms Doppler Spread is obtained.

From the Delay Power Density Spectrum, $P_h(\tau)$, also known as delay power profile of the channel, the average excess delay, rms delay spread and excess delay spread are usually extracted. These parameters are used as guidelines for wireless system design. The average excess delay is defined as,

$$\bar{\tau} = \frac{\int_0^\infty \tau P_h(\tau)d\tau}{\int_0^\infty P_h(\tau)d\tau}$$

(3.8)

For a discrete model of the $P_h(\tau)$ i.e. $\sum_j P(\tau_j)$, the average excess delay may be written as

$$\bar{\tau} = \frac{\sum_j \tau_j P(\tau_j)}{\sum_j P(\tau_j)}$$

(3.9)

The rms delay spread is the square root of the second central moment of the power delay profile and is defined as:

$$\sigma_\tau = \sqrt{\bar{\tau^2} - (\bar{\tau})^2}$$

(3.10)

where

$$\overline{\tau^2} = \frac{\sum_j \tau_j^2 P(\tau_j)}{\sum_j P(\tau_j)} \tag{3.11}$$

The maximum excess delay is defined as the time delay over which the multipath normalized power falls to X dB below the maximum. Figure 3.3 defines these parameters.

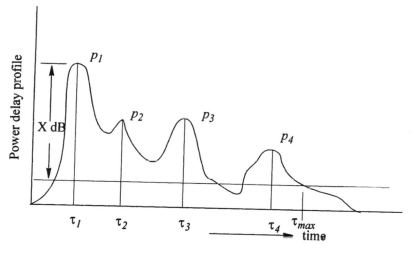

Figure 3.3 Definitions of channel time spread parameters

In a similar manner, we can use the Doppler Power Spectral Density, $P_H(v)$, to obtain frequency domain characteristics. Using relations similar to those in (3.8) - (3.11), we can define channel parameters like mean Doppler, and rms Doppler Spread.

The channel impulse response and the channel transfer function form a Fourier transform pair. As we have seen earlier that the impulse response provides clues to the shapes of the magnitude and phase responses of the channel. A larger number of paths and greater rms delay create larger number of peaks and troughs in the magnitude frequency response[†]. The average range of frequencies over which the magnitude response remains relatively flat is known as the coherence bandwidth, B_c, of the channel. It is a statistical measure but it provides an estimate of the maximum distortion free transmission rate over the channel. Alternatively, we can say that spectral components

---

[†]. *The phase response of the channel is considered to remain linear over the channel bandwidth. Any deviation of the phase response from linearity will also result in ISI.*

within the coherence bandwidth have high amplitude correlation. Different values of correlation coefficients have been used to define the coherence bandwidth. For a correlation coefficient value of 0.5, the coherence bandwidth is given by:

$$B_c = \frac{1}{2\pi\sigma} \qquad (3.12)$$

For higher correlation values, say 0.9, the coherence bandwidth is much narrower. Lee [3] has suggested that the coherence bandwidth is approximately given by:

$$B_c \approx \frac{1}{50\sigma} \qquad (3.13)$$

where σ is the channel delay spread in seconds and B_c in Hz.
Several theoretical and empirical models for channel impulse response have been suggested in literature.

EXAMPLE E-3.1.1

In an impulse response measurement experiment in a microcellular environment, the impulse response was found to consist of six paths with dB strengths and arrival times in microseconds as (0,0), (-3, 0.3), (-11,0.6), (-6, 1.2), (-8, 1.5) and (-12, 3.3).
(a) Evaluate the mean and rms delay of the channel.
(b) Find the channel transfer function.
(c) What is the coherence bandwidth of the channel?

Solution

(a) Average delay

$$\bar{\tau} = \frac{\sum_j \tau_j P(\tau_j)}{\sum_j P(\tau_j)} = \frac{2.9515}{2.723} = 1.0839 \qquad (E-3.3.1)$$

and the mean square delay

$$\overline{\tau^2} = \frac{\sum_j \tau_j^2 P(\tau_j)}{\sum_j P(\tau_j)} = \frac{3.98845}{2.7323} = 1.4647 \qquad (E-3.3.2)$$

The channel delay spread is calculated using (3.10) is 0.538 µsec.

(b) The channel impulse response is given by

$$h(\tau) = \delta(\tau) + 0.5\delta(\tau - 0.3 \times 10^{-6}) + 0.079\delta(\tau - 0.6 \times 10^{-6}) + \qquad \text{(E-3.3.3)}$$
$$0.251\delta(\tau - 0.251 \times 10^{-6}) + 0.83\delta(\tau - 1.5 \times 10^{-6}) + 0.063\delta(\tau - 3.3 \times 10^{-6})$$

The transfer function is given by the Fourier transform of the above equation.

(c) The channel coherence bandwidth, given by (3.12) is calculated as approximately 296 KHz.

3.1.2 Delay and Frequency Spread Models

The importance of delay and frequency characteristics can not be overstressed because these have profound impact on system design and consequently on the type of services that the channel can offer. In the past, most of the outdoor applications were limited to narrowband communication, that is the transmission bandwidth was in general narrower than the channel coherence bandwidth. With growing interest in wideband applications over wireless channels, several researches made wideband measurements in the outdoor environments. These measurements resulted in several empirical models. These are described in 3.1.2.1. In the indoors environment, wideband applications, particularly wireless LANs are becoming common and a large volume of research is dedicated to the characterization of wideband indoors wireless channel.

3.1.2.1 Outdoor Frequency and Delay Spread Models

The design of a digital communication system starts with a laboratory environment study on the alternatives in modulation. To do so, analytical or simulation methods are used. Development of communication system models, including that of the channel block is an essential step before system are evaluated. In the absence of reliable models, the performance of digital mobile communication can only be studied by expensive field trials.

Several attempts have been made to develop channel models [4],[5]-[10]. These models were proposed from echo-envelope data, under the assumption that the carrier phase is uniformly distributed over $(0, 2\pi]$. The global statistical model is based on multipath data over a large inhomogeneous area, and does not include the effects of the nonstationarity associated with multipath statistics. Turin *et al* [4] suggested that if a signal, $\Re\{s(t) exp(j\omega_o t)\}$, $t \in (-\infty, \infty)$, is sent then $\Re\{\rho(t) exp(j\omega_o t)\}$, $t \in (-\infty, \infty)$ is received, where

$$\rho(t) = \sum_{k=0}^{\infty} S^{fk} a_k s(t - t_k) exp(j\theta_k) + n(t) \tag{3.14}$$

and $n(.)$ is low pass complex valued white noise; $20 \log_{10} a_k$ and $20 \log_{10} S$ have normal distribution with means μ_k and variances σ_k^2 respectively. θ_k is uniformly distributed over $(0-2\pi)$; $\{t_k - t_0\}_1^{\infty}$ is a modified Poisson sequence with mean arrival rate determined by the empirical probability of occupancy curves. All parameters a_k, θ_k, and S are independent; and the set $\{f_k\}^{\infty}$ is determined empirically. Thus, the kth path strength $S^{fk} a_k$ is lognormal as required, viz.,

$$20 \log(S^{fk}) a_k = f_k(20(\log S)) + 20 \log a_k \tag{3.15}$$

the function $20 \log_{10} S^{fk} a_k$ is normal with mean $(f_k \mu + \mu_k)$ and variance $f_k^2 \sigma^2 + \sigma_k^2$.

Wideband channel soundings have been reported in [6][7][8][9][10]. Bajwa [9], through extensive channel sounding experiments, extended the dispersion model to include Doppler shifts, previously ignored by Turin *et al* [4]. From the experimental data, scatter plots for average delay $\bar{\tau}$ against delay spread s and coherence bandwidth for both urban and suburban areas were obtained, which resulted in the following empirical expressions for $\bar{\tau}$, the average delay in microseconds, and B_c, the coherence bandwidth in kHz:
For suburban areas:

$$\bar{\tau} = 0.143 + 0.88\sigma$$
$$B_c = 1420\sigma^{-1.57} \text{ or } 1000\sigma^{-1} \tag{3.16}$$

For urban areas:

$$\bar{\tau} = 0.227 + 0.82\sigma$$
$$B_c = 730\sigma^{-1.42} \text{ or } 1000\sigma^{-1} \tag{3.17}$$

The difference between the values for coherence bandwidth in urban and suburban environments results from different modes of propagation. For multipath effects resulting primarily from scattering, the coherence bandwidth B_c is inversely proportional to the delay spread σ. When propagation is through scattering only, the value of B_c is given by

$$B_c = 1000\sigma^{-1} \tag{3.18}$$

When a line-of-sight component is present, the index in the inverse relationship is smaller, so that the coherence bandwidth is less dependent on the delay spread. For example, if the measured delay spread in a suburban area is say 2 microseconds, the corresponding coherence bandwidth using (3.16) is estimated to be 478 kHz. For the same delay spread in an urban area, the estimated bandwidth is 272 kHz. The corresponding values from (3.12) and (3.13) are calculated to be 79.5 and 10 kHz respectively. It is seen that depending on which relationship we use, there could be appreciable difference between estimates of this parameter. This suggests that an exact relationship between the delay spread and the coherence bandwidth does not exist and the values predicted by these empirical relations are just estimates.

Extensive fields trials have also been directed towards the measurement of the channel impulse response primarily to obtain delay-Doppler spreads of the channel [4],[5]-[10]. Some discrete channel impulse response models also give statistics on the number of resolvable paths in CIR. The complex impulse response is generally obtained by exciting the channel with a pseudorandom sequence and finding its low-pass equivalent. The experimental results provide delay and doppler profiles of the channel, which are useful in the estimation of correlation or coherence bandwidth. Some work on the temporal variations in the impulse response is also reported in the literature [39]. Based on this and related models, stored channel and hardware simulators have been developed [11][12].

3.1.2.2 Delay-Frequency Models for Indoors Wireless Channels

The channel impulse response models are based on number of paths, their magnitudes and locations. These parameters are generally considered to be derived from random processes, each having a distinct distribution. A theoretical model proposed by Jakes included explicitly the role of delays between rays arriving from different directions [13]. In his model all the signals constituting a ray have similar delays but different phase relationships. The measured results indicate that an exponential distribution of the delay spread gives the best approximation [14]. However, other distributions are also used. When the angle of the incident power is uniformly distributed, the function $p(\sigma, T)$ can be expressed as

$$p(\sigma, T) = \frac{1}{\sigma} exp\left(-\frac{T}{\sigma}\right) \tag{3.19}$$

where σ is a measure of the time delay spread. The number of paths and the overall length of impulse response depend on the environment; the path arrival is usually modeled as a Poisson, and the path strength (magnitude) is assumed to vary randomly with either Rayleigh, Nakagami, Ricean or lognormal distribution.

3.1.2.3 Path-Arrival Models

The path arrival process was largely neglected except in [4], [15]-[19], where Standard Poisson (SM) and Modified Poisson (MM) models have been used to characterize path arrivals. In this body of work, two major approaches to the statistical modelling of the path arrival process are discussed. One of them focuses on the outdoor mobile radio channels and is reported by Turin *et al* [4], Suzuki [15], and Hashemi [16],[17]. The other is due to Saleh *et al* [18], and pertains to the indoors mobile radio environment. Recently Hashemi [19] has applied modifications to the outdoor model in order to characterize the indoors environment.

3.1.2.3.1 Turin's Model

Turin hypothesized a Poisson arrival process. However, it was discovered that for a small Poisson parameter λ, the empirical distributions were more clustered around a mean than suggested by Poisson distribution. This was interpreted to reflect the tendency of the near line of sight paths to arrive in groups. Moreover, as arrival rate, λ, increases, the fit between the empirical and Poisson distribution improves, possibly suggesting that late arriving paths in the sequence are more accurately modeled by the Poisson arrival assumption [4].

Based on these observations, a modified Poisson process (or a so called Δ-K model) was suggested, which takes into account the possibility that paths may arrive in groups, reflecting the grouping property of buildings (or reflectors) in the environment. The channel can be represented by two states S_1 and S_2 for the man arrival rates λ and $K\lambda$ respectively. The process starts in S_1 with mean arrival rate of λ. When a path arrives, the state changes to S_2 with the mean arrival rate $K\lambda$ for a time interval of $(t, t+\Delta)$. If no path arrives in this interval, a transition back to the state S_1 is made. For $K = 1$ or $\Delta = 0$, Δ-K model reverts to a standard Poisson process.

3.1.2.3.2 Saleh's Model

The statistical arrival model proposed by Saleh [18] is based on the physical reality that measured rays arrive in clusters. The arrival of the first rays of successive clusters are modeled as a poisson arrival process with some fixed mean arrival rate, Λ. The rays within each cluster of rays are also assumed to arrive with Poisson arrival process with another fixed mean arrival rate, λ. In general, $\lambda \gg \Lambda$, which means that each cluster may consist of many rays.

The arrival time of lth cluster is denoted by T_l, where $l = 0, 1, 2, ...$ and the arrival time of the kth ray in the lth cluster is assumed to be τ_{kl} where $k = 0, 1, 2, ...$ Also, for the first cluster, $T_0 = 0$ and $\tau_{0l} = 0$. Then according to the model, T_l and τ_{kl} are described by the following independent inter-arrival exponential probability density functions,

$$P(T_l | T_{l-1}) = \Lambda e^{-\Lambda(T_l - T_{l-1})} \qquad l > 0 \qquad (3.20)$$

and

$$P(\tau_{kl} | \tau_{(k-1)l}) = \lambda e^{-\lambda(\tau_{kl} - \tau_{(k-1)l})} \qquad k > 0 \qquad (3.21)$$

It is seen that the Saleh's model is similar to Turin's but the latter model is more general since it takes into account clustering as well as homogeneity of the environment. The main difference between the two models is the way variability in the mean arrival rate is modeled. In the first case, it is modeled in a stochastic manner, whereas in the second case a deterministic model is used.

The standard Poisson model assumes that the multipath are the result of randomly located obstacles. According to Poisson distribution, in a given time duration T, the number of paths, l, occur with a probability

$$P(l) = \frac{\mu^l e^{-\mu l}}{l!} \qquad (3.22)$$

where μ the Poisson parameter (number of arrivals in T). μ and mean arrival rate at time t, $\lambda(t)$, are related as

$$\mu = \int_T \lambda(t) dt \qquad (3.23)$$

The inter-arrival rate is distributed exponentially as

$$f_X(x) = \lambda e^{-\lambda x}; \qquad x > 0 \qquad (3.24)$$

In the indoor environment scatterers are not randomly located; the result is bunching of paths in clusters.

Standardization bodies like Joint Technical Committee (JTC) has recommended several models. Exponentially decaying power and equal power path models have been more used in development of GSM and TIA systems respectively. Details on some of these models are given in Appendix D.

3.1.2.4 Indoors Measurements and Statistical Models

Measurement and statistical model development of indoor radio channel saw a surge of activity over a ten year period between 1986 and 1996 [19]-[26]. In [19] measurements at 1.5 GHz are reported. The channel is observed to be very slowly varying, with the delay spread extending to a range up to about 200 nanosecond and rms delay spread of up to about 50 nanoseconds. The received signal is observed to arrive in clusters. The contributing rays have independent uniform phases, and independent Rayleigh amplitudes with variance that decays exponentially with cluster and ray delays. The clusters and the rays within the cluster form Poisson arrival processes with different but fixed rates.

Hashemi [20] wrote a tutorial paper that outlined the principles of radio communications in indoor environment. The channel is modeled as a linear time-varying filter at each location in the three dimensional space. The paper discusses several models for ray arrivals. These include standard Poisson Model, Modified Poisson - Δ-K Model, and Modified Poisson - nonexponential interarrivals. Two other point process models are also suggested. The first model is Gilbert's burst model, the second is pseudo-Markov models, which is used to spike trains from nerve cells.

In [21] time delay spread and signal level measurements of 850 MHz were made over inside to outside radio paths at two residential locations. Rms delay spreads of up to 420 nanosecods were encountered. When a direct path was present, this improved the delay spread to 325 nanoseconds, and even to 100 nanoseconds.

In [24] and [25] statistical modeling and simulation of rms delay spread is taken as topic of discussion. It is found after extensive measurements that the number of multipath components in each impulse response estimate is a normally distributed random variable with a mean value that increases with increasing antenna separation. A modified Poisson distribution shows a good fit to the data, amplitudes are lognormally distributed over both local and global areas. The rms delay spread over large area is normally distributed. The amplitudes of the adjacent paths in the same impulse response were found to be uncorrelated.

In [23] and [26] results of measurements at 58 and 60 GHz are reported. In [23], a worst case rms delay spread of 98 nanosecond is reported. This will support a transmission rate of 2 Msymbols/s. In [26] experimental results on the polarization dependence of multipath propagation are reported. The measurements were conducted in a conference room within a modern office building. The floor area was about 90 m^2, and the height of the ceiling was 2.6 m. The walls were of concrete, which were covered by plasterboard. The results indicated that the use of circular polarization effectively sup-

---

pressed multipath delayed waves. This phenomenon is due to difference in the reflection characteristics of the wall for circular and linear polarization. The use of circular polarization is found to be effective in reducing BER due to multipath propagation in high speed transmission channels.

3.2 Doppler Spread and Coherence Time

The foregoing section discussed the time spreading properties of mobile radio channels. Since either one or both transceivers forming the radio link could be in motion, the impulse response becomes time variant. Consider a link between a stationary base station and a mobile terminal travelling at a constant speed. While travelling, the mobile terminal is assumed to continuously change its orientation relative to the base station, thereby altering the velocity vector joining the mobile with the base station. Suppose a pure tone of f_o Hz is transmitted. The received frequency at the mobile is given by:

$$f = f_o \pm f_D \qquad (3.25)$$

where f_D is the Doppler frequency given by

$$f_D = \frac{v}{\lambda}\cos\psi = f_m \cos\psi \qquad (3.26)$$

where v, λ, and ψ are speed of the vehicle in meters/sec, wavelength in meters and the angular direction of the mobile travel relative to the line joining the transmitter with the receiver. Since f_D is continuously changing by virtue of changing orientation of travel, a transmitted pure tone will be received as a tone drifting randomly between $f_o - f_m$ and $f_o + f_m$, where f_m is the maximum Doppler frequency. This band of frequencies is known as Doppler spread. Doppler spread is indicative of the rate at which channel changes. If the bandwidth of the signal is much wider than $2f_m$, the degrading impact of Doppler spread on signal reception is negligible. The time over which the channel impulse response does not change noticeably is known as channel coherence time. The coherence time, T_c, is inversely proportional to Doppler spread and is given by

$$T_c \approx \frac{9}{16\pi f_m} \qquad (3.27)$$

for time correlation coefficient of 0.5 [27].

3.3 Physically Motivated Radio Channel Models

The above discussion on the channel impulse response does not give an insight into understanding the time and frequency domain behaviors of the channels. So far, we have determined that communication channels e.g. mobile radio channels can be characterized in either time or frequency domain. In time domain, we examined the channel impulse response and found rms delay spread and coherence bandwidth as characterization parameters. In frequency domain, we characterized the channel in terms of Doppler spread and channel coherence time. Coherence bandwidth and coherence time lead to the definitions of frequency and time selectivity respectively. However, our interest also lies in interpretation of the results in relation to the environmental features. Let us first examine what efforts have been made towards the development of physically motivated models of mobile radio channels.

Many measurement on mobile radio channels were made to characterize the mobile radio channels from physics of propagation and several channel models were proposed [10], [28]-[38]. Many researchers used the measured results in proposing mobile channel models suitable for both analog and digital transmission. Ossanna proposed a reflection model for signal propagation in mobile environments [28]. In his model, large randomly located vertical reflectors were assumed to contribute to the signal at the receiver. This model was successful in predicting a sharp cutoff frequency equal to the maximum Doppler frequency. However, the model could not accurately predict the presence of high energy low frequency components in the fading spectra. Also, the model did not take into account the effects of shadowing, intermodulations of concurrent reflections, and non-random reflector orientations. Nylund [29] confirmed one aspect of Ossanna's model through his measurements, and showed that within a small area the signal envelope follows a Rayleigh distribution.

Gilbert examined several models suitable for mobile environments based on scattering of radio waves [30]. These models considered a large number of waves arriving uniformly from all directions. Based on a comprehensive analysis of the mechanism of multipath propagation, Clarke proposed an extension of Gilbert's model that also made Ossanna's model a special case [31]. Clarke suggested that the failure of the latter to predict relatively high energy components at low frequencies was due to large changes in ground level which were not included in Ossanna's model. Reudink used Clarke's model to derive expressions for average fade rate and duration [32]. Gans applied Clark's model to the prediction of antenna propagation effects and the development of diversity schemes [34]. Hansen and Meno included in their model the effect of large-scale ground level variation and proposed a super-imposed Rayleigh and lognormal model for mobile fading [35]. This was demonstrated experimentally by Turin *et al* [4]. Work by Rice led to an extension of the model in the situation where one plane wave or a line of sight component may dominate the received signal [36]. In this case, the probability density function was shown to follow Ricean distribution. Many researchers investigated the statistics of the measured envelope and compared the observed distribution with Rayleigh, Ricean, Nakagami and other distributions [37],[38]. Note that all these models are based on the assumption that the signal arrives uniformly from all directions in the azimuth plane. The possibility that the signal may arrive from some preferred directions was not considered, though it seems to be more realistic. This aspect will be considered later in the chapter.

3.3.1 Development of a Model

Consider an information bearing signal, *s(t)*, transmitted over a mobile radio channel is written as:

$$s(t) = \Re e[u(t)exp(j2\pi f_c t)] \tag{3.28}$$

where *u(t)* is a complex valued low-pass equivalent waveform and f_c is the carrier frequency. The signal transmitted from the base station is assumed to reach the mobile via a large number of paths following reflections and diffractions from various randomly located and oriented obstructions in the propagation path. The radio wave is assumed to be vertically polarized[†], that is the *E*-vector is perpendicular to the ground plane and all the component waves are travelling horizontally in the azimuth plane. When the signal is measured over an interval significantly shorter than the average distance between the prevailing terrain irregularities, the standard deviation of the measured signal mean remains approximately constant. It has been shown by Lee [42] that for a spatial interval of measurement approximately equal to 40λ, the standard deviation of the mean is about 1 dB. This distance is usually used to obtain signal local mean. The received signal *r(t)* is considered to be made of a very large number of wave components, N $(N \rightarrow \infty)$, with amplitude A_i and delay τ_i. The received signal *r(t)* is given by:

---

† *Analysis for horizontal polarization is similar.*

$$r(t) = \Re\left\{\sum_{i}^{N} A_i u(t - \tau_i(t)) exp(j2\pi f_c(t - \tau_i(t)))\right\}$$

$$\text{(3.29)}$$

$$= \Re\left\{\sum_{i}^{N} g_i(t) u(t - \tau_i)\right\}$$

where $g_i(t)$ is the complex gain associated with the *i*th path. Note that $g_i(t)$ is time varying as perceived by the mobile receiver because of its movement relative to its surroundings. The time variant impulse response of the channel is written as:

$$h(\tau, t) = \sum_{i} g_i(t)\delta(\tau - \tau_i(t))exp(j\phi_i(t)) \qquad \text{(3.30)}$$

where $g_i(t)$, $t_i(t)$ and $\phi_i(t)$ are the time variant attenuation factor, propagation delay and phase of the *i*th path, respectively. The parameter i is the path number.

The value of $h(\tau, t)$ at each time delay, τ_i, is a random process whose statistics are assumed to remain stationary over the time interval corresponding to a distance of approximately 40λ. Equations (3.29) and (3.30) provide starting points for development of several models. A complete characterization of the received signal requires the description of five parameters (i, $g_i(t)$, $t_i(t)$, $\phi_i(t)$ and $u(t)$), the first four of these are due to propagation through the medium. The channel characterization very much depends on the signal bandwidth whether wider or narrower than the channel coherence bandwidth. The first step is to consider a signal whose bandwidth is much narrower than the channel coherence bandwidth. In this case, since all the signal frequencies are expected to be affected in a similar manner, it may be simpler to consider transmission of a single sinusoid.

3.3.1.1 Channel Model for Narrowband Transmission: Single Sinusoid Transmission

When an unmodulated carrier at a frequency of ω_c radians/sec is transmitted over a mobile radio environment, a large number of contributing waves, each with a random amplitude and an independent phase, construct the received signal. The received signal[†] may be written as:

$$y(t) = \Re\left\{ \sum_{i=-\infty}^{\infty} a_i e^{j((\omega_c + \omega_i)t + \phi_i)} \right\}$$

(3.31)

$$= \sum_{i=-\infty}^{\infty} a_i \cos(\omega_i t + \phi_i)\cos\omega_c t - \sum_{-\infty}^{\infty} a_i \sin(\omega_i t + \phi_i)\sin\omega_c t$$

Equation (3.31) may be rewritten as a sum of two lowpass random processes, X_1, X_2, modulated in quadrature,

$$y(t) = X_1(t)\cos\omega_c t - X_2(t)\sin\omega_c t$$

(3.32)

where $X_1(t)$ and $X_2(t)$ may be considered as sum of a large number of sinusoids with random frequencies, phase and amplitudes. These may be approximated by a complex Gaussian random process. Strictly speaking, the condition of the composite signal being complex Gaussian holds only under the central limit theorem (when the number of contributing waves approaches infinity), but it has been shown that if the number of contributing waves is seven or more, the resulting distribution is very nearly Gaussian [43]. Note that we are considering only small phase differences between the paths; in other words the inter-path delay is infinitesimally small ($\tau_i \rightarrow 0$) and paths with large (significant) delays are absent. The above discussion leads to the probability densities of X_1 and X_2 as:

$$p_{X_1}(x_1) = \frac{1}{\sigma\sqrt{2\pi}} exp\left(-\frac{x_1^2}{2\sigma^2}\right)$$

(3.33)

$$p_{X_2}(x_2) = \frac{1}{\sigma\sqrt{2\pi}} exp\left(-\frac{x_2^2}{2\sigma^2}\right)$$

and mean and mean square of X_1 and X_2 are given by

$$\langle X_1 \rangle = \langle X_2 \rangle = 0$$

$$\langle X_1^2 \rangle = \langle X_2^2 \rangle = \sigma^2$$

(3.34)

---

†. *The transmitted radio wave is assumed to be vertically polarized. The analysis presented here pertains to received electric field component, E_z being vertically polarized. Since the direction of the vehicle is not along either x or y axis in the x-y plane, two other field components, H_x and H_y exist. The analyses for H_x and H_y are similar to that for the electric field component. For more detail see [13].*

In writing (3.34), we have assumed that

$$\langle \sum_i a_i^2 \rangle = 1.$$

(3.35)

Also, it can be shown that X_1 and X_2 are uncorrelated i.e. $\langle x_1 x_2 \rangle = 0$, as well as jointly Gaussian [44]. Using transformation from Cartesian to polar coordinates, the joint probability density function for the signal envelope and its phase may be written as [45]:

$$p(r, \phi) = \frac{r}{2\pi\sigma^2} exp\left(-\frac{r^2}{2\sigma^2}\right); \qquad r \geq 0, -\pi \leq \phi \leq \pi$$

(3.36)

The above equation suggests that $p(r,\phi) = p(r).p(\phi)$, i.e. envelope and phase are independent random variables.

$$p(r) = \frac{r}{\sigma^2} exp\left(-\frac{r^2}{2\sigma^2}\right); \qquad r \geq 0$$

$$p(\phi) = \frac{1}{2\pi}; \qquad -\pi \leq \phi \leq \pi$$

(3.37)

Figure 3.4 shows the probability density functions for the envelope for $\sigma = 1$ and 2.

Two important moments of a random process, $p_X(x)$, are its mean m and variance μ, which are defined as:

$$m = \langle x \rangle = \int_{-\infty}^{\infty} x p_X(x) dx$$

$$\mu = \langle x^2 \rangle - \langle x \rangle^2 = \int_{-\infty}^{\infty} x^2 p_X(x) dx - m^2$$

(3.38)

The means and variance of the preceding random variables are given in Table 3.1.

The cumulative distribution function is given by

$$P(r \leq R) = \int_0^R \frac{r}{\sigma^2} exp\left(-\frac{r^2}{2\sigma^2}\right) dr = 1 - exp\left(-\frac{R^2}{2\sigma^2}\right)$$

(3.39)

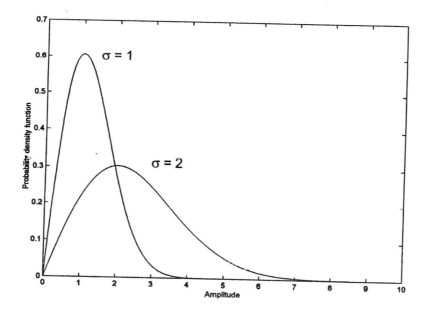

Figure 3.4 Rayleigh probability density function for fading envelope

---

Table 3.1	*Statistical parameters of envelope and phase*

Random Variable	Mean (m)	Mean Square $\langle x^2 \rangle$	Variance μ
x_1 or x_2	0	σ^2	σ^2
r	$\sigma(2/\pi)^{1/2}$	$2\sigma^2$	$(2-\pi/2)\sigma^2$
ϕ	0	$\pi^2/3$	$\pi^2/3$

Equation (3.39) also gives the probability of signal outage, i.e the fraction of the time when the signal falls below a certain threshold, R. It must be emphasized here that the above model is developed after assuming that at the receiver the signal arrives equally from all directions (in azimuth) and the line of sight propagation is absent. In actual mobile environment, this assumption may not remain valid since either a line of sight component is present or the signal does not arrive equally from all directions. When the base station is visible from the mobile unit and a line of sight propagation exists, the received signal is modeled as a sum of the direct path and scattered signal components. The received signal amplitude in this case follows Ricean distribution, which is written as,

$$p(r) = \frac{r}{\sigma^2} exp\left(-\frac{r^2 + Q^2}{2\sigma^2}\right) I_0\left(\frac{rQ}{\sigma^2}\right) \tag{3.40}$$

where $Q \cos 2\pi f_c t$ represents the line of sight component, r the scattered signal contribution with a mean power σ^2, and $I_0(.)$ is the modified Bessel function of the first kind with index zero. One parameter of interest is the ratio of direct path power to the scattered signal power, known as the Ricean parameter, K i.e. $K = Q^2/(2\sigma^2)$. The degree with which the signal fades is inversely proportional to the value of K.

3.3.1.2 Other Non-Conventional Channel Models

Rayleigh and Ricean distributions are derived from the physics of signal propagation in an environment consisting of randomly located signal scatterers. These models may be regarded as physically motivated models. Besides these models, several other statistical models have been used to characterize mobile radio channels. Nakagami [46] proposed the m-distribution (now known as Nakagami distribution) as a suitable model for mobile radio channels. His proposal was based on the observation that the experimental data has a better fit to an m-distribution than to Rayleigh distribution. This was confirmed by measurements reported in [47]. The Nakagami distribution also emulates Rayleigh, Ricean and lognormal distributions quite nicely and therefore is now regarded as a more general model. The other advantage of the Nakagami distribution resides in its ability to model more severe fading than is possible with the Rayleigh model (when $\frac{1}{2} < m < 1$). A Nakagami probability density function is given by:

$$p(r) = 2\left(\frac{m}{2\Omega}\right)^m \frac{1}{\Gamma(m)} r^{2m-1} exp\left(-\frac{mr^2}{2\Omega}\right) \tag{3.41}$$

The standard deviation Ω is defined as:

$$\Omega = 2\left(\bar{r}\frac{\Gamma(m)}{\Gamma\left(m+\frac{1}{2}\right)}\right)^2 \tag{3.42}$$

where $\Gamma(.)$ is the Gamma function, 2Ω is the mean signal power, and $m \geq 1/2$. $E[r]$ is the mean defined as

$$E[r] = \bar{r} = \frac{\Gamma\left(m + \frac{1}{2}\right)}{\Gamma(m)} \sqrt{\frac{\Omega}{m}} \tag{3.43}$$

Note that for $m = 1$, the above distribution reduces to Rayleigh. The parameter $1/m$ is usually referred to as the amount of fade (AF). As m increases, the channel fading reduces and when $m = \infty$, the fading disappears completely. In other words, as m increases, it emulates Rice factor, K, of Ricean distribution.

The parameters of the two distributions may be used to develop relationship between the Ricean and Nakagami distributions. It can readily be obtained if one rewrites the distribution in (3.40) in terms of its second moment Ω and Ricean parameter K, i.e.

$$f(r) = \frac{2r(1+K)}{\Omega} e^{-\left(K + \frac{(1+K)r^2}{\Omega}\right)} I_0\left(2r\sqrt{\frac{K(1+K)}{\Omega}}\right) \tag{3.44}$$

The nth moment of distribution in (3.40) is,

$$E\left\{r^n\right\} = (P_r)^{\frac{n}{2}} \Gamma\left(1 + \frac{n}{2}\right) {}_1F_1\left(-\frac{n}{2}, 1; \frac{P_s}{P_r}\right) \tag{3.45}$$

where P_r and P_s are power in scattered component and specular component respectively. After some manipulations, we find,

$$m = \frac{1}{\left[1 - \frac{K^2}{(1+K)^2}\right]} = 1 + \frac{K^2}{1 + 2K}. \tag{3.46}$$

The foregoing discussion reveals that the Rayleigh model is based on a scattering model and has a strong physical motivation. This is also true for Ricean model where a line of sight component is present with a scattered signal. It is, however, difficult to find a similar physical motivation for Nakagami channel. This may be a reason that the Nakagami model has not been widely used in the mobile communications related literature, however this trend is changing, and a number of publications appeared in the past decade e.g. [33].

3.4 Spectral Theory of Signal Reception

The transmission of a single sinusoid in changing environment produces envelope and phase variations due to continually changing the base-mobile geometry. The received frequency also changes continuously due to time variant Doppler frequency. Gans [34] studied the spectral theory of mobile reception. Our interest lies in finding the received power spectrum density. When closely positioned scatterers surround the receiver, it can the be assumed that the power arrives equally from every direction. It should be emphasized, however, that arrival of equal power from all direction is for analytical simplification and this assumption may not represent the field reality accurately. Thus, from the spatial point of view, the probability of receiving signal in a differential angle $d\alpha$ around angle α is given by $P(\alpha)\, d\alpha$, where $P(\alpha)$ is the probability density function of the angle of arrival. For an antenna gain $G(\alpha)$ in the direction of α, the received power is $\sigma^2 G(\alpha)P(\alpha)d\alpha$, where σ^2 is the average power received by an isotropic antenna. Obviously, the total received power is given by,

$$P_r = \int_0^{2\pi} \sigma^2 G(\alpha)P(\alpha)d\alpha \tag{3.47}$$

Consider a mobile travelling in the direction of α at a speed v m/sec. The received frequency is $f_c + v/\lambda \cos \alpha$. Suppose the power spectral density of this random spectrum is $S(f)$ and the differential frequency received in a differential angle $d\alpha$ around the angle α is df. Then, equating the received powers in $d\alpha$ (spatial) and in the frequency band df, we have

$$S(f)df = [\sigma^2 P(\alpha)G(\alpha) + \sigma^2 P(-\alpha)G(-\alpha)]d\alpha. \tag{3.48}$$

Note that the signal arriving at the mobile from angular directions α and $-\alpha$ have the same Doppler frequency as shown in Figure 3.5.
We also note,

$$f = \frac{v}{\lambda}\cos\alpha + f_c \tag{3.49}$$

$$df = \sqrt{f_m^2 + (f-f_c)^2}|d\alpha|$$

$$\cos\alpha = \frac{f-f_c}{f_m} \tag{3.50}$$

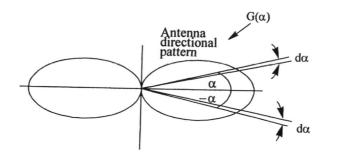

Figure 3.5 Antenna gain pattern

$G(\alpha)$ is the directional pattern of the antenna and may taken to be an even function so that $G(\alpha) = G(-\alpha)$ as shown in Figure 3.5. Also it can be assumed that the distribution of the angle of arrival is a uniform function $P(\alpha) = P(-\alpha) = 1/2\pi$. The general expression of the spectrum is therefore given by:

$$S(f) = \frac{\sigma^2}{\sqrt{f_m^2 - (f-f_c)^2}} |P(\alpha)G(\alpha) + P(-\alpha)G(-\alpha)| \qquad (3.51)$$

The field spectra shape is different for the three field components because $G(\alpha)$, the radiation pattern of the antennas, is different for the field components E_z, H_x and H_y. For E_z it is equal to 1.5. Substituting for $G(\alpha)$ and $P(\alpha)$ in equation we find the spectra for the E_z field component as:

$$S_{EZ}(f) = \frac{3\sigma^2}{\omega_m} \left[1 - \left(\frac{f-f_c}{f_m}\right)^2\right]^{-1/2} \qquad (3.52)$$

Expressions for the other two field components can be found using (3.46)-(3.50)[†]. The spectra of the field component E_z is shown for two values of maximum Doppler, 20 *Hz* and 30 *Hz* in the range of $|f| \le 29.99$ and 19.99, in Figure 3.6. A sharpest increase in the spectral density is evident at 20 and 30 Hz where it goes to infinity.

It easy to see that if $G(\alpha)$ and $P(\alpha)$ are not uniformly distributed over 2π, the frequency range and shape of the received spectrum will be different. We can use this property to our advantage to reduce the Doppler spectrum to a narrow range. The following example illustrates this.

[†]. *Note that H_x and H_y are found by dividing E_z by the intrinsic impedance of the free space and including with it the direction of the field components. The intrinsic impedance of the free space is 120π.*

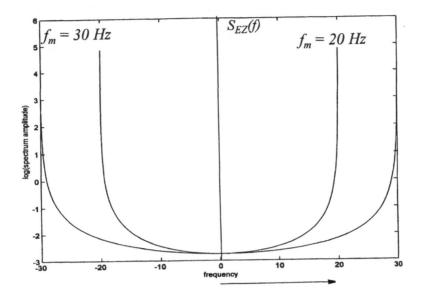

Figure 3.6 Power Spectra of the E_z field component

EXAMPLE E-3.4.1 Doppler Frequency Variation with Angle of Arrival

Plot the spectrum of the signal received by an antenna whose beam-width is $10°$ and is directed (a) perpendicular to and (b) along the motion of a mobile, which is travelling at 50 km/hr. The transmitted signal is a single sinu-soid at 821 MHz. The directional gain of the antenna is G_o within an angle $|\alpha| < \beta/2$ and $G(\alpha) = 0$ otherwise. Assume antenna gain $G_0 = 30$.

Solution:

(a) When the antenna is directed perpendicular to the vehicular direc-tion of travel, Doppler frequency is confined to $f < f_m \sin \beta/2$ and $S(f)$ is given by:

$$S(f) = \frac{G_o}{f_m \sqrt{1 - \left(\frac{f}{f_m}\right)^2}} = \frac{30}{38 \sqrt{1 - \left(\frac{f}{f_m}\right)^2}} \qquad \text{(E-3.3.3)}$$

$f_m = v/\lambda = 38$ Hz; $\alpha/2 = 5°$, giving

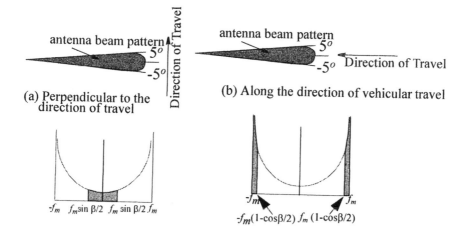

Figure E-3.4.1 Spectra of signals for different directions of vehicular travel

$$S(f) = \frac{30}{38\sqrt{1 - \sin^2(\beta/2)}} = \frac{30}{38\cos(\beta/2)} \qquad \text{(E-3.3.4)}$$

The spectrum is shown in Figure 3.4.1.

(b) When antenna is directed along the direction of travel of the vehicle,

$$S(f) = \frac{2G_o}{f_m\sqrt{1 - \left(\frac{f}{f_m}\right)^2}}; \quad 1 > \left|\frac{f}{f_m}\right| > 1 - \cos\frac{\beta}{2}$$

$$\qquad \text{(E-3.3.5)}$$

$$= \frac{2G_o}{f_m\sin\beta/2} = 18.12$$

which is in the range of 37.85 to 38 Hz. The spectrum is shown in Figure E-3.4.1. Thus, directing the mobile antenna in the direction of its travel will result in higher Doppler frequency as compared to the case when the antenna is directed perpendicular to the vehicle travel. This example demonstrates that directive antennas could play an important role in reducing the effects of fading.

3.4.1 Field Components Correlation Functions

In the previous sections, we observed that a signal transmitted over a mobile channel is subject to random changes in amplitude, frequency and phase. The dependence of these random changes in time and space are of importance in system design, where the task of system designers is to seek methods or techniques so that the impacts of these randomly varying parameters are minimized. If the field components of the signal are found to be uncorrelated at some points in space, then picking these components at these points and combining them in certain manner could yield an output that does not fade as severely as the components. The process of combining two or more independently fading versions of the signal in a certain way is known as diversity reception. Diversity plays an important role in combating signal fading in mobile radio communications. The cross-correlation of the field components can be found using standard technique [48] e.g.

$$\Re_{xy}(\tau) = \int_{-\infty}^{\infty} \int_{-\infty}^{\infty} xyp(x,y)dxdy \qquad (3.53)$$

where x and y are two random variables and $p(x,y)$ is their joint probability density function.

It can be shown that the cross-correlations $\Re_{EzHx}(\tau)$, and $\Re_{HxHy}(\tau)$ are zero for all values of τ. Furthermore, the crosscorrelation between E_z and H_y, $\Re_{EzHy}(\tau)$, is zero when $\tau = 0$ [13]. Thus, we could install two loop antennas at any distance and combine the outputs of these antennas to produce simple space diversity system. Similarly, installation of a vertical dipole and a loop at the same location gives an alternate diversity in polarization.

3.5 Envelope and Phase Correlation Functions

In the previous section, it was demonstrated that a signal propagating through mobile radio channel undergoes both envelope and phase fluctuations. Since frequency and phase modulated analog and digital signals are extensively used in mobile wireless communications, it might appear that channel effects on signal envelope would be of little importance. However, since the signal amplitude represents instantaneous signal power, therefore its fluctuations will result in a randomly varying signal to noise or signal to interference ratio. It will be shown in Chapter 4, that the fading of the envelope results in a significant degradation in performance. We conclude that correlation properties of the envelope are also important from the point of view of creating uncorrelated versions of the signal so that diversity reception could

---

be implemented. Furthermore, the signal phase fluctuations result in degradation in digital transmission systems particularly in the case of phase modulation. It is therefore pertinent to study the correlation properties of the phase as well.

The auto-correlation of Rayleigh faded envelope, $r(t)$, is given by [49]; [also derived in Appendix A]:

$$\Re_r(\tau) = \langle r(t)r(t+\tau) \rangle = \frac{\pi}{2}\mu\left[1 + \frac{J_o^2(\omega_m\tau)}{4}\right] \tag{3.54}$$

where μ is average power output power of a vertical dipole, ω_m is the maximum Doppler frequency in radians/sec, and τ is the delay. The autocovariance of the envelope is given by

$$L_e(\tau) = \frac{\pi}{8}\mu J_o^2(\omega_m\tau) \tag{3.55}$$

It can be seen that $L_e(\tau)$ reduces to zero when $\omega_m\tau$ is approximately equal to 2.4 or alternatively the signal envelope received at two points spaced by a distance $d = v\tau$ or 0.38λ apart will be uncorrelated. In [47], autocovariance of the signal received over mobile radio channel has been measured in several areas of Ottawa-Hull, Canada. The results are shown in Figure 3.7. It is seen that the autocovariance follows that is given in (3.55). The autocovariance function crosses the abscissa around 0.38λ. A small discrepancy from the theoretical results is due to some experimental error.

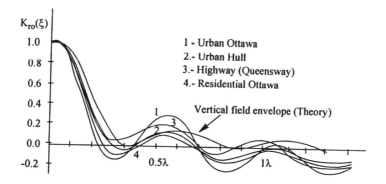

Figure 3.7 Autocovariance function of fast fading by area

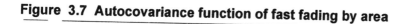

EXAMPLE E-3.5.1   Envelope Correlation Function

The autocorrelation coefficient of the envelope of an E_z field signal for a τ second time delay is given by:

$$L_e(\tau) = \frac{\pi}{8}\mu J_o^2(\omega_m\tau) \qquad\qquad \text{(E-3.5.1.1)}$$

Two antennas are installed d meters apart on top of a vehicle that travels at a speed of v meters/sec. If the frequency of operation is 900 MHz, find:

(a) The distance, d, when the signals received by the two antennas are perfectly uncorrelated.

(b) The distance when correlation coefficient is 0.57.

Solution:

(a) We first normalize $L_e(\tau)$ such that $\mu = 8/\pi$. When $L_e(\tau) = 0$, $J_o(\omega_m\tau) = 0$. We find that $\omega_m\tau = 2\pi(d/\lambda)$, assuming speed of the mobile, v, remains uniform over the observation period. From Bessel function tables, $J_o(2.41) = 0$. This gives a value of $d = 0.3836\ \lambda$. The wavelength at 900 MHz is 0.333 meter. The distance is calculated to be 12.79 cm.

(b) Now, for $L_e(\tau) = 0.57$, the value of $\omega_m\tau$ from the tables is read to be 1.4. This gives a value of spacing between the antennas $d = 0.223\ \lambda$ or 7.43 cm.

3.5.1 Average Level Crossing Rate (ACR)

The received signal envelope experiences fades whenever the terminal moves. In addition to the motion of the mobile, the movements of surrounding bodies can also disturb the spatial standing waves thereby causing fading. Rayleigh distribution has been found to be an appropriate model to describe the fading envelope. It is also of interest to know the average number of times the signal fades below a threshold and the average length of time that the signal stays below it.

An examination of a time samples of the received envelope would reveal that shallower fades occur more frequently than very deep fades, the latter are relatively rare. The signal fade rate may be characterized by the number of times the envelope crosses a threshold level in a unit time. Since the threshold level may be set to an acceptable level of reception, therefore level crossing rate may be considered to be synonymous with the quality of reception. The level crossing statistics are used in determining outage rate, the rate at which reception quality becomes unacceptable. It is evident that the level crossing rate is directly related to the speed of the mobile since a faster moving mobile will cut the spatial standing waves more rapidly. Several researchers have attempted to model the crossing rate statistically but in the absence of the second order statistics, modeling is largely empirical [13],[51]. The difficulty in modeling lies because an infinite number of signal envelope waveforms can yield Rayleigh distribution. Time statistics therefore are not unique, and only the average fading rate or level crossing is possible.

The derivation of an expression for the average crossing rate is simple. Consider a range of random variable R representing envelope between r and $r+dr$. Let R within this range has a slope of r'. Note that the envelope can cross the threshold with a positive or a negative slope. For either of slopes, the count will be identical. However, we will count only level crossings with positive slope. The time spent in the interval dr for the slope r' is dr/r'. The number of crossings in a unit time is thus equal to the expected amount of time the signal remains in the interval dr with a slope r', divided by the time taken for one crossing at that slope. i.e.

$$\frac{E[t]}{\frac{dr}{r'}} = \frac{p(r,r')dr\,dr'\,dt}{\frac{dr}{r'}} = p(r,r')r'\,dr'\,dt \qquad (3.56)$$

The number of crossings in time T, for given slope r' is

$$N_{T,\,r'} = p(r,r')r'\,dr'\,T = Tp(r,r')r'\,dr' \qquad (3.57)$$

The average crossing rate per second for a given slope r' is

$$\frac{Tp(r,r')r'\,dr'}{T} = p(r,r')r'\,dr' \qquad (3.58)$$

The average crossing rate per second for all possible (positive) slopes, i.e. $r' = 0$ to ∞, is

$$N_R = \int_0^\infty p(r,r')r'\,dr' = \int_0^\infty p(R,r')r'\,dr' \qquad (3.59)$$

where R is the threshold level.

Using $p(r,r')$ in (C.6) in Appendix C, which indicates that r and r' are independent variables. N_R, the Average Crossing Rate (ACR) for a threshold R, is thus.

$$N_R = \int_0^\infty \frac{R}{\sigma^2}exp\left(-\frac{R^2}{2\sigma^2}\right)\frac{r'}{\sqrt{2\pi\mu_2}}exp\left(-\frac{r'^2}{2\mu_2}\right)dr' \qquad (3.60)$$

Define $R_{r.m.s} = (2\sigma^2)^{1/2} = (\langle r^2 \rangle)^{1/2}$, and

$$\rho = \frac{R}{\sqrt{\langle r^2 \rangle}} = \frac{R}{R_{rms}} = \frac{R}{\sqrt{2}\sigma} \qquad (3.61)$$

$$N_R = \sqrt{\frac{\mu_2}{\sigma^2}}\rho \, exp(-\rho^2) \qquad (3.62)$$

For E_z, the average crossing rate is given by:

$$N_{REz} = \sqrt{2\pi}f_m \rho \, exp(-\rho^2) \qquad (3.63)$$

The maximum average crossing rate occurs at -3 dB relative to R_{rms} i.e. $\rho = R_{r.m.s.} /(\sqrt{2})$. Figure 3.8 shows the average crossing rates for Rayleigh faded channel for three values of f_m.

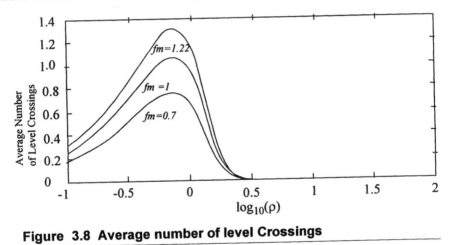

Figure 3.8 Average number of level Crossings

EXAMPLE E-3.5.2 Average Crossing Rates and Mobile Terminal Speed

(a) In a mobile radio communication experiment the mobile unit is travelling in a downtown area at a speed of 40 km/hr. The experimenter finds that the receiver threshold is crossed on the average of 20,000 in 16 minutes. Given that the operating frequency of the transmitter is 850 MHz, find the mean signal relative to the rms level.

(b) The mean signal level remains unchanged from part (a) and the number of crossings drops to 14,523 in an observation time of 24 minutes. Estimate the speed at which the mobile is travelling.

Solution:

(a) Vehicle speed = 11.11 m/sec; f_m = 31.48 Hz. Number of crossings per second = 20.83. $N_{REz} = \sqrt{2\pi}f_m\rho exp(-\rho^2) = 20.83$. This gives a value of ρ = 1.2456. R = 1.2456 R_{rms} or 1.9075 dB above R_{rms}

(b) Using ρ = 1.2456, and N_R = (14523)/(24 x 60) = 10.0854 crossing / second, we find f_m from $N_{REz} = \sqrt{2\pi}f_m\rho exp(-\rho^2)$ to be 15.246. Thus, $v = f_m\lambda$ = 5.38 m/sec or 19.37 km/hr.

3.5.2 Average Fade Duration (AFD)

The average duration of fades is the average length of time for which the signal envelope remains below threshold of reception. More recently, the average fade duration (signal outage) has gained acceptance as a performance measure along with either average signal to noise ratio in analog or bit error rate in digital communications. ACR and AFD may be used to determine the impact of fading on digital transmission systems particularly on currently widely used packet transmission operating in fading mobile radio channels. In these systems, average fade duration may serve as a guide to the design of packet radio system. For example, in an adaptive packet transmission system, where the packet length changes with the vehicular speed, estimation of average fade duration is needed.

Figure 3.9 shows a sample waveform of a typical envelope of a received signal. The signal consists of cyclic occurrence of fades and non-fades. Let τ_i be the duration of ith fade. The average fade duration at a threshold level, R is the average amount of time that the signal remains below the threshold level R. During an observation time interval of T seconds, the total time that the signal remains below R is equal to $\Sigma \tau_i$.

The fraction of time that the signal remains below threshold level R is

$$\frac{1}{T}\sum_i \tau_i \tag{3.64}$$

where τ_i is the ith segment of time over which the signal remained below the threshold. (3.64) also gives the probability that signal r is below the reference level R, i.e.

$$P(r < R) = \frac{1}{T}\sum_i \tau_i \tag{3.65}$$

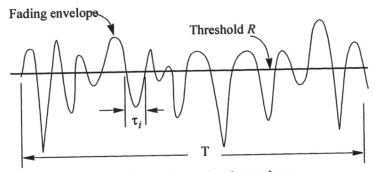

Figure 3.9 Sample waveform of received envelope

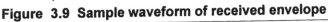

The probability that $r < R$ is given by the product of the average fade duration and the number of crossings in a unit time. Average Fade Duration (AFD) $\bar{\tau}$ is therefore given by,

$$\bar{\tau} = \frac{1}{N_R} P(r < R) \tag{3.66}$$

$$
\begin{aligned}
P(r < R) &= \int_0^R p(r)dr \\
&= \int_0^R \frac{r}{\sigma^2} exp\left(-\frac{r^2}{2\sigma^2}\right)dr \\
&= (1 - exp(-\rho^2))
\end{aligned}
\tag{3.67}
$$

where

$$\rho^2 = \frac{R^2}{2\sigma^2} \tag{3.68}$$

The average fade duration is:

$$\bar{\tau} = \frac{1}{\rho}\sqrt{\frac{\sigma^2}{\mu_2}}(exp(-\rho^2) - 1) \tag{3.69}$$

For the envelope of E_z field component, the average fade duration is given by,

$$\bar{\tau}_{Ez} = \frac{exp(\rho^2) - 1}{\rho f_m \sqrt{2\pi}} .$$

(3.70)

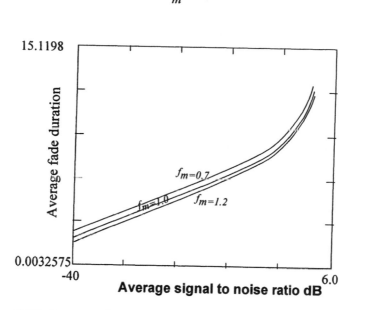

Figure 3.10 Average fade duration on Rayleigh faded envelope

    Several attempts have been made to model fade duration statistically [50],[51],[52]. DaSilva [50] reported that the cumulative distributions of fade duration and interfade (non-fade) interval vary with the threshold level. The results in [50] indicate that fade duration statistics closely approximate an exponential distribution when the threshold is set at the signal average. The interfade interval is also found to follow exponentially distribution over a threshold range of -10 to -30 dB relative to the average envelope [50].

    Using a simulated mobile radio channel, Bodtmann and Arnold [51] measured fade depths over a wide range channel outputs and extended the analytical results for shallow fades between 0.08 dB and 5 dB. The measured fade duration distribution is shown in Figure 3.11. For the threshold level of 10 dB or more below the median signal level, all of the distributions have identical shapes. For deep fades, the distributions quickly reach an asymptotic slope of (fade duration)$^{-3}$. For shallower fades however, the shape of distribution changes due to peaks in the probability density function of the fade duration. The reasons for these peaks can be explained as follows. The Doppler spectrum has peaks displaced by $\pm f_m$ from the carrier. These cause a strong quasi-periodic component in the fading envelope at around $2f_m$. It follows from Fourier series analysis that fade duration will have some degree of periodicity with period $1/(2f_m)$; consequently the fade duration probability density function will have peaks at multiples of this period. Various field measurements of average fade duration and level crossing rate have been found to substantiate the theoretically predicted values [47],[49].

The average crossing rate and the average fade duration may be altered substantially if the receiver uses a directional antenna. Narrow beam directive antennas limits the Doppler spectrum to a range much narrower than $\pm f_m$ produced by an isotropic antenna. More details on this will be given in Chapter 8.

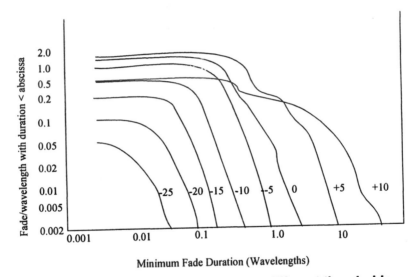

Figure 3.11 Measured fade durations at different thresholds

EXAMPLE E-3.5.3 Estimation of Average Packet Length

Estimate average packet length for transmission rate of 4800 bits/sec over a mobile channel that links a base station with the communication equipment abroad a train, which is travelling at 80 km/hr. The link operates at 946 MHz. The detection threshold of the receiver is set at -2 dB relative to rms signal level.

Solution:

The train speed is 80 km/hr, which is equivalent to 22.22 m/sec. The frequency of operation is 946 MHz which corresponds to the wavelength $\lambda=$

0.3171 meter. The threshold ρ is set at -2 dB relative to the R_{rms} or $\rho = R/R_{rms}$ = 0.7943.

$$\bar{\tau} = \frac{P(r \geq R)}{N_R} = \frac{\int\limits_R^\infty p(r)dr}{N_R}; \quad but \qquad \text{(E-3.4.3.1)}$$

$$p(r) = \frac{r}{\sigma^2}exp\left(\frac{r^2}{2\sigma^2}\right) \qquad \text{(E-3.4.3.2)}$$

and

$$\bar{\tau} = \frac{exp(-\rho^2)}{\sqrt{2\pi}f_m exp(-\rho^2)} = \frac{1}{\sqrt{2\pi}f_m\rho} \qquad \text{(E-3.4.3.3)}$$

$f_m = v/\lambda = 70.07$ Hz, or average fade duration is 0.1423 seconds. Thus, the average packet duration is $0.1423 \times 4800 = 683$ bits.

Rice [52] derived an asymptotic expression for the distribution function for the duration of deep fades as:

$$F(\mu, R) = \frac{2}{\mu}I_1\left(\frac{2}{\pi\mu^2}\right)exp\left(\frac{-2}{\pi\mu^2}\right) \qquad \text{(3.71)}$$

where μ is the fade duration normalized to its mean value.

$$\mu = \tau_\lambda\sqrt{\tau_\lambda R} \qquad \text{(3.72)}$$

and $I_1(z)$ is a Bessel function of imaginary argument. To obtain the distribution of the number of fades per wavelength, $F(\mu, R)$ must be multiplied by the number of fades of depth R, $N_\lambda(R)$:

$$N_\lambda(\mu, R) = N_\lambda(R)F(\mu, R) \qquad \text{(3.73)}$$

3.5.3 Signal Phase Characterization

The widespread use of angle (frequency and/or phase) modulation in mobile systems suggests that the study of phase characteristics of the channel is also important. In a FM receiver, the information is obtained by taking the derivative of the signal phase. However, additional noise is produced because of the presence of random FM as result of channel fading. The random FM noise distorts the information signal. On the other hand, when the data encodes the carrier phase, the presence of random FM determines the error rate. It is clear that the signal phase characteristic also determines the rate at which data can be transmitted using phase modulation.

3.5.3.1 Random Frequency Modulation

The mobile terminal when roams in the service area changes the velocity vector to the base station randomly. Such motion produces randomly varying Doppler frequency at the receiver. These frequency variations are known as random frequency modulation. It can be shown that the Random FM defines the limit to performance of mobile communication systems when the signal to noise ratio is high. The probability density $p(\theta')$ is obtained by integrating $p(r, r', \theta, \theta')$ over r, r' and θ. From Rice [45] the joint density function in the four random variables r, r', θ, θ' of a Gaussian process is written as (see Appendix C):

$$p(r, \theta, r', \theta') = \frac{r^2}{(2\pi)^2 \sigma^2 v^2} exp\left[-\left(\frac{r^2}{2\sigma^2} + \frac{r^2\theta'^2 + r'^2}{2v^2}\right)\right] \tag{3.74}$$

and $p(\theta')$ is derived as (Appendix C).

$$p(\theta') = \frac{1}{2}\frac{\sigma}{v}\left(1 + \frac{\sigma^2}{v^2}\theta'^2\right) \tag{3.75}$$

From (3.75), we find that the mean square value of random FM, $<\theta'^2>$, is infinite. This means that random FM power can become arbitrarily large unless it is controlled by restricting the bandwidth which is the case for all practical systems. The probability density of random FM conditioned on the received envelope, R, determines the rms bandwidth at the selected value of R. For a given value of R, the conditional probability density $p(\theta'|R)$ has Gaussian distribution with a standard deviation of μ_2 ($v=\mu_2^2$ in (3.75)) i.e.

$$p(\theta'|R) = \frac{p(\theta', R)}{\sqrt{2\pi v}} = \frac{R}{\sqrt{2\pi v}}exp\left(-\frac{R^2\theta'^2}{2v}\right) \tag{3.76}$$

It can be seen that whenever the signal envelope experiences a deep fade, the frequency deviation due to random FM increases proportionally. For example, a fade depth of 20 dB will produce a frequency deviation of 10 ω_m. This is shown in Figure 3.12. This phenomenon creates difficulties when coherent digital modulation is attempted over fast fading mobile radio channels. For a carrier recovery scheme to work properly, a narrow band noise limiting loop filter is desirable. However, whenever a deep signal fade occurs, the instantaneous frequency deviation could potentially reach as high as 30 times the maximum theoretical Doppler frequency. To keep the phase-locked loop in lock, the loop filter bandwidth defining the lock range must be widened. This is contradictory to the above mentioned requirement of narrow-band loop filter. A possible solution to this problem lies in the use of a frequency divider stage preceding the phase locked loop.

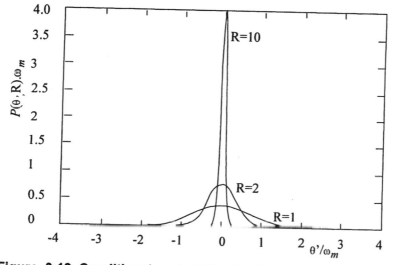

Figure 3.12 Conditional probability distribution of random FM

3.5.3.2 Power Spectrum of Random FM

At the output of the FM receiver, the signal is accompanied by so called random FM noise, which is caused by channel fading. It will be shown in Chapter 4 that random FM limits the performance of FM systems under fading conditions. To calculate the noise power due to random FM, and estimate the output signal to noise ratio, we need random FM noise power spectral density.

The power spectrum of θ' is obtained by taking the Fourier transform of the autocorrelation function $\Re_{\theta'}(\tau)$ i.e.

$$S_{\theta'}(f) = \int_{-\infty}^{\infty} \Re_{\theta'}(\tau)exp(-j\omega\tau)d\tau = 2\int_{0}^{\infty}\Re_{\theta'}(\tau)cos(\omega\tau)d\tau \qquad (3.77)$$

where $\Re_{\theta'}(\tau)$ is given by (Appendix C).

Jakes [13] evaluated the spectrum by separating the time range of integration, τ, into three regions so that appropriate approximations can be applied to solve Bessel and logarithmic functions in (3.77). The sum of the spectra obtained for the three time ranges gives the resulting spectrum. This is shown in Figure 3.13. In the limit when frequencies approaching infinity, the random FM power spectrum approaches an asymptotic form given by:

$$\lim_{f \to \infty} S_{\theta'}(f) \cong \left(\frac{\mu_2}{\sigma^2} - \frac{\mu_1^2}{\sigma^4} \right) f^{-1} \qquad (3.78)$$

where μ_1, and μ_2 are the first and the second central moments and σ is the average power received by an isotropic antenna. These are defined by the following equation.

$$\mu_n = (2\pi)^n \int_{f_c - f_m}^{f_c + f_m} S_i(f)(f - f_c)^n df \qquad (3.79)$$

where $S_i(f)$ is the input spectrum of the field component. The asymptotic form of (3.78) has been found to give a reasonably accurate estimate of the random FM power spectrum over the audio frequency range of 300 to 3000 Hz.

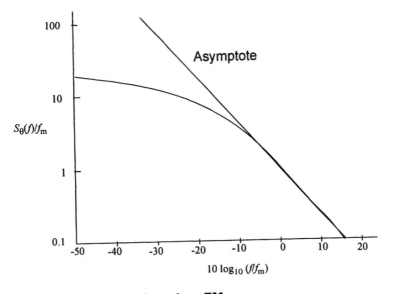

Figure 3.13 Spectrum of random FM

EXAMPLE E-3.5.4 Impact of Vehicular Speed on Random FM

In the previous section, it has been demonstrated that the random FM power spectral density can be approximated by an asymptotic relationship given by

$$\lim_{f \to \infty} S_{\theta'}(f) \cong \left(\frac{\mu_2}{\sigma^2} - \frac{\mu_1^2}{\sigma^4} \right) f^{-1} \qquad \text{(E-3.5.4.1)}$$

A mobile radio system operates at 900 MHz. Find the relative levels of random FM for the following cases:

(a) For vehicular speeds of 40 km/hr and 100 km/hr when the baseband range is 300 to 3000 Hz,

(b) How much random FM power increases when the high end of the baseband frequency is increased from 3000 to 7500 Hz. The vehicular speed in both cases is 40 Km/hr.

(c) Explain why does an increase in speed of the vehicle has a greater effect on the random FM power than does an increase in the bandwidth?

Solution:

(a) From Section 3.5, the relation between μ_n and σ is,

$$\mu_n = (2\pi)^n \int_{f_c - f_m}^{f_c + f_m} S_i(f)(f - f_c)^n df$$

$$= \sigma^2 \omega_m^n \frac{1.3.5.....(n-1)}{2.4.6...n} \qquad \text{(E-3.5.4.2)}$$

With $\sigma^2 = (3E_o^2/4)$ and ω_m equals 209 and 476 rad/sec for 40 and 100 km/hr respectively. $\mu_2 = \sigma^2\omega^2/2$ and $\mu_1 = 0$.

Substituting in (E-3.5.4.1), we have: for 40 km/hr $S_{\theta'}(f) = 0.5$ $(209.44)^2 f^1$, and for 100 km/hr $S_{\theta'}(f) = 0.5 (476)^2 f^1$. This gives $P_{rf40}/P_{rf100} = 0.1936$ or -7.13 dB. This means that power due to random FM at 100 km/hr is 7.13 dB higher than that at 40 km/hr.

(b) Now the baseband range has been increased by a factor of 2.5, i.e. from 3000 to 7500. $P_{rf40} = 0.5 (209.44)2 \log_e (7500/300) = 70598$, and $P_{rf40} = 0.5 (209.44)2 \log_e(3000/300) = 50501$. This shows that increasing the baseband bandwidth by a factor of 2.5, the change in random FM power is only +1.45 dB.

(c) The random FM power spectral density is proportional to the square of the vehicle speed, on the other hand it is proportional to the loga-

---

rithm of the bandwidth ratio. Thus, vehicle speed has a greater effect on the random FM power.

3.5.4 Envelope and Phase Correlation Bandwidth

The model presented in Section 3.3.1.1 is valid only when a single sinusoid is transmitted. As mentioned earlier, in practical systems single sinusoid is hardly used for information transmission. The channel over which all components of the transmitted spectrum fade in unison is termed flat fading channel. The flat fading channels are also known as time selective channels. These channels usually transmit information over a single path, which is scattered by the scatterers local to the receiver. In other words the channel impulse response is in the form of a solo slightly time stretched impulse devoid of other distinct paths. In contrast to this, propagation in urban areas is characterized by highly dispersive multipath effects, caused by reflections and scattering from terrain irregularities including buildings that are located at relatively greater distances from the receiver.

Each distinct path thus received is scattered by the local scatterers. These multipath effects manifest themselves frequency selective fading, which appears as intersymbol interference. Intersymbol interference occurs when differential time delays between the received replicas of the transmitted signal are significant, usually $T_b/8$ or more. The received time dispersed signal replicas contribute to distortion of spectral components. Moreover, these replicas also fade independently because different set of scatters affect each path. Thus, the signal spectral components undergoing temporal variation in magnitude result is an irreducible error rate. Irreducible error rate implies that the error rate does not decrease with an increase in signal to noise ratio. This is because an increase in the transmit power proportionally increases power of the multipath components. The performance of digital communication systems can be improved with the use of channel equalizers [53][54][55]. A detailed discussion on this topic is presented in Chapter 9.

The channel dispersion reduces the channel coherence bandwidth. A signal occupying bandwidth less than coherence bandwidth suffers negligible distortion but when the signal bandwidth exceeds the coherence bandwidth, severe intersymbol interference results. The channel correlation bandwidth is measured by transmitting two sinusoids separated by frequency ω_s and evaluating the correlation between their received envelopes or phases to obtain respectively the envelope or phase coherence bandwidth. A correlation coefficient of 0.5 is usually used in defining the coherence bandwidth. Other values of correlation coefficients have also been used.

To determine coherence bandwidth, joint probability distribution of envelopes (r_1, r_2), and phases (θ_1, θ_2), due to two sinusoid separated by a frequency range is given by [Appendix A, Equation (A.14)]:

$$p(r_1, r_2, \theta_1, \theta_2) = \frac{r_1 r_2}{(2\pi)^2 \mu^2 (1 - \zeta^2)} \cdot exp\left[-\frac{\left(r_1^2 + r_2^2 - 2\zeta r_1 r_2 \cos(\theta_2 - (\theta_1 + \phi)) \right)}{2\mu(1 - \zeta^2)} \right] \tag{3.80}$$

$$p(r_1, r_2) = \frac{r_1 r_2}{\mu^2 (1 - \zeta^2)} exp\left[-\frac{r_1^2 + r_2^2}{2\mu(1 - \zeta^2)} \right] I_0\left(\frac{r_1 r_2}{\mu(1 - \zeta^2)} \right)$$

where

$$I_0(x) = \frac{1}{2\pi} \int_0^{2\pi} e^{x(x \cos\theta)} d\theta \tag{3.81}$$

and the envelope correlation function is given by:

$$\Re(r_1, r_2) = \int_0^\infty dr_1 \int_0^\infty r_1 r_2 p(r_1, r_2) dr_2 \tag{3.82}$$

r_1, r_2 are defined as envelopes of sinusoids of frequency f_1 at t, and f_2 at $t+\tau$ respectively. Similar definitions for θ_1 and θ_2 apply.

We consider two cases. The first is when a single sinusoid is transmitted and correlation between the envelopes at times spaced by τ seconds is measured. The second case relates to determination of coherence bandwidth. Consider that a single sinusoid is transmitted, then r_1, r_2, are the envelops and θ_1, and θ_2 are the phases of the received signal at times t and $t + \tau$ respectively.

From Appendix A, Equation (A.28) the autocorrelation function is given by:

$$\Re(r_1, r_2) = \frac{\pi}{2} \mu^2 \left(1 + \frac{J_0^2(\omega_m \tau)}{4} \right) \tag{3.83}$$

Figure 3.14 shows the autocorrelation function of the received envelope as a function of Doppler frequency or vehicular velocity. This can also be regarded as the cross correlation function of the envelopes of the signals received by two antennas separated by a distance $d = v/\tau$. Reference [47]

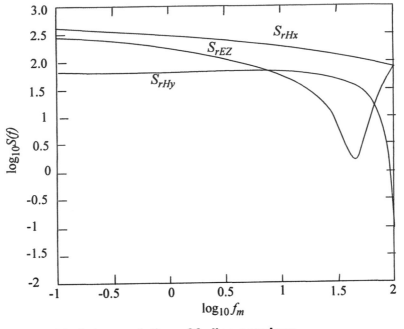

Figure 3.14 Autocorrelation of fading envelope

describes the measurement of autocovariance of fast and slow fading envelope, $S_{rEz}(f)$, and Figure 3.15 shows the measured results; the shapes of the measured autocovariance functions are similar to those predicted from theory. There are, however, two observations to make. First, the peaks of the function $J_o^2(\omega_m)$ occur closer together than predicted. The inaccuracies in the calibration of the spatial scale (i.e. measurement of wheel radius) have been cited as one reason for this discrepancy. Second, the measured autocovariance function makes excursions below zero, while $J_o^2(\omega_m \tau)$ is a positive valued function. This may be due to the filtering process that removes the slow fading component. In practical field measurements, there is no clear distinction between the slow and fast fading components i.e separating the two could be quite difficult. Too short a filter length will result in a mean with large variance, and an excessively long filter may result in some loss of the slowly varying component. In a similar manner, correlation functions for the phase angles of the corresponding signal phasors, $\Re(\theta_1, \theta_2)$, is found as:

$$\Re(\theta_1, \theta_2) = \langle \theta_1, \theta_2 \rangle = \int_{-\pi-\pi}^{\pi \ \pi} \theta_1 \theta_2 p(\theta_1, \theta_2) d\theta_1 d\theta_2 \qquad (3.84)$$

We define coherence bandwidth based on the signal phase as the bandwidth over which the phase correlation coefficient drops to 0.5. It is estimated that this bandwidth is nearly half of that based on the envelope correlation function [13]. The Fourier transform of these correlation functions i.e (3.83) and (3.84) provide the power spectral density of the envelope and phase respectively.

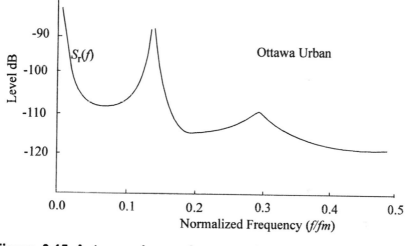

Figure 3.15 Autocovariance of measured envelope

EXAMPLE E-3.5.5 Estimation of Correlation Bandwidth

In an experiment, the channel mean delay spread is measured to be 4.3 microseconds. Find the correlation bandwidth if the delay spread follows an exponential distribution.

Solution:

The delay spread is given by $p(\sigma,T) = (1/\sigma) \, exp(-T/\sigma)$, where $\sigma = 4.3$ μsec. This model is based on equal power being received from all directions. Correlation coefficient $= \rho_e(S,\tau) = [J_o(\omega_m\tau)]^2 / (1+S^2\sigma^2)$.

For estimation of coherence bandwidth, assume $\tau = 0$ or $J_o^2(\omega_m\tau) = 1$. Thus $0.5 = 1/(1+S^2\sigma^2)$ or $S = 232,558$ rad/sec. This gives $\Delta f = 37$ kHz. This is equivalent to the classical definition of $S = 1/\tau$. However, the resulting figure of 37 kHz is much lower than the usually accepted figure.

3.6 M-B and M-M Propagation Models

The preceding discussions apply to communications on the forward channel from an elevated base station to a mobile surrounded by closely located scatterers. It was assumed that there is little or no spatial dispersion of radio waves prior to reaching the scatterers, but an extensive local scattering does exist. As the mobile unit moves, the local scatterers strongly affect the received signal characteristics such as its strength and correlation function.

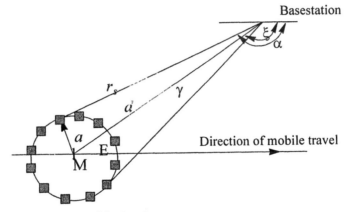

a is the radius of the local scatterers

Figure 3.16 Mobile to base (M-B) scattering modeling

On the reverse channel, when the mobile transmits, the signal is scattered by the environmental features surrounding the mobile. The presence of scatterers in close proximity of the transmitter results in dispersion of the signal over wider angles. It can be seen that propagation characteristics describing the base-to-mobile and mobile-to-base links may be different. It is likely that channel reciprocity may not be strictly valid.

Figure 3.16 shows a typical mobile to base link. We use the technique developed in Section 3.3.1.1, with one assumption that the geometry forming a ring of radius 'a' surrounding the mobile unit does not change appreciably over the measurement time even when the mobile moves. This is an obvious simplification, but its validity is justified if the measurement time is short. Because the scatterers are present only in the vicinity of the mobile unit, the arrival of radio plane waves at the scatters is confined to a small azimuth angle γ and for large link distance, angle γ also does not change appreciably. Let the distribution of radiated power from the mobile antenna in the azimuth angle γ be $p(\gamma)$. The power incident on scatterers within a differential angle γ will be $Kp(|\gamma|)\, d\gamma$, where K accounts for propagation loss. This incident power is assumed to have scattered uniformly over angle α[†]. Thus, neglecting the multiplicative factors affecting the loss between the scatterers and the base station, the received power at the base station over $d\alpha$ of the angle α is:

$$p(\gamma)d\gamma = p(\alpha)d\alpha \qquad (3.85)$$

Following the technique in Section 3.3.1.1, this is the amount of power in the differential frequency range of df for a given power spectral density $S(f)$.

$$S(f)df = p(\alpha)d\alpha = p(\gamma)d\gamma \qquad (3.86)$$

An inverse transform of $S(f)$ will result in complex correlation function $\Re(\tau)$ [13]

$$R(\tau) = g(\tau) + jh(\tau) = \int_{-\infty}^{\infty} exp(j2\pi(f - f_c))S(f)df \qquad (3.87)$$

where f is given by $f = f_c + f_m \cos \alpha$. However, here α depends on γ through the relation

$$\alpha(\gamma) = acos\left[\frac{\cos\xi + k\cos(\gamma+\xi)}{\sqrt{1 + 2k\cos\gamma + k^2}}\right]; \quad k = \frac{a}{d} \qquad (3.88)$$

Thus, if ξ is the angle of the mobile to the base station, we have, for uniform distribution of radiated power from the mobile,

$$c(\tau) = \int_{-\pi}^{\pi} exp(j\omega_m \cos(a(\gamma)))d\gamma \qquad (3.89)$$

It has been shown that decorrelation distance at the base station is several orders of magnitude larger than the one found at the mobile unit. The decorrelation distance also changes with the orientation of the mobile travel. For example, when the mobile travels directly towards the base station, the decorrelation distance becomes very large. The decorrelation distance

---

†. *This assumption is used to simplify analysis.*

becomes much smaller when the mobile moves perpendicular to the line joining the mobile and the base station.

Direct mobile-to-mobile communication without the intervention of a
mobile is one of several requirements of law enforcing agencies. Applications
of mobile to mobile link include communications between rescue squads,
ambulances, and military or security vehicles in cases of emergency or disas-
ters. This mode of communication is quite common in private land mobile
systems. Except for communications in the Citizen's band, this mode of communications is not allowed in the public cellular systems. It is though possible
that this mode of communications will be introduced sometime in the future.
For example, a proposed Personal Radio Communication System requires no
intervention from the base station for calls between mobile units that are
located within a small locality [56]. However, Wireless Adhoc Networks
seem to be more promising application of mobile to mobile communications.

A mobile-to-mobile propagation model is needed to evaluate mobile
to mobile communication systems. The major difference between mobile to
mobile and the mobile-to-base lies in the locations of antennas, their heights
and surrounding scatterers. In this mode antennae heights on either side of the
link are low, both the transmitter and the receiver are surrounded by rings of
scatterers. A mobile to mobile propagation model based on a time variant
channel transfer function $H(f,t)$ is proposed in [59]. The signal envelope follows Rayleigh distribution. The space-time correlation function $\Re(d,\tau)$ is
shown to be:

$$R(d, \tau) = \sigma^2 J_0(\beta V_2 \tau) J_0(\beta V_1 \tau + \beta d) \qquad (3.90)$$

where d is the distance between the terminals, and τ is the time separation. In
the model it is assumed that transmitted and received powers are equally distributed over 2π radians. The power spectrum of the complex envelope,
obtained by taking the Fourier transform of (3.90), is given by

$$S(f) = \frac{\sigma_1^2}{\pi f_m \sqrt{a}} \left[\frac{1+a}{2\sqrt{a}} \left(\sqrt{1 - \frac{f^2}{(1+a)f_m^2}} \right) \right] \qquad (3.91)$$

where a is the ratio of transmitter and the receiver speeds i.e. V_2/V_1.

3.7 Wideband Channel Models

The channel characteristics presented in Sections 3.4.1 and 3.6 apply
to either unmodulated sinusoid or narrowband signals. A greater number of

wideband applications of mobile wireless communications in practically all types of environments are now emerging. Recently, five air interface standards have been approved for use in the third generation (3G) systems designed to offer multimedia services. The maximum data rates for vehicular, pedestrian, and indoor channels have been specified to be 144 kbits/sec, 384 kbits/sec, and 2Mbits/sec. Three of these systems will use Code Division Multiple Access (CDMA) and the other two will use wideband TDMA.

Section 3.5.4, introduced the concept of coherence bandwidth to make a distinction between wide and narrowband applications. A signal whose bandwidth exceeds the coherence bandwidth of the channel is defined as wideband signal. In time domain, the channel data bearing capability is measured by the channel delay spread, which is inversely proportional to the channel coherent bandwidth. Though symbol duration equal to delay spread determines the theoretical upper limit to the transmission rate, in practice, however, channel delay spread in excess of one eighth of symbol duration starts causing performance degradation. It is imperative, therefore, to characterize time varying wideband channels.

The wideband channel may be characterized by its time variant impulse response, $h(\tau;t)$, which is obtained by exciting the channel with impulses or very narrow pulses as discussed in Section 3.1. The discussions in the preceding sections were concentrated on time-frequency channel models. The presence of scatterers in the vicinity of either mobile terminal or base station leads to dispersion of the signal in many directions. If we create narrow beams and direct them to the signal source, the received signal could be freed from interference. The propagation models that include the angle of arrival of the signal are called time-space-frequency models, because the signal arriving from any direction can be with frequency dispersion. Since these models involve properties of antennas, therefore we defer the discussion on this topic to Chapter 8.

3.8 Channel Models for Simulations

The models presented in sections 3.3, 3.7, and 3.9 form the basis of several channel simulators which have been used to evaluate systems under controlled conditions. The channels simulators fall in two distinct categories: hardware and software.

The hardware simulators have been mostly based on filtered quadrature Gaussian noise sources as shown in Figure 3.17. The filter cutoff frequency, f_m, is indicative of the vehicle speed. The design of filters with very sharp cutoff at maximum Doppler frequency, f_m, pose major difficulty. These filters are generally realized with the use of digital signal processors. These filters are generally realized with   of digital signal processors. It has been suggested that this realizatio   , reasonable results as far as the envelope characteristics are concerned but the Doppler spectrum is different from the desired [57].

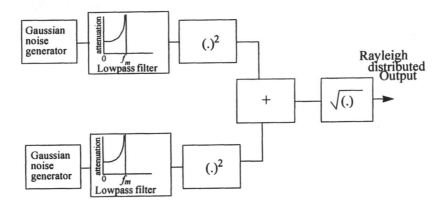

Figure 3.17 Schematic of a simulator based on filtered quadrature Gaussian Process

A model based on sum of n sinusoids suggested by Jakes [13] is now widely used in simulations. A schematic of the reference simulator known as Jakes' model is shown in Figure 3.18. In this realization, N_o oscillators ($N_o \geq 8$) with frequency shifts within the range of Doppler shift $\omega_m \cos 2\pi n/N$, ($N = 4N_o+2$; $n = 1,2,...,N_o$) plus one frequency representing maximum Doppler shift, ω_m, are used to generate signal frequencies from a carrier ω_c (0 in the case of baseband generation). The phases β_n and the value α are chosen as close as possible to give a uniform distribution over 0 to $1/2\pi$. The oscillators outputs are summed after appropriately adjusted in magnitude. This is given by

$$R(t) = \sqrt{2} \sum_{n=1}^{N} C_n \cos(\omega_c t + \omega_m t \cos A_n + \Phi_n) \qquad (3.92)$$

C_n is chosen to be 1, i.e. equal strength sinusoids, and $\Phi_n = -2\pi(f_c+f_m)\tau_n$.

A typical fading envelope generated by sum of N sinusoids is shown in Figure 3.19. The carrier frequency is chosen to be 800 MHz and the mobile speed is 50 km/hr.

This model can be extended to produce frequency selective fading with the use of up to N_o Jakes' simulators. The signals on each path is faded independently by selecting path phases given by the following equations. The

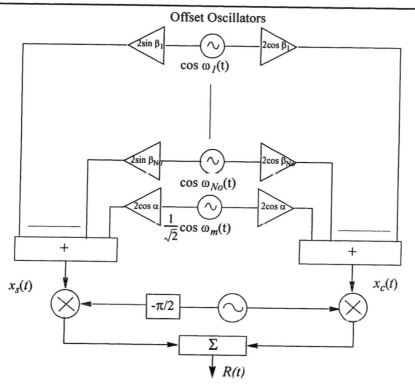

Figure 3.18 Schematic of a simulator based on sum of sinusoids

nth oscillator is given an additional phase shift $\gamma_{nj}+\beta_{nj}$ and using the gains given in the case of a single simulator. The path delays are selected according to a chosen impulse response model.

$$\beta_n = \frac{\pi n}{N_o + 1}$$

$$\gamma_{nj} = \frac{2\pi(j-1)}{N_o + 1}, \qquad n = 1, 2, ..., N_o$$

(3.93)

In [58] it is demonstrated that the signal produced by Jakes' model is not wide sense stationary (WSS), therefore it needs to be modified. It is further shown that introduction of random phases in the low frequency oscillators eliminates, for small values of time, the correlation between signals. Recently, the use of multi-element antennas are recommended for wideband signals to remove interference. The presence of scatterers around the mobile terminal smears the angle of arrivals at the receiver array. The spread in arrival angle due to the scatterers around the mobile can also be incorporated

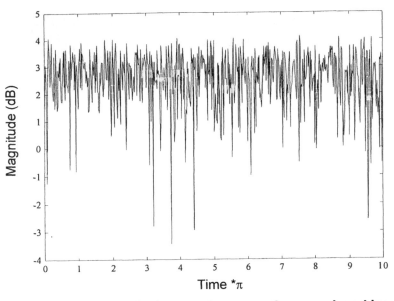

Figure 3.19 A typical fading envelope waveform produced by Jakes' sum of *N* sinusoid simulator

in the simulation model to appropriately correlate the multiple fading waveform for the array.

A real time hardware fading simulator based on fast signal processor has been reported in [12]. This programmable real time simulator covers 250 to 1000 MHz band and is dual channel with programmable correlation between the channels. In addition to hardware simulators, stored channel simulators have also been used [38]. The software simulators are more versatile, but possess a major drawback of being slow. Several channel software simulation modules are now integrated with several commercially available communications simulation packages.

3.9 Markovian Models for Digital Transmission

Several channel models presented in the previous sections are useful in the evaluation of digital transmission systems and provide guidelines for selection of system design parameters. If we view the communication system from end to end, we note that in the case of digital transmission, the receiver output information data is contaminated by random and burst errors, the pattern of which is a function of the channel characteristics. Rather than model individual components of a communication system like transmitter, radio channel, and the receiver separately, it is possible to lump all the system blocks into one block that describes the overall behaviour of the digital data system in terms of errors bursts and error free periods. The length of these periods and their occurrences are random. These models, known as Markovian channel models, are quite useful in predicting the error performance of channels and design of error control schemes. Several models to represent Mobile Digital Channel (MDC) have also been proposed [11],[38],[40],[41], the simplest being Gilbert's model which is based on a two state Markov chain. Other models extend Gilbert's model to those with more than two states. The extended models are useful in the modeling the behaviour of those channels that produce random as well as burst errors.

Models proposed by Gilbert [11], and Fritchman [60], [61] have been applied to mobile radio systems. Out of these Fritchman's model is considered to be the most appropriate for mobile radio channels. Before we go into details of this model, it is worthwhile to provide a short introduction to the background of these models.

The mobile radio channels suffer from deep fades that occur randomly in time and have random duration and depth. During the period that the channel is not in fade, the signal strength at the receiver input is sufficiently high and the data is delivered practically error free. However, when the channel is in deep fade the receiver tend to make errors in succession producing an error burst. The length of the fade and its depth determine the character of the error burst. The channel therefore can be modeled by a two state Markovian model, that is a good (error free) state and bad (error burst) state. Observing this, Gilbert proposed a two state Markovian model [11]. This is shown in Figure 3.20. In the state G, the error probability is assumed to be zero while there is high probability of making errors (approx. 0.5) in the bad state. The Gilbert model assumes that during the G state, when the channel is not in fade, no errors are made. However, even in good state the channel noise will force some random errors. Elliot [62] modified the Gilbert Model to allow occurrence of random errors in the G state. The channel still has two states but in good state there is finite probability of making errors.

The Markovian channel models are developed from the statistics of error and error gap processes. These processes are defined in the following. Consider a simple representation of a digital channel shown in Figure 3.21. In this representation x_i and y_i are the input and the output symbols respectively. It is assumed that the input sequence $\{x_i\}$ and noise (or other channel impairment) sequence $\{n_i\}$ are statistically independent. An error is said to occur when noise sample is greater than the symbol sample and is represented by a

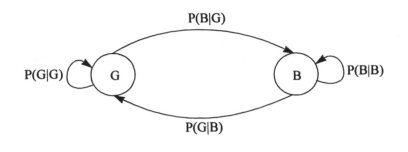

Figure 3.20 Two state Markovian model

1, whereas when signal sample is greater than the noise sample no error is made and a 0 represents such occurrence. A sequence of zeros between two errors is called an error gap. The length of the gap is defined as one plus the number of zeros in an error file. For all positive integers m, error gap probability mass function (EGPMF) is defined as probability of having a gap of length m. The error gap distribution is defined as the probability of having a gap of length greater than m and is represented as:

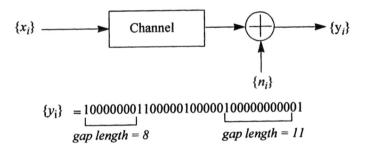

Figure 3.21 A simple representation of digital channel

$$P(\text{gaplength} = m) = P(0^{m-1}1|1)$$

$$P(\text{gaplength} > m) = P(0^m|1)$$

$$= \sum_{k}^{\infty} P(0^k 1|1)$$

(3.94)

From the above definitions, it is observed that $P(0^m|1)$ is a monotonically decreasing function of m and

$$P(0^m 1|1) = P(0^1|1) - P(0^{m+1}|1) \tag{3.95}$$

In general, the gap process is a renewal process, successive gap lengths are uncorrelated and the process is uniquely defined by EGPMF. However, in most of the channels with memory, gap process is non-renewal process and the successive gap lengths have some correlation between them. The Gilbert's model need three channel parameters, q_1, q_2, and p, which can be measured. q_1 and q_2 are the transition probability for B to G state and from G to B state respectively. p is the probability of making error in B state. The following probabilities apply:

$$P(G) = \frac{q_1}{q_1 + q_2} \qquad P(B) = \frac{q_2}{q_1 + q_2}$$

$$P(1) = \frac{pq}{q_1 + q_2} \qquad P(1|1) = p \tag{3.96}$$

$$P(111) = p^2 P(1) \qquad P(101) = p(P(1)(1 - p + q_1 q_2))$$

$$\frac{P(111)}{P(101) + P(111)} = \frac{p}{1 + q_1 + q_2}$$

Fritchman developed a very general case of Markovian model for channels with memory [60]. In his model, the channel has N states which are divided into two groups: K error free states in group A and $(N-K)$ error states in group B as shown in Figure 3.22.

The state sequence $\{z_i\}$ of the system corresponds, at each instance i, to a state of the Markov chain (states 1 to N). At each discrete time i, the error bit e_i is equal to 0 or 1 depending on whether the state z_i belongs to group A or B.

$$e_i = 0; \text{ if } z_i \in A$$
$$= 1; \text{ if } z_i \in B \tag{3.97}$$

For real (physical) channels a simplified version where transitions within the chain are permitted from a state belonging to one group to a state within the other group. Transitions to and from the same state are also allowed. In many cases a string of 1's distribution can be modeled by a single exponential. This leads to further simplification of the model since we can now consider only one error state ($N-K = 1$). For this model, the probability

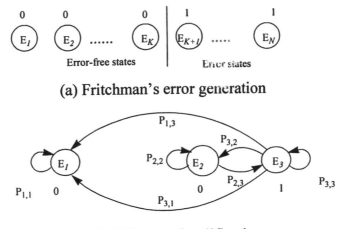

(a) Fritchman's error generation

(b) Gilbert's simplification

Figure 3.22 Error generation methods for digital channel

$P(0^m|1)$ of obtaining an interval (a gap) length greater or equal to m bits is given by:

$$P(0^m|1) = \sum_{i=1}^{K} \frac{P_{N,i}}{P_{i,i}}(P_{i,i})^m$$

$$= \sum_{i=1}^{K} \alpha_i \beta_i^m$$

(3.98)

where α_i and β_i are the Fritchman's model parameters which are used to calculate the state transition probabilities, i.e.

$$P_{i,i} = \beta_i, \quad i = 1, 2, ..., K$$

$$P_{1,N} = 1 - \beta_i$$

$$P_{N,i} = \alpha_i \beta_i$$

$$P_{N,N} = 1 - \sum_{i=1}^{K} \alpha_i \beta_i$$

(3.99)

Fritchman's parameters can be estimated by using curve fitting techniques for a given error gap distribution (EGD). Chouinard et.al. [63] used measured results to study the distribution of transmission errors in mobile radio channels. A gradient algorithm was used for curve fitting.

The reported results show that the error burst length distribution $(P(1^m|0))$ can be described using a unique error state (N-K=1). The error burst length distribution is written as:

$$P(1^1|0) = \alpha_N \beta_N^m \qquad (3.100)$$

or in logarithm scale

$$\begin{aligned} \log_{10} P(1^m|0) &= \log_{10} \alpha_N + m \log_{10} \beta_N \\ &= (a_N + m b_N) \end{aligned} \qquad (3.101)$$

3.10 Summary

This chapter examined in detail the micro-characteristics of the mobile radio channel. The emphasis has been on the analysis of scattering effects and channel models. The study was divided into two main parts - wideband characterization and narrowband (single carrier). The channel impulse response formed the basis of wideband channel characterization. The path arrival in a random environment was defined. Several models were discussed. The probability density function of the received envelope and phase was derived for narrowband channel characterization. The theoretical basis for envelope and phase correlation and coherence bandwidth was established and coherence bandwidths were defined. The concepts of random frequency modulation was introduced. It was observed that the transmission path geometry between the base station and the mobile unit introduces a significant degree of non-reciprocity. Models for wideband mobile radio channels were introduced. Methods to simulate mobile fading channels were described. Finally, digital channel models based on error performance were introduced. All of these considerations laid the groundwork for design of systems that allow efficient signal transmission and reliable reception in the mobile radio environment. The next Chapter 4 will consider the issues related to the performance of analog and digital transmission systems.

Problems

Problem 3.1:

(a) Show that a channel whose magnitude response is uniform but the phase response is not linear produces ISI.

(b) Derive and expression for the channel impulse response is given by

$$H(f) = (1 + \cos 2\pi fT)e^{-j\sin\pi(f/T)} \tag{P-3.1.1}$$

Problem 3.2

(a) Define mean excess delay, rms delay spread, and maximum excess delay of a mobile radio channel.

(b) A power delay profile is shown in Figure P-3.2. Determine the channel parameters as defined in (a).

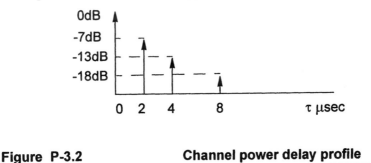

Figure P-3.2   **Channel power delay profile**

Problem 3.3

The power delay profile shown in Figure P-3.2 is obtained after averaging 1000 profiles. All four paths have been shown to fade with exponential distribution with the path averages shown. Determine the probability that all four paths have powers within 1 dB.

Problem 3.4

Repeat Problem 3.3 for the case when the path amplitude fade with Nakagami distribution. The corresponding path power fade with Gamma distribution, which is given by

$$p(s) = \frac{m^m s^{m-1} \exp\left[-\frac{m}{\bar{s}}s\right]}{\Gamma(m)\bar{s}^m} \qquad \text{(P-3.4.1)}$$

where m is the Nakagami parameter, $\bar{s}$ is the mean power.

Problem 3.5

During its travel from a sub-urban area towards the city core over a distance of 3 km. The measured channel delay spread changes linearly over this distance from 2.1 μseconds to 6.7 μseconds. How does the coherence bandwidth and supportable data rate changes over this distance?

Problem 3.6

(a) Starting from (3.41) derive the relationship between m and K given in (3.46).

(b) Plot Ricean and Nakagami distributions and demonstrate that with the value chosen according to (3.46) Nakagami distribution indeed emulate the Ricean distribution.

Problem 3.7

In a mobile radio communication scenario, the presence of a certain geometry of the obstacles results in a non-uniform angle-of-arrival distribution at the receiver. It is estimated that $p(\alpha)$, the arrival distribution, is given by the following:

$$\begin{aligned}
p(\alpha) &= \frac{3}{5\pi}; \quad 0 \le |\alpha| \le \frac{\pi}{4} \\
&= \frac{2}{5\pi}; \quad \frac{\pi}{4} \le |\alpha| \le \frac{\pi}{2} \\
&= \frac{4}{5\pi}; \quad \frac{\pi}{2} \le |\alpha| \le \frac{3\pi}{4} \\
&= \frac{1}{5\pi}; \quad \frac{3\pi}{4} \le |\alpha| \le \pi
\end{aligned} \qquad \text{(P-3.7.1)}$$

The mobile antenna has a gain of 1.5. Because the antenna is installed on the rear fender, the antenna directional pattern is not uniform but has a pattern is shown Figure P-3.7. Plot the Doppler spectrum of the received carrier when the operating frequency is 850 Mhz and the mobile is travelling at 60 km/hr.

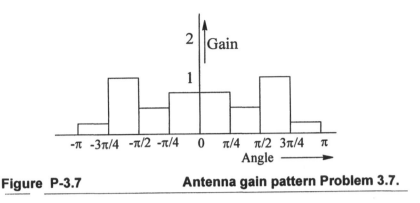

Figure P-3.7   **Antenna gain pattern Problem 3.7.**

Problem 3.8

(a) What is the channel coherence bandwidth?

(b) In terms of symbol period, T_s, the signal bandwidth B_s, channel coherence bandwidth B_c and rms delay spread of the channel, state the conditions for

      (i) a flat fading channel

      (ii) a frequency selective fading channel

      (iii) a slow fading channel

      (iv) a fast fading channel.

Problem 3.9

a) What is the correlation coefficient $\rho(\tau)$ and the corresponding separation distance for the mobile received signal by a system having the following parameters? Transmission frequency is 845 MHz and the mobile is travelling at 50 Km/hr. Mean signal power is -89 dBm and the mean signal voltage is -83 dBm. Correlation function $R(\tau)$ for τ seconds separation is -84 dBm.

(b) Explain why the physical separation of two co-located base station antennas is larger for in line propagation than for broadside propagation for the same correlation coefficient function.

Problem 3.10

Assume that the carrier frequency is 850 MHz and the audio band extends from 300 to 3300 Hz. What is the average power ratio between two random FM signals one received while the mobile is travelling at 60 km/hr and the other for the mobile speed of 25 km/hr?

Problem 3.11

Mobile to mobile communication without a central base station finds applications in rescue, emergency services as well as in forward line military communications. In this mode local scatterers surround both the transmitter and the receiver. The time space correlation has been derived to be,

$$\Re(d, \tau) = \sigma^2 J_o(\beta V_2 \tau) J_o(\beta V_1 \tau + \beta d) \qquad \text{(P-3.8.1)}$$

where σ^2 is half the rms signal power, d is the distance in meters between the two mobiles and τ is the time separation. Plot $\Re(d,\tau)$ for $d =0.2$, 1.0, and 2 Km and $\tau =1, 2, 5$ μsec. Assume $V_1 =15$ Km/hr and $V_2 = 20$ Km/hr ($\beta = 2\pi/\lambda$, $f= 821$ MHz).

Problem 3.12

The autocovariance function of the received envelope due to vertical electric field E_z is given by

$$L_e(\tau) = \frac{\pi}{8} b_o J_o^2(\omega_m \tau) \qquad \text{(P-3.1.2)}$$

where b_o is the average power of the envelope. It is observed that $L_e(\tau)$ is a function of vehicular speed. If the minimum $L_e(\tau)$ required for a desired diversity gain is found to be 0.25 (when $b_o\pi/8 =1$) then find the minimum distance between antennas to satisfy the above condition when the mobile speed varied between 15 and 100 Km/hr. The transmission frequency is 850 MHz.

Problem 3.13

The envelope of E field at a frequency of 834 MHz is received by a mobile unit travelling at 47 km/hr. Assume that the reception threshold of receiver is set at -13 dB relative to the envelope rms power. (a) Find relative frequency of error bursts. (b) Estimate the average length of error free packet. (c) Repeat (a) if the mobile speed has changed to 100 Km/hr. (d) Estimate the length of the packet when a better receiver with a reception threshold of - 20 dB relative to rms is used.

Problem 3.14

(a) If the received envelope power is -80dBm and the fade margin -18 dB, what percentage of time the received signal is in fade?
(b) Repeat (a) for envelope power of -70 dB and fade margin of 14 dB.

Problem 3.15

The envelope of an E field signal at a frequency of 870 MHz is received by a mobile unit travelling at 70 Km/hr. The mean signal strength at the receiver is measured to be -9 dBm. If the envelope of the fading signal is Rayleigh distributed, find the average fade duration and cross rates for threshold level of -13 dBm.

Problem 3.16

Repeat Problem 3.14 when the signal fades with Nakagami statistics.

Problem 3.17

Repeat Problem 3.14, for lognormal faded signal.

References

[1] R.C. Dixon, *Spread Spectrum Systems*, 2nd Edition, J. Wiley and Sons, New York, 1984.

[2] A.U. H. Sheikh, *Delay Spread Measurement in Hong Kong*, Report to British Council, Research Grant No. H-ZF25, the Hong Kong Polytechnic University, Department of Electronic and Information Engineering, August 1999.

[3] W.C.Y. Lee, *Mobile Cellular Telecommunication Systems*, McGraw Hill Publishers, New York, 1989.

[4] G. L. Turin, F.D. Clapp, T.L. Johnston, S.B. Fine and D. Lavry, "A statistical model of urban multipath propagation," *IEEE Trans. on Veh. Tech.*, vol. VT-21, pp.1-9, February 1972.

[5] D. Black and D. O. Reudink, "Some characteristics of mobile radio propagation at 836 mhz in the philadelphia area," *IEEE Tran. on Veh. Tech.*, vol. VT-21, pp.45-51, May 1972.

[6] D. C. Cox and R. P. Leck, "Distribution of multipath delay spread and average excess delay for 910-mhz urban mobile radio paths," *IEEE Trans. Antennas and Propagation.*, vol. AP-23, pp.206-213, March 1975.

[7] D. C. Cox, "910 MHz urban mobile radio propagation: multipath characteristics in New York City," *IEEE Trans. on Comm.*,vol. COM-21, pp. 1188-1194, Nov. 1973.

[8] D. C. Cox and R. P. Leck, "Correlation bandwidth and delay spread multipath propagation statistics for 910-mhz urban mobile channels," *IEEE Trans. on Communication,* vol. COM-23, pp. 1271-1280, Nov. 1975

[9] A. S. Bajwa, and J. D. Parsons, "Small-area characterization of uhf urban and suburban mobile radio propagation," Proc. IEE, vol. 129 Pt. F, pp. 102-109, April 1982.

[10] A. S. Bajwa and J. D. Parsons, "Large area characterization of uhf multipath propagation and its relevance to the performance bounds of mobile radio systems," *Proc. IEE*, vol. 132, Part F, pp. 99-106, April 1985.

[11] E. N. Gilbert, "Capacity of a burst noise channel," *Bell System Technical Journal*, vol., 39, pp. 1253-1265, September 1960.

[12] R. Goubran, H.M. Hafez, and A.U. Sheikh, "Implementation of a real-time mobile channel simulator using a DSP chip," *IEEE Trans. on Instrumentation and Measurements*, vol. 40, pp. 709-714, August 1991.

[13] W.C. Jakes (Ed.), *Microwave Mobile Communications*, J. Wiley, 1974.

[14] D.C. Cox, "Delay doppler characteristics of multipath propagation at 910 mhz in a suburban mobile environment," *IEEE Trans. on Antennas Propagat.*, vol. AP-20, Sep 1972. 41

[15] H. Suzuki, "A statistical model for urban radio propagation," *IEEE Trans. on Communications*, vol. COM-25, no. 7, July 1977.

[16] H. Hashemi, "Simulation of the urban radio propagation channel," *IEEE Trans. on Veh. Tech.*, vol. VT-28, no. 3, August 1979.

[17] H. Hashemi, *Simulation of the Urban Radio Propagation Channel*, Ph.D. Dissertation, University of California, Berkeley, 1977.

[18] A.M. Saleh et.al., "A statistical model for indoor multipath propagation," *IEEE Journal of Selected Areas in Communications*, vol. SAC-5, no. 2, pp. 128-137, February 1987.

[19] H. Hashemi, "The indoor radio propagation channel," *Proceedings of IEEE*, vol. 81, no. 7, pp. 941-968, 1993.

[20] D. M. J. Devasirvatham, "Time delay spread and signal level measurements of 850 mhz radio waves in building environments," *IEEE Trans. on Antenna and Propagation*, vol. AP-34, no. 11, pp. 1300-1305, November 1986.

[21] D. Molkdar, "Review on radio propagation into and within buildings," *IEE Proceedings*, vol. 138, no. 1, pp. 61-73, February 1991.

[22] P.F.M. Smulders and A. G. Wagemans, "Wideband indoor radio propagation measurements at 58 ghz," *Electronics Letters*, vol. 28, no. 13, pp. 1270-1272, June 1992.

[23] H. Hashemi, "Impulse response modeling of indoor radio propagation channels," *IEEE-JSAC*, vol. 11, no. 7, pp. 967-978, September 1993.

[24] H. Hashemi and D. Tholl, "Statistical modeling and simulation of the RMS delay spread of indoor radio propagation channels," *IEEE Trans. of Veh. Tech.*, vol. 43, no. 1, pp. 110-120, February 1994.

[25] T. Manabe, K. Sato, H. Masuzawa, K. Taira, T. Ihara, Y. Kasashima and K. Yamaki, "Polarization dependence of multipath propagation and high speed transmission characteristics of indoor millimeter-wave channel at 60 GHz," *IEEE Trans. of Veh. Tech.*, vol. 43, no. 1, pp. 268-274, May 1995.

[26] R. Steele (*Ed.*), *Mobile Radio Communications*, IEEE Press, 1994.

[27] J. F. Ossana, "A Model for Mobile Radio Fading Due to Building Reflections: Theoretical and Experimental Fading Waveform Power Spectra," *Bell System Technical Journal*, vol. 43, pp 2935-2971, November, 1964.

[28] H. W.Nylund, "Characteristics of small area signal fading on mobile circuits in the 150 MHz band," *IEEE Trans. on Veh. Tech.*, vol. VT-17, pp. 24-30, October 1968.

[29] E. N. Gilbert, "Energy reception for mobile radio," *Bell System Technical.Journal.*, vol. 44, pp. 1779-1803, October 1965.

[30] R. H. Clarke, "A statistical theory of mobile-radio reception," *Bell System Technical Journal*, vol. 47, pp. 957-1000, July 1968.

[31] D. O. Reudink, "Properties of mobile radio propagation above 400 MHz," *IEEE Trans. on Veh. Tech.*, vol. VT-23, pp. 143-159, November 1974.

[32] U. Charash, "Reception through Nakagami fading channels with random delays," IEEE Trans. on Comm., vol. 27, no. 4, pp. 657-670, 1979.

[33] M. J. Gans, "A power spectral theory of propagation in the mobile radio environment," *IEEE Trans. on Veh. Tech.*, vol. VT-21, pp. 27-38, February 1972.

[34] F. Hansen and F. I. Meno, "Mobile fading- rayleigh and lognormal superimposed", *IEEE Trans. on Veh. Tech.*, vol. VT-26, pp. 332-335, November 1977.

[35] S. O. Rice, "Statistical properties of a sine wave plus random noise," *Bell System Technical Journal*, vol. 27, pp. 109-157, January 1948.

[36] H. Suzuki, "A Statistical model for urban radio propagation channel," *IEEE Trans., on Communications*, vol. COM-25, pp. 673-679, July 1977.

[37] T. Aulin, "Characteristics of a digital mobile radio channel," *IEEE Trans. on Veh. Tech.*, vol. VT-30, pp. 45-53, May 1981.

[38] H. Hashemi, "Simulation of the urban radio propagation channel," *Proc. 1977 National Telecommunication Conference*, Paper 38-1.

[39] L. Kittel, "Radio signalling design and discrete channel modelling for mobile automatic telephone systems," *Mobile Radio systems and Techniques*, IEE Conference Publication No.238, September 1984.

[40] J. P. A. Adoul, "Error intervals and cluster density in channel modelling," *IEEE Trans. on Information Theory*, vol. IT-, pp. 125-129, Jan. 1974.

[41] W.C.Y. Lee and Yeh, "On the estimation of the second order statistics of lognormal fading in mobile radio environment," *IEEE Trans. on Communications*, vol.COM-22, pp. 869-873, June 1974.

[42] T. C. Kelly and J. E. Ward, *Investigation of Digital Mobile Radio Communications*, U.S. Department of Justice, Grant No. NI-71-129-G, October 1973.

[43] M. Schwartz, *Information Transmission, Modulation and Noise*, Third Edition, McGraw Hill, New York, 1980.

[44] S. O. Rice, "Mathematical analysis of random noise," *Bell System Technical Journal*, vol. 23, pp. 282-332

[45] M. Nakagami, "The *m*-distribution, a general formula of intensity distribution of rapid fading," in *Statistical Methods in Radio Wave Propagation*, W.G. Hoffman Ed. Pergamon Press, Oxford, 1960.

[46] T. Knight, *Characteristics of Mobile Radio Channel*, M.Eng. Thesis, Carleton University, Ottawa, Canada, 1985.

[47] W. R. Young, Jr. and W. L. Root, *Introduction to Random Signal and Noise*, McGraw Hill Book Co. New York, pp.170, 1958

[48] W. C. Jakes, Jr. and D. O. Reudink, "Comparison of mobile radio transmission at UHF and X-bands," *IEEE Trans. Veh. Tech.*, vol. VT-16, pp. 10-14, Oct. 1967.

[49] J. A. S. DaSilva, *Performance Analysis of Land Mobile Packet Radio Systems*, Ph.D. Thesis, Department of Systems and Computer Engineering, Carleton University (Report No. SC80-3), 1980.

[50] W. F. Bodtmann and H. W. Arnold, "Fade duration statistics of a rayleigh distributed Wave," *IEEE Trans. on Comm.*, vol. COM-30, pp 549-553, March 1982.

[51] S. O. Rice, "Distribution of duration of fades in radio transmission: Gaussian noise model," *Bell Syst. Tech. Journal*, vol.37, pp. 581-635, May 1958.

[52] P. Monsen, "Feedback equalization for fading dispersive channels," *IEEE Tran. on Infor. Theory*, vol. IT-17, pp. 56-64, 1971.

[53] B. Glance and L.J. Greenstein, "Frequency-selective fading effects in digital mobile radio with diversity combining," *IEEE Trans. on Comm.*, vol. COM-31, Sept. 1983.

[54] N.W.K.Lo, D.D. Falconer, and A.U.H. Sheikh, "Adaptive equalization and diversity combining for mobile radio using interpolated channel estimates," *IEEE Trans., Veh. Tech.*, vol. 40, pp. 636-645, August 1991.

[55] D.C. Cox, "Universal portable radio communications," *IEEE Trans. Commun.*, vol. 38, pp. 1272-1280, Aug. 1980.

[56] G. L. Stüber, *Principles of Mobile Communications*, Second Edition, Kluwer Academic Publishers, Boston, 2001.

[57] M. F. Pop and N. C. Beaulieu, "Limitations of sum-of-sinusoids fading channel simulators," *IEEE Trans. Commun.*, vol. 49, no. 4, pp. 699-708, April 2001.

[58] A.S. Akki and F. Haber, "A statistical model of mobile-to-mobile land communication channel," *IEEE Trans. Veh. Tech.*, vol. VT-35, pp. 2-7, 1986.

[59] B.D. Fritchman, *A binary Channel Characterization Using Partitioned Markov Chains with Applications to Error Correcting Codes*, Ph.D. Dissertation, Lehigh University, Bethlehem, PA. 1966.

[60] A.N. Kanal and A.R.K. Sastry, "Models for channels with memory and their applications to error control," *Proceedings of IEEE*, vol. 66, July 1978.

[61] E.O. Elliot, "Estimates of error rates for codes on burst noise channels," *Bell System Technical Journal*, vol. 42, pp. 1977-1997, 1963.

[62] J.V. Chouinard, M. Lecours, and G.Y. Delisle, "Estimation of Gilbert's and Fritchman's models parameters using gradient method for digital mobile radio channels," *IEEE Trans. Veh. Tech.*, vol. 37, pp. 158-166, August 1988.

Chapter 4

SIGNAL TRANSMISSION OVER MOBILE RADIO CHANNELS

The mobile radio channels were characterized for signal strength and its variations in Chapter 2 and for channel impulse response, transfer function and signal microstructure in Chapter 3. The macro- and micro-characterization provided us with sufficient knowledge to proceed with the analyses and designs of signal transmission systems over mobile radio channels. The undesirable influence of the environment dependent multipath propagation on signal parameters makes the mobile radio link unreliable. The degradation in the quality of the received signal is measured using one or more quality measures like signal to noise ratio, bit error rate (BER) and signal outage. The first two measures are well known, the third signal outage is defined as the probability with which the signal falls below the reception threshold. The reception threshold is selected as the minimum of signal level that provides adequate reception quality. It could also be defined in terms of maximum probability of error. The relatively poor signal to noise ratio during signal outages renders communication difficult if not impossible.

In this chapter, we develop the above mentioned signal quality measures. The aim is to understand how the mobile radio channel impairments contribute to degradation of the signal quality. The goal of a system designer is to develop innovative signal processing techniques so that susceptibility to channel impairments is minimized.

As a framework for discussion, we need to consider different forms of signals produced by signal sources, and their processing for transmission. The sources considered here are capable of generating a variety of signals like voice, data, compressed video and image signals. The signals are either in analog or digital format. The analog signals are prepared for transmission by filtering and digital signal by encoding. The analog signals are also formatted into digital signals with the use of sampling, and encoding process. Amplitude, frequency, phase or any combination of these three types can be used to modulate a carrier. Note that analog representation almost always implies a voice source, while a digital baseband signal may pertain to voice or data or image.

An important measure of the received signal is that of carrier-to-noise ratio $(C/N)_i$. In the case of analog representation, signal quality may be gauged by the receiver output signal-to-noise ratio $(S/N)_o$. For digital signals the equivalent measure is the average bit error rate (BER). Signal outage is also used to evaluate system performance.

These observations underscore the structure of this chapter. In the first part, Sections 4.1 to 4.3, the output signal to noise ratio, $(S/N)_o$ is examined in relation to the input carrier to noise ratio, $(C/N)_i$. Both frequency and amplitude-modulated signals are considered. In Sections 4.4 through 4.7, the performance of digital transmission systems using different modulation formats is examined in terms of BER.

Angle modulation is extensively used over mobile radio channels and for several decades frequency modulation had been the choice modulation method for such signals. Even today several private land mobile systems use frequency and phase modulation. These methods also constitute the foundation of a large variety of digital modulation formats. It is therefore reasonable to study transmission of frequency modulated signals over mobile channels as such a study facilitates understanding the impact of multipath fading on digitally modulated signals.

In Section 4.1, we consider the impact of channel impairments on FM signals. Frequency modulation signals are robust in the presence of fading since the information lies in the zero crossing of the carrier rather than the envelope. The impact of fading on the signal envelope can be removed with the use of a level limiter. The 'capture effect' another noteworthy property of FM lets the receiver to 'capture' the stronger received signal and suppress the weaker signals. The bandwidth to power exchange is another feature of FM. A larger deviation of the modulated signal increases the output signal to noise ratio without increasing the transmit power. This is particularly important in mobile systems where low power mobile transmitters are preferable in order to make the terminal light and avoid frequent battery charging.

The use of wide band FM is not favored because of spectrum inefficiencies associated with FM modulation and a desire to conserve spectrum. FM has some other limitations; the upper limit set by the noise due to random phase fluctuations to the performance of FM systems is an important one. In addition, the receiver is captured by noise when forced to operate below threshold during deep signal fades. Hence, the performance of an FM receiver is greatly dependent on the probability distribution of the signal fades.

While searching for a modulation method that uses the spectrum efficiently, one may focus on to amplitude modulation (AM). At first sight, one would rule out AM because channel fading severely affects the carrier envelope. However some years ago, advances in signal processing techniques revived some interest in the use of single sideband AM (SSB-AM) because of its high spectral efficiency. The interest diminished because of increased interest in digital transmission and complexities associated with implementation difficulties in using SSB-AM for digital transmission. We shall discuss AM and SSB-AM briefly.

While the use of digital transmission over mobile wireless channels has been under investigation for the past thirty years [1]-[10], only during the past decade, it became the dominant mode of transmission. Beginning with the second generation cellular systems, all future mobile radio systems are expected to be digital. These systems have better spectrum efficiency, reliability, and security because of the use of error control coding and encryption. Digital formats facilitate integration of voice, data and video signals. The analysis of digital systems show that their performance differ significantly from that of analog systems. Section 4.4.2 examines several baseband digital waveforms and pre-filtering schemes that conserve bandwidth. The requirements of a digital modulation system are considered in Section 4.4. The application of AM, FM and PM modulation techniques to digital signals is discussed in this section. Because of the abrupt level shifts that characterize the baseband signal, both this signal and its modulated counterpart occupy a fairly broad frequency spectrum. The need to maintain spectral efficiency, avoid interference, and provide resistance to fading, results in the use of the constant envelope-continuous phase (CECP) class of digital modulation. The effects of fading and shadowing on the error performance are analyzed in section 4.5.

4.1 Frequency Modulation of Analog Signals

In this section, we consider reception of frequency or phase modulated analog signals over wireless mobile channels characterized by multiplicative Rayleigh fading and additive noise. The objective of this section is to determine the output signal-to-noise ratio for a given input carrier to noise. The solution to the problem is approached in three stages. First we determine baseband signal-to-noise ratio in the absence of fading. This is followed by an examination of the baseband noise spectrum under conditions of no signal, strong signal, and at the boundary between them that is when the transmitted signal and the thermal noise components are comparable. Finally, an analytical expression is derived for the output signal-to-noise ratio in the presence of fading.

A schematic of a typical FM link is shown in Figure 4.1. The baseband information bearing message $m(t)$ is passed through a pre-emphasis filter. The function of the pre-emphasis filter is to emphasize weaker higher frequency signal components so that these components do not suffer adversely at the signal detection stage. More details are given in Section 4.1.1. The output of the pre-emphasis filter is passed to either a direct or indirect FM modulator.

In the past it was common practice to use an indirect narrowband phase modulator (NBPM) [11] from which the FM signal was derived. The NBFM was followed by a series of multipliers and mixers to produce FM modulated signal at a desired carrier frequency and deviation. Direct modulators in the form of phase locked loops are now increasingly in use as concerns on the stability of oscillators have largely been alleviated. In any case, the transmitted signal is written as:

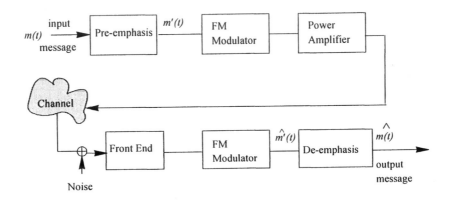

Figure 4.1 A typical FM radio link

$$s(t) = Re[A \exp(j(\omega_c t + \mu(t)))] \tag{4.1}$$

where

$$\mu(t) = \int_{-\infty}^{t} m'(\lambda)d\lambda \tag{4.2}$$

and $m'(t)$ is the pre-emphasized signal.

The signal while propagating through the channel encounters multiplicative fading and noise is added at the receiver. The received signal is therefore of the form:

$$s'(t) = \sum_{i=0}^{\infty} A_i(t)\exp(-j\beta v \cos(\phi_i - \alpha))s(t) + n_i(t) \tag{4.3}$$

where $A_i(t)$ the attenuation of ith transmission path, ϕ_i is its angle of arrival and α is the angle of the direction of the travel of the vehicle (Figure 4.2). $n_i(t)$ is the noise at the receiver input.

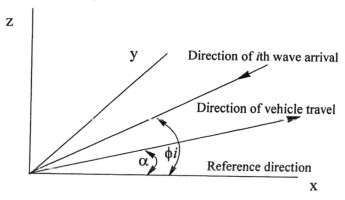

Figure 4.2 Relationship between the direction of vehicle travel and wave arrival

Since

$$\sum_{i=0}^{\infty} exp(-j\beta v\cos(\phi_i-\alpha)) \qquad (4.4)$$

is a narrowband Gaussian process, equation (4.2) can be written in terms of the envelope and phase as:

$$s'(t) = Re[r(t)exp(j\psi_r(t))exp(j(\omega_c t + \mu(t)))] + n_i(t) \qquad (4.5)$$

Equation (4.5) provides a general expression for the received frequency modulated signal over fading channels. In the absence of fading, we consider only one path and $A_i(t)$ is no longer a function of time but is constant i.e. $A_i = A$. Furthermore, since terminal mobility causes fading, the speed of terminal causing fading is considered to be zero in this case.

4.1.1 S/N in the Absence of Fading

We begin with some modification to (4.5) on the basis that multipath fading is absent, that is the received signal consists of a single path and its envelope is constant. Rice [12] has shown that the baseband noise power spectral density at the receiver output remains unaffected in the absence or presence of modulation. In determining the output noise power, we use this observation and assume that the carrier is unmodulated i.e. modulation signal $\mu(t) = 0$. However, when our interest is to estimate output signal power, we shall reintroduce modulation. The received RF signal (carrier plus noise), in the absence of fading, is then given by:

$$s'(t) = Re[Aexp(j\omega_c t)] + n_i(t)$$
$$= A\cos\omega_c t + x_c(t)\cos\omega_c t - x_s(t)\sin\omega_c(t) \qquad (4.6)$$
$$= R(t)\cos(\omega_c t + \theta_n(t))$$

where $x_c(t)$ and $x_s(t)$ are the in-phase and quadrature components of the noise, $\theta(t) = \tan^{-1}[x_s(t)/(x_c(t)+A)]$ and R is the received signal envelope.

Equation (4.6) may also be used to write the IF signal by replacing the centre frequency ω_c by ω_i, provided the receiver front end is linear and has a sufficiently wide bandwidth. Without losing generality, in the following analysis, however, we retain the subscript c. The spectrum at the IF filter output or the input to the limiter discriminator is shown in Figure 4.3. The carrier power is $A^2/2$ and the noise power is given by

$$N = \frac{\eta}{2}\int_{-\infty}^{\infty}[G(f-f_c) + G(f+f_c)]df \qquad (4.7)$$

where $\eta/2$ is the two sided input noise power spectral density and $G(f)$ is the receiver power gain function, which is magnitude square of the filter transfer function.

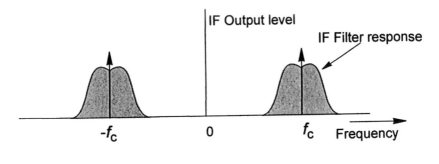

Figure 4.3 IF filter output spectrum

The IF carrier to noise ratio, $(C/N)_i$ is given by:

$$\rho = \frac{A^2}{2\eta \int_0^\infty G(f-f_c)df} \qquad (4.8)$$

To compute output $(S/N)_o$, we need to understand how the receiver processes the input signal to deliver the output. The FM receiver differentiates the input phase, $\theta(t)$, to deliver output voltage or current (see Figure 4.4). The envelope, therefore, does not possess any information related to the signal. However, if envelope fluctuations are retained, the magnitude of the output will vary randomly as well. Thus, FM receivers use limiters, which remove the envelope fluctuations to deliver a constant magnitude having a value C. The constant envelope frequency modulated signal is then operated upon by a discriminator, which is modeled by a differentiator, an envelope detector, and a low pass filter. The discriminator is assumed to have a gain k.

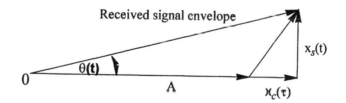

Received signal envelope

$\theta(t)$

0

A

$x_s(t)$

$x_c(\tau)$

Figure 4.4 Phasor representation of signal plus noise

The output of the limiter is:

$$s_1(t) = C\cos(\omega_c t + \theta(t)) \qquad (4.9)$$

and the differentiator output is given by:

$$s_d(t) = -kC(\omega_c + \theta'(t))\sin(\omega_c t + \theta(t)) \qquad (4.10)$$

where $\theta(t) = -\tan^{-1}(x_s(t)/(A+x_c(t)))$. The envelope detector output is given by:

$$s_d(t) = -kC(\omega_c + \theta'(t)) \qquad (4.11)$$

where $\theta'(t)$ is the differential of the phase angle $\theta(t)$, and ω_c represents a dc output, which is blocked by the baseband filter. The computation of $(S/N)_o$ is simple when $\rho \gg 1$. We shall consider this first.

The discriminator output under $\rho \gg 1$ is

$$kC\theta'(t) = kC\frac{d}{dt}\left(\text{atan}\frac{x_s(t)}{A}\right) \cong kC\frac{x_s'(t)}{A}; \qquad \text{for } X_c, X_s \ll A \qquad (4.12)$$

Since $x_s(t)$ is a Gaussian distributed random process, the above condition is satisfied almost all the time. The output noise power can be determined from the power spectral density of $x_s(t)$, which is given by [13]

$$W_{x_s}(f) = 4\int_0^\infty g(t)\cos(2\pi f\tau)d\tau \qquad (4.13)$$

where $g(\tau)$, the impulse response of the filter is defined as

$$g(\tau) = \int_0^\infty \eta G(f)\cos(2\pi(f-f_c)\tau)d\tau \qquad (4.14)$$

Thus,

$$W_{x_s}(f) = \eta[G(f_c+f) + G(f_c-f)] \qquad (4.15)$$

The power spectral density, $S_{\theta'}(f)$, of kCx'_s/A is given by

$$S_{\theta'}(f) = k^2 C^2 \frac{W_{xs}(f)}{A^2}|j2\pi f|^2 \qquad (4.16)$$

which is plotted in Figure 4.5. To simplify calculations, we assume that the IF filter centered at IF and baseband filter have ideal frequency response within a bandwidth of $2B$ and W Hz respectively. The output noise power is given by:

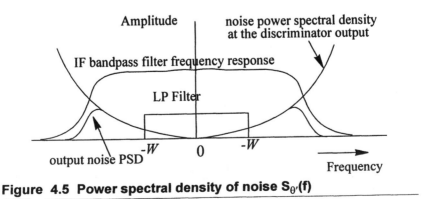

Figure 4.5 Power spectral density of noise $S_{\theta'}(f)$

$$N_o = \frac{2}{3}k^2 C^2 \left[\frac{2\pi}{A}\right]^2 W^3 \tag{4.17}$$

It is seen that the output noise power spectral density being proportional to the square of the frequency affects the higher frequency components of the information-bearing signal more adversely. To avoid excessive degradation of the signal to noise ratio of the higher frequency components, pre-emphasis circuit is used to enhance the higher frequency components at the transmitter and the emphasized components are de-emphasized at the receiver.

Now, we turn our attention to the output signal noise power. We assume that the carrier is modulated by an information bearing signal. The recovered baseband signal i.e. the detector output, for $\rho \gg 1$, is given by

$$s_d(t) = -kC(\omega_c + \alpha m(t)) \tag{4.18}$$

Thus, the output signal power is

$$S_0 = \langle s_d(t)^2 \rangle = (kC\alpha)^2 \overline{m(t)^2} \tag{4.19}$$

Two important observations are made from (4.17) and (4.19). Equation (4.19) suggests that the output signal power is proportional to the square of the deviation. Furthermore, (4.17) suggests that the output noise power is independent of the IF bandwidth as long as the IF bandwidth is much wider than twice the signal (or baseband filter) bandwidth. This implies that for reasonable input carrier-to-noise ratio, the deviation ratio in the transmitted signal can be increased without increasing the transmit power. However, the bandwidth expansion cannot be increased without a limit because of two reasons; firstly, to conserve bandwidth and to conform to the regulations on FM transmission and secondly, spreading of the signal power over wider bandwidth with the use of larger deviation ratio results in higher noise power and lower carrier to noise ratio. As the bandwidth is increased further the decreasing C/N reaches a certain threshold below which the noise appears in the form of clicks. As shown in Figure 4.6, these clicks are produced as a result of sudden phase change of the order of 2π. Further reduction of the C/N degrades the baseband S/N very rapidly.

Combining (4.17) and (4.19), the output signal to noise ratio is given by

$$\left(\frac{S}{N}\right)_o = \frac{3}{8\pi^2}(A\alpha)^2 \frac{\overline{m(t)^2}}{W^3} \tag{4.20}$$

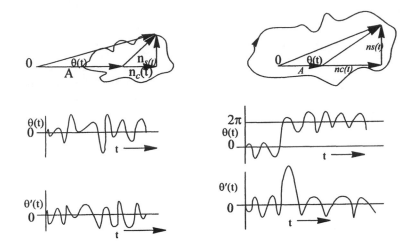

Figure 4.6 Formulation of noise clicks in the FM receiver

The expression (4.20) applies only when the input carrier to noise ratio $\rho \gg 1$. This condition is not satisfied for all times because of the presence of fading. Thus, we need to analyze the behaviour of the FM receiver when the input carrier-to-noise ratio is low.

4.1.2 Threshold Performance

We now analyze the performance of the FM receiver when the input carrier to noise ratio is close to or below threshold. We seek a general expression for the output noise spectral density so that we could extend the results to the fading case. Rewriting expression for $\theta(t)$ from (4.6), we obtain

$$\theta(t) = \text{atan}\left(\frac{n_s(t)}{n_c(t) + A}\right) \tag{4.21}$$

Note that in this case we have not ignored $n_c(t)$ in the denominator since A may not be very large compared to $n_c(t)$. The discriminator output is then given by:

$$\theta'(t) = \frac{(A + n_c(t))n_s'(t) - n'_c(t)n_s(t)}{(A + n_c(t))^2 + (n_s(t))^2} \tag{4.22}$$

The difficulty arises when we try to find correlation function, $\Re_{\theta'(\tau)}(\tau)$ of $\theta'(t)$. This involves averaging over the statistics of eight random variables.

The output noise spectrum is obtained by taking the Fourier transform of the correlation function, $\Re_{\theta'(\tau)}(\tau)$. To circumvent this difficulty, Rice [14] approximated the output noise spectrum by a sum of three noise spectral terms under the assumption that the IF filter response is Gaussian. The assumption of using Gaussian IF filter is not critical as long as the output baseband filter bandwidth is much narrower than the IF filter bandwidth. The three noise terms mentioned above are related to three received carrier to noise conditions; first when the received carrier to noise ratio is high, second when it is very low or near zero and third when it is around receiver's operating threshold. The output noise spectrum is then written as:

$$S_n(f) = S_1(f) + S_2(f) + S_3(f) \tag{4.23}$$

where $S_1(f)$ is the spectrum of the thermal noise that captures the receiver in the absence of a carrier. $S_2(f)$ is the spectrum of quadratic noise, given by equation (4.25). This component represents the noise when the input carrier to noise ratio is sufficiently high; in this case noise component, $S_2(f)$, becomes insignificant for very large input carrier levels. To calculate the signal to noise ratio between these extremes, a correction term, $S_3(f)$ is introduced. Rice [12] presented an exact expression for the spectrum of noise clicks when the receiver is operating below or about threshold level. Davis [15] however decided to fit an empirical expression to approximate $S_1(f) + S_3(f)$ within the frequency range 0 to W Hz, where W is the cutoff frequency of the baseband filter. These two expressions, one due to Rice ($S_c(f)$) and the other due to Davis ($S_D(f)$) are given in (4.24):

$$S_1(f) + S_3(f) = S_c(f) = 4\pi B\sqrt{2\pi}(1 - erf\sqrt{\rho})$$

$$S_1(f) + S_3(f) = S_D(f) = \frac{8\pi B exp(-\rho)}{\sqrt{2}(\rho + 2.35)} \tag{4.24}$$

where ρ is the input carrier to noise ratio.

The quadratic noise power spectral density, $S_2(f)$, is identical to the power spectral density of the output noise defined as $S_{\theta'}(f)$ in the previous section. However, equation (4.16) becomes invalid in the absence of the carrier; in this case, the noise spectral density goes to infinity. This difficulty is overcome by rewriting it as:

$$S_2(f) = [1 - exp(-\rho)]^2 S_{\theta'}(f) \tag{4.25}$$

Thus, the total output noise spectral density, using Davis's empirical expression, is given by:

$$S_n(f) = S_2(f) + S_D(f)$$

$$S_n(f) = [1 - exp(-\rho)]^2 S_{\theta'}(f) + \frac{8\pi B exp(-\rho)}{\sqrt{2}(\rho + 2.35)} \tag{4.26}$$

For a Gaussian IF filter, the output noise spectral density is given by:

$$S_n(f) = \frac{2\pi f[1 - exp(-\rho)]^2}{B\rho} exp\left(\frac{-\pi f^2}{B^2}\right) + \frac{8\pi B exp(-\rho)}{\sqrt{2}(\rho + 2.35)} \tag{4.27}$$

The choice of a Gaussian IF filter does not alter the output baseband noise power spectral density because its bandwidth is considerably larger than the baseband filter bandwidth in a typical voice application, *e.g.* 3 kHz against 30 kHz. Over the baseband frequency range of interest (*0-W*), the noise power spectral density essentially remains flat, so that the output noise power is

$$N_{\theta'}(\rho) = \int_{-W}^{W} S_n(f) df \tag{4.28}$$

Suppose, the recovered modulation signal power S is present, i.e.

$$S = \overline{m(t)^2}; \alpha = 1 \tag{4.29}$$

It is now possible to calculate the output signal-to-noise ratio $(S/N)_o$. The output signal in absence of noise is given by:

$$s_{do}(t) = s_d(t)(1 - exp(-\rho)) \tag{4.30}$$

The output baseband signal power, S_o ($k = C = 1$ from (4.19) and (4.29)) is given by:

$$S_0 = [1 - exp(-\rho)]^2 S \tag{4.31}$$

The output signal to noise ratio is therefore

$$\left(\frac{S}{N}\right)_o = \frac{[1 - exp(-\rho)]^2 S}{\int_{-W}^{W} S_n(f) df} = \frac{[1 - exp(-\rho)]^2 S}{\frac{a[1 - exp(-\rho)]^2}{\rho} + \frac{8\pi B W exp(-\rho)}{\sqrt{\rho + 2.35}}} \tag{4.32}$$

where

$$a = \frac{4\pi^2 W^3}{3B}\left[1 - \frac{6\pi}{10}\left(\frac{W}{B}\right)^2 + \frac{12\pi^2}{56}\left(\frac{W}{B}\right)^4 + \dots\right] \tag{4.33}$$

A plot of $(S/N)_o$ against $(C/N)_i$ is shown in Figure 4.7.

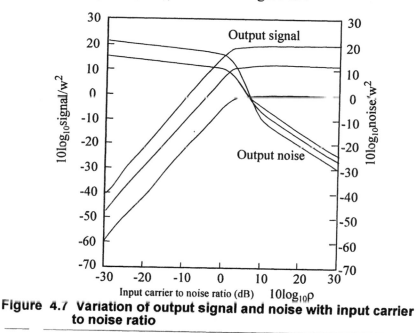

Figure 4.7 Variation of output signal and noise with input carrier to noise ratio

The output signal to noise ratio plots clearly show the threshold effect. For low CNR, not much improvement in the output S/N is seen. However, as input CNR increases, a relatively strong improvement in S/N is noticed. To determine the effect of changes in the input CNR on the output SNR, calculation of S in the foregoing equation is required. For this, we use Carsons's rule, which states that the transmission bandwidth is given by:

$$B = 2(W + f_d) = 2W(1 + \beta_i) \tag{4.34}$$

where β_i is the worst case tone modulation index and f_d is the maximum deviation. From the above,

$$f_d = \frac{B - 2W}{2} \tag{4.35}$$

and signal power is given by $k(2\pi f_d)^2$, $k < 1$. We assume that the signal level is chosen arbitrarily such that the rms deviation is 10 dB below the peak, giving a value for $k = 0.1$. Such an assumption is suitable for modulation signals with large peak to rms ratio in order to avoid over modulation, e.g. speech signals. Thus, the input signal power, S, is given by [13]

$$S = 0.1(2\pi f_d)^2 = \frac{\pi^2(B-2W)^2}{10} \tag{4.36}$$

This result is based on the assumption of Gaussian modulation with a uniform spectrum and rms frequency deviation σ_m. The IF bandwidth is chosen according to the expression so that modulation peak deviation is lower than σ_m +10 dB. The spectral components then lie within the maximum deviation defined by Carson's rule for sinusoidal modulation. Now, we can modify (4.32) to get expression for the output signal to noise ratio in terms of transmission and baseband bandwidths.

EXAMPLE E-4.1.1   Signal to Noise Ratio at a FM Receiver Output

Estimate the output signal to noise ratios of a FM receiver for input C/N ratios of 1, 5, 10 dB. The signal bandwidth is 3 kHz and the peak modulation index is 10. Assume that the IF filter has Gaussian response with bandwidth, B_{IF}, much wider than the modulation signal bandwidth.

Solution:

We use Equations (4.32) and (4.33) to solve the problem. To avoid over modulation, we assume that $k = 0.1$, in which case S is given by:

$$S = \frac{\pi}{10}(B-2W)^2 \tag{E-4.1.1}$$

To estimate the noise bandwidth B_N. we assume the 3dB bandwidth of the IF filter to be equal to 15 kHz. Then B_N is estimated to be $f_{3dB}/0.6738 =$ 22.26 kHz. For various values of carrier to noise ratio the output signal to noise ratio is calculated by substituting system parameters in equations (4.32) and (4.33).

Table E-4.1.1	*Output signal to noise ratio variation with input carrier to noise ratio*

(C/N)$_i$ dB	1	5	10
(S/N)$_o$ dB	-7.7	4	33

It is seen that the output signal to noise ratio improves rapidly when the input carrier to noise ratio is increased from 5 dB to 10 dB - an increase of input CNR of 5 dB results in SNR improvement of 29 dB. Clearly, a sharp threshold effect is demonstrated.

4.1.3 Performance in Rayleigh Fading

In the previous two sections, expressions for output signal to noise ratio for high, low, and medium values of input career to noise ratios were derived. The input conditions were assumed to remain static. However, in the presence of fading, the CNR changes dynamically; whenever the CNR becomes strong, it captures the receiver while at other times it goes through deep troughs at which the noise captures the receiver. In other words, the signal dynamics force the receiver to operate above and below threshold alternately thereby randomly varying the output signal to noise ratio. Clearly, the degree of fading affects the 'quality' of the received signal.

In voice communications, it has been shown by subjective tests that the ratio of average signal power to average noise power is a reasonable measure of quality. Davis [15] used averaging of instantaneous output signal, $s_{do}(t)$, and output noise over fade statistics in order to obtain average output signal-to-noise ratio. The difference between the fluctuating output signal and the averaged signal is a zero mean random process, which is uncorrelated with the averaged signal and contributes an additional noise called the signal suppression noise. The term signal suppression noise is used because wider fluctuations add to the noise power thereby reducing the output signal to noise ratio as if suppressing the signal. It is also assumed that the signal and the noise conditions do not change significantly over the averaging time - a reason why this technique of calculation of S/N is called quasi-static. This is reasonable assumption since the channel fade rate is much slower than the information bearing signal. If the averaged output signal is given by $\langle s_{do}(t)\rangle$, then the signal suppression noise, $n_s(t)$, is written as:

$$n_s(t) = s_{do}t - \langle s_{do}t\rangle \tag{4.37}$$

The envelope, $r(t)$, of the signal at the IF filter output is Rayleigh distributed i.e.

$$p(r) = \frac{2r}{r_o^2}exp\left(-\frac{r^2}{r_o^2}\right); \qquad 0 \leq r \leq \infty \tag{4.38}$$

where $r_o^2 = \langle r^2\rangle$ is the mean square value of r averaged over the fading statistics. The instantaneous SNR (averaged over 1 RF cycle) is given by:

$$\gamma = \frac{r^2}{2N} \tag{4.39}$$

where N is a constant noise power, and the *pdf* for the instantaneous *SNR* is given by the exponential distribution.

$$p(\gamma) = \frac{1}{\Gamma} exp\left(-\frac{\gamma}{\Gamma}\right) \tag{4.40}$$

The signal power S_0 is then calculated as follows:

$$\Gamma = \langle \gamma \rangle = \frac{\langle r^2 \rangle}{2N} \tag{4.41}$$

$$\overline{s_{d0}(t)} = \langle s_{do}(t) \rangle = \langle s_o(t)(1 - exp(-\gamma)) \rangle$$

$$\langle s_{do}(t) \rangle = s_o(t) \int_0^\infty \frac{1 - exp(-\alpha)}{\Gamma} exp\left(-\frac{\alpha}{\Gamma}\right) d\alpha = s_o(t)\left(\frac{\Gamma}{1+\Gamma}\right) \tag{4.42}$$

$$\overline{S}_0 = \langle \langle s_{do}(t) \rangle^2 \rangle = \frac{\Gamma^2}{(1+\Gamma)^2} \langle s_{do}(t)^2 \rangle = \frac{S\Gamma^2}{(1+\Gamma)^2} \tag{4.43}$$

where $S = \langle S_{do}(t)^2 \rangle$, and is defined earlier in (4.36). Signal suppression noise $n_s(t) = s_{do}(t) - \langle s_{do}(t) \rangle$ is written as

$$n_s(t) = s_o(t)(1 - exp(-\gamma)) - s_o(t)\left(\frac{\Gamma}{1+\Gamma}\right) \tag{4.44}$$

and

$$N_s = \langle [s_o(t)(1 - exp(-\gamma))]^2 - \left[s_0(t)\left(\frac{\Gamma}{1+\Gamma}\right)\right]^2 \rangle$$

$$= \frac{S\Gamma^2}{(1+\Gamma)^2} \tag{4.45}$$

The total noise power is

$$\overline{N} = \int_0^\infty \left[\frac{a}{\alpha}(1 - exp(-\alpha^2)) + \frac{8\pi BW}{\sqrt{2}(\alpha + 2.35)} exp(-\alpha) + \right. \tag{4.46}$$

$$\left. \frac{1}{1+\Gamma} exp(-\alpha^2) S\left(\frac{1}{\Gamma} + exp\left(-\frac{\alpha}{\Gamma} \right) \right) \right] d\alpha$$

Note that α is dummy variable representing ρ.

$$\overline{N} = S\left(\frac{1}{2\Gamma+1} - \frac{1}{(1+\Gamma)^2} \right) + \frac{a}{\Gamma} \log\left(\frac{(1+\Gamma)^2}{1+2\Gamma} \right) + \tag{4.47}$$

$$8\pi BW \sqrt{\frac{\pi}{2\Gamma(1+\Gamma)}} exp\left(\frac{2.35(1+\Gamma)}{\Gamma} \right) erfc\left(\sqrt{\frac{2.35(1+\Gamma)}{\Gamma}} \right)$$

In addition to the above noise components, random FM noise is also present in the baseband output. It is assumed that the signal baseband frequencies of interest (300-3000 Hz) are much larger than expected Doppler shift (0-100 Hz) at 900 MHz for a reasonable vehicular speeds. The random FM power spectral density can then be given by an asymptotic approximation given below:

$$S_{\theta'}(f) \cong \frac{f_m^2}{f}; \qquad f > f_m$$

$$\overline{N_{\theta'}} = \int_{-W}^{W} (2\pi)^2 \frac{f_m^2}{f} |H(f)|^2 df \tag{4.48}$$

where $H(f)$ is the transfer function of the baseband filter, which could also include the de-emphasis circuit. For simplicity we assume that the baseband circuit is an ideal rectangular filter of bandwidth W, and an impulse response of type

$$h(\tau) = \frac{\sin(2\pi W\tau)}{\pi\tau} \tag{4.49}$$

The output signal to noise ratio SNR is then given by:

$$\frac{\overline{S}}{\overline{N}} = \frac{\overline{S}}{\overline{N} + \overline{N_{\theta'}}} \tag{4.50}$$

The output signal to noise ratio vs the input *CNR* is shown in Figure 4.8 for several values of the ratio B/W.

The curves for finite *B/W* has two distinct regions. At low values of γ, the output signal to noise ratio increases faster than a corresponding increase in γ indicating that FM improvement is still working. At relatively higher values of γ, there is a linear increase in the output *S/N* with increase in the mean carrier to noise ratio, suggesting that FM advantage in improving *S/N* is lost due to channel fading. We will see in Chapter 7 that this advantage can be restored to a large extent by use of diversity.

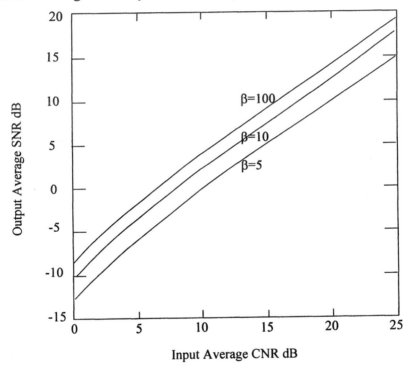

Figure 4.8 Output signal to noise ratio in the presence of fading

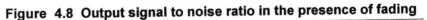

EXAMPLE E-4.1.2 Signal to Noise Ratio at FM Receiver Output in the Presence of Fading

Estimate the output signal to noise ratio of an FM receiver operating under fading conditions for mean *C/N* ratios of 1, 5, and 10 dB. The system parameters are given as baseband bandwidth = 3 kHz, peak modulation index = 10. The IF bandpass filter is assumed to have Gaussian response.

Solution:

Using (4.45) and (4.47), we calculate the signal and the noise power. Note that we have ignored the noise power due to random FM. The signal power is calculated from

$$\overline{S}_o = S\frac{\Gamma^2}{(1+\Gamma)^2} \qquad S = \frac{\pi^2(B-2W)^2}{10} \qquad \text{(E-4.2.2.1)}$$

Substituting values for ρ_o, B, W and S in the above equation we find the output signal to noise ratio (S_o/N_o) for various values of C/N.

Table E-4.1.2		*Signal to Noise Ratio in Presence of Fading*	
C/N dB	1	5	10
S/N dB	-8.85	-3.50	1.66

EXAMPLE E-4.1.3   Limiting Signal to Noise Ratio Under Fading

In mobile communications random FM is caused by signal fades. The limiting output SNR is given by:

$$SNR_{rfm} = \frac{(B-2W)^2}{20\left(\frac{v}{\lambda}\right)^2 ln10} \qquad \text{(E-4.2.3.1)}$$

Under these conditions, the SNR can not be improved by increasing CNR. Assuming that the maximum modulation frequency is 3 KHz, the vehicular speed is 55 Km/hr, and the carrier frequency is 850 MHz, what is the ratio $B/(2W)$ for $SNR_{rfm} < 30$ dB.

Solution:

By direct substitution of the parameter values, we get $B/(2W) < 2.548$.

4.2 DSB-AM of Analog Signals

Prior to the invention of frequency modulation, double sideband amplitude modulation (DSB-AM) was routinely used in mobile communications. Even today, some mobile systems operating at high frequency (HF) use DSB-AM. In DSB-AM, the information bearing signal modulates the envelope of the RF carrier. The chief advantage of DSB-AM lies in its relatively narrow RF bandwidth, which is equal to twice the information signal bandwidth. Unfortunately, the modulated signal envelope is highly susceptible to channel fading and noise - a reason why AM having bandwidth advantage, is not used in mobile communications. It is still possible to use amplitude modulation, if some corrective action is taken to counter the impact of fading. Since amplitude modulation is generally unsuited to mobile communications, we shall compare the relative merits of AM and FM for historical reasons. The AM signal is given by

$$s(t) = A[1 + \mu m(t)]\cos\omega_c t \qquad (4.51)$$

where $\mu < 1$ is the modulation index and $m(t)$ is the modulation signal. The AM detector output signal-to-noise ratio is

$$\frac{S_o}{N_o} = \frac{A^2 \langle \mu^2 m(t)^2 \rangle}{2\eta W} \qquad (4.52)$$

where η is the single sided noise power spectral density in Watts/Hz and W is the baseband bandwidth. To compare it with FM, consider that for both systems, an output signal-to-noise ratio of 30 dB is required. For both FM and AM, the rms modulation is chosen such that signal peaks do not cause over modulation. In this case $\langle \mu^2 m^2(t) \rangle$ is taken to be 0.1. Figure 4.9 illustrates the trade off of AM carrier power against FM bandwidth which takes place as FM output signal-to-noise ratio is increased by increasing frequency deviation. To maintain an equal signal-to-noise ratio, AM power must be increased, resulting in a significant power penalty over FM. It is seen that beyond an IF-to-base bandwidth ratio of 8, the FM system begins to lose its advantage over AM; degradation of carrier-to-noise ratio causes the receiver to operate in the threshold region. In any case, increased bandwidth in itself is no panacea since it exacerbates the effects of spectral congestion. The capture effect in FM systems, however, is extremely useful in the rejection of interference, particularly in mobile systems, which are based on extensive frequency reuse.

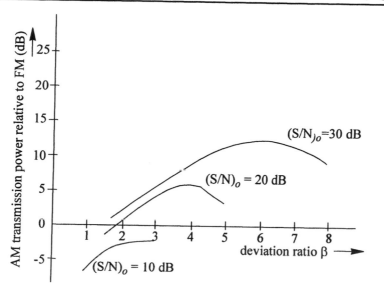

Figure 4.9 Trade off between AM and FM Systems

4.3 SSB-AM

An alternative to DSB-AM is single sideband (SSB) modulation. In SSB one sideband carrying redundant information is suppressed. Thus, the transmission bandwidth is halved or spectrum efficiency is doubled. Theoretically, within the standard 30 kHz FM channel bandwidth, five or even six single sideband channels can be accommodated. Note that such advantage is realized in analog voice communications. It is difficult to use SSB for digital transmission. Thus, SSB in mobile communications does not have good prognosis for use over mobile radio channels because all future mobile communication systems are expected to be digital.

Jakes [13] considered the case of a conventional DSB-AM system with a rms modulation index of 0.1. It was shown that the magnitude of the signal suppression noise resulting from channel fading can be as high as three times the magnitude of the signal itself. Similar results are obtained for conventional SSB systems. Here the received signal is the product of the complex Gaussian process of the multipath fading medium $r(t)exp(j\omega t)$ and the transmitted signal. If the spectrum of $r(t)$ lies well below the lowest frequency of the baseband information bearing signal, the signal suppression noise can be reduced by more than 10 dB with the use of a high pass filter having a cutoff frequency greater than the maximum Doppler frequency. Since, the effect of fading is multiplicative, the inter-modulation products thus produced cannot be removed by simple filtering. In this case, $r(t)exp(j\omega t)$ due to channel fading should be extracted to undo the channel-induced impairments.

One method of correcting the complex envelope is to transmit a pilot and extract gain and phase of the complex channel fading envelope. The main requirement is to locate the pilot as close as possible to the signal spectrum so that it maintains high correlation with the latter. The difficulty arises when the spectrum of the Doppler affected pilot overlaps the received signal spectrum; in this case the recovery of the faded pilot becomes impossible. These two requirements are of course conflicting, because as separation is increased to avoid overlap, the fading on the pilot and the signal become increasingly uncorrelated while decreasing the separation increases the chances of spectrum overlap. In order to achieve 20-dB output signal to signal-suppression noise,[†]an amplitude correlation between the pilot and the signal frequencies (over the band of signal frequencies e.g. 300 to 3 kHz) of 0.9999 is required [13]. For example, if a single sideband modulated carrier occupying 3 kHz bandwidth is transmitted over a channel having one microsecond delay spread and an output signal to signal-suppression noise of 20 dB is desired, then the pilot and the signal spectral components (e.g. 300 Hz and 3.3 kHz) must be separated by less than 2 kHz.

The foregoing discussion suggests that the design of a SSB system for mobile channels requires solutions to two significant problems: where to locate the pilot and how to reduced inter-modulation products? A variety of solutions are possible.

Almost all single sideband systems proposed to date make use of a low level pilot tone (-7.5 to -15 dB relative to peak envelope power) for automatic frequency and gain control. The following systems will be described briefly.

    a) Diminished pilot carrier SSB;

    b) Pilot tone-in-band SSB (PTIB) and Transparent tone-in-band (TTIB);

    c) Pilot tone-above-band SSB;

    d) Amplitude companded single side band system (ACSSB).

---

† *When the channel goes into a deep fade, the correction circuit producing the channel inverse must increase its gain. During a very deep fade the correction circuit gain must be limited to a certain maximum value in order to reduce noise enhancement. In this case an optimum gain magnitude limiting correction circuit is usually used.*

4.3.1 Diminished Pilot Carrier SSB

A diminished carrier SSB system was developed by Philips Research Laboratory [16] where the partially suppressed carrier is used as a pilot tone by the receiver for automatic gain control (AGC) and automatic frequency control (AFC). The SSB signal is generated with the use of a Hilbert transformer, which can be implemented using digital signal processors.

The major advantage offered by the diminished carrier SSB lies in its transparency with respect to all speech and data formats. No restrictions are imposed on the baseband signal and theoretically no amplitude or phase distortion is introduced. Disadvantages include an asymmetric pull-in range of the AFC circuit during initial synchronization, distortion of the pilot, possible adjacent channel interference, and low correlation between the pilot and the higher frequency components lying at the higher edge of the message spectrum. Distortion of the pilot occurs because of the high differential group delay of the filter relative to the signal spectra, which degrades the performance of AFC [17] and AGC [18] circuits.

4.3.2 Pilot Tone-in-Band (TIB) SSB

The pilot tone-in-band SSB, developed at the University of Bath, England, follows the most radical approach in satisfying the requirement of a high correlation between the pilot and the signal. In addition, the suggested solution achieves a large symmetrical pull-in range in the frequency control circuit [19]. A low-level pilot transmitted with the message signal provides a reference to the receiver AGC and AFC circuitry. In this approach, a part of the signal spectrum is removed to create a notch of width equal to the maximum expected Doppler spread. A tone of 10 to 15 dB below the peak signal level is inserted in this notch. This is shown in Figure 4.10.

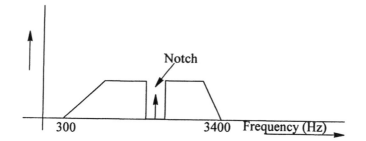

Figure 4.10 Pilot tone inband single side band spectrum

Since a portion of the signal spectrum is removed to make a room for the notch, the system is no longer transparent to all signal types; that is some signals are affected more than the others. However, due to susceptibility to fast fading, the instantaneous random FM could reach several times the maximum Doppler frequency as shown in Section 3.4.3.1. If the notch is not wide enough, there is danger that the speech and pilot spectra overlap. In this case, it becomes difficult to recover the pilot cleanly and the intelligibility of the recovered speech is severely degraded. On the other hand, a wider notch eliminates a significant portion of the signal spectrum. This causes an unacceptable level of signal distortion in the case of analog speech communications and error bursts in digital communications. The long error bursts in digital transmission will render data transmission highly unreliable. To alleviate these problems, the transparent tone-in-band (TTIB) system was proposed in [20].

The basic concept of TTIB is relatively simple but the system implementation is quite complex. The baseband signal spectrum (300-3000 Hz) is split into two approximately equal frequency segments (300 to 1700 Hz and 1700 to 3000 Hz) with the use of suitable filters. The upper segment is shifted upwards in frequency in order to make a room for a low level reference pilot, which is inserted in the centre of the resulting gap. This is shown in Figure 4.11.

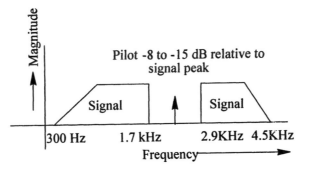

Figure 4.11 TTIB Location of tone in relation to signal spectrum

This scheme has high pilot-signal correlation, avoids an asymmetrical pull-in range, and is also suitable for data transmission [21]. The transmitter configuration is shown in Figure 4.12(a), and that of the receiver in Figure 4.12(b).

The spectral manipulations in the transmitter (*a* to *f*) and receiver (*f* to *l*) are shown in Figure 4.13. The composite signal is transmitted using conven-

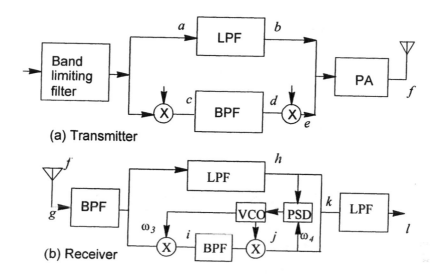

Figure 4.12 TTIB Transmitter and Receiver configurations

tional SSB techniques. At the receiver, the final stages of audio processing remove the pilot signal for use in the AGC and AFC circuits. A complementary downward frequency translation of the upper half of the audio spectrum is performed, thereby generating the original baseband signal, which is shown in Figure 4.14. TTIB thus gives the user a complete 300-3000 Hz channel while retaining all of the system advantages of tone-in-band SSB transmission. For UHF or higher frequencies the gap width can be increased. In this way the faded pilot can be extracted unambiguously for use in feedforward signal regeneration (FFSR) shown in Figure 4.14(b).

A mathematical description of the received faded signal $y_i(t)$ can be used to illustrate the salient features of FFSR operation:

$$y_i(t) = Ex(t)\cos(\omega_1 t + \omega_p t + y(t)) + Sx(t)\cos(\omega_1(t) + \omega_s(t) + y(t)) \quad (4.53)$$

where $x(t)$ = random amplitude modulation, $y(t)$ = random phase modulation, $\omega = \omega_p$ the angular frequency of the pilot, ω_s = the angular frequency of the audio signal component, E and S are corresponding amplitudes of the pilot and the signal respectively, and ω_I= IF frequency of the receiver.

The action of FFSR is to generate a control signal $c(t)$ at ω_2 the second intermediate frequency. This is given by:

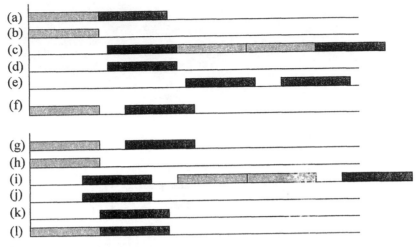

Figure 4.13 Spectral Manipulation in TTIB

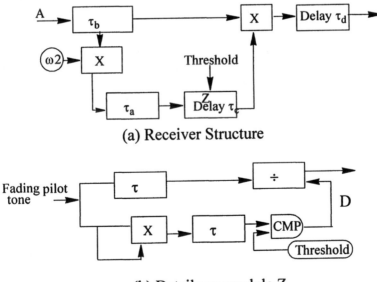

(a) Receiver Structure

(b) Details on module Z

Figure 4.14 Feedforward Signal Reconstruction

$$\chi(t) = \frac{C}{x(t)} \cos(\omega_2 t + y(t)) \tag{4.54}$$

where C is a constant.

$y_i(t)$, the received signal is linearly mixed using $\chi(t)$, to get the output signal, $y_o(t)$, which is free of the unwanted random amplitude and phase variations, $x(t)$ and $y(t)$, i.e.

$$y_o(t) = \frac{EC}{2} \cos(\omega_p t + (\omega_1 - \omega_2)t) + \frac{SC}{2} \cos(\omega_s t + (\omega_1 - \omega_2)t) \tag{4.55}$$

If the receiver is configured so that ω_1 and ω_2 are equal in the above expression, the signal is demodulated to the baseband.

$$g(t)_A = Ex(t)\cos(\omega_1 t + \omega_p t + \omega_e t + y(t)) + Sx(t)\cos(\omega_s t + \omega_e t + y(t)) \tag{4.56}$$

If ω_1 and ω_2 are not equal, the pilot and signal frequencies will be in error by ω_e.

$$g(t)_B = Ex(t)\cos((\omega_1 - \omega_2)t + \omega_p t + \omega_e t + y(t)) \Big|_{t = t - \tau_s} \tag{4.57}$$

Providing that the amplitude of the pilot envelope is above a predetermined threshold level, the block Z generates at point C a control signal of the form.

$$g(t)_C = \frac{M}{x(t)} \cos((\omega_1 - \omega_2)t + \omega_p t + \omega_e t + y(t)) \Big|_{t = t - \tau_s - \tau_d} \tag{4.58}$$

where M is a constant.

This control signal is now used to linearly mix down a delayed version of the input signal to yield a regenerated signal at point D, provided the time delays in the upper and lower signal paths are matched i.e. $\tau_b = \tau_b + \tau_c$. The signal at point D is given by:

$$g(t)_p = \frac{ME}{2} \cos(\omega_2 t) + \frac{MS}{2} \cos(\omega t + (\omega_2 - \omega_p)t) \Big|_{t = t - \tau_s - \tau_d} . \tag{4.59}$$

If $\omega_2 = \omega_p$ the resulting signal is correctly demodulated to baseband. For pilot carrier SSB systems, $\omega_2 = \omega_p = 0$ therefore the first mixer and the oscillator in the control path can be removed.

4.3.3 Tone-Above-Band SSB

While TTIB is transparent to various forms of data, its practical implementation is more complex than simple tone-in-band systems. A variation on the diminished tone system is found in tone-above-band SSB (TAB). This variation is also transparent to the signal to be transmitted. Here, the pilot tone is placed above the audio spectrum. Such an arrangement, while relatively simple to implement, makes the system more susceptible to adjacent channel interference and yields an asymmetrical pull-in behavior from the automatic frequency control circuitry. Further, the pilot tone is placed in a region of high differential group delays with respect to most of the audio band, resulting in degraded performance of the AGC and AFC circuits.

4.3.4 Amplitude Companded Single Side-Band (ACSSB)

ACSSB is a modified form of tone-above-band (TAB) system. The baseband signal is compressed and spectrally shaped for companding purposes prior to the addition of a pilot. The composite signal is then modulated to generate ACSSB signal. The pilot provides AGC, AFC, squelch, and selective call functions. The system performs equally well or even better than commercial FM systems but uses only one fifth of the bandwidth.

The normal SSB requires precise tuning (within about 20 Hz) for good quality voice reproduction. The AGC signal in a standard SSB receiver is obtained by averaging the voice power over several word lengths. This does not provide adequate correction to rapid fading; in addition the background noise is enhanced to an unacceptable level. ACSSB solves these problems by frequency modulating the pilot using a subcarrier just above the voice band. In land mobile radio version, the voice bandwidth extends to 3100 Hz and the pilot is centered at 3300 Hz. This pilot signal serves as a reference for the receiver AFC circuitry. The selective calling function can be incorporated by producing a locally generated pilot having a high degree of correlation with the pilot with the received signal, thus locking the receiver to the calling signal.

The standard SSB is inherently noisier than FM. The demodulated signal-to-noise ratio is nearly the same as the C/N, thus there is no S/N improvement. ACSSB uses amplitude companding, a technique similar to that used in Dolby noise reduction scheme. Compandors compress the large amplitude variations of the speech signal on transmission, so that low level signal components are raised in amplitude above the noise level. On reception, the amplitude range is expanded to its normal level, which quietens down the background noise during weak syllabi and word gaps. The use of amplitude companding results in approximately uniform SNR over the frequency band. The resulting improvement in SNR is similar to that due to FM capture effect in the presence of interference.

High spectral efficiency is an added advantage of ACSSB over conventional SSB when used in cellular communications networks. The spectrum efficiency of the cellular systems is measured in terms of the number of mobiles per MHz/km^2. In this measure, the channel reuse distance and the channel bandwidth are equally important. A typical carrier to interference ratio for FM is about 8 dB; the normal ratio for a conventional SSB system is between 20 and 22 dB. Amplitude companding of the information system decreases the effect of co-channel interference in SSB by about 15 dB. Since ACSSB requires the co-channel reuse distance no greater than that for FM (5-7 dB for ACSSB against 8 dB for FM) and it uses only one fifth of the FM bandwidth, therefore ACSSB has spectrum efficiency five-fold higher than that of a conventional FM system.

4.3.5 Performance of SSB Systems

Performance results of SSB systems over fading channels have not been widely reported. McGeehan *et al* [21] have described the results of field trials on subjective evaluation of the quality of the recovered speech. An experimental TTIB SSB system operating at 942 MHz with channel band-width of 6.25 kHz was used. The quality of speech for SSB was compared with that of a conventional 25 kHz bandwidth FM system. The tone-above-band system used a diminished carrier, -10 dB relative to peak envelope power, positioned at 3.9 kHz with a signal band extending from 300 Hz to 3.4 kHz. At the receiver, Feed Forward Signal Reconstruction (FFSR) was used to remove the effect of fading. Equal peak envelope powers for both the SSB and FM systems were used for fair comparison. Subjective tests for the two systems indicated that the performance of the SSB system with feedforward signal reconstruction was superior to that of the FM system for all signal lev-els.

4.4 Digital Transmission over Mobile Channels

In contrast to the situation some twenty years ago, digital transmission is now the most important mode of signal transmission over mobile radio channels. The use of digital transmission over mobile channels accelerated when transition from first generation analog system to second-generation dig-ital cellular was made in 1993. At present all second-generation mobile wire-less systems are digital. The family of third generation cellular systems recently standardized will use digital transmission methods. Four primary reasons are behind the transition from analog to digital transmission. First, the law enforcement agencies, transportation, and emergency services started to access computer databases in order to make their operation more efficient. For security reasons, these applications require encryption whose implemen-tation is much easier for digital signal formats. Second, the very large growth in the user base of the existing VHF/UHF mobile network required that spectrum efficiency be maximized, in this respect digital technology is much more efficient than analog. For example, routine messages such as a dispatcher's voice call message for a particular taxi may take up to four seconds, but the same message (coded with 100% overhead for protection against errors) can be transmitted digitally at a rate of 2400 bits/sec in three hundred milliseconds. This gives approximately thirteen-fold increase in the channel utility. The third reason is related to reliability, cost and size. The implementations of digital communication systems using VLSI subsystems are small in size, consume much lower power, and are highly reliable. The digital transmission systems also facilitate multiplexing of several data sources of different rates. The network control information can be interleaved in a common data frame with ease. A fourth reason driving the development of digital transmission is the move towards ISDN and Broadband ISDN as well as wireless ATM for high-speed data transmission over mobile networks. For instance, within the context of recently developed cellular radio systems, mixed speech, data, and image transmission is becoming a primary requirement. The 3G systems will provide variable rate services with data rates between 9 kbit/sec to 2 Mbits/sec. It is difficult to conserve spectrum with the use of conventional methods of digital encoding such as PCM, DPCM and delta modulation. In this regard research in speech and image compression techniques is becoming very important for spectrum conservation. Recent research into speech encoding shows that speech with adequate quality can be transmitted using 4.8 kbits/sec or more. It is now possible to digitally encode speech at 2.4 kbits/sec. However, the quality of the speech is marginal. Details on a form of speech encoding are discussed in Section 12.8. Similar efforts are being made to encode good quality moving picture at a rate 2 Mbits/sec or lower.

4.4.1 Basic Principles of Digital Communications

The desirable characteristics of a digital transmission system suitable for mobile communications applications are:

(a) **Good spectral efficiency:** Radio spectrum is a finite non-renewable resource; a limited amount of it is allocated to mobile radio communications. Also, a channel used by a user at certain location precludes the use of same resource to another user at the same location. The limitation of spectrum resource requires that a maximum number of users are accommodated within a given spectrum range. This means that maximum information should be packed into minimum of bandwidth. For example, in the case of digital FM modulation sudden changes in the phase of the RF signal due to abrupt changes in the level of the baseband waveform give rise to expansion in the occupied bandwidth. Thus, smoothing the baseband data pulses results in more spectrum efficient modulated signal. Furthermore, using spectrum efficient multiplex access schemes such as TDMA or CDMA also saves spectrum.

(b) **Constant envelope:** The mobile radio channel is prone to fading, therefore the envelope of the modulated signal should not bear any message information that is the envelope of the modulated signal should remain constant. If imbedding of information onto the envelope is unavoidable, the receiver should use some technique to recover information from the faded envelope.

(c) **Robust modulation method:** As the channel fading is expected, the chosen modulation method should be robust in the presence of fading, that is the bit error performance should not degrade significantly with a small reduction in the received carrier level. Furthermore, the chosen modulation method should be not be too susceptible to impairments of mobile radio channels, e.g. fading, shadowing and interference.

(d) **Easy implementation:** The terminal cost is an important consideration. A simple modulator and detector implementation results in a low cost terminal.

(e) **Availability of timing information:** One of the functions of the digital receiver is to recover symbol timing accurately. The transmitter should use data waveforms that aids such recovery. In mobile environment, this requirement becomes even more desirable because of the presence of multi path fading.

Figure 4.15 shows the principal components of a typical digital mobile radio transmission system. We shall consider each system block and discuss how the above mentioned characteristics are realized in a practical digital system design. The data source is assumed to produce impulses of strength +1 or -1, representing binary 1 and 0 respectively. The source generates data symbols at a rate known as symbol rate. The source output is encoding into a certain format that imbeds above mentioned desirable properties in the waveform.

4.4.2 Baseband Signalling Waveforms

The choice of the pulse shape has a significant influence on the spectrum efficiency and system reliability, therefore the data pulse should be shaped appropriately before it is modulated. The impact of the pulse shape on the system reliability is measured by the degree of accuracy with which receiver is able to recover symbol timing. If the symbol timing information is imbedded in the data pulse then recovery of timing information at the receiver becomes easier. A selection of signalling pulses are shown in Figure 4.16. The salient features of these signalling pulses are discussed in the next section.

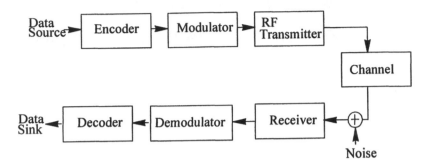

Figure 4.15 Principal components of a typical digital transmission system

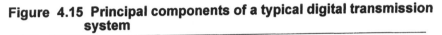

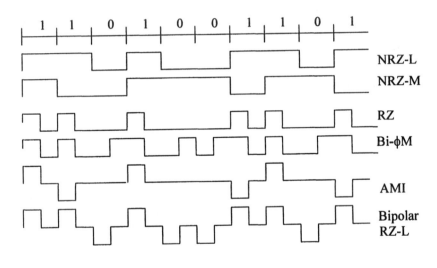

Figure 4.16 Baseband Digital Waveforms

4.4.2.1 Power Spectrum of Signalling Waveforms

A data sources is considered to generate random impulse train of certain periodicity. For example, a binary source generates an impulse of height a_k at kth time interval. The data waveform is produced by passing these pulses through a filter, whose response to these impulses produces the desired data pulse waveforms. The power spectrum and the bandwidth of the random sequence can be obtained by multiplying the power spectrum of the random impulse train with the magnitude square of the transfer function of the pulse shaping filter. We derive a general expression for the power spectrum of the impulse train in terms of symbol time and amplitude of a random impulse train.

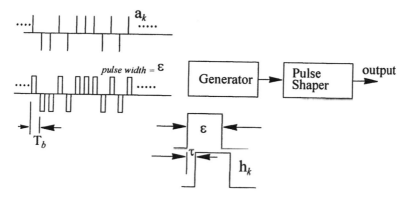

Figure 4.17 Calculation of Auto-correlation and Spectral density

Because the width of an impulse is infinitesimally small reduces to zero in the limit. It is difficult to determine the autocorrelation of a random impulse train unless we approximate the impulses by equivalent rectangular pulses of height a_k and width ε so that area under the pulse is equal to the impulse strength i.e. $h_k \varepsilon = a_k$. Once the random train autocorrelation is obtained, we reduce the pulse width to approximately zero (≈ 0) and take Fourier transform to get the power spectrum of random impulse train. Referring to Figure 4.17, for $\tau < \varepsilon$, we can write for a sequence of N pulses ($N \rightarrow \infty$).

$$\Re_{\hat{x}}(\tau) = \frac{\Re_{\hat{x}}(0)}{\varepsilon T_b}\left(1 - \frac{|\tau|}{\varepsilon}\right) \qquad ; |\tau| < \varepsilon$$

$$\Re_{\hat{x}}(0) = \frac{1}{N}\sum_k a_k^2$$

(4.60)

Thus, for a pulse train the auto-correlation function is given by

$$\Re_x(\tau) = \frac{1}{T_b}\sum_{n = -\infty}^{\infty} \Re_n(\tau)\delta(\tau - nT_b)$$

(4.61)

The power spectral density, $S_x(f)$ is the Fourier transform of the auto-correlation function of (4.61).

$$S_x(f) = \frac{1}{T_b} \sum_{n=-\infty}^{\infty} \Re_n e^{-j2\pi f n T_b}$$

$$= \frac{1}{T_b}\left[\Re_0 + 2 \sum_{n=1}^{\infty} \Re_n \cos(n\omega T_b)\right] \tag{4.62}$$

If the impulse train ($\varepsilon \to 0$) is passed through a filter, which has an impulse response $p(t)$ or magnitude transfer function $|P(f)|$, then the power spectral density at the output of this filter is

$$S_x(f) = \frac{|P(f)|^2}{T_b}\left[\Re_0 + 2 \sum_{n=1}^{\infty} \Re_n \cos(n\omega T_b)\right]. \tag{4.63}$$

Equation (4.63) gives a general expression for the power spectral density of a random pulse train. $\Re_0$ is the autocorrelation function when delay is zero and $\Re_n$ is the autocorrelation function for a delay of n bits. In the following, we consider several line codes and find their power spectra.

4.4.2.1.1 Polar Signalling

In polar signalling a '1' is transmitted by a pulse $ap(t)$ and a zero by a pulse $-ap(t)$. The probability of occurrence of 1 and 0 is equal i.e. $P(0) = P(1) = 0.5$, and $a_k^2 = a^2$. We assume that $p(t) = 1$ for $0 \le t \le T_b$. The spectrum properties may be determined from the auto-correlation function as follows. We first consider the case when NRZ-bipolar signalling (full pulse) is used. We use (4.63) to calculate spectrum and for this we need $\Re_0$ which is given by

$$\Re_0 = \operatorname*{Lim}_{N \to \infty} \frac{1}{N}\sum_k a_k^2 = a^2 \tag{4.64}$$

We also need $\Re(\tau)$, which not only depends on the time shift but also on the combination of symbols. $\Re_j\Re_n$ is either a^2 or $-a^2$. Out of N symbols half of them will be a^2 and the other half $-a^2$. Therefore $\Re_n = 0$, for $n \ge 1$. For a full pulse (period T_b), $P(f)$ is given by

$$P(f) = \int_{-T_b/2}^{T_b/2} 1.exp(-j2\pi ft)dt \tag{4.65}$$

which is $T_b sinc(fT_b)$. Thus, using (4.63), The pulse power spectrum is

---

$$S_x(f) = \frac{|P(f)|^2}{T_b} \sum_{n=-\infty}^{\infty} \Re_n exp(-j2\pi n f T_b) = a^2 T_b sinc^2(f T_b) \qquad (4.66)$$

The power spectrum of the pulse train is shown in Figure 4.18. A similar technique may be used to find the power spectrum for half pulse, i.e. pulse width $T_b/2$. Obviously, it will result in a spectral width twice that for the full pulse. We note that the spectrum peak occurs at frequency = 0, requiring dc coupling in the modem. Also, the spectrum is devoid of any spectrum lines, which if present could have aided timing recovery at the receiver.

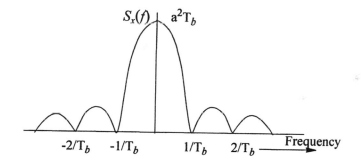

Figure 4.18 Power spectrum of full pulse polar signalling

4.4.2.1.2 NRZ Unipolar

In the NRZ-unipolar signalling, 1 is transmitted as a pulse of duration T_b and 0 as no pulse. Again $P(1) = P(0) = 1/2$. and a_k^2 is either a^2 or 0, and $\Re_0$ is therefore $a^2/2$. $\Re_n$ has four possibilities

$a_k \, a_{k-n} = 0. \, 0 = 0$

$a_k \, a_{k-n} = a. \, 0 = 0$

$a_k \, a_{k-n} = 0. \, a = 0$

$a_k \, a_{k-n} = a. \, a = a^2$

Out of N calculations of $\Re_n$, $N/4$ have value of a^2 and the other $3N/4$ values are zero. Thus, $\Re_n$ is

$$\Re_n = E\{a_k a_{k-n}\} = \frac{a^2}{4} \qquad (4.67)$$

Using $P(f)$ from (4.65), we get the power spectrum of NRZ-unipolar pulse train as

$$S_x(f) = T_b \text{sinc}^2(fT_b) \left(\sum_{n=-\infty}^{\infty} \Re_n \exp(-j2\pi nfT_b) \right)$$

$$= T_b \text{sinc}^2(fT_b) \left(\frac{a^2}{4} + \sum_{n=-\infty}^{\infty} \frac{a^2}{4} \exp(-j2\pi nfT_b) \right)$$

(4.68)

Note that for $n = 0$, $\Re_0 = a^2/2$, half of which is written within the summation and half is brought outside. We can write this in an alternative form using Fourier series. We know that in time domain the following relationship holds.

$$\sum_{m=-\infty}^{\infty} \delta(t - mT_b) = \frac{1}{T_b} \sum_{n=-\infty}^{\infty} e^{j2\pi nT_b f}$$

(4.69)

We can write (4.68) as

$$S_x(f) = T_b \text{sinc}^2(fT_b) \left(\frac{a^2}{4} \left(1 + \frac{1}{T_b} \sum_{n=-\infty}^{\infty} \delta(f - nf_s) \right) \right)$$

(4.70)

The first term in parenthesis represents a continuous spectrum whereas the second term represents a discrete spectrum consisting of tones of strength $1/T_b$, located at multiples of f_s, where $f_s = 1/T_b$. These impulses are suppressed by the function $\text{sinc}(fT_b)$, which is zero at multiples of f_s. Therefore, the clock recovery becomes difficult. However, if we reduce the pulse width to $T_b/2$, clock timing recovery becomes possible. Figure 4.19 shows the power spectra of full and half length pulses.

4.4.2.1.3 Bi-polar Pseudoternary (Alternate Mark Inversion (AMI))

In bi-polar pseudoternary or AMI signalling 0 is represented by a no pulse and 1 is represented by $+a$ or $-a$ depending on whether the previous 1 was $-a$ or $+a$. There are thus three levels $+a$, 0, $-a$, a reason why it is known as pseudoternary. Considering equal probability for 1 and 0, we can write

$P(a_k = a) = 1/4$

$P(a_k = -a) = 1/4$

$P(a_k = 0) = 1/2$

Thus $\Re_0 = a^2/2$. For $n = 1$, we have (0,0), (0,1), (1,0), and (1,1) possible combinations. Considering this, it is easy to see that

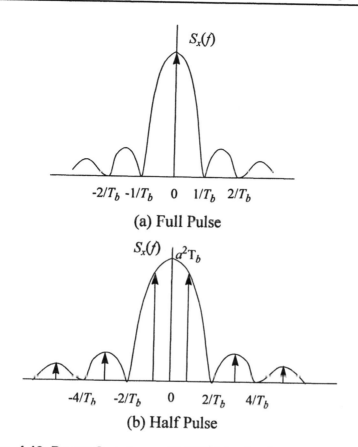

(a) Full Pulse

(b) Half Pulse

Figure 4.19 Power Spectrum of NRZ Unipolar Signalling

$$a_k a_{k+1} = 0; \qquad \text{for} \frac{3}{4} N$$

$$a_k a_{k+1} = -a^2; \qquad \text{for} \frac{1}{4} N \tag{4.71}$$

Negative sign appears because of bipolar inversion. $\Re_l$ is calculated as

$$\Re_1 = \underset{N \to \infty}{\text{Lim}} \frac{1}{N} \left[\frac{N}{4}(-a^2) + \frac{3N}{4}(0) \right] = -\frac{a^2}{4} \tag{4.72}$$

and

$$\Re_n = \underset{k}{\text{Lim} N \to \infty} \frac{1}{N} \sum a_k a_{k+n} \tag{4.73}$$

Since products $+a^2$, $-a^2$, and 0 are formed and there are equal number of a^2 and $-a^2$, therefore $\Re_n = 0$. The power spectrum is therefore given by

$$S_x(f) = \frac{|P(f)|^2}{T_b} \left[\Re_0 + 2 \sum_{n=1}^{\infty} \Re_n \cos(2\pi n f T_b) \right] \tag{4.74}$$

$$S_x(f) = a^2 \frac{|P(f)|^2}{T_b} \operatorname{sinc}^2(\pi f T_b) = a^2 T_b \operatorname{sinc}^2(\pi f T_b) \sin^2(\pi f T_b)$$

The power spectrum of AMI signalling is shown in Figure 4.20. It is seen that AMI does not use excessive bandwidth. It also does not contain a dc component, which is of an advantage in design of modems. The presence of alternate mark inversion in this signalling format gives an ability to the receiver to detect single errors. It can be seen that timing clock can be recovered by rectifying the signal. The disadvantages of AMI include its non-transparency and higher power requirement

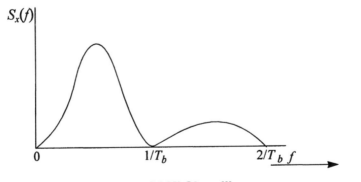

Figure 4.20 Power spectrum of AMI Signalling

4.4.2.1.4 Duobinary Signalling

Duobinary signalling is similar to pseudo-ternary, but its bandwidth is only half that of latter. A 0 is transmitted by no pulse and 1 is transmitted by $p(t)$ or $-p(t)$ depending on the polarity of the previous pulse and number of zeros between them. A '1' is encoded by the same pulse as was used to encode the preceding 1 provided an even number of 0's separate the two 1's, otherwise polarity of the preceding pulse representing 1 is reversed. An example on duobinary encoding is shown in Figure 4.21.

Data   1 1 1 0 0 0 1 1 0 1 0 0 1

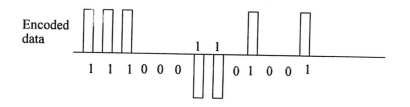

Figure 4.21 Duobinary encoding example

To calculate $\Re_0$, we observe that on the average half of the pulses have amplitude $a_k = 0$ and the remainder have the amplitude either a or $-a$. Hence

$$\Re_0 = \lim_{N \to 0} \frac{1}{N}\left[\frac{N}{2}(\pm a)^2 + \frac{N}{2}(0)^2\right] = \frac{a^2}{2} \tag{4.75}$$

To determine $\Re_1$, we note that for $a_k \, a_{k-1}$ is zero for the last three out of four possibilities 11, 10, 01, and 00. $N/4$ combinations are non zero because of duobinary rule that the a bit sequence 11 can only be encoded by two consecutive pulses of the same polarity. Thus,

$$\Re_1 = \lim_{N \to \infty} \frac{1}{N}\left[\frac{N}{4}(1) + \frac{3N}{4}(0)\right] = \frac{a^2}{4} \tag{4.76}$$

To calculate $\Re_2$ we enumerate 8 equally likely combinations i.e. 111, 101, 110, 100, 010, 001, and 000. In the last six combinations either $a_k = 0$ or $a_{k+2} = 0$, that is $a_k \, a_{k+2} = 0$. Therefore,

$$\Re_2 = \lim_{N \to \infty} \frac{1}{N}\left[\frac{N}{8}(a^2) + \frac{N}{8}(-a^2) + \frac{3N}{4}(0)\right] = 0 \tag{4.77}$$

Similarly, $\Re_n$ is zero for $n > 1$. The power spectrum is given by

$$S_x(f) = a^2 \frac{|P(f)|^2}{2T_b}(1 + \cos(2\pi f T_b)) = \frac{|P(f)|^2}{2T_b}[\cos(f T_b)]^2$$

$$= a^2 \frac{T_b}{4} \operatorname{sinc}^2\left(\frac{f T_b}{2}\right)[\cos(f T_b)]^2$$

(4.78)

4.5 Bandpass Digital Transmission Systems

Baseband signalling is used where data signals do not need a carrier for signal transportation. Since mobile communications always need RF modulation, therefore we need to discuss bandpass digital modulation. The baseband data is modulated on to RF frequency using an appropriate modulation method prior to transmission over radio channels. Since, the modulated signal occupies a finite RF bandwidth and therefore it is known as bandpass modulation. Many bandpass digital modulation methods ranging from simple amplitude shift keying to multi-level modulation schemes are analyzed extensively in the literature on digital communications. However, not all of those modulation techniques are suitable for transmission over mobile radio channels. In the following sections, we review basic principles of bandpass digital modulation schemes and provide necessary background to the reader so that basic principles can be extended for use in wireless environment. We begin with analyses of the performance of several bandpass digital modulation methods over Gaussian noise channels. The analyses are then extended to include the channel impairments such as fading, frequency selectivity, and random FM. The impact on interference on performance will be discussed in Chapter 7.

4.5.1 Amplitude Shift Keying

As noted in Section 4.2, encoding of information in the envelope of the carrier should be avoided as far as possible unless some form of envelope correction is used in the receiver. Non-coherent amplitude shift keying (NCASK) suffers significant degradation due to fading and is therefore is rarely used over mobile radio channels [13]. However, coherent ASK has a distinct advantage in its simplicity of combining diversity branches. Bitler and Stevens [22] used subjective voice evaluation methods to demonstrate that a four branch space diversity system using amplitude shift keying can provide an acceptable performance at data rates up to 50 kbits/sec. Since, coherent binary ASK is functionally equivalent to BPSK (Binary Phase Shift Keying), therefore we defer its analysis to the next section.

4.5.2 Binary Phase Shift Keying

In binary phase-shift keying (BPSK) the transmitted signal is a sinusoid of fixed amplitude but the carrier phase is either zero and π for data symbols '1' and '0' respectively. The transmitted signal is given by:

$$s_{1BPSK}(t) = \sqrt{2P_s}\cos(\omega_c t); \qquad \text{when } b(t) = 1$$

$$s_{0BPSK}(t) = \sqrt{2P_s}\cos(\omega_c t + \pi); \quad \text{when } b(t) = 0 \tag{4.79}$$

where $b(t)$ is the information symbol of duration T_b seconds, ω_c is the carrier, and P_s is the signal power. Thus, BPSK signal is similar to bi-polar ASK. At the receiver, the input signal, contaminated by noise is given by:

$$s_i(t) = s_{iBPSK}(t) + n(t); \qquad i = 0, 1 \tag{4.80}$$

where $n(t)$ additive white Gaussian noise (AWGN) with (double sided) power spectral density $N_o/2$ watt/Hz. At the receiver, the received signal is multiplied by a locally generated waveform $[s_1(t)-s_0(t)]$, where $s_1(t)$ and $s_0(t)$ are the signalling waveforms for '1'and '0' respectively. The output of the multiplier is then passed through an integrator whose output is sampled every T_b seconds i.e. every bit interval. The receiver output signal when $s_1(t)$ is transmitted and output noise are given by

$$s_o(T) = \int_0^{T_b} s_1(t)[s_1(t) - s_0(t)]dt$$

$$n_o(T) = \int_0^{T_b} n(t)[s_1(t) - s_0(t)]dt \tag{4.81}$$

At the receiver $[s_1(t) - s_0(t)] = 2A\cos\omega_c t$ is generated. Since $s_1(t) = -s_0(t)$, therefore $s_o(T_b) = A^2 T_b$. Other alternative receiver structures, which can be used are shown in Figure 4.22.

For a bit duration of T_b seconds, the probability of error is given by [23]:

$$P_e = \frac{1}{2}erfc\left(\sqrt{\frac{A^2 T_b}{2N_o}}\right) = \frac{1}{2}erfc\left(\sqrt{\frac{E_b}{N_o}}\right) = Q\left(\sqrt{\frac{2E_b}{N_o}}\right) \tag{4.82}$$

where $erfc(x)$ is defined as

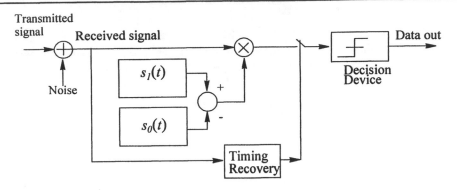

Figure 4.22 Receiver structures for BPSK

$$erfc(x) = \frac{2}{\sqrt{\pi}} \int_{x}^{\infty} \exp(-u^2)du \qquad (4.83)$$

and $Q(x)$ is Q-function[†] defined as

$$Q(x) = \frac{1}{\sqrt{2\pi}} \int_{x}^{\infty} \exp\left(-\frac{u^2}{2}\right)du \qquad (4.84)$$

It can be seen that $erfc(x)$ is related to $Q(x)$ by the relations

$$erfc(x) = 2Q(\sqrt{2}x) \text{ or} \qquad (4.85)$$

$$Q(x) = \frac{1}{2}erfc\left(\frac{x}{\sqrt{2}}\right) \qquad (4.86)$$

The bit error rate is plotted in Figure 4.23.

    For coherence demodulation the receiver local oscillator of the BPSK receiver must be synchronized to the received signal carrier. One of the method to recover carrier phase is called squaring and filtering method. In this method the received signal is squared and passed through a narrowband filter centered at $2\omega_0$. A frequency divider then produces the reference phase. Note that in the presence of fading, the recovery of the carrier phase becomes diffi-

---

[†]. *$Q(x)$ is bounded by* $\left(1 - \frac{1}{x^2}\right)\frac{1}{x\sqrt{2\pi}}e^{-x^2/2} < Q(x) < \frac{1}{x\sqrt{2\pi}}e^{-x^2/2}$.
For x>3, both these bounds are tight.

cult because of the presence of random FM. Furthermore, phase ambiguities may also result. Versions of BPSK, Differential PSK (DPSK) and differentially encoded PSK (DEPSK) circumvent the problem of phase ambiguity by representing carrier phase transitions in terms of the phases for the preceding and the current symbol.

EXAMPLE E-4.5.1 Probability of Error for BPSK Transmission

(a) Calculate error rates Coherent BPSK for E_b/N_0 of 3, 6, and 12 dB.
(b) In BPSK, the error rate increases when the local carrier is not coherent with the incoming carrier. What is the degradation in the error rate if the receiver has a phase error of 30°.

Solution

(a) E_b/N_0 of 3, 6, and 8 dB results in 2, 4, and 6.31 on linear scale. The error rate is given by

$$ P_e = \frac{1}{2} erfc \sqrt{\frac{A^2 T_b}{2\eta}} = \frac{1}{2} erfc \sqrt{\frac{E_b}{N_0}} \tag{E-4.5.1} $$

Using Matlab, we find the error probability to be 2.3×10^{-2}, 2.35×10^{-3}, 2×10^{-4} respectively.

(b) When the phase error of 30°, the error rate is degraded since effective E_b/N_0 is now weighted by cosine of the phase error. The new error rates are 3.1×10^{-2}, 4.25×10^{-3}, and 4.5×10^{-4} respectively.

4.5.3 Differential Phase-Shift Keying

In DPSK, the phase of the modulated symbol depends upon the present data symbol and the encoder output resulting from the preceding symbol. Assuming at a certain instant of time, kT_b, if the encoder output $v(t-kT_b)$ is A, then for $b(kT_b) = 1$, $v(t-(k+1)T_b) = A$ and for $b(kT_b) = 0$, $v(t-(k+1)T_b) = -A$. This waveform is then applied to a balanced modulator to produce an output,

$$ \Phi_k(t) = \mp A \cos(\omega_c t) \tag{4.87} $$

This means that the modulator output phase may change by π radians depending on the bit stream at the encoder output. The demodulator circuit shown in Figure 4.24 recovers the transmitted signal.
The multiplier output is

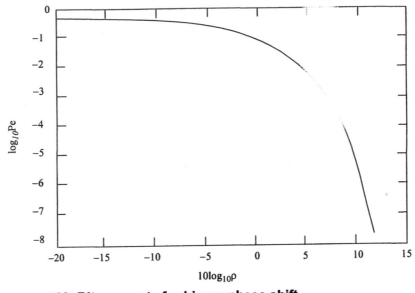

Figure 4.23 Bit error rate for binary phase shift

$$\frac{v(t)}{A}\frac{v(t-T_b)}{A}\cos(\omega_c t + \phi)\cos(\omega_c(t-T_b)+\phi)$$

$$= \frac{v(t)v(t-T_b)}{2A^2}\left[\cos(\omega_c T_b) + \cos\left(2\omega_c\left(t-\frac{T_b}{2}\right)\right)+2\phi\right] \tag{4.88}$$

The low pass filter output is

$$\frac{v(t)v(t-T_b)}{A^2}\cos(\omega_c T_b) \tag{4.89}$$

This output is maximized when $\omega_c t = \pm1$.

The transmitted data is given by the product $v(t)v(t-T_b)$. The DPSK system has the advantage of avoiding the need for a complex circuitry for carrier recovery. Its main disadvantage lies in its higher error rate caused by the use of a noisy reference signal to determine data bit polarity. Also, this noisy reference results in an error propagation to the succeeding bits, i.e. in DPSK the errors usually occur in pairs.

The probability of error is given by [24]

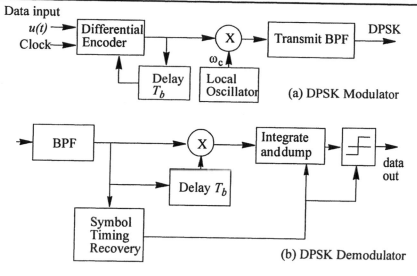

Figure 4.24 DPSK Transmitter and Demodulator

$$P_e = \frac{1}{2}exp\left(-\frac{A^2 T_b}{2N_o}\right) = \frac{1}{2}exp\left(-\frac{E_b}{N_o}\right) \qquad (4.90)$$

From this expression it can be seen that DPSK suffers a penalty of approximately 3dB relative to BPSK. In the case of DPSK recovery of the exact signal phase is not necessary, since the phase is derived from the previous detected signal. The important assumption is that the phase does not change over the duration of two bits. This is of a definite advantage in the case of slowly fading channel.

4.5.4 Binary Frequency Shift Keying (BFSK)

In binary frequency shift keying (BFSK) a '1' is represented by a carrier of frequency ω_1 and a '0' by another carrier of frequency ω_0. The frequency separation $\omega_1 - \omega_0$ is chosen such that the signals at ω_1 and ω_0 are orthogonal. For coherent FSK, the minimum frequency separation is equal to $1/2T_b$. This separation becomes double for non-coherent FSK. Since the frequency rather than the envelope of the carrier contains the information, fluctuations in the envelope have little effect on the error performance provided the carrier level remains high. As expected both frequency and phase fluctuations can degrade the system performance. Figure 4.25 shows a typical coherent BFSK receiver configuration. The chief advantage of frequency shift keying lies in its simpler implementation

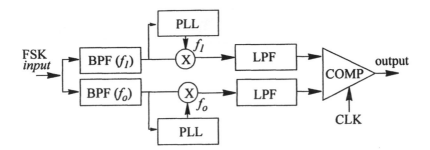

Figure 4.25 A typical CFSK Detector

In the absence of fading, the received signal is given by:

$$y(t) = Re[a_k h(t - kT_b)exp(\omega_c \pm \omega_i t)] + n(t) \tag{4.91}$$

The mark filter (designed to detect a '1') will produce an output

$$y_1(t) = Re[(a_k h(t - kT_b) + n_1(t))exp(j\omega_1 t)] \tag{4.92}$$

The space filter (designed to detect a '0') has an output consisting of noise only and is given by:

$$y_2(t) = Re[n_2(t)exp(j\omega_2 t)] \tag{4.93}$$

The coherent detector at the output of the filters gives the baseband output as:

$$u_1(kT_b) = a_k + n_{c1}(kT_b)$$
$$u_2(kT_b) = n_{c2}(kT_b) \tag{4.94}$$

where $n_{c1}(kT_b)$ and $n_{c2}(kT_b)$ are the in-phase components of the noise in the two branches.

If the transmitted signal is a binary '1', and $u_1(kT_b) < u_2(kT_b)$, then the receiver has made an error. The probability of error, given that a '1' has been transmitted, is given by:

$$P(e|1) = P(u_1 < u_2) = P(u_1 - u_2 < 0) = P(n_{c1} + a_k - n_{c2} < 0) \tag{4.95}$$

Similarly, when a '0' is transmitted the probability of making an error is given by:

$$P(e|0) = P(u_2 < u_1) = P(n_{c1} - a_k - n_{c2} > 0) \tag{4.96}$$

In (4.95) and (4.96), we require the probability distribution of n_{c1}-n_{c2}. Since the noise components n_{c1} and n_{c2} are independent Gaussian samples with variance σ^2, their difference is also Gaussian, with variance $2\sigma^2$. The probability of error is then given by:

$$P_e = \frac{1}{2}[P(e|1) + P(e|0)] = P(n_{c1} - n_{c2} < -a_k) \tag{4.97}$$

If $a_k = A$, then

$$P_e = \int_{-\infty}^{A} P_n(\lambda) d\lambda = \frac{1}{2} erfc\left(\frac{1}{\sqrt{2}} \frac{A}{\sqrt{2}\sigma}\right) = \frac{1}{2} erfc\left(\sqrt{\frac{\gamma}{2}}\right) \tag{4.98}$$

where $\gamma = (A^2/2\sigma^2)$.

The probability of error is plotted in Figure 4.26.

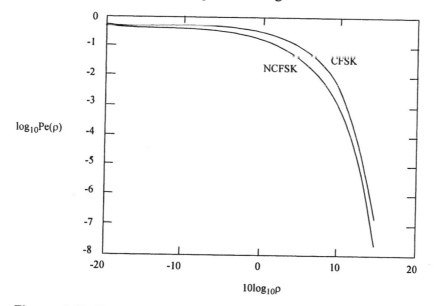

Figure 4.26 Error probability for BFSK and NCFSK

In the case of non-coherent detection, the receiver makes use of envelope detector instead of a coherent detector i.e. multiplier and integrator. When a '1' is transmitted, the mark filter will provide an output which is the envelope of the carrier at frequency ω_1 with additive noise, the resulting envelope has Ricean distribution [25]. The space filter output will deliver the envelope of the noise only, which is Rayleigh distributed. Thus, when '1' is transmitted we have:

$$p(r_1) = \frac{r_1}{\sigma^2} exp\left(-\frac{r_1^2 + A^2}{2\sigma^2}\right) I_0\left(\frac{r_1 A}{\sigma^2}\right); \qquad 0 \le r_1 \le \infty$$

(4.99)

$$p(r_2) = \frac{r_2}{\sigma^2} exp\left(-\frac{r_2^2}{2\sigma^2}\right); \qquad 0 \le r_2 \le \infty$$

An erroneous decision is made when '1' is transmitted and $r_1 < r_2$. The probability of this error is given by:

$$P_e = Prob(r_1 < r_2) = \int_0^\infty p(r_1) \int_0^\infty p(r_2) dr_1 dr_2$$

(4.100)

$$= \frac{1}{2} exp\left(-\frac{\gamma}{2}\right)$$

This is plotted in Figure 4.26. It has been shown that coherent FSK has an advantage of approximately 3dB over non-coherent detection for low signal-to-noise ratios, and that this advantage gradually decreases with an increase in γ. On the other hand, the receiver for non-coherent FSK is simpler and does not require phase tracking loop, the performance of which may vary because of rapid phase fluctuations resulting from rapid channel fading.

4.6 Performance of *M*-ary Signalling

In *M*-ary signalling, the transmitted waveform is obtained by encoding k bits into M different symbols, where $M = 2^k$. The *M*-ary signalling can use multiple carrier phases, multiple carrier amplitudes, multiple frequency tones, or their combinations. The actual transmission rate over the channel reduces by factor, k. However, since the transmitted symbols are closely located in the signal constellation for the same transmit power, the error rate usually degrades. The degradation is offset by bandwidth efficiency. M-ary signalling is extensively used in microwave links or any other high capacity transmission scheme.

4.6.1 Performance of MPSK

In this modulation method, the transmitter maps a sequence of k bits unto a particular phase out of possible $M = 2^k$ phases. The carrier with this phase shift is transmitted. The error distance is given by $A \sin(\pi/M)$. The symbol error probability is given by

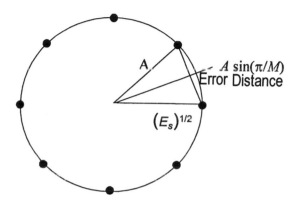

Figure 4.27 Signal constellation of 8-PSK signal

$$P_e \approx \frac{2}{\log_2 M} Q \left(\sin\left(\frac{\pi}{M}\right) \sqrt{\frac{2E_b \log_2 M}{N_o}} \right)$$

(4.101)

The above expression is derived as follows. Symbol error rate is given by

$$P_s = Q \left(\frac{a_1 - a_2}{2\sigma_0} \right)$$

(4.102)

where σ_0 is the standard deviation of the output noise power and $a_1 - a_2$ is the distance between the received symbols are the output of the detector. The distance between the symbol points at the input of the decision device in this case is $2A\sin(\pi/M)$ and the standard deviation of the output noise power is $\sigma_s = \sqrt{N_o/T_s}$. T_s is the symbol time. We rewrite (4.102) as

$$P_s = Q\left(\frac{2A\sin\left(\frac{\pi}{M}\right)}{2\sqrt{N_o/T_s}}\right) \tag{4.103}$$

Symbol energy E_s is given by $A^2 T_s/2$, or $A = (2E_s/T_s)^{1/2}$. Thus, symbol error is

$$P_s = Q\left(\sin\left(\frac{\pi}{M}\right)\sqrt{2\frac{E_s}{N_o}}\right) \tag{4.104}$$

and

$$E_s = \log_2 M E_b \tag{4.105}$$

where E_b is bit energy. Thus, symbol error probability is given by

$$P_s = Q\left(\sin\left(\frac{\pi}{M}\right)\sqrt{\frac{2\log_2 M E_b}{N_o}}\right) \tag{4.106}$$

A symbol consisting of $N = \log_2 M$ bits is in error if one or more errors occur in N bits, i.e.

$$P_s = \binom{N}{1}(P_b)^1(1-P_b)^{N-1} + \binom{N}{2}(P_b)^2(1-P_b)^{N-2} + \dots + \tag{4.107}$$
$$\binom{N}{j}(P_b)^j(1-P_b)^{N-j} + \dots + \binom{N}{N}(P_b)^N$$

where P_b is bit error probability and $\binom{J}{K} = J!/(K!(K-J)!)$.

Since bit error probability is quite small, therefore its higher powers can be neglected, i.e. finding more than one error in N bits is extremely rare. Bit error probability is therefore approximately given by

$$P_b = \frac{1}{N}P_s = \frac{1}{\log_2 M}Q\left(\sin\left(\frac{\pi}{M}\right)\sqrt{\frac{2\log_2 M E_b}{N_o}}\right) \tag{4.108}$$

We can also calculate signal to noise ratio of M-ary signalling in terms of bit energy to noise power spectral density as,

$$\frac{S}{N} = \log_2 M\left(\frac{E_b}{N_o}\right) \qquad \text{for } M > 2 \tag{4.109}$$

The above result can be extended to an important modulation scheme known as Quadrature Phase Shift Keying, which defines a four point constellation. For QPSK, $M = 4$, thus using (4.101), we can write for P_e as,

$$P_e = Q\left(\sqrt{\frac{2E_b}{N_o}}\right) \tag{4.110}$$

4.6.2 Performance of MFSK

In MFSK, the signal points are located on each of M orthogonal axes at a distance of

$$d = \sqrt{2E_s} \tag{4.111}$$

Each signal point has M-1 equidistant neighbors. For a $M = 3$, the locations of signal points is shown in Figure 4.28.

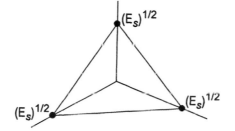

Figure 4.28 Location of signal points in 3-FSK transmission

Thus, each signal has the same conditional probability of error. The receiver consists of M correlators and the decision is made on the basis of the largest output from the correlators. A schematic of the receiver is shown in Figure 4.29. The probability of error can be found in two ways. One is more intuitive but give upper bound error rate. The exact analysis is much more difficult, therefore will not be considered here.

The upper bound error rate means that the error rate for M-ary FSK is lower or equal to a value given by the bound. Here we use Union Bound. The probability of error for coherently detected binary FSK is given by

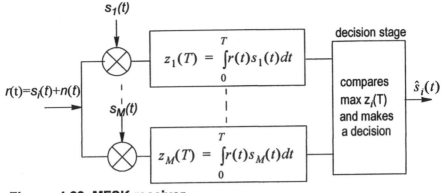

Figure 4.29 MFSK receiver

$$P_b = Q\left(\sqrt{\frac{E_b}{N_o}}\right) \qquad (4.112)$$

and the symbol error rate $P_e(M)$ is given by

$$P_E(M) \leq (M-1)Q\left(\sqrt{\frac{E_s}{N_o}}\right) \qquad (4.113)$$

where $E_s = E_b \log_2 M$. For non-coherent detection, the symbol error rate is given by

$$P_E(M) \leq \frac{1}{M} \exp\left(-\frac{E_s}{N_o}\right) \sum_{j=2}^{M} (-1)^M \binom{M}{j} \exp\left(-\frac{E_s}{jN_o}\right) \qquad (4.114)$$

For $M>7$, there is not much difference between the coherent and non-coherent detection error rate. This is given by

$$P_E(M) < \frac{M-1}{2} \exp\left(-\frac{E_s}{2N_0}\right) \qquad (4.115)$$

EXAMPLE E-4.6.1

Compare the bit error performance of coherent detectors for 8-PSK and 8-FSK for E_s/N_o of 10 dB.

Solution

We use (4.108) and (4.113) for comparison between the performances. The symbol error rates for the two systems are calculated to be 4.457×10^{-2} and 5.754×10^{-3}. The corresponding bit error rates are 5.57×10^{-3} and 7.19×10^{-4}.

4.6.3 Spectrally Efficient Modulation Methods

The rectangular pulses associated with the data stream occupy a broad spectrum, a property that has a distinct disadvantage when a premium is placed on bandwidth as is the case in mobile communications. Through filtering of the baseband data stream, sudden changes in the modulated signal phase can be smoothed out, resulting in considerable savings in spectral occupancy. Garrison [29] derived analytically the spectrum requirements of various filtered versions of FSK signals. Bandwidth reduction is obtained through the use of continuous phase frequency shift keying (CPFSK), a subclass of constant envelope continuous phase modulation signals. Members include duobinary MSK [30], Gaussian Filtered MSK (GMSK) [31], Tamed FM (TFM) [32], and generalized TFM (GTFM) [33],[34]. All of these signals have excellent spectral characteristics. In general a CPFSK signal can be written as:

$$s(t) = A\cos(2\pi f_c t + \psi(t)) \tag{4.116}$$

where

$$\phi(t) = \pi h \int_{-\infty}^{t} \sum_k a_k g(t - kT_b) dt \tag{4.117}$$

is the transmitted phase. The a_k is the kth data symbol out of the set $\{a_k\}$, $g(t)$ is the pulse shape or transmit filter impulse response, and h is the modulation index. In its most general form, the pulse shape $g(t)$ is rectangular and when $h = 0.5$, a MSK signal is generated i.e.

$$g(t) = \frac{1}{T_b}; \qquad 0 < t < T_b$$
$$= 0; \qquad \text{elsewhere} \tag{4.118}$$

For duobinary MSK (with $h = 0.5$)

$$g(t) = \frac{1}{2T_b}; \qquad -T_b < t < 0$$

$$= \frac{1}{2T_b}; \qquad 0 \le t \le T_b \qquad\qquad (4.119)$$

$$= 0; \qquad \text{elsewhere}$$

For TFM with $h = 0.5$

$$g(t) = \frac{1}{4T_b}; \qquad -T_b < t < 0$$

$$= \frac{1}{2T_b}; \qquad 0 < t < T_b \qquad\qquad (4.120)$$

$$= \frac{1}{2T_b}; \qquad T_b < t < 2T_b$$

$$= 0; \qquad \text{elsewhere}$$

Another way of differentiating between duobinary MSK and TFM is to consider them as having the same filter response but differently coded data streams. The polynomial $f(D) = (1 + D)/2$ is used for duobinary MSK, and $f(D) = \{(1 + D)/2\}^2$ for TFM [32].

4.6.4 Minimum Frequency Shift Keying (MSK)

Minimum frequency shift keying (MSK) can be regarded as a form of continuous phase digital FM with a modulation index of ½ [26] or a frequency shift keying when the separation between the signalling carriers is half of data rate. The main advantage of MSK lies in a considerable reduction in the sideband spillover, which significantly lowers the occupied bandwidth and adjacent channel interference. The frequency is shifted between f_H and f_L such that the two signals produced are mutually orthogonal over one bit interval, i.e.,

$$\int_0^{T_b} \sin(\omega_H t)\sin(\omega_L t)dt = 0 \qquad\qquad (4.121)$$

A frequency discriminator as well as differential or coherent phase detectors can be used to detect MSK signals. Coherent detection is superior to other detection methods when BER performance is considered; the performance is equivalent to an antipodal BPSK signal (in which a 1 is represented by $s(t)$ and 0 by $-s(t)$) when maximum Doppler frequency is not high [27]. Figure 4.30 shows the coherent receiver configuration. The receiver separates the even and odd bit streams $u_1(t)$ and $u_2(t)$. The correlation period is $2T$. One can derive an expression for bit error rate for MSK with coherent detection in a manner similar to that used for QPSK, since the signal constellation and receiver structure for both are similar.

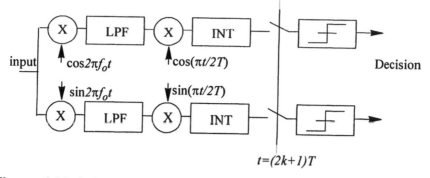

input

$\cos 2\pi f_o t$ $\cos(\pi t/2T)$ Decision

$\sin 2\pi f_o t$ $\sin(\pi t/2T)$

$t = (2k+1)T$

Figure 4.30 Coherent MSK receiver

The orthonormal unit vectors of the coordinate system are given by,

$$u_H(t) = \sqrt{\frac{2}{T_s}} \sin(\omega_H t)$$

$$u_I(t) = \sqrt{\frac{2}{T_s}} \sin(\omega_L t)$$

(4.122)

The minimum distance between the four possible signal vectors is give by

$$d = \sqrt{2E_s} = 2\sqrt{E_b},$$

(4.123)

which is identical to the case of QPSK. The probability of error for QPSK, and hence MSK, is [24]:

$$P_{eb}(MSK) = P_{eb}(QPSK) = \frac{1}{2}erfc\left(\sqrt{\frac{E_b}{N_o}}\right)$$

(4.124)

The symbol error rate is therefore given by

$$P_e = erfc\sqrt{\frac{E_b}{N_o}}$$

(4.125)

For differential and discriminator detection of MSK, the expressions for the probability of error are much more complex. Simon and Wang [28] have analyzed the performance of one bit differential detection of narrow-band digital

FM signals. It is shown that for modulation index of 0.5 (MSK) the error rate for the two types of detectors is equivalent.

4.6.5 Gaussian Filtered Minimum Frequency Shift Keying (GMSK)

The concept of using a pre-modulation low pass filter to shape the spectrum of a minimum frequency shift keying (MSK) signal while retaining its constant envelope property has been investigated by several researchers [30],[31],[35]-[39]. The requirement on the filter are:

(a) narrow bandwidth and sharp cutoff,

(b) lower time side lobes in the impulse response,

(c) preservation of filter output pulse area to assure the phase shift of $\frac{1}{2}\pi$.

A Gaussian filter satisfies the above characteristics [31]. The spectral efficiency of GMSK has been calculated by Murota [31] in terms of the $B_b T$ product (ratio of filter bandwidth to signalling rate). It is seen that a decrease in value of $B_b T_b$ results in more compact spectra. When $B_b T_b = 0.2$, the spectra of GMSK and TFM are equivalent. For a normalized bandwidth $(f-f_c)/T_b$ of 1.0, this represents a suppression of approximately 40 dB over the spectrum of MSK.

Coherent detection of GMSK offers superior performance but has the usual drawback of requiring the recovery of the RF carrier phase. This is a difficult task in the mobile environment, therefore either discriminator or differential detection is preferred. As $B_b T_b$ product decreases, demodulation becomes even more difficult because phase recovery becomes more and more difficult. Furthermore, the error rate degrades as bandwidth-time product is reduced. A rigorous mathematical analysis of the performance for both one-bit and two-bit differential detection has been reported in [38],[39] and discriminator detection is discussed in [40]. However, the expressions are too complex to be useful for system design. Much simpler BER performance is given in [31] as

$$P_e = Q\left(\sqrt{\frac{2\alpha E_b}{N_o}}\right) \tag{4.126}$$

where α is a constant related to BT as

$$\alpha \cong \begin{cases} 0.68 & \text{for } BT = 0.25 \\ 0.85 & \text{for } BT = \infty \end{cases} \tag{4.127}$$

4.6.6 Duobinary MSK

In duobinary signalling the signal bit stream is encoded in such a way that it introduces a well-controlled intersymbol interference. The extent of the ISI is known at the receiver and the resulting signal occupies a reduced bandwidth. Figure 4.31 shows a duobinary MSK transmitter.

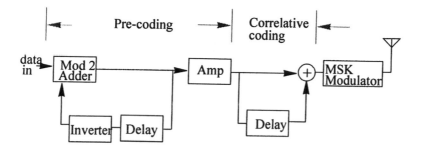

Figure 4.31 Block diagram of duobinary coded MSK transmitter

The signal $s(t)$ is assumed to be of the form

$$s(t) = \cos(\omega_o t + \phi(t)) \tag{4.128}$$

In which

$$\phi(t) = x_k + \frac{\pi}{4T_b} v_k \tag{4.129}$$

Gupta and Elnoubi investigated the incoherent detection of doubinary MSK signals [42]. Relative to MSK, duobinary MSK is spectrally more efficient but occupies a wider spectrum than TFM.

The received signal model

$$e(t) = s(t) + i(t) + n(t) \tag{4.130}$$

is used, where $n(t)$ is narrowband Gaussian noise. Let θ be the total phase angle resulting from the signal phase and noise. Then $\theta(t)$ is the output of the discriminator and is obtained as follows:

$$e(t) = X(t)\cos(\omega_o t) - Y(t)\sin(\omega_o t)$$

$$\theta'(t) = \frac{X(t)Y'(t) - X'(t)Y(t)}{X(t)^2 + Y(t)^2} \tag{4.131}$$

Using Equation (C.4) for joint probability distribution function of X, X', Y, and Y', we get the probability density function $p(\theta')$ which is given by

$$p(\theta') = \frac{1}{2}\frac{\sigma_2}{\sigma_1}(1-\rho^2)\left[\theta'^2 - 2\left(\frac{\sigma_2}{\sigma_1}\right)\rho\theta' + \left(\frac{\sigma_2}{\theta'}\right)^2\right] \qquad (4.132)$$

The probability of error may be derived as

$$P_e = \frac{1}{2}\left[\int_\gamma^\infty p(z|\text{space})dz + \int_0^\gamma p[z|\text{mark}]dz\right] \qquad (4.133)$$

in which $z = |\theta'|$ and γ is the detection threshold.

For limiter-discriminator detection, the worst case performance is given by:

$$P_e = \frac{1}{2} - \frac{3}{4}\left[\frac{1}{\sqrt{1 + 32 f_D^2 T_b^2}} - \frac{1}{\sqrt{9 + 32 f_D^2 T_b^2}}\right] \qquad (4.134)$$

where f_D is the Doppler frequency.

4.6.7 TFM and GTFM

Tamed Frequency Modulation (TFM) was first proposed by Jager and Dekker [32]. Other researchers later investigated the performance of Tamed and Generalized Tamed Frequency Modulation (GTFM) [34],[42]. The essential differences in the phase trellis (the way the modulated signal phase changes from bit to bit) for MSK (sinusoidal smoothed phase), and TFM are shown in Figure 4.32.

If the data signal $a(t)$ is specified by

$$a(t) \sum_{n=-\infty}^\infty a_n g(t - nT_b); \qquad \text{with } a_n = \pm 1 \qquad (4.135)$$

with $a_n = \pm 1$, then for TFM the phase is defined at the bit intervals as

$$\phi(mT_b + T_b) - \phi(mT_b) = \frac{\pi}{2}\left(\frac{a_{m-1}}{4} + \frac{a_m}{2} + \frac{a_{m+1}}{2}\right) \qquad (4.136)$$

where $\phi(0) = 0$ if $a_o. a_1 = 1$ and $\phi(0) = \pi/4$ if $a_o.a_1 = -1$

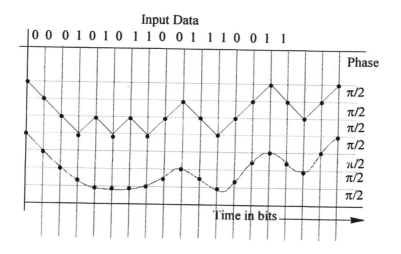

Figure 4.32 Phase behavior of MSK and TFM

Chung [34] examined Generalized Tamed Frequency Modulation (GTFM), a broader class of TFM, where the major alteration has been in the definition of phasing, which is now given by:

$$\phi_i(mT_b + T_b) - \phi_i(mT_b) = \frac{\pi}{2}(Ca_{m-1} + Ba_m + Ca_{m+1}) \qquad (4.137)$$

where C and B satisfy the condition *(2C+B)* = 1. In GTFM, the maximum phase change in phase $\phi_i(t)$ during one symbol time is restricted to $\pm\pi/2$.

Several demodulator structures have been proposed for GTFM. These include quadrature coherent detection (used for MSK), limiting discriminator detection and maximum likelihood sequence estimation. Theoretical expressions for BER are difficult to obtain; in [32] bit error rates for four phase and TFM are compared and empirical expression for BER using TFM is given as,

$$P_{eTFM} = \frac{1}{4}erfc(\sqrt{\rho}) + \frac{1}{4}erfc(\sqrt{0.69\rho}) \qquad (4.138)$$

where ρ is the signal to noise ratio at the input of the receiver with a bandwidth of $1/T_s$, T_s being symbol time.

4.7 Performance Over Impaired Channels

In the foregoing sections, an overview on the performance of various digital transmission schemes in the presence of AWGN was given. In addition to AWGN, the mobile wireless channels have generally more than one impairment e.g. frequency selectivity, fading, and shadowing. The frequency selectivity is the result of multipath propagation where inter-path delay could be a significant portion of the symbol period ($>T_b/4$). Transmission of a very high data rate (in excess of 10 Mbits/sec for wireless LAN application) over relatively static indoor mobile channel with a delay spread of around 300 nanosecond is a good example of operation in the presence of frequency selectivity. These channels are more space variant than time variant. A relatively narrowband signal propagating through wideband indoor radio channel experiences flat fading where all the spectral components fade in unison. A moderately wideband signal may also may also experience both frequency selectivity and fading in outdoor mobile radio channels. In this section, we discuss in some detail the performance of digital transmission systems over impaired channels.

4.7.1 Performance in Frequency Selective Channels

For high speed digital transmission, the channel delay spread can not be neglected since errors are produced by inter-symbol-interference (ISI). For high speed data transmission over channels that do not fade but exhibit frequency selectivity which leads to inter-symbol-interference. A system with fixed terminals using indoor radio channel devoid of fading is an example of this scenario. The average bit error rate depends on the received signal modulation pulse shape produced as a result of the transmit and receive filters, the transmitted data sequence and the channel RMS delay spread. Calculation of error rate due to ISI is quite difficult. A simplification to the analysis may be used by assuming statistical characterization of ISI. For example, Adachi [43] made assumption that ISI resembles zero mean Gaussian noise and derived results for QDPSK with differential detection.

In the bit error calculations, an equivalent SNR is first derived and it is then used in the average BER expression for differential detection, which is given by

$$P_e = \frac{1}{2}\left[1 - \frac{\Gamma_s}{\sqrt{\Gamma_e^2 + 4\Gamma_e + 2}}\right] \tag{4.139}$$

where Γ_e is the equivalent SNR, which is given by

$$\Gamma_e = \frac{\int\limits_{-\infty}^{\infty} h^2(-\tau)\xi(\tau)d\tau}{\sum\limits_{\substack{k \neq 0 \\ k = -\infty}}^{\infty} \int_{-\infty}^{\infty} h^2(-\tau - kT)\xi(\tau)d\tau} \qquad (4.140)$$

where $h(.)$ is the composite transmit and receive filter response and $\xi(.)$ is the channel delay power profile.

For small rms delay spreads ($\tau_{rms}/T < 0.1$), (4.140) can be simplified further to

$$P_e \approx \frac{1}{\Gamma_e} = \left\{ \sum\limits_{\substack{k \neq 0 \\ k = -\infty}}^{\infty} [Th'(kT)]^2 \right\} \left(\frac{\tau_{rms}}{T}\right)^2 \qquad (4.141)$$

where $h'(kT)$ is the derivative of $h(t)$ at $t = kT$.

An alternative approach is presented in [44]. In general it is difficult to compute the effect of ISI because of the presence of the past history. The exact analysis requires the complete past history of the transmitted sequence. One could, however, simplify analysis by truncating the history or even to a single ISI term, when it produces the worst case ISI, i.e. it is equal but opposite to the signal but occurs one bit later. If we assume that the output due to a certain chosen bit is only affected by the preceding bit, then we can write the mean value of $y(t)$, the output of a single pole Butterworth filter, at the end of this interval is written as,

$$E[y(t)] = \sqrt{\frac{E_b}{T_b}}(1 - 2\exp(-\omega t))$$
$$\qquad (4.142)$$
$$E[y(nT_b)] = \sqrt{\frac{E_b}{T_b}}(1 - 2\exp(-4B_N T_b))$$

The error probability is then given by

$$P_e = Q\left(\frac{\overline{y(nT_b)}}{\sigma_y}\right) = Q\left(\frac{E_b}{N_o}\frac{\left(1 - e^{\left(\frac{-4B_N T_b}{}\right)}\right)}{\sqrt{B_N T_b}}\right) \qquad (4.143)$$

The choice on the bandwidth B_N can be made by minimizing the error rate. It is shown that the minimum error results in

$$P_e = Q\left(\sqrt{1.118\frac{E_b}{N_o}}\right) \qquad (4.144)$$

4.7.2 Performance Over Flat Fading Channels

In the previous section, it was shown that channel frequency selectivity places an upper limit on the performance of digital systems. This upper limit is a function of the channel delay spread. When the transmission bandwidth is narrower than the channel coherence bandwidth, the signal experiences flat fading, that is the signal spectral components experience correlative or time selective fading. Rayleigh or Nakagami statistics are considered to be appropriate models of flat fading. The impact of fading is to randomly change the carrier to noise ratio, C/N, at the receiver input. The C/N degrades when the channel goes into fade and it improves as the signal comes out of fade. Thus, the input C/N varies randomly with time. This is illustrated in Figure 4.33.

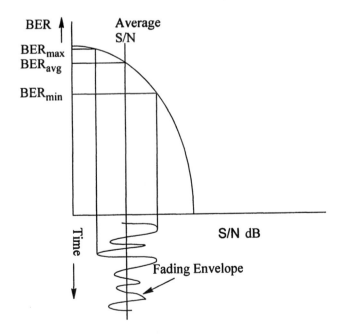

Figure 4.33 Variation of BER with signal fading

In addition to flat fading, the signal is affected by shadows created by large obstacles that obstruct the propagation path. If the channel fade rate is much slower than the transmission rate, the quasi-static approach is justified in calculation of BER. The general method of evaluating the performance of a digital transmission system is to average the error rate over the statistics of C/N. The method that French [45] developed is used here to find an expression for input C/N when Rayleigh fading is accompanied by lognormal shadowing. This is used later used to average the instantaneous BER.

It was shown in Chapter 3 that the envelope of the received signal follows Rayleigh distribution with probability density function

$$p\left(\frac{r}{\bar{r}}\right) = \frac{\pi r}{2(\bar{r})^2} exp\left(-\frac{\pi r^2}{4(\bar{r})^2}\right) \tag{4.145}$$

the error probability is given by:

$$P(e) = \int_0^\infty p(\gamma)P(e|\gamma)d\gamma \tag{4.146}$$

where γ is instantaneous CNR, $P(e|\gamma)$ is the conditional probability of error and $p(\gamma)$ is given by

$$p(\gamma) = \frac{1}{\Gamma}exp\left(-\frac{\gamma}{\Gamma}\right) \tag{4.147}$$

and Γ is average CNR. Equation (4.146) is a general expression for BER that can now be applied to several modulation schemes discussed earlier.

The error probability under fading conditions has been analyzed extensively [35]-[50]. Not surprisingly, fading causes a significant degradation in error probability. For example, results indicate that to achieve an error rate of 10^{-3} for CFSK under fading, the mean C/N must be increased by nearly 19 dB from that needed in a steady additive Gaussian noise channel to counter the effects of fading. Calculations show that under fading performance differences between various modulation schemes are not significant when the input average signal to noise ratio is high. For example, to achieve an error rate of 10^{-3}, we require the same mean carrier to noise ratio for both NCFSK and one bit differential detection MSK.

Frequency modulation when used in a fading environment is accompanied by random FM. This limits the resulting S/N. If the inverse of the Doppler frequency is more than a fraction of the bit duration i.e $\geq T_b/4$, a significant increase in the irreducible error rate due to random FM is observed. The irreducible error rate is a function of $f_D T_b$, where f_D and T_b are the Doppler frequency and symbol time respectively. This performance limit is of great concern when data communication systems operate over mobile radio channel.

Hata and Miki [35] have analyzed the performance of two bit differential detection over mobile radio channels. The probability of error in terms of input signal envelope of the detector is given by:

$$P_e(R) = \frac{1}{4}\left[erfc\left(\frac{R\rho}{\sqrt{2\pi}\sqrt{1-\rho^2}}\right) + erfc\left(\frac{R\rho k^2}{\sqrt{2\sigma}\sqrt{1-\rho^2 k^2}}\right)\right] \tag{4.148}$$

where R is the envelope at the detector input, σ^2 is the mean signal power and ρ is given by

$$\frac{\Gamma J_o(4\pi f_D T)}{1+\Gamma} \tag{4.149}$$

$k = \cos^2(\pi\tau/2T)$ and Γ is the mean carrier to noise ratio.

The error probabilities for several modulation schemes when not limited by random FM are shown in Figure 4.34. It is seen that there is not much difference in performance between many modulation schemes when these operate under fading condition. Thus, to realize superior qualities of a modulation method under fading conditions, it becomes important to use diversity and forward error control techniques. These are discussed in Chapter 7.

EXAMPLE E-4.7.1 Probability of Error for DPSK and CFSK

Differential phase shift keying (DPSK) and non-coherent frequency shift keying are two suitable techniques to transmit data over mobile channels. Under steady signal conditions the error rates for the two modulation schemes are given by $P_e = 0.5 \exp(-\gamma)$ and $P_e = \exp(-\gamma/2)$ respectively and γ is defined as carrier to noise ratio.

(a) Drive an expression for P_e, when the signal envelope fades with Rayleigh distribution i.e.

$$p(r|\bar{r}) = \frac{\pi r}{2(\bar{r})^2}\exp\left(-\frac{\pi r^2}{4(\bar{r})^2}\right) \tag{E-4.7.1.1}$$

(b) Estimate average carrier to noise ratio for $P_e = 10^{-3}$ under steady and fading channel conditions.

Solution:

(a) From (4.147) and (4.146), The probability of error for the two modulation schemes is derived as:

$$P_e = \int_0^\infty \frac{1}{2}exp(-\gamma)\frac{1}{\Gamma}exp\left(-\frac{\gamma}{\Gamma}\right)d\gamma = \frac{1}{2(1+\Gamma)}; \qquad \text{for DPSK}$$

(E-4.7.1.2)

$$P_e = \int_0^\infty \frac{1}{2}exp\left(-\frac{\gamma}{2}\right)\frac{1}{\Gamma}exp\left(-\frac{\gamma}{\Gamma}\right)d\gamma = \frac{1}{1+\Gamma}; \qquad \text{for NCFSK}$$

where $\gamma = <r^2>/2N$ and $\Gamma = <\gamma>$.

(b) By direct substitution in (E-4.7.1.2), we get $\Gamma = 26.98$, 29.99 dB for DPSK and NCFSK respectively. It shows 3 dB superiority of DPSK over NCFSK.

The error probability of GMSK in the presence of Rayleigh fading and Gaussian noise is obtained by averaging (4.126) over the channel fading statistics. The average BER is given by

$$P_e = \frac{1}{2}\left(1 - \sqrt{\frac{\alpha\Gamma}{1+\alpha\Gamma}}\right) \cong \frac{1}{4\alpha\Gamma} \qquad (4.150)$$

α is a constant whose values are given in (4.127).

4.7.2.1 Performance in Rayleigh Faded Shadowed Channels

It was also shown that variations in the local mean follow a lognormal distribution. i.e.

$$p(\bar{r}_d) = \frac{1}{\sqrt{2\sigma^2}}exp\left(-\frac{(\bar{r}_d - m_d)^2}{2\sigma^2}\right) \qquad (4.151)$$

where σ is the standard deviation (in dB) of the local mean.

The density function of the received signal envelope is obtained by combining equation (4.145) and (4.151) to yield

$$p(r) = \int_{-\infty}^\infty p\left(\frac{r}{\bar{r}_d}\right)p(\bar{r}_d)d\bar{r}_d = \sqrt{\frac{\pi}{8\sigma^2}}\int_{-\infty}^\infty \frac{r}{r^2}exp\left(-\frac{\pi r^2}{4r^2}\right)exp\left(-\frac{(\bar{r}_d - m_d)^2}{2\sigma^2}\right)d\bar{r}_d \quad (4.152)$$

Also,

$$p\left(\frac{r}{r_d}\right) = p\left(\frac{r}{\bar{r}}\right) \tag{4.153}$$

Replacing r with $10^{(rd/20)}$, we get:

$$p(r) = \sqrt{\frac{\pi}{8\sigma^2}} \int_{-\infty}^{\infty} \frac{r}{10^{\left(\frac{r_d}{20}\right)}} exp\left(-\frac{\pi r^2}{4 \times 10^{\left(\frac{r_d}{20}\right)}}\right) exp\left(-\frac{(r_d - m_d)^2}{2\sigma^2}\right) dr_d \tag{4.154}$$

Because of the difficulty of finding a closed form solution, a numerical evaluation of the above integral is usually sought. A simpler case is that in which the fading follows a Rayleigh distribution only because most signal degradation results from fast fading. Under these conditions the above integral reduces to form in (4.145), and (4.146).

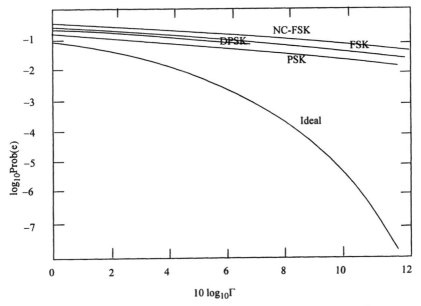

Figure 4.34 Error probability for several modulation schemes under Rayleigh Fading

4.7.2.2 Performance in the Presence of Frequency Selective Fading

The static condition of channel frequency selectivity changes into dynamic channel selectivity when either end of the mobile radio link becomes mobile. This is a typical case of high speed data transmission over mobile communications channels. We observe that the difficulty of analyzing performance of a particular modulation scheme is further compounded by the presence of fading and shadowing. The usual technique is to use diversity and equalization to counter the degradation effect of the fading and ISI. Details on diversity and equalization are presented in Chapter 7.

In [51] the bit error probability of a matched filter in a Rayleigh fading multipath channel in the presence of Interpath and ISI is derived. The channel is assumed to be slowly fading and is discrete multipath. The received signal is expressed as

$$r(t) = \sum_{i=1}^{L} f_i(t)s(t - \tau_i) + n(t) \tag{4.155}$$

where $\{f_i(t)\}$ form a set of complex zero mean Gaussian wide-sense stationary time variant coefficients of the channel. For practical reasons, the number of interfering bits may be assumed to be finite. τ_i is assumed to be constant. The channel is known to the receiver and thus a filter matched to the received bit waveform can be assumed. The output of the matched filter forming the decision variable is used to find the probability of error. It is shown that the probability of error conditioned on interfering bits, b_n, $n \neq k$ is

$$P(\{b_n, n \neq k\}) = \sum_{i=1}^{2L} \prod_{i=1}^{2L} \frac{1}{1 - \lambda_l/\lambda_i} \tag{4.156}$$
$$\lambda_i < 0 \quad i \neq l$$

where $\{\lambda_i\}$ are the eigenvalues of the matrix $Q(k)M$. The matrix $M = E[x(k)z^H(k)]$ is the covariance matrix of the vector, $z(k)$. In the present application the matrix M is the Hermitian matrix, i.e.

$$M = \begin{bmatrix} F & 0 \\ 0 & R_n \end{bmatrix}_{2L \times 2L} \tag{4.157}$$

where the matrix $F = E[f(k)f^H(k)]$ is the covariance matrix of coefficient vector $f(k)$, and the matrix $R_n = E[n(k)n^H(k)] = 4N_o[R(\tau_i-\tau_j)]_{L \times L}$ is the covariance matrix of the noise vector, $n(k)$. The average bit error probability is obtained by averaging $P(\{b_n, n \neq k\})$ over all possible sequences $\{b_n, n \neq k\}$. If the transmit pulse is zero outside the interval $[0,T_b]$ and the maximum delay difference is

$$\max_{i,j}\{\tau_j - \tau_i\} T_b \qquad (4.158)$$

The average bit error probability is given by

$$P_e = \frac{1}{4}\sum_{\substack{i = -1 \\ i \neq 0}}^{1}\sum_{\substack{j = -1 \\ j \neq 0}}^{1} P(b_{k-1} = i, b_{k+1} = j) \qquad (4.159)$$

The results show that the effect of IPI and ISI strongly depend on the bit waveform.

4.8 Summary

In this chapter, transmission of analog and digital transmission over mobile channels were examined. The transmission methods were analyzed by using quasi-static technique, which assumes that the channel characteristics vary relatively much slower than the signal information rate. It was shown that channel fading does not allow imbedding of information in the signal envelope. Thus, amplitude modulation can not be used without additional signal processing at the receiver. The needed signal processing has been shown in the case of single sideband modulation. In analog FM transmission, the presence of fading wipes out the usual advantage of enhancing signal to noise ratio by increasing modulation index.

The performance analyses of several systems show that in general signal fading has a disastrous effect on the performance of all transmission methods. The analyses of digital systems were presented using step by step approach. First, performance in the presence of noise was considered, the other channel impairments like channel frequency selectivity, fading and shadowing were introduced gradually. The presence of frequency selectivity results in irreducible error rate that increases with an increase in the channel delay spread. When frequency selectivity and multipath fading are combined as impairments, the system suffers further degradations. The usual advantage of one digital transmission scheme over the other is more or less lost because of fading and the decision on the particular type of modulation is then taken on the basis of complexities and cost of transmitter and receiver implementations.

Problems

Problem 4.1:

What is the average power ratio between two random FM signals, one received while the mobile unit is travelling at 70 Km/hr and the other at 100 km/hr. The carrier frequency of 870 MHz is frequency modulated by an audio signal covering the bandwidth 300 to 3400 Hz.

Problem 4.2

In an analog mobile cellular system, the base station transmits power P such that the average signal strength at a distance of 200 meters from it is measured to be -40 dBm. The terrain surrounding the base attenuates the signal at a rate of 46 dB per decade. The mobile receiver has the following characteristics:

(i) Noise bandwidth = 20 kHz

(ii) Receiver Noise Figure = 4 dB

(iii) Operating Frequency = 850 MHz.

Assume that handover takes place when the baseband S/N drops to 18 dB. Estimate the radius of the cell for the above link. The transmission bandwidth is given by $B = 2(W+f_d)$, where f_d is the peak deviation. The modulation signal amplitude is chosen such that it gives rms deviation is 10 dB below the peak in order to avoid over-modulation.

Problem 4.3

In Problem 4.2, include random FM and estimate the new cell radius if noise due to random FM is also taken into account. The average vehicular speed within the cell is 38 Km/hr.

Problem 4.4

(a) It was noted in this chapter that the output baseband signal to noise ratio under high input carrier to noise ratio is limited by random FM, and is given by,

$$SNR_{rfm} = \frac{(B-2W)^2}{20\left(\frac{v}{\lambda}\right)^2 \ln 10} \qquad (P\text{-}4.1.1)$$

where W is the audio bandwidth $(W_2 - W_1)$ is the baseband bandwidth in Hz. Compare the limiting baseband signal to noise ratio for the following cases:

(i) Case A: $W_2 = 4300$, $W_1 = 340$, $V = 100$ Km/hr and Frequency = 450 MHz

(ii) Case B: $W_2 = 3400$, $W_1 = 300$, $V = 70$ Km/hr and Frequency = 821 MHz.

(b) In a certain system, the baseband signal to noise ratio requirement of 18 dB is satisfied when the IF carrier to noise ratio, γ, is 10 dB. Assuming that $W = 3$ kHz, and $B = 25$ kHz, what value of γ will meet 18 dB output signal to noise ratio for same value of W.

Problem 4.5

Estimate output signal to noise ratio of a FM system for *C/N* ratios of 1,5, and 10 dB. The system uses a modulation index of 10. Assume the input signal bandwidth of 3000 Hz. The receiver IF filter has Gaussian response 3 dB bandwidth B_{IF} much wider than the input signal bandwidth.

Problem 4.6

(a) What are main differences between unipolar and bipolar codes? How RZ and NRZ formats affect these codes?

(b) Manchester code is a special form of NRZ codes. Explain how binary codes are represented by Manchester code. Sketch Manchester encoded waveform for a binary sequence 100011100110101.

Problem 4.7

(a) Several definitions of signal bandwidth are used. Explain the differences in following definitions of the signal bandwidths.

(i) Absolute bandwidth.
(ii) Null to null bandwidth.
(iii) Half power bandwidth.
(iv) 6-dB bandwidth.

(b) In the channel capacity is given by

$$C = B\log_2\left(1 + \frac{S}{N}\right) \qquad \text{(P-4.7.1)}$$

Calculate the theoretical maximum data capacity of a channel bandwidth of 200 kHz and signal to noise ratios of 15 and 12 dB and compare your answer to the GSM standard of channel rate of 270.8333 kb/sec in a bandwidth of 200 kHz.

Problem 4.8

A linearly modulated signal is represented by

$$s(t) = \Re\left[v(t)e^{j2\pi f_c t}\right]$$

(P-4.8.1)

where $v(t)$ is the lowpass equivalent. If $v(t)$ is a random pulse train given by

$$v(t) = \sum_{n=-\infty}^{\infty} I_n g(t - nT)$$

(P-4.8.2)

where $g(.)$ is the transmission pulse, I_n is the information sequence and T is the symbol time. Show that the power spectrum of $v(t)$ is given by

$$\Phi_{vv}(f) = \frac{\sigma_i^2}{T}|G(f)|^2 + \frac{\mu_i^2}{T^2}\sum_{m=-\infty}^{\infty}\left|G\left(\frac{m}{T}\right)\right|^2\delta\left(f - \frac{m}{T}\right)$$

(P-4.8.3)

where σ_i^2 is the variance of the information sequence and μ_i is its mean value.

Problem 4.9

(a) Determine the differential encoded sequence for an initial bit stream of 01110011010010001 for transmission as DPSK signal, if the initial encoded bit is 1, and the initial encoded bit is zero. From the results obtained comment on the relationship between the output sequence and the value of the initial encoded bit.

(b) Give reasons why DPSK is preferred over CPSK for digital signalling over mobile radio channels despite its inferior BER performance.

Problem 4.10

The probability of error for MSK and QPSK is the same. What are the significant differences between the two modulation schemes with regard to bandwidth efficiency?

Problem 4.11

The $\pi/4$-shifted QPSK is a modified version of QPSK. Explain the $\pi/4$-QPSK constellation with the help of a diagram. What is the maximum phase change that can occur in this modulation scheme?

A bit stream 110001110010101000 is to be sent using $\pi/4$-QPSK. If the initial phase $\theta_o = \pi/4$ and the left most bits are sent first, determine the phase θ_k during transmission.

Problem 4.12

In a digital mobile communication system operating over fading channels, the probability of error is obtained by averaging $P_e(\gamma)$ over the statistics of γ which is defined as the instantaneous (averaged over 1 cycle) carrier to noise radio.

(a) Find the statistics of γ, if the received envelope has Rayleigh distribution.

(b) Extend your results when the median signal strength is lognormal distributed with standard deviation of σ dB.

(c) Estimate error rate in the urban area for FSK signalling in an urban environment given that $\sigma = 12$ dB. The receiver error rate may be given by $P_e = 0.5 \, Q \, exp \, (-Kr)$ where $Q = 48$ and $K = 2$ and r is the signal amplitude.

Problem 4.13

The outage in digital communications is defined as the probability of error rate exceeding a certain threshold. In a QPSK system operating in a Rayleigh fading environment, the reception threshold is set at 10^{-4}.
(a) Find average signal to noise ratio for an outage probability of 10^{-2}.
(b) If we increase the power determined in (a) by 3 dB, what is the improvement in the outage probability?
(c) Using the power determined in (b), determine the outage probability when the threshold is changed to 10^{-3}.

References

[1] E.J. Bruckert and J.H. Sangster, "The effects of fading and impulsive noise over a land mobile radio channels," *IEEE Veh. Tech. Conf. Record*, PP. 14-16, December,1969.

[2] R.W. Johnson, "Field tests of mobile radioteleprinter communications," *National Symposium on Law Enforcement Science and Technology*, vol. II, 1968.

[3] R.C. French, "Mobile radio data transmission - error performance," *Proc. IERE Conf. on Civil Land Mobile Radio*, Teddington, Middlexsex, UK, pp. 93-100, November 1975.

[4] T.C. Kelly and J.E. Ward, *Investigation of Digital Mobile Radio Communications*, U.S. Department of Justice, Law Enforcement Assistance Administration, National Institute of Law enforcement and Criminal Justice, October 1973.

[5] MRDS Project, "*Study report on characterization of digital mobile radio systems*," Cantel Engineering Associates, Vancouver, B.C.,Canada,March 1976.

[6] P.J. Cadman, R.L. Brewster, "Data transmission over VHF and UHF land mobile radio channels," *Proc. IEE*, vol. 130 Part F, pp. 527-531, October 1983.

[7] R.C. French, "Mobile radio data transmission in the urban environment," *Proc. ICC*, pp. 27-15-27-20, 1976.

[8] K. Hirade and K. Murota, "A study of modulation for digital mobile telephony," *IEEE Veh. Tech. Conf.*, pp. 13-19, 1979.

[9] G.A. Arrendondo and J.I. Smith, "Voice and data transmission in a mobile radio channel at 850 MHz," *IEEE Trans. on Veh. Tech.* VT-26, pp. 88-93, February 1977.

[10] P.M. Petrovic, "Digitized speech transmission at vhf using existing fm mobile radios," *IEEE Trans. on Veh. Tech.* vol. VT-31, pp. 76-88, May 1982.

[11] A.B. Carlson, *Communication Systems, An Introduction to Signals and Noise in Electrical Communication*, McGraw Hill Book Co., 1986.

[12] S.O. Rice, "Noise in FM receivers," *Symposium of Time Series Analysis Proceedings*, M. Rosenblatt(ed.) Wiley, New York, 1963.

[13] W.C. Jakes (Ed.), *Microwave Mobile Communications*, J. Wiley, New York, 1974.

[14] S.O. Rice, "Statistical properties of a sine wave plus random noise," *Bell System Technical Journal*, vol. pp. 109-157, January 1948.

[15] B.R. Davis, "FM noise with fading channels and diversity," *IEEE Trans. on Veh. Tech.* vol. COM-19, pp. 1189-1199, December 1971.

[16] R. Wells, "SSB for VHF mobile radio at 5 kHz channel spacing," *Proc. IERE Conf. on Radio Receivers and Associated Systems*, Southampton, pp. 29-36, 1976.

[17] J.P. McGeehan, and A. Lymer, "Acquisition performance of 2nd order phase-locked loops in the presence of time delay for switched and swept signal conditions,"

[18] J.P. McGeehan and D.F. Burrows, "Performance limits of feedforward automatic gain control in mobile radio receivers," *Proc. IEE* Part F, pp. 385-392, 1981.

[19] J.P. McGeehan, and J.P.H. Sladen, "Comparative adjacent-channel selectivity performance of phase locked pilot tone SSB mobile radio receivers with particular reference to long-, short and split loop configurations," *Proc. IEE*, vol. 129, Pt. F. pp. 439-446, December 1982.

[20] J.P. McGeehan and A. Lymer, "Problem of speech pulling and its implementation for the design of phase locked ssb radio systems," *Proc. IEE*, vol. 128 Part F, pp. 361-369, 1981.

[21] J.P. McGeehan and A.J. Bateman, "Phase-locked transparent tone-in-band (ttib): a new spectrum configuration particularly suited to the transmission of data over mobile radio networks," *IEEE Trans. on Comm.* vol. COM-32, pp. 81-87, January 1984.

[22] J.S. Bitler and C.O. Stevens II, "A UHF mobile telephone system using digital modulation: preliminary study," *IEEE Trans. on Veh. Tech.*,. Tech, vol. VT-22, No.3, pp. 78-81, August 1973.

[23] H. Taub and D.L. Schilling, *Principles of Communication Systems*, McGraw Hill Book Company, New York, 1986.

[24] M. Schwartz, W.R. Bennet and S. Stein, *Communication Systems and Techniques*, McGraw Hill Book Company, New York, 1966.

[25] P.F. Panter, *Modulation, Nose and Spectral Analysis*, McGraw Hill Book Company, New York, 1965.

[26] R. de Buda, "Coherent demodulation of frequency shift keying with low deviation ratio", *IEEE Trans. on Comm.* pp. 429-435, June 1972

[27] S. Pasupathy, "Minimum shift keying: a spectrally efficient modulation", *IEEE Comm. Magazine*, vol. 17, pp. 14-22, July 1979.

[28] M.K. Simon, C. Wang, "Differential *Versus Limiter Discriminator Detection of Narrowband* FM", *IEEE Trans. on Comm.* vol. COM-31, pp.1227-1234, November 1983.

[29] G.J. Garrison, "A Spectral density analysis for digital FM", *IEEE Trans. on Comm.* vol. COM-23, pp. 1228-1243, November 1975.

[30] G.S. Deshpande, P.H. Wittke, "Correlative encoded digital FM", *IEEE Trans. on Comm.*, vol. COM-29, pp. 156-162, February 1981.

[31] K. Murota and K. Hirade, "GMSK modulation for digital mobile radio telephony", *IEEE Trans. on Comm.* vol.COM-29, pp. 1044-1050, July 1981.

[32] F.Jager and C. Dekker, "Tamed frequency modulation, a novel method to achieve spectral economy in digital transmission", *IEEE Trans. on Comm.*, vol. COM-26, pp. 534-542, May 1978.

[33] K.S. Chung, L.E. Zegers, "Generalized tamed frequency modulation", *Phillips Journal of Research*, pp.165-177, 1982.

[34] K.S. Chung, "Generalized tamed frequency modulation and its applications for mobile radio communications", *IEEE J. Selected Areas in Communications*, vol. JSAC-2, pp. 487-497, July 1984.

[35] T. Miki and M. Hata, "Performance of 16kbits/s GMSK transmission with postdetection selection diversity in land mobile radio", *IEEE Trans. on Veh. Tech.*, vol. VT-33, pp. 128-133, August 1984.

[36] M. Hata and T. Miki, "Performance of MSK high-speed digital transmission in land mobile radio channels", *Proc. IEEE Globecomm*, pp. 16.3.1-16.3.6, 1984.

[37] K. Daikoku, K. Murota, and K. Momma, "High speed digital experiments in 920 MHz urban and suburban mobile radio channels", *IEEE Trans. on Veh. Tech.* vol. VT-21, pp. 70-75, May 1982.

[38] M.K. Simon and C. C. Wang, "Differential detection of Gaussian MSK in a mobile radio environment", *IEEE Trans. on Veh. Tech.* vol. VT-33, pp. 307-320, Nov. 1984.

[39] S.M. Elnoubi, "Analysis of GMSK with differential detection in land mobile radio channels", *IEEE Trans. on Veh. Tech.*, vol. VT-35, pp. 162-167, November 1986.

[40] S.M. Elnoubi, "Analysis of GMSK with discriminator detection in mobile radio channels", *lIEEE Trans. on Veh. Tech.*h. vol. VT-35, pp. 71-76, May 1986.

[41] S. Elnoubi, and S. Gupta, "Error rate performance of non-coherent detection of duobinary coded msk and tfm in mobile radio communication systems", *IEEE Trans. on Veh. Tech.* vol. VT-30, pp. 62-76, May 1981.

[42] F.Adachi, "Analysis of DPSK error rate due to multipath delay spread," *Electronics Letters*, vol. 28, no. 7, pp. 623-624, 1992.

[43] K. Simon, S.M. Hinedi and W.C. Lindsey, *Digital Communications Techniques: Signal Design and Detection*, Prentice Hall, Upper Saddle River, N.J., 1995.

[44] R.C. French, "Error rate predictions and measurements in the mobile radio data channel", *IEEE Trans. on Veh. Tech.* vol. VT-27, pp. 110-116, August 1978.

[45] J.G. Proakis, *Digital Communications*, McGraw Hill, 1983.

[46] J-E. Stjernvall, J. Uddenfeldt, "Gaussian MSK with different demodulators and channel coding for mobile telephony", *ICC Conf. Record*, pp. 1219-1222, 1984.

[47] K. Hirade, K. Murota, M. Hata, "GMSK transmission performance in land mobile radio", *Globecomm Conf. Record*, pp. B3.4.1-B3.4.6., 1982.

[48] K. Hirade, M. Ishizuka, F. Adachi, and K. Ohtani, "Error rate performance of digital fm with differential detection in land mobile radio channels", *IEEE Trans. on Veh. Tech.*, VT-28, pp. 204-212, 1979.

[49] S. Ogose, K. Murota, and K. Hirade, "A transmitter diversity for msk with two bit differential detection", *IEEE Trans. on Veh. Tech.* vol. VT-33, pp. 37-43, 1984.

[50] V. Kaasila, and A. Mämmelä, "Bit error probability of a matched filter in a rayleigh fading multipath channel in the presence of interpath and intersymbol interference," *IEEE-Trans. on Communications*, vol. 47, no. 6, pp. 809-812, 1999.

Chapter 5

INTERFERENCE, DISTORTION AND NOISE

Interference, noise along with other causes of signal distortion such as intersymbol interference and intermodulation products constitute unwanted but ubiquitous aspects of any radio mobile radio system whether frequencies are reused or not. When two transmitters, located within the same geographical area for the purpose achieving higher spectrum efficiency, are assigned identical carrier frequency as is the case in simulcast system[†] or in two geographically distinct areas separated by a distance d in cellular systems, co-channel interference results. The desired and the interfering signals may travel to the receiver via the same or different paths; in the latter case the two signals fade independently with or without identical distributions. The impact of co-channel interference on the system performance is in the form of degradation in the signal to interference plus noise. The impact is either worsening of the symbol error or outage probability. The former was discussed in detail in Chapter 4, the latter will be examined in Sections 5.1, 5.2, 5.3, and 5.4. The impact of interference on the output signal to noise plus interference ratio is discussed in Section 5.7.1. Adjacent channel interference, considered in Section 5.7.2, arises when transmitters tuned on neighboring channels operate in the same geographical area. The degradation in the system performance results due to either assignment of several neighboring frequencies in a small area or frequency reuse. Thus, to protect the desired signal from the unwanted interfering signal and achieve higher efficiency, a proper frequency assignment is crucial.

The Intermodulation products resulting from intermixing of signals at different frequencies in non-linear elements of the receiver or transmitter may fall within the signal bandwidth thus causing interference. Interference due to intermodulation products is briefly discussed in Section 5.5. In Section 5.6 near-to-far ratio effect, which is particularly important in spread spectrum communications, is discussed. The transmission of digital signals in the presence of interference is considered in Section 5.8. Performance of π/4-QDPSK in presence of channel fading and interference is provided in considerable

---

†. *Strictly speaking in a simulcast system, two transmitters are tuned to the same frequency and carry the same information from two neighboring locations. Thus, the signals from the two transmitters reach a receiver via different paths and may result in improved performance because of independent fading.*

detail because this particular modulation format has been adopted in the North American Digital Cellular (DAMPS/TIA IS-54 or IS-136) System. The details on DAMPS/TIA IS-54 or IS-136 are given in Chapter 12. Earlier we examined the mobile radio channel interference and distortion in the absence of noise. Gaussian (thermal) noise was considered in Chapter 4 in relation to evaluation of various modulation formats. In the actual radio environment there are other sources of noise which may have characteristics other than Gaussian. Although, certain forms of non-Gaussian noise are not of much consequence for frequencies beyond 800 MHz. We include information on impulse noise in mobile environment in an attempt to quantify its characteristics in Section 5.9 for completeness.

5.1 Interference and Frequency Reuse

Mobile radio communication necessitates extensive frequency reuse in order to meet the growing demand without infusion of additional spectrum and incurring unacceptable levels of interference. When terrain details of the coverage area are known, a fixed transmitter-receiver geometry can be used to predict signal strength over distance. This information may be used to find carrier to interference ratio for a given link. Thus, for each locale, a minimum frequency reuse distance can be estimated for a desired threshold of carrier-to-interference ratio. In this regard, we need reasonably accurate propagation models for particular terrain features. In Chapter 2, several simple models based on smooth earth propagation proposed in [1] and [2] were detailed. However, these simplistic models do not apply either to a typical mobile radio situation or to any other practical radio communication system where terrain characteristics do not remain uniform within the coverage area. In most cases, the prevailing terrain conditions may let the signal propagate across the designated boundaries of the coverage area, as hard coverage boundaries are seldom found in mobile communications. The boundaries being soft, in general these could result in areas of significant mutual co-channel interference. We will start with a simple model to estimate the frequency reuse distance and then extend the model for more complex situation when the fading is introduced in following section.

5.1.1 Frequency Assignment in Fading

Frequency assignment is an important aspect of mobile radio systems. In order to achieve high spectrum efficiency, carrier frequencies must be reused in elementary areas located as close as possible. The physical separation between base stations operating at identical frequencies determines the level of mutual interference at mobiles in the cells reusing frequency. When vehicles move in multipath propagation, the desired and the interfering signals experience random and independent fading. The quality of reception can be measured by estimating the signal to interference ratio, which is a result of a random process with two degrees of freedom. One can estimate this quality metric in the form of average signal to interference ratio. The other quality measure of probability that the interfering signal exceeds the desired signal is also useful. This measure is known as the signal outage probability.

In the section that follows, signal outage probability in the presence of co- or adjacent channel interference will be examined. It is clear that the outage probability will depend strongly on the statistical properties of channels through which the desired and interfering signals propagate. The outage probability can be used as an indicator of performance margin for a particular transmission scheme, frequency reuse distance, and finally the spectrum efficiency of the system. The performance margin is defined as the difference between the received signal to interference and the required signal to interference ratio that could be achieved when both the desired signal and the interference fade independently. The analysis is later extended in defining spectrum utilization and a point is made that propagation models play important role in determining overall spectrum utilization. The parameters of a certain channel model include distance related signal attenuation factor, signal fading statistics, and the modulation method.

5.2 Co-channel Interference and Outage

Consider a scenario where a frequency 'f', assigned to a base station that provides coverage to cell A, is reused in six cells known as co-channel cells, each located at a distance D from cell A. A mobile roaming within the cell A will be affected by the signals from six neighboring base stations[†]. The desired signal and the six co-channel interfering signals fade independently as all signals arrive at the mobile via different multipath environments. The instantaneous carrier to interference ratio at the input of the mobile terminal determine the quality of reception. Since the desired and the interference both fade continuously and randomly, the instantaneous signal to interference ratio is not very practical for design purposes. Instead, the average signal to interference level is used as a meaningful design parameter. Even this parameter alone does not uncover a full picture; we need other parameters to appreciate fully the impact of fading on the signal to noise ratio. Because of fading, the

---

† The closest reuse cells are called tier 1 cells. Interference also occurs from tier 2, 3,... cells that number 12, 18, etc. However, increased distance related attenuation makes the impact of interference from outer tiers negligible. In most analyses, interference from only first tier is considered.

signal to interference will fall below a certain minimum *C/I* known as the reception threshold. Of course, the fraction of time for which carrier to interference ratio *C/I* remains above the reception threshold, can also provide a performance criterion. However, when *C/I* fades below this threshold the receiver performance degrades significantly and at times the communications becomes unintelligible. Obviously, the cellular system operator will not allow signal to interference to fade below the threshold often. How *C/I* variation depends on the statistical descriptions of the signal and interference? A little reflection will show that the statistical description of the propagation channels is the starting point to find the statistical description of *C/I*. In addition to the statistical models given in Chapter 3, we shall also use the distance dependent attenuation models discussed in Chapter 2.

Our objective is to determine signal to interference plus noise ratio in cellular environment, where number of interference signals and the distances of the interference sources play important roles. Thus, we divide the signal and interference modeling into two distinct cases: one for macrocellular radio systems and second for microcellular systems. The rationale behind such division is the distinctiveness of propagation mechanisms in microcells and macrocells cells. In the macro cellular scenario, signal and interferences fade with Rayleigh or Rayleigh-lognormal statistics. In the small cells, the desired signal most likely fade with Ricean statistics and the interference arriving from the distant cochannel cells follow Rayleigh statistics. It is interesting to note that Nakagami statistics can emulate Rayleigh, Ricean and lognormal statistics well - a reason why Nakagami distribution is now considered to be more general channel model. We shall, however, present an outage analysis in microcellular environment using traditional models, which are based on Rayleigh, Ricean and lognormal distributions. It should be noted that our results using Rayleigh, Ricean, and lognormal may be used to map into the case of Nakagami model provided some mapping between them is found.

5.2.1 Co-channel Interference Models

As mentioned above, the microcell co-channel interference modeling is different from those for macro or larger cells because of distinctiveness of propagation conditions in these environments lead to different fading statistics for the desired signal and the co-channel interfering signals. Since, the line of sight path between the mobile and the interference base stations is unlikely, the interfering signals are most likely to follow Rayleigh statistics. Furthermore, as we noted in Chapter 2, the interfering signals also experience lognormal shadowing. To complete the picture, we need to include the impact of shadowing into the model. First, we shall analyze Ricean/Rayleigh and Ricean/Rayleigh-plus-lognormal scenarios to develop statistical models for signal to interference ratio. This will be followed by the derivation of outage probabilities.

5.2.1.1 Ricean-Rayleigh Model

The Ricean-Rayleigh is found in microcellular systems where the signal from the serving base station has a specular component to the mobile while the interfering base stations propagate through a Rayleigh fading channels. We follow the analysis given in [3].

Figure 5.1 shows the conceptual model for signal to interference analysis in cellular environment. In Figure 5.1(a) collection of cells is shown where a mobile is subjected to interference from cochannel cells. In Figure 5.1(b), it is shown that the desired signal propagates through L_0 paths, and each interfering signal propagates through L_I paths. Suppose $v(t)$ is the signal transmitted from the serving base station and $v_i(t)$ $i=1,2,..,I$ are I interfering signals. The ith interfering signal arrives at the mobile via L_i paths. The receiver input, $u(t)$, is then written as,

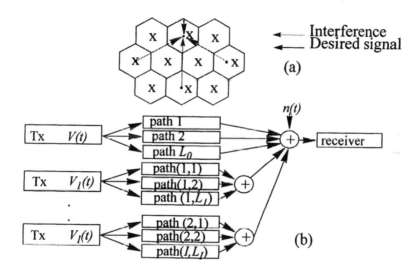

Figure 5.1 Co-channel interference in microcellular systems: (a) an interference scenario (b) interference modelling

$$u(t) = \sum_{l=1}^{L_0} \alpha_l v(t - t_l) + \sum_{i=1}^{I} \sum_{l=1}^{L_i} \beta_{i,l} v_i(t - t_{i,l}) + n(t) \qquad (5.1)$$

The path l of the desired signal having an amplitude α_l is received with a delay t_l. Similarly, the lth path of interfering signal i has an amplitude $\beta_{i,l}$ and delay $t_{i,l}$.

In the Ricean-Rayleigh model, we consider that signal paths from the desired mobile fade with Ricean statistics, while the interference signal from the cochannel cells arrive through paths that follow Rayleigh distribution. The pdf of α_l is, therefore,

$$p_{\alpha_1}(\alpha_1) = \frac{\alpha_1}{X_1} exp\left(-\frac{\alpha_1^2 + s^2}{2X_1}\right) I_0\left(\frac{\alpha_1 s}{X_1}\right) \tag{5.2}$$

The signal power X_l measured over 1 RF cycle is $\frac{1}{2}\alpha_1^2$. The pdf of X_l is given by chi-square pdf with two degrees of freedom i.e.

$$p_{X_1}(x_1) = p_{\alpha_1}(\alpha_1)\frac{d\alpha_1}{dx_1} = \frac{1}{X_1}exp\left(-\frac{2x_1 + s^2}{2X_1}\right) I_0\left(\frac{\sqrt{2x_1}s}{X_1}\right) \tag{5.3}$$

$I_0(.)$ is the modified Bessel function of the first kind and order zero. There are a total of $\sum_{i=1}^{I} L_i$ co-channel interference components, all subject to Rayleigh fading, that is lth path of ith interferer has Rayleigh statistics,

$$p_{i,l}(\beta_{i,l}) = \frac{\beta_{i,l}}{Y_{i,l}}exp\left(-\frac{\beta_{i,l}^2}{2Y_{i,l}}\right) \tag{5.4}$$

and the corresponding interference signal power of the lth path of ith signal, $y_{i,l}$, has exponential distribution with a mean power $Y_{i,l}$.

$$p_{Y_{i,l}}(y_{i,l}) = \frac{1}{Y_{i,l}}exp\left(-\frac{y_{i,l}}{Y_{i,l}}\right) \tag{5.5}$$

Out of L_0 desired signal paths, L_0-1 are components due to frequency selective multipath, each of these is subject to Rayleigh fading. The pdfs of the amplitude α_l and power x_l of the these ISI components are written as:

$$p_{\alpha_i}(\alpha_i) = \frac{\alpha_i}{X_i}exp\left(-\frac{\alpha_i^2}{2X_i}\right) \tag{5.6}$$

and

$$P_z(z) = \frac{z^{L-1}}{X_2^L(L-1)!} exp\left(-\frac{z}{X_2}\right) \tag{5.7}$$

respectively for $2 \leq l \geq L_0$ and X_l is the mean power. The total interference power, z, is the sum of L interference powers. Assuming that all the interference components have different mean powers, the pdf of total interference power is given by [3]:

$$P_z(z) = \sum_{i=1}^{L} z_i^{L-2} exp\left(-\frac{z}{Z_i}\right) \prod_{\substack{j=1 \\ j \neq i}}^{L} \frac{1}{Z_i - Z_j} \tag{5.8}$$

and the pdf of the signal to interference ratio, defined as $r = x/z$, is given by [3]:

$$P_r(r) = \int_0^\infty \frac{z_1}{X_1} exp\left(-\frac{z_1^2 + s^2}{2X_1}\right) I_0\left(\frac{\sqrt{2rzs}}{X_1}\right) \sum_{i=1}^{L} z_i^{L-2} exp\left(-\frac{z}{Z_i}\right) \prod_{\substack{j=1 \\ j \neq i}}^{L} \frac{1}{Z_i - Z_j} dz$$

$$= \sum_{i=1}^{L} \frac{b_i}{(r+b_i)^2}\left(1 + \frac{ar}{r+b_i}\right) exp\left(-\frac{ab_i}{r+b_i}\right) \prod_{\substack{j=1 \\ i \neq j}}^{L} \frac{b_j}{b_j - b_i} \tag{5.9}$$

where $b_i = X_l/Z_i$.

The pdf of signal to interference ratio can be simplified under the assumption that all interfering components have equal average power. It is given by:

$$P_r(r) = \frac{L}{b}\left(\frac{b}{r+b}\right)^{L+1} exp\left(-\frac{ab}{r+b}\right) \sum_{i=0}^{L} C_L^{L-i} \frac{1}{i!}\left(\frac{ar}{r+b}\right)^i \tag{5.10}$$

where $a = \frac{s^2}{2X_1}$, $b = \frac{X_1}{X_2}$, and C_i^{L-1} is a binomial coefficient, i.e. $\frac{L!}{i!(L-1)!}$.

In modeling the interference from other sources two approaches are possible. First, when the phasors representing the interfering signals vary significantly fast and cannot maintain coherence with each other and therefore are added noncoherently. In this case the pdf of the joint interference is given by multiple convolution of the power pdfs of the interference signals. Second, when the interference phasor hardly change during the observation time, the phasors are added on the basis of their amplitudes rather than power. In this case, all interference components add up to one Rayleigh phasor. The pdf of the total interference power is,

$$P_z(z) = \frac{B_L}{(r + B_L)^2}\left(1 + \frac{ar}{r + B_L}\right)\exp\left(-\frac{aB_L}{r + B_L}\right) \tag{5.11}$$

where

$$B_L = X_1' \left(\sum_{i=1}^{L} z_i\right) \tag{5.12}$$

5.2.1.2 Ricean-Rayleigh-Lognormal Model

In a practical cellular radio system, signal propagation path is usually obstructed by large objects that create shadows. Thus, each path resulting from multipath propagation is subject to Rayleigh fading and lognormal shadowing. However, it can be safely assumed that the desired Ricean signal is devoid of shadowing[†]. Thus, all $L_0 - 1 + \sum_{i=1}^{L} L_i$ paths experience superimposed Rayleigh fading and lognormal shadowing. The contributions from all Rayleigh faded paths add up in voltage to result in a Rayleigh phasor. The sum of lognormal random variables is well approximated by another lognormal random variable. The pdf of interference is given by [3]:

$$p_Z(Z) = \frac{C}{Z}\exp\left(-\frac{(\log Z + A)^2}{B}\right) \tag{5.13}$$

where

$$C = \frac{\log e}{\sqrt{2\pi}\sigma}, A = -\log Z_0 + \frac{\sigma^2}{2\log e} \text{ and } B = 2\sigma^2. \tag{5.14}$$

Z_0 is the area mean power of the composite interference and σ is its standard deviation in Bels.

The pdf of the signal to interference ratio is given by:

[†]. *In certain circumstances, the plane earth propagation can give rise to signal strength lobes, which can be confused with the presence of shadowing.*

$$p_r(r) = C \int_0^\infty \frac{(a+1)rV+1}{(rV+1)^3} \exp\left\{-\frac{1}{B}\left[\log(bV) + \frac{B}{4}\ln 10\right]^2 - \frac{a}{rV+1}\right\}dV \quad (5.15)$$

where

$$a = \frac{x}{2X_1} \quad \text{and} \quad b = \frac{X_1}{Z_0}. \quad (5.16)$$

For noncoherent phasor addition (power sum), (5.15) is simplified by approximating the total interference power by a lognormal random variable. It should be noted that it is very difficult to find an exact expression for the noncoherent sum of powers. The simplified expressions for the pdf of the total interference and signal to interference power ratio are:

$$p_z(z) = \frac{C}{z}\exp\left[-\frac{(\log z + A)^2}{B}\right] \quad (5.17)$$

and

$$p_r(r) - \frac{C}{X_1}\int_0^\infty \left\{\exp-\left[\frac{2rz+s^2}{2X_1} + \frac{\log z + A^2}{B}\right]\right\}I_0\left(\frac{\sqrt{2rzs}}{X_1}\right)dz \quad (5.18)$$

The probability density functions for signal to noise ratios may now be used to find outage probability, which may be used to define the cell dimensions. Table 5.1 lists several possible microcell co-channel interference models. The main differences between various models arise when the interference signals with equal or unequal powers are added coherently or non-coherently.

Table 5.1		Microcell co-channel Interference Model

Model	Interference Addition	Interference Power
Ricean/Rayleigh	Non-coherent	equal mean
		unequal mean
	Coherent	
Ricean/Rayleigh-plus-log-normal	Coherent	
	Non-coherent	

5.3 Signal Outage Analysis

In cellular environment, in addition to bit error performance the outage probability is used as a measure of the service quality. The outage is defined as the probability that the received signal or signal to interference falls below a certain threshold, which is normally selected to give acceptable quality of service. The BER criterion is commonly used in noise limited environments but outage definition related to signal to interference is used in fading and interference environment.

The outage probability is defined as,

$$P_{out} = Prob(r < R_t)$$

$$= \int_0^{R_t} p_r(r)dr \qquad (5.19)$$

where R_t is the threshold level. If all the interference components have equal power, we use (5.10) and (5.19) to get:

$$P_{out} = \frac{R_I}{R_I + b}\exp\left(-\frac{ab}{R_I + b}\right) \sum_{m=0}^{k} \frac{b^k}{(R_I + b)^k} \sum_{m=0}^{k} C_k^{k-m} \frac{1}{m!}\left(\frac{aR_I}{R_I + b}\right)^m \qquad (5.20)$$

If the interference components have mutually different mean powers, the outage probability is obtained from (5.9) and (5.19).

$$P_{out} = \sum_{i=1}^{I} \frac{R_I}{R_I + b_i}\exp\left(-\frac{ab_i}{R_I + b_i}\right) \prod_{j=1, j \neq i}^{L} \frac{b_j}{b_j - b_i} \qquad (5.21)$$

Using (5.20), the outage probabilities for Rice Factors of $a = 0$ and $a = 5$ are plotted against signal to interference ratio for the case of equal mean interference power in Figure 5.2. The number of contributing interferers is six, each having two paths. The threshold level is fixed at $R_t = 5$. It is observed that as soon as the signal to interference increases beyond the selected threshold level, the outage decreases rapidly.

5.3.1 Medium to Large Cell Co-channel Interference Modeling

The above analysis, though applicable to microcellular environment under certain propagation conditions, is directly applicable to medium to large cells. Since in medium to large cells direct line of sight is not expected, we may simply take a, the Ricean factor as 0. Then, from (5.20), we have

---

$$P_{out} = \frac{R_t}{R_t + b} \sum_{k=0}^{L-1} \frac{b^k}{(R_t + b)^2} \tag{5.22}$$

Using the parameters similar to those in Figure 5.2, Figure 5.3 compares the outage probability of a medium/large cell system ($a = 0$) with that of microcell ($a = 5$). It is seen that outage probabilities are lower than in the case of microcells.

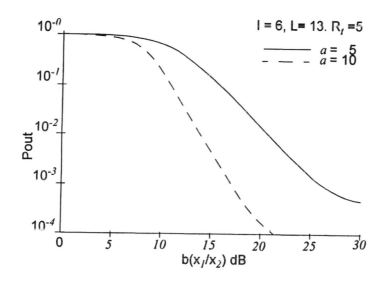

Figure 5.2 Outage probability: Ricean/Rayleigh model with equal mean interference power

5.3.2 Noncoherent and Coherent Interferers

The expression given in (5.20) assumes that the powers of interferers are added. However, in a very slow fading environment the addition of interferers can be considered coherent. The outage probability is derived as:

$$P_{out} = \int_0^{R_I} \frac{B_L}{(r + B_L)^2}\left(1 + \frac{ar}{r + B_L}\right)\exp\left(-\frac{aB_L}{r + B_L}\right) dr$$

$$= \frac{R_I}{R_I + B_L}\exp\left(-\frac{aB_L}{R_I + B_L}\right) \tag{5.23}$$

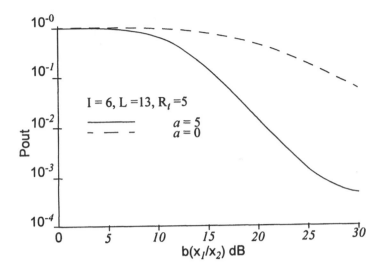

Figure 5.3 Outage probability: Comparison of microcell and medium large cell systems

The outage results are shown in Figure 5.4. It is observed that the outage probabilities are not significantly different for the two cases considered.

5.3.3 Effect of Frequency Selective Multipath Fading

In the presence of frequency selective multipath, self interference due to multipath propagation adds an additional interference. There are a total of $L_o - 1 + \sum_{i=1}^{I} L_i$ interferers out of which $L_o - 1$ are due to self interference and I are the cochannel interferers. It should be noted that the desired signal power is weakened due to dispersion of its energy into multipaths. The outage probability is therefore degraded. The results are shown in Figure 5.5.

5.3.4 Effect of Shadowing

The results derived for outage probabilities in the Ricean/Rayleigh environment can be extended to include lognormal shadowing. Assuming that the interference signals are added non-coherently, the outage probability in a microcell environment for the case of Ricean/Rayleigh-plus-lognormal fading is found to be,

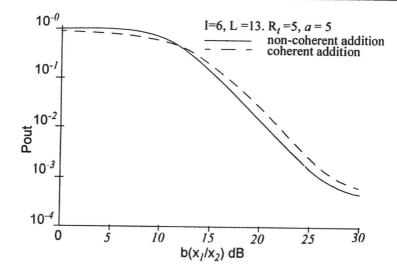

Figure 5.4 Outage probability: comparison of coherent and non-coherent addition of interference components

$$P_{out} = C \int_0^R \int_0^\infty \frac{(a+1)rV+1}{(rV+1)^3} \exp\left\{-\frac{1}{B}\left[\log(bV)+\frac{B}{4}\ln 10\right]^2 - \frac{a}{rV+1}\right\} dVdr$$

$$= CR_i \int_0^\infty \frac{1}{R_iV+1} \exp\left\{-\frac{1}{B}\left[\log(bZ)+\frac{B}{4}\ln 10\right]^2 - \frac{a}{R_iV+1}\right\} dV$$

(5.24)

The numerical results for $a = 5$ and $R_I = 5$ for several shadowing depths are presented in Figure 5.6. The case of no shadowing is also presented for comparison. It is seen that the presence of shadowing does not significantly affect the outage probability. The impact of shadowing worsens as the signal to interference ratios increases. It seems that the desired signal experiences heavier shadowing in comparison with the composite interference. The contributing interfering signals experience independent fading and shadowing, therefore when these signals combine, the composite interference signal becomes much steadier due to path diversity. Under low signal to interference ratios and heavier shadowing, the outage probability decreases. It is because of lower level of the resultant composite interference.

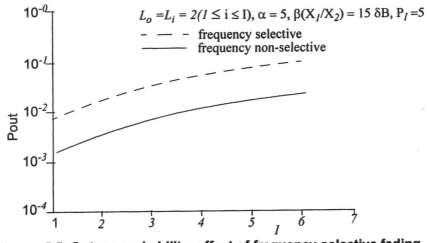

Figure 5.5 **Outage probability: effect of frequency selective fading**

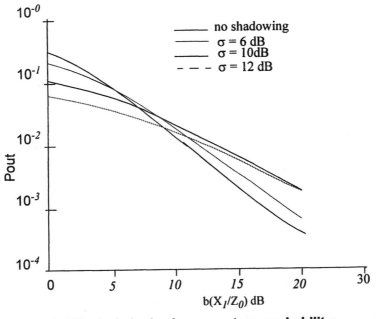

Figure 5.6 **Effect of shadowing on outage probability:**

5.4 Calculation of Reuse Distance

The co-channel reuse distance, U, is defined as the ratio of the distance between the centers of the nearest co-channel neighboring cells to the cell radius:

$$U = \frac{d_i + r_c}{r_c} \qquad (5.25)$$

where d_i is the distance from the mobile located at the edge of the cell to the interfering base station and r_c is the radius of the cell. We can use a propagation loss model, say in [4], to calculate the mean power of the received signal i.e.

$$P = \frac{C_p}{d^\alpha (1 + d/g)^\beta} \qquad (5.26)$$

where C_p is a constant, d is the distance from the transmitter, g is the breakpoint in the path loss curve, α is the basic attenuation rate and β is the attenuation rate beyond g, which is approximately 200 m. In large cells the distance of the mobile unit is much larger than the turning point, therefore using (5.26) we find the mean power of a received interference as

$$P_i = \frac{C_p g^\beta}{d_i^{\alpha + \beta}} \qquad (5.27)$$

then

$$\frac{P_d}{P_i} = (U - 1)^{\alpha + \beta} \qquad (5.28)$$

where P_d is the mean power of the received signal.

For microcells where r_c may be much greater than g,

$$\frac{P_d}{P_i} = (U - 1)^{\alpha + \beta} \left[\frac{r_c/g}{1 + r_c/g} \right]^\beta \qquad (5.29)$$

Equations (5.28) and (5.29) may be combined with (5.22), (5.23) and (5.24) to get outage probabilities versus the reuse distance U. These are shown in Figure 5.7 to Figure 5.9.

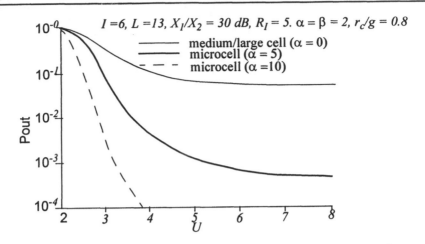

**Figure 5.7 Outage probability versus reuse distance: comparison
of microcell systems and medium/large cell systems**

It is seen that the outage probability decreases with an increase in reuse distance. When frequency selectively is present, analysis reported in [3] indicates that the outage probability becomes asymptotic to an irreducible outage, that is increasing the reuse distance does not decrease P_{out}.

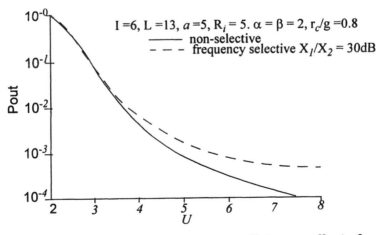

**Figure 5.8 Outage probability versus reuse distance: effect of
frequency selective multipath fading**

The outage probability in Figure 5.7 decreases with an increase in the reuse distance. However, a decrease in the outage probability becomes minimal with a further increase in the reuse distance; frequency selective multi-path fading is the probable cause of this. The effect of frequency selectivity is shown in Figure 5.8. In the absence of frequency selectivity the outage keeps decreasing with an increase in the reuse distance. On the other hand, the presence of frequency selectivity leads to irreducible outage.

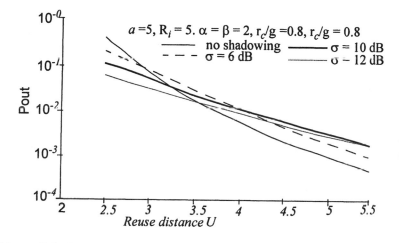

Figure 5.9 Outage probability versus distance: effect of shadowing

Figure 5.9 shows the outage probability versus the reuse distance for microcellular environment for different degree of shadowing. In order to achieve an acceptable outage probability the reuse distance should be increased beyond 4 for deeper shadowing. For lower reuse distance $U < 3$, the impact of shadowing is reversed.

5.5 Intermodulation Interference

In addition to co-channel and adjacent channel interference, cellular radio may be affected by intermodulation products if the frequencies within the coverage area are not assigned properly. The intermodulation interference may arise either at the receiver or at the transmitter. Unlike adjacent channel interference, intermodulation products cannot be removed by simple filtering. Receiver intermodulation products are produced under two conditions: (i) frequencies of interfering signals are integrally related to the desired frequency as indicated below, and (ii) the interfering signal has sufficient amplitude to produce an intermodulation product greater than the receiver threshold. Third and fifth order receiver intermodulation are produced for the following frequency relationships.

a) two signal third order $\quad F_0 = 2F_1 - F_2$;

b) three signal third order $\quad F_0 = F_1 - F_2 + F_3$;

c) two signal fifth order $\quad F_0 = 3F_1 - 2F_2$;

d) three signal fifth order $\quad F_0 = 2F_1 - 2F_2 + F_3$.

where F_1, F_2 and F_3 are the frequencies of interfering signals, and F_0 is the frequency of the desired signal. The FCC's Advisory Committee for Land Mobile Radio Services [5] rates the powers of the interfering signals as:

a) two signal third order product

$$IM = 2A + B + 10 - 60\log_{10}(dF);$$

b) three signal third order product

$$IM = A + B + C - 81\log_{10}(dF);$$

c) two signal fifth order product

$$IM = 2A + 3B - 57 - 135\text{Log}_{10}(dF);$$

d) Three signal fifth order product

$$IM = A + 2B + 2C - 132 - 195\log_{10}(dF),$$

where A, B, C are the received power from the interfering signals at frequencies F_1, F_2 and F_3 respectively. dF is the average of the several offsets from the received frequency.

The intermodulation products (IMP) are produced when the desired and interfering signals enter a nonlinear device such as a frequency mixer, therefore their effect can be reduced by making improvements to the linearity of the receiver stages. Proper frequency allocations within the cluster area also helps. There are other causes of intermodulation products, the most important being the intermodulation products produced at the transmitter. Transmitter intermodulation products are produced when a signal from one transmitter leaks into another transmitted having some nonlinearity. In such cases, the major sources of intermodulation products are AM-PM conversion through amplification, antenna mismatching, and interaction (coupling) between co-located antennas. These intermodulation products are radiated along with the desired signal produced at the transmitter locations. Thus, it becomes very difficult to separate the IMP from the desired signal it at the receiver. This may be responsible for performance degradation.

The main cause of AM-PM conversion distortion is power amplifier nonlinearities. Linearization of power amplifier will avoid this problem. The amplifier linearity can be achieved either by using a feedforward amplifier or linear amplification using nonlinear components.

　　In the feedforward amplifier technique the input signal is passed through two paths; one through the nonlinear amplifier and the other that provides a reference input signal. After an appropriate level adjustment, the reference signal is subtracted from the level adjusted power amplifier output. This provides an output consisting of intermodulation products. This output is then linearly amplified and subtracted from the power amplifier output to give an intermodulation free signal. The system is shown in Figure 8.10.

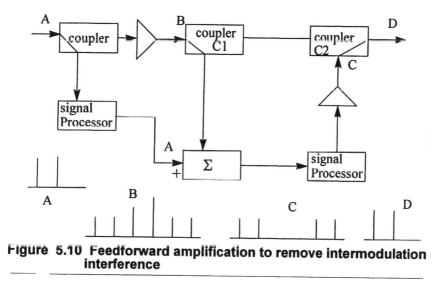

Figure 5.10 Feedforward amplification to remove intermodulation interference

　　The second technique of linear amplification using non-linear components (LINC) was proposed by Cox [6]. This technique has been used in several practical systems [7][8]. In this system the signal to be amplified is separated into envelope and phase components. The information is phase modulated in such a way that the phase angle is proportional to the inverse sine of the envelope voltage. The phase modulator output is split and its conjugate formed. The inverse sine modulated carrier and its conjugate are used along with the phase information derived from the LINC input signal to provide two constant envelope signals to a sub-system consisting of two nonlinear amplifiers and a combiner. This system is shown in Figure 5.11 and the circuit operation is given as follows.

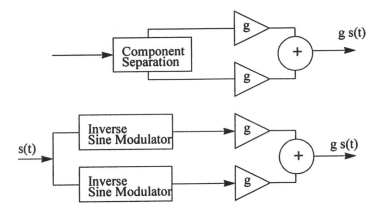

Figure 5.11 Linear amplification with non-linear components

$$s(t) = P\cos(\omega t + \theta)\sin(\psi(t))$$

$$= \frac{P}{2}\cos(\omega t + \theta)\sin(\psi(t)) + \frac{P}{2}\cos(\omega t + \theta)\sin(\psi(t))$$

$$= \frac{P}{2}\sin(\omega t + \theta)\cos(\psi(t)) - \frac{P}{2}\sin(\omega t + \theta)\cos(\psi(t)) \qquad (5.30)$$

$$= \frac{P}{2}\sin(\omega t + \theta + \psi(t)) - \frac{P}{2}\sin(\omega t + \theta - \psi(t))$$

From the block diagram it is evident that the performance of LINC depends on the inverse sine modulator, which is found to be sensitive to loop delays. This restricts the bandwidth of the signals which can be linearly amplified. It has been shown experimentally that for a 10 MHz bandwidth, distortion components in excess of 40dB below the desired signal components can be achieved. For wider bandwidth, automatic phase and gain equalization of the signals on the two branches are required.

5.6 Near-to-Far-End Ratio Interference

Consider two mobiles, A and B, not co-located but both transmit equal power at the same frequency. Consider distances of mobiles A and B from the base station be d_1 and d_2 $(d_2 > d_1)$ respectively. The signal received at the base station from mobile A will dominate that from B because of substantially lower distance related attenuation. With a fourth power propagation law, the ratio of the powers from the two mobiles, known as near-to-far-end ratio, is given by:

$$R_{rf} = \frac{d_2^4}{d_1^4} = \left(\frac{d_2}{d_1}\right)^2 \tag{5.31}$$

In cellular systems using FDM and TDM, a frequency or a time slot is not shared by another user in the cell, therefore near-to-far-end ratio interference is not a problem. Furthermore, the assignment of frequencies or time slots are based on the minimization of total interference. When the adjacent channel interference is involved, then R_{rf} is reduced by a factor equal to the IF filter attenuation of the adjacent channel. This is usually between 65 and 70 dB.

In systems such as direct sequence spread spectrum, where all the mobiles within a cell share the same carrier frequency, near-to-far-end ratio interference is of a major concern and efforts are made so that none of the signals arriving at the base station dominates the others. This essentially means that the transmit power of mobiles are controlled. To do so, the base station measures the powers from each mobile within the cell and orders every mobile to adjust its power.

Only recently, the near-far phenomenon in a fading environment received considerable attention. The power received over fading environment is continuously changing thereby requiring that the transmitter adjusts its power continuously to compensate for fading as well the distance. The power control in fading environment requires adjustments to the output power at a rate faster than the channel fade rate. Usually power control at both ends of the link is required. Several schemes to deliver accurate power control have been proposed; the detail on power control will be provided in Chapter 12. The near-far problem may be analyzed using techniques described in Section 5.5.3. It has been shown that under fading conditions, the degradation due to near end to far end interference in the absence of power control is not as severe as that under steady signal conditions. The near far problem in steady channel can be analyzed by considering that the mobiles are distributed in service area with a certain distribution. However, when fading is present, the probability of outage i.e. interference power > desired signal power, depends on the channel fading characteristics in addition to user distribution.

The near end to far end interference can be reduced by adequate frequency separation of desired and undesired signals. Such interference may also be reduced considerably by the attenuation slope of the IF filter. Given a required near end to far end interference ratio of, say, N dB, one can determine the necessary frequency separation. Let S be the frequency separation and X be the slope of the IF filter transfer function in dB/octave. For equal transmitted powers (P dBW) of the desired and the undesired transmitter situated at distances of d_1 and d_2 ($d_1 > d_2$), the near end to far end ratio is given by:

$$P_{rf} = \frac{10^{\frac{P}{10}}}{d_1^{-\alpha}} \left(10^{\frac{(P - x\log_2 s)}{10}} \frac{}{d_2^{-\alpha}} \right)^{-1}$$

(5.32)

or

$$P_{rf} = \left(\frac{d_2}{d_1} \right)^{\alpha} 10^{\frac{x\log_2 s}{10}}$$

(5.33)

The near end to far end ratio in dB is then given by

$$R_{rf} dB = x\log s + 10\alpha\log\frac{d_2}{d_1}$$

(5.34)

In (5.32) and (5.34), α is the propagation constant whose value may vary between 2 and 5.

EXAMPLE E-5.6.1

A base station receiver capable of providing adjacent channel isolation of 70 dB is receiving signals at two adjacent channels. The desired signal source is located 4 Km away. An interference margin of 10 dB is required. What is the minimum distance that a second mobile can transmit its signal without causing interference to the desired signal? Assume path loss model (i) $d^{3.5}$ (ii) $d^{4.2}$, where d is the distance between the transmitter and the receiver.

Solution

(i) Let p_o be the 1 meter intercept for both transmitters. $d_1 = 4000$ meters, IF filter isolation is 10^{-7}, and protection margin in 10. Thus, we can write

$$\frac{P_o}{(d_1)^{3.5}} = 10 \times \frac{P_o}{(d_2)^{3.5}} \times 10^{-7} \qquad \text{(E-5.6.1)}$$

Substituting for $d_1 = 4000$ meters, we solve for $d_2 = 150$ meters.
(ii) Using $a = 4.2$ instead of 3.5, we get $d_2 = 258$ meters.

5.7 Impact of Interference on Modulated System

In the previous sections, we presented a general methodology to evaluate outage probability due to co-channel interference in the presence of different channel fading and shadow conditions but in the absence of any modulation. The results obtained above are applicable in determining cell dimensions and frequency reuse patterns. The performance of a system from the users' perspective is related to the quality of reception. The quality of reception can be defined in several ways; output signal to noise ratio, outage probability, and error rate are possible measures. The transformation of input carrier to noise ratio to the output signal to noise ratio or symbol error rate takes place when the signal passes through the receiver. This will be discussed in the following sections.

5.7.1 Co-channel Interference in Analog Systems

In the following sections, we start with analog frequency modulation because of historical reasons as well as this case provides us with an insight into how channel impairments impacts the signal amplitude and phase of the receive signal. Since FM is no longer in predominant use for mobile communications, we shall treat this subject rather briefly.

5.7.1.1 Signal and Interference Conditions Steady Channel Conditions

Capture effect is an important characteristics of FM receiver; it is the ability of the FM receiver to suppress interference and noise provided signal to noise or signal to interference ratio is high. Since, in the shadowed environment, the mean signal level varies with lognormal statistics, the receiver will lose capture with a finite, though small, probability whenever the signal drops below the reception threshold. The loss of signal capture results in an output that is largely due to interference. To quantify this phenomenon, this section will examine the effect of co-channel interference on FM receiver output signal-to-interference ratio, first under steady state conditions and then in the presence of fading. Our analysis follow the method of Prabhu et.al. [9][10].

Consider the schematic on an FM receiver shown in Figure 5.12. The receiver consists of a linear front end, a down convertor to IF, a limiter/IF amplifier and a discriminator followed by a low pass filter. The input to the receiver is composed of a desired signal *s(t)* and an interference signal *i(t)*, and thermal noise *n(t)*. Both signal and interference are frequency or phase modulated,

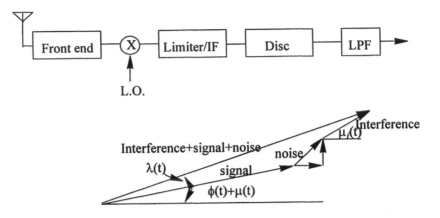

Figure 5.12 Schematic of an FM receiver and received signal phasor diagram

The two signals are represented by:

$$s(t) = \cos(\omega_c t + \phi(t) + \mu(t))$$
$$s(t) = Re[j(\omega_c(t) + \phi(t) + \mu(t))]$$

(5.35)

and

$$i(t) = R\cos(\omega_c t + \omega_d t + \phi_i(t) + \mu_i(t))$$
$$= Re[Rexp(j(\omega_c t + \omega_d t + \phi_i(t) + \mu_i(t)))]$$

(5.36)

where ω_c is the carrier frequency and interference lies ω_d radians from it; ω_d is usually small in case of co-channel interference. $\phi(t)$ and $\phi_i(t)$ are the phase modulation for the desired and interference signal respectively. $\mu(t)$ and $\mu_i(t)$ are arbitrary phases for the desired and interfering signals. R is the amplitude of the interfering signal relative to the desired signal, which is assumed to be unity.

The input to the receiver is

$$e(t) = s(t) + i(t) + n(t)$$

(5.37)

Ignoring noise, $n(t)$, we can write (5.37)

$$e(t) = Re[j(\omega_c(t) + \phi(t) + \mu(t)) + Rexp(j(\omega_c t + \omega_d t + \phi_i(t) + \mu_i(t)))] \quad (5.38)$$

$$e(t) = Re\{1 + Rexp(j(\omega_d(t) + \phi_i(t) + \mu_i(t) - \mu(t) - \phi(t)))\} \quad (5.39)$$
$$\{[exp(j(\phi(t) + \mu(t)))](exp(j\omega_c t))\}$$

The receiver is assumed to be linear from the antenna to the discriminator input, which is given by equation (5.38) with ω_c replaced by the IF frequency ω_{if}. The difference between the phases of the local oscillator frequency and the received carrier frequency is considered to remain constant. Equation (5.39) is then written as:

$$e(t) = Re\{1 + Rexp(j(\omega_d(t) + \phi_i(t) + \mu_i(t) - \mu(t) - \phi(t)))\} \quad (5.40)$$
$$\{[exp(j(\phi(t) + \mu(t)))](exp(j\omega_{if} t))\}$$

The function of PM receiver[†] is to detect the signal phase. Thus, it is convenient to write equation (5.40) in the form $Re\{K\ exp(jv)\}$ so that the receiver output is easily recognized and obtained. Thus, we can write

$$1 + Rexp(j(\omega_d(t) + \phi_i(t) - \phi(t) + \mu_i(t) - \mu(t))) = exp(x) \quad (5.41)$$

Then

$$x = ln(1 + Rexp(j(\omega_d(t) + \phi_i(t) - \phi(t) + \mu_i(t) - \mu(t)))) \quad (5.42)$$

Since $R < 1$, therefore $|R\ exp[j(\omega_d t + \phi_i(t) - \phi(t) + \mu_i(t) - \mu(t))]| < 1$ and we can use the relation

$$\lambda(t) = Im\{x\}$$
$$= \sum_{k=1}^{\infty} \frac{R^k}{k}(-1)^{k+1} \cos[\omega_d t + \phi_i(t) - \phi(t) + \mu_i - \mu] \quad (5.43)$$

Thus, in the output of the phase detector, the component due to interference will be proportional to $\lambda(t)$ in PM system and to its differential in FM. Without loss of generality we consider the constant of proportionality to be 1.

To estimate the output signal-to-interference ratio we could find the mean square value of $\lambda(t)$. However, since baseband filtering is usually involved in the final stage of any receiver, it may be more intuitive to first estimate the power spectrum of $\lambda(t)$. The output power spectral density of the

---

† *The FM receiver output is differential of the received phase angle.*

interference is then given by the product of the power spectrum of $\lambda(t)$ and magnitude square of the baseband filter transfer function. The power spectrum is given by the Fourier transform of the autocorrelation of $\lambda(t)$. We first average $\lambda(t)$ over $\mu_i(t)$ and $\mu(t)$ which are independent random processes with uniform distributions over the interval $[0, 2\pi]$. The autocorrelation function $\Re_\lambda(t)$ is given by:

$$\Re_\lambda(\tau) = \lim_{T \to \infty} \frac{1}{2T} \int_{-T}^{T} E[\lambda(t)\lambda(t+\tau)]_{\mu_i,\mu} dt \tag{5.44}$$

$$E[\lambda(t)\lambda(t+\tau)]_{\mu_i,\mu} = \sum_{k=1}^{\infty} \frac{1}{2k^2} R^{2k} \tag{5.45}$$

$$\cos[k(\omega_d t + \phi_i(t) - \phi_i(t+\tau) + \phi(t+\tau) - \phi(t))]$$

and $\Re_\lambda(t)$ is then written as

$$\Re_\lambda(\tau) = \sum_{k=1}^{\infty} \frac{1}{2k^2} R^{2k} \tag{5.46}$$

$$\lim_{T \to \infty} \frac{1}{2T} \int_{-T}^{T} \cos[k(\omega_d t + \phi_i(t) - \phi_i(t+\tau) + \phi(t+\tau) - \phi(t))] dt$$

By letting $k[\phi_i(t) - \phi(t)] = \psi(t)$, and $\omega_d = 0$ (co-channel interference case), we obtain

$$\Re_\lambda(\tau) = \sum_{k=1}^{\infty} \frac{1}{2k^2} R^{2k} \lim_{T \to \infty} \frac{1}{2T} \int_{-T}^{T} \cos[\psi(t+\tau) - \psi(t)] dt \tag{5.47}$$

where $\psi(t)$ is a random variable whose probability density function is to be determined. It is evident that the pdf of $\psi(t)$ depends on the type of modulation signal. We can consider a sinusoid phase modulated by a random process, $\psi(t)$ i.e., $A \exp(j[\omega_c t + \psi(t) + \mu(t)])$. The autocorrelation function, $\Re_\lambda(\tau)$ is given by [9],[10],[11],

$$\Re_\lambda(\tau) = \sum_{k=1}^{\infty} \frac{1}{2} \frac{R^{2k}}{k^2} \exp(-K_\psi(0) + K_\psi(\tau)) \tag{5.48}$$

If $K_\psi(0)$ and $K_\psi(\tau)$ are known, then autocorrelation function, $\Re_\lambda(\tau)$, and its power spectrum density $S_\lambda(f)$ can be found. This mathematics gets quite involved if $K_\psi(\tau)$ is chosen arbitrarily. Consider a relatively simple case where the modulation signal has a rectangular power spectrum similar to that of speech signals. This is shown in Figure 5.13. The idea is to find the impact of relative interference, R, on the output S/I, when the desired and the interference signals have similar power spectrum. In (5.48), the auto-correlation function of the sum of desired and interference signals, $K_\psi(\tau)$, is needed to estimate $S_\lambda(t)$.

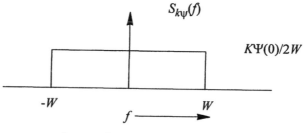

Figure 5.13 Power spectrum of modulation signal

Consider now an input, which is composed of signal and interference, each with a rectangular power spectrum of Figure 5.13. $K_\psi(0)$ is then the intensity of the input, and the power spectrum density is given by

$$S_{k\psi}(f) = \frac{K_\psi(0)}{2W}; \quad |f| < W$$
$$= 0; \quad elsewhere$$

(5.49)

and

$$K_\psi(\tau) = \int_{-\infty}^{\infty} S_{k\psi}(f)\exp(j2\pi f\tau)df = K_\psi(0)\frac{\sin(j2\pi W\tau)}{2\pi W\tau}$$

(5.50)

and $S_\lambda(f)$ is written as

$$S_\lambda(f) = \int_{-\infty}^{\infty} \Re_\lambda(\tau)\exp(-j2\pi f\tau)d\tau = \frac{1}{2}\sum_{k=1}^{\infty}\frac{R^{2k}}{k^2}S_k(f)$$

(5.51)

where

$$S_k(f) = \int_{-\infty}^{\infty} \exp(-K_\psi(0) + K_\psi(\tau)) \exp(-j2\pi f\tau) d\tau$$

$$= \int_{-\infty}^{\infty} \exp\left(-K_\psi(0)\left[1 - \frac{\sin 2\pi W\tau}{2\pi W\tau}\right]\right) \exp(-j2\pi f\tau) d\tau \tag{5.52}$$

With $2\pi W\tau = p$, and $\zeta = f/W$

$$S_k(f) = \frac{1}{2\pi W} \int_{-\infty}^{\infty} \exp\left(-K_\psi(0)\left[1 - \frac{\sin p}{p}\right]\right) \exp(-j\zeta p) dp \tag{5.53}$$

Expanding

$$exp\left[-K_\psi(0)\left(1 - \frac{\sin p}{p}\right)\right] \tag{5.54}$$

we obtain

$$\exp(-K_\psi(0)) \sum_{m=0}^{\infty} \frac{K_\psi(0)^m}{m!} \left(\frac{\sin p}{p}\right)^m \tag{5.55}$$

$S_k(\zeta)$ is given by:

$$S_k(\zeta) = \exp(-K_\psi(0))\left[\delta(f) + \frac{1}{2\pi W} \sum_{m=1}^{\infty} \frac{K_\psi(0)^m}{m!} \int_{-\infty}^{\infty} \left(\frac{\sin p}{p}\right)^m \exp(-j\zeta p) dp\right] \tag{5.56}$$

Consider terms within the integral

$$\int_{-\infty}^{\infty} \frac{\sin p}{p} \exp(-j\zeta p) dp = F_1(\zeta) = \pi; \quad |\zeta| < 1$$

$$= 0; \quad \text{otherwise}$$

$$= \pi[u_{-1}(\zeta + 1) - u_{-1}(\zeta - 1)] \tag{5.57}$$

where $U_{-1}(x)$ is the step function.
Also,

$$\int_{-\infty}^{\infty} \left(\frac{\sin p}{p}\right)^m \exp(-j\zeta p) dp$$

$$= \frac{m\pi}{2^{m-1}} \sum_{k=0}^{N} (-1)^k \frac{(|\zeta| + 1 - 2k)^{m-1}}{k!(m-k)}; \quad 0 \le |\zeta| < 1, m \ge 2$$

$$= 0; \quad \text{otherwise} \tag{5.58}$$

where

$$N = \text{INTEGER}\left\lceil \frac{l + |\zeta|}{2} \right\rceil \tag{5.59}$$

$\text{INTEGER}\lceil x \rceil$ is the largest integer not greater than x.
$F_m(\zeta)$ is continuous for $m \geq 2$ and is discontinuous at $\zeta = f/W = 1$ [9]. For large m, $F_m(\zeta)$ can be approximated as:

$$F_m(\zeta) = \left(\frac{6\pi}{m}\right)^{\frac{1}{2}} exp\left(-\frac{3\zeta^2}{2m}\right) \tag{5.60}$$

$$S_k(f) = exp(-K_\psi(0))\left[\delta(f) + \frac{1}{2\pi W} \cdot \sum_{m=1}^{\infty} \frac{K_\psi(0)^m}{m!} \cdot F_m\left(\frac{f}{W}\right)\right] \tag{5.61}$$

Substituting (5.58) in (5.54), we get an expression for $S_k(f)$. Recall that this is kth component of the spectrum $S_\lambda(f)$, as indicated in (5.49).

The power spectral density has two significant characteristics: an impulse at $\zeta = 0$ and a discontinuity at $\zeta=1$. The presence of an impulse indicates that worst case interference occurs at $f = 0$ and thus provides a lower bound on output C/I. The discontinuity suggested in (5.59) in the power spectrum density is due to shape of the signal power spectrum. The above equations have been derived for the a power spectrum density $S_{K\psi}(f)$ flat over $2W$.

Now, we extend the results to composite signal and interference from another FM transmission obtained by modulating a baseband signal having a power spectrum similar to that of desired signal. Assume that $\phi(t)$ and $\phi_i(t)$ are independent stationary Gaussian processes with flat two sided power spectra over the frequencies $-W$ to $+W$. The assumption that the modulation is a Gaussian process is only a rough approximation to the statistics of a single voice signal. The statistics of the speech may deviate from the Gaussian model, but with little impact on the results. Φ^2 and Φ_i^2 represent the mean square modulation index of the desired and the interference signal respectively. At the output of an FM receiver the power is proportional to the mean square modulation index Φ^2. Without loss of generality it can be assumed that the constant of proportionality is unity. Thus, the total power of the signal is equal to the mean square modulation index. The two-sided power spectrum density $S_\phi(f)$ is then given by:

$$S_\phi(f) = \frac{\Phi^2}{2W}; \quad |f| \le W \tag{5.62}$$

$$= 0; \quad \text{otherwise}$$

The power spectrum density of the interference signal is given by a similar expression. The receiver baseband output consists of signal and interference. Thus,

$$K_\psi(t) = \frac{1}{T \to \infty T} \int_{-T}^{T} k^2 \left(\Phi^2(t) + \Phi_i^2(t) \right) dt \tag{5.63}$$

We use the above result for the case when mean square modulation indices are equal i.e. $\Phi^2 = \Phi_i^2$. The result shows that the output co-channel interference power spectral density depends on the IF interference to desired signal power ratio R^2.

For $R<<1$, only the first term in equation (5.48) is significant, and $S_\lambda(f)$ is proportional to R^2. This proportionality is maintained even if R^2 approaches unity provided f/W is small. Also, it is seen that the baseband spectrum is quite flat in the range $|f|<W$; thus the total baseband signal-to-interference ratio is approximately equal to the ratio of their respective power spectrum densities. The worst case occurs at $f = 0$ where the interference is greatest because of the presence of an impulse.

When $R>1$ the receiver output will be heavily distorted since the receiver is then captured by the interference. The signal is heavily contaminated with noise, and is unintelligible or the listener will hear interference. In this case, the total receiver output may be considered to be made up of interference. The foregoing analysis can be applied to this case, as well that in which the roles of signal $\phi(t)$ and interference $\phi_i(t)$ are reversed, with R replaced by $1/R$. Thus for $R > 1$, the power spectrum density of the baseband interference is the sum of $S_\lambda(f)$ and $S_{\phi i}(f)'$. The interference power spectrum density is given by [11]:

$$S_I(f) = S_\lambda(f); \quad R < 1$$
$$= S_\lambda(f)' + S_{\phi i}(f); \quad R > 1 \tag{5.64}$$

where $S_\lambda(f)'$ is calculated by replacing R with $1/R$ and $\phi(t)$ with $\phi_i(t)$.

In Chapter 4, we observed that output power is proportional to the mean square modulation index. Thus, it is seen that under steady (non-fading) channel conditions, the signal to interference ratio is proportional to Φ^3. We also note that if both the interference and the desired signal are located at the same centre frequency, the maximum baseband interference occurs at zero frequency. If there is an offset between the two frequencies of $\pm kf_d$, then $S_k(f)$

is also shifted by the same amount, resulting in the presence of delta function, previously present at $f=0$, at $\pm kf_d$. When the shift is large compared to W, the interference peak will shift outside the baseband filter.

EXAMPLE E-5.7.1

(a) At a particular location in a cellular system, the mobile is receiving input carrier to co-channel interference power ratio $R^2 = 5$. Calculate baseband *SIR* when the modulation index of both the interference and desired signal is $\Phi = 2$ radians.

(b) The modulation index of the interfering signal is reduced to $\Phi_i = 1.4$. to improve *SIR*. Estimate the improvement in *SIR*.

<u>**Solution**</u>

(a) Substituting in the following equation, $r = 1/R = 0.4472$

$$\left(\frac{S}{I}\right)_{min} = \frac{\Phi^3}{r^2\sqrt{3/\left(8\pi\left[1+(\Phi_i/\Phi)^2\right]\right)}}, \qquad \Phi^2 + \Phi_i^2 > 2.35, r^2 < 1 \qquad (E\text{-}5.7.1)$$

we get 163.74 or 9.77dB.

(b) When the modulation index of the interfering signal is reduced to 1.4, the $(S/I)_{min}$ is calculated using (E-5.7.1) to be 499.33 or 27 dB. The result shows that by reducing the modulation index of the interference signal by a small amount substantial gain in $(S/I)_{min}$ is obtained.

5.7.2 Adjacent Channel Interference

In the previous section, it was shown that when $R^2 < 1$, and a frequency offset f_d exists between the desired and the interference signal, the peak of the baseband output interference also shifts by f_d, thus the technique of analyzing the problem of co-channel interference can also be used for adjacent channel interference analysis. When $R^2 > 1$, the analysis is more difficult in the case of adjacent channel interference but the output baseband interference may not play as critical role as that in case of co-channel interference because the adjacent channel interference is heavily attenuated between 60 to 70 dB by the IF filter. Thus, the condition of $R^2 > 1$ will not necessarily occur except under unusual conditions e.g. independent fading of the signal and interference. This condition occurs only on the reverse, mobile to base, channel if considered a single cell system. For cellular geometry both the forward and reverse links are affected. It is also noted that the probability of $R^2 > 1$ happening is very small. To handle the case of $R^2 > 1$, we need an alternate technique to estimate the output interference power. This requires a reasonable model to represent the interference centered at the adjacent channel frequency. A Gaussian noise model may be used for the composite adjacent interference (from several sources). In this case, the composite interference in the form of Gaussian noise is centered at the adjacent channel frequency f_n. To the FM receiver the adjacent channel interference can be taken as a heavily attenuated offset Gaussian noise. We have seen in Chapter 3 that, when the carrier frequency coincides with the receiver centre frequency, the baseband noise has three components with power spectrum densities given by [11]:

$$W(f) = exp(-2\gamma)W_N(f) + (1 - exp(-\gamma))^2 W_c(f) + (1 - exp(-\gamma))^2 W_{\theta'}(f) \quad (5.65)$$

where $W_N(f)$ is the baseband output power spectrum densities in the absence of a signal, $W_c(f)$ is the click noise spectrum, and $W_{\theta'}(f)$ is the baseband noise spectrum when the IF signal-to-noise ratio is large ($\gamma \gg 1$).

When the noise is centered at an frequency offset f_n, the instantaneous output of the discriminator whose function is to deliver an output proportional to the rate of change of the input phase angle between the centre frequency f_c and the offset $2\pi(f_n - f_c)$. Thus, expressions for the power spectrum densities of the three noise components are modified to include the effect of frequency offset. These are [11]:

$$W_N(f) = 4\pi^2 (f_n - f_c)^2 \delta(f) + \frac{8\pi B}{\sqrt{4.7}}$$

$$W_c(f) = 8\pi^2 (f_n - f_c)^2 exp(-\gamma) \qquad (5.66)$$

$$W_{\theta'}(f) = 2(2\pi f)^2 S_\lambda(f)$$

These components must be modified further to account for the bandpass characteristics of the IF filter. In other words, the adjacent channel interference components must be attenuated by the IF filter. Because of the presence of term $(f_n - f_c)^2$, adjacent channel interference would be worse than co-channel interference except for the effect of the IF filter, which attenuates the adjacent channel signals by as much as 70 dB. This large attenuation reduces the severity of the adjacent channel interference problem in steady channel. However, when the radio channel fades, the interference signal has a finite probability of exceeding the desired signal even after being heavily attenuation by the IF filter.

5.7.3 Effect of Fading on Co-channel Interference

In Section 5.7.1, we have seen that under steady channel conditions, signal to interference at the output of the receiver is proportional to the cube of the modulation index, therefore in the absence of fading the output signal-to-interference ratio can be improved simply by increasing the modulation index.

However, when the signal fades, the output S/I degrades appreciably. The question arises whether it is still possible to improve the situation by increasing the modulation index. We can use the quasi-static analysis technique similar to the one used in Section 4.2.3. to evaluate noise performance in fading environment.

We start with the assumption that the desired and the interfering signals undergo independent Rayleigh fading because of their propagation to the receiver via different propagation paths. On a short term basis, their relative strengths vary considerably, with the result that signal to interference ratio goes through large fluctuations. Thus, it is appropriate to determine the average signal to average interference ratio. Signal strength can be viewed as a random variable with a certain mean value but varying much slower than the observation time T, that is it can be assumed to remain constant over this time. The fluctuations around this mean may be modeled as a zero mean random variable known as signal suppression noise. The output signal power and interference power depend on the value of R^2, the interference to signal power ratio which continuously vary because of channel fading. The probability density function of R^2 is given by an F-distribution with two degrees of freedom [10]:

$$P_{R^2}(\alpha) = \frac{\Gamma}{(1+\Gamma\alpha)^2} \qquad (5.67)$$

where Γ is the ratio of the average power of the desired signal to the average power of the interfering signal at the IF stage. When the receiver is captured by the signal (i.e., the instantaneous value of R^2 is less than 1), full strength modulation appears at the baseband output. On the other hand when the instantaneous value of R^2 exceeds 1, the receiver being captured by the interference provides an output in the form of heavily distorted unintelligible signal which may be regarded as noise for all practical purposes.

The output of the receiver due to the desired signal may be obtained by taking an average over the statistics of R^2. This is given by:

$$\overline{\phi_0(t)} = \langle \phi_0(t) \rangle$$

$$= \phi(t) \int_0^1 P_{R^2}(\alpha) d\alpha \qquad (5.68)$$

$$= \frac{\Gamma \phi(t)}{1 + \Gamma}$$

The signal suppression noise, $n_\phi(t)$, is

$$n_\phi(t) = \phi_0(t) - \overline{\phi_0(t)}$$

$$= \frac{1}{1 + \Gamma} \phi(t); \quad R^2 < 1 \qquad (5.69)$$

$$= \frac{-\Gamma}{1 + \Gamma} \phi(t); \quad R^2 > 1$$

The baseband output interference power is made of two components - the signal suppression noise and the worst case interference given by (E-5.7.1). The averaged noise power is given by [11]:

$$\overline{N_\phi} = \int_0^1 \frac{\Gamma}{(1 + \Gamma\alpha)^2} \left\{ \alpha \sqrt{\frac{3}{8\pi\left(\Phi^2 + \Phi_i^2\right)}} + \frac{1}{(1 + \Gamma)^2} \Phi^2 \right\} d\alpha + \qquad (5.70)$$

$$\int_1^\infty \frac{\Gamma}{(1 + \Gamma\alpha)^2} \left\{ \Phi_i^2 + \frac{\Gamma^2}{(1 + \Gamma)^2} \Phi^2 \right\} d\alpha$$

which results in [11]

$$\overline{N_\phi} = \frac{1}{1 + \Gamma} \left\{ \left[\left(1 + \frac{1}{\Gamma}\right) \log(1 + \Gamma) - 1 \right] \times \sqrt{\frac{3}{8\pi\left(\Phi^2 + \Phi_i^2\right)}} + \Phi_i^2 + \frac{\Gamma\Phi^2}{1 + \Gamma} \right\} \qquad (5.71)$$

The output signal power is

$$\overline{\Phi^2} = \langle \phi(t)^2 \rangle = \frac{\Gamma^2}{(1 + \Gamma)^2} \Phi^2 \qquad (5.72)$$

The ratio of output signal power to averaged output interference shows that for different values of Γ the curves tend to remain flat expect for some advantage below $\Gamma = (\log_e \Gamma)^{1/3}$. This means that there is no advantage in increasing the modulation index in the presence of fading. In other words, the Γ^3 advantage in the absence of fading in lost when the signal is transmitted over fading environment.

EXAMPLE E-5.7.2

Given the parameters as in part (a) of Example E-5.7.1, with mobile system operating in a Rayleigh fading environment, what will be the output *SIR* if mean carrier to noise ratio is given to be 18 dB.

Solution

Dividing (5.72) by (5.71), we obtain an expression for output *SIR*,

$$\frac{\overline{\Phi^2}}{\overline{N_\phi}} = \frac{(\Gamma^2/(1+\Gamma)^2)\Phi^2}{\frac{1}{1+\Gamma}\left\{\left[\left(1+\frac{1}{\Gamma}\right)\log(1+\Gamma)-1\right]\times\sqrt{\frac{3}{8\pi\left(\Phi^2+\Phi_i^2\right)}}+\Phi_i^2+\frac{\Gamma\Phi^2}{1+\Gamma}\right\}} \qquad \text{(E-5.5.1)}$$

Substituting for $\Gamma = 63$ (18dB), $\Phi_i = \Phi = 2$, we get $SIR = 28.66$ of 14.57 dB.

5.7.4 ACI in Presence of Rayleigh Fading

Consider two signals located at adjacent frequencies transmitted from different base stations and received by a same mobile unit. It has been shown previously (Section 3.8) that the time delay spread of the two signals determines the correlation of the received amplitudes as a function of frequency separation. The probability density function of the signal to adjacent channel interference ratio ρ after the IF filter is given by:

$$P_\rho(\alpha) = \frac{d}{d\rho}P[\alpha > \rho]$$

$$= -\frac{d}{d\alpha}P[r_2 > ar_1]\frac{d\alpha}{d\rho} \qquad (5.73)$$

where $a = (\rho/G)^{1/2}$ and G is the relative power gain of the IF filter for the desired signal and the adjacent channel interferer. The value of G depends on the shape of the IF filter pass band. Using $P[r_2 \geq ar_1]$, the equation (5.73) can be written in an alternative form [11].

$$P_\rho(\alpha) = \frac{(1 - \lambda^2)\left(1 + \frac{\alpha}{G}\right)}{G\left[\left(1 - \frac{\alpha}{G}\right)^2 - 4\lambda^2 \frac{\alpha}{G}\right]^{\frac{3}{2}}}$$ (5.74)

where

$$\lambda^2 = \frac{J_0^2(\omega_m \tau)}{\sqrt{1 + s^2 \sigma^2}}$$ (5.75)

It is seen that the probability of adjacent channel interference depends on the relative gain G and the correlation coefficient between the desired and the interference envelopes. For a fixed relative filter gain, G, the interference decreases with a decrease in the correlation coefficient which in turn is related to frequency separation. For a fixed channel spacing it depends on the band-pass characteristics of the IF filter. A larger value of G (i.e. sharper bandpass characteristic of IF the filter) will result in smaller value of $p_p(\alpha)$, but also in a greater receiver cost. Relaxation in the filter tightness will result in greater susceptibility to adjacent channel interference. In cellular mobile systems where user equipment cost is of paramount importance, adjacent channel interference can be reduced more effectively by an appropriate frequency allocation scheme.

To analyze the impact of adjacent channel interference on the output-signal-to-interference ratio, the quasi static approach can also be used. As noted in Chapter 4, the fluctuating signal at the output is divided into an average signal and a signal suppression noise. These are given by [11]:

$$\overline{\phi(t)} = \int_0^\infty P_p(\alpha)(1 - \exp(-\alpha))\phi(t)d\alpha$$ (5.76)

$$= \phi(t)[1 - f(G, \lambda)]$$

where

$$f(G, \lambda) = (1 - \lambda^2) \int_0^\infty \frac{\exp(-Gx)(1 + \alpha)}{[(1 + x)^2 - 4\lambda^2 x]^{3/2}} d\alpha$$ (5.77)

The output signal power is:

$$S = \langle \phi(t)^2 \rangle = \Phi^2 [1 - f(G, \lambda)]^2 \tag{5.78}$$

Signal suppression noise $\phi_{ns}(t)$ is given by [11]:

$$n_\phi(t) = (1 - exp(-\rho))\phi(t) - \overline{\phi(t)} = [f(G, \lambda) - exp(-\gamma)]\phi(t) \tag{5.79}$$

Signal suppression noise power is given by:

$$N_{s\phi}(\rho) = \Phi^2 [f(G, \lambda) - exp(-\gamma)]^2 \tag{5.80}$$

The total output interference power, $N_\phi(\rho)$, for a given carrier to interference ratio, ρ, may be written as:

$$N_\phi(\rho) = \Phi^2 [f(G, \lambda) - exp(-\gamma)]^2 + \int_{0.1W}^{W} \frac{W(f)}{4\pi^2 f^2} df \tag{5.81}$$

The second term in (5.81) represents interference power due to random FM. The signal-to-noise plus interference ratio in the presence of adjacent channel interference is obtained by averaging $N_\phi(\gamma)$ over the statistics of γ, the carrier to interference ratio [11].

$$\overline{N_\phi} = \int_0^\infty N_\phi(\gamma) p_\gamma(\gamma) d\rho$$

$$= \int_0^\infty \frac{(1 - \lambda^2)\left(1 + \frac{\gamma}{G}\right) N_\phi(\gamma)}{G\left[\left(1 + \frac{\gamma}{G}\right)^2 - 4\lambda^2 \frac{\gamma}{G}\right]^{3/2}} d\gamma \tag{5.82}$$

$$\frac{\overline{S}}{\overline{N_\phi}} = \frac{\Phi^2 [1 - f(G, \lambda)]^2}{\int_0^\infty \frac{(1 - \lambda^2)\left(1 + \frac{\gamma}{G}\right) p_\gamma(\gamma)}{\sqrt{G\left[\left(1 + \frac{\gamma}{G}\right)^2 - 4\lambda^2 \frac{\gamma}{G}\right]^{3/2}}} d\gamma + \int_{0.1W}^{W} \frac{W(f)}{4\pi^2 f^2} df} \tag{5.83}$$

If the interference and the desired signal were approximately equal before IF filtering and the radio channels (for signal and interference) were steady, the combined effect of IF filtering and demodulation will considerably reduce the effect of the interference. However, in the presence of fading there is a finite probability that the interfering signal is greater than the desired signal ($\rho < 1$) with a result that output interference is higher than that predicted under steady state conditions. Thus, resulting average signal-to-noise ratio may be lower than that expected under steady signal conditions.

5.8 Digital Transmission in Interference

It is recognized that mobile radio system performance is often limited by the presence of interference, therefore system performance in the presence of interference is an important topic of investigation. In the past, however, since mobile radio systems have been primarily analog, digital systems performance in interference infested mobile radio channel has received relatively little attention. Substantial research has been done on phase modulated systems over stationary channels [12]-[15]. There have been some investigations of packet errors when identical signals are transmitted from more than two transmitters, i.e. simulcast systems for wide area paging [16][17]. The performance of phase shift keying (PSK) and differential phase shift keying (DPSK) systems in the presence of co-channel interference over steady channel has been discussed by Rosenbaum [18][19]. System performance depends on the number of interfering signals; it appears that an upper bound of error probability is obtained when composite interference from several sources is modeled as Gaussian noise.

The effect of fading on digital transmission in the presence of interference is difficult to model accurately. Arredondo [16] analyzed radio paging errors in multiple transmitter mobile systems. French [17] measured error probabilities for simulcast system and concluded that the probability of error depends on the frequency difference and time delays between the simulcast transmitters. In the following subsections, we analyze performance of several digital transmission schemes suitable for use in mobile systems; particular emphasis will be placed on two widely used modulation schemes, GMSK and $\pi/4$-QDPSK.

5.8.1 Performance in Interference and Fading

Fading of the desired signal and interference represents a typical mobile environment. The desired signal may be represented by either Rayleigh or Ricean distribution, the interference signals also fade with either of these two distributions. Other types of distributions are not considered here, though Nakagami distribution is a good candidate. Although multiple interferers beset the desired signal, we consider that interferers form an equivalent interfering signal.

In Section 5.3, a general methodology was presented to find distributions of signal to interference ratio for several mobile radio scenarios. The bit error rate is obtained by averaging the bit error rate in the presence of interference over the statistics of the interference. The average bit error rate for DPSK modulation under the condition that the desired signal fades with Ricean statistics and the interfering signals have identical Rayleigh statistics. In the case when all interfering signals have the same average power and fade independently Gamma pdf of the form [32]

$$p_Y(y) = \frac{y^{I-1}}{Y_1^I(I-1)} \exp\left(-\frac{y}{Y_1}\right) \tag{5.84}$$

The desired signal fades with Ricean statistics, i.e.

$$p_X(x) = \frac{1}{X}\exp\left(-\frac{2x+s^2}{2X}\right) I_0\left(\frac{\sqrt{2xs}}{X}\right) \tag{5.85}$$

The pdf of the signal to interference ratio, $r = x/y$, is found as [33]

$$p_R(r) = \int_0^\infty y p_x(ry) p_y(y) dy \tag{5.86}$$

$$= \frac{1}{b_1}\left(\frac{b_1}{r+b_i}\right)^{I+1} \exp\left(-\frac{ab_1}{r+b_1}\right) \sum_{i=0}^{I} C_I^{I-i}\frac{1}{i!}\left(\frac{ar}{r+b_i}\right)^i$$

where $b_1 = X/Y_1$, $a = s^2/(2X)$ and C_I^{I-i} is a binomial coefficient, i.e. $I!/[i!(I-i)!]$. When all interferers have mutually different mean powers, i.e. $Y_i \neq Y_j$ when $i \neq j$, the pdf of signal to noise ratio, r, is given by [32]

$$p_r(r) = \sum_{i=1}^{I} \frac{b_i}{(r+b_i)^2}\left(1+\frac{ar}{r+b_i}\right)\exp\left(-\frac{ab_i}{r+b_i}\right) \prod_{\substack{j=1 \\ j \neq i}}^{I} \frac{b_j}{b_j-b_i} \tag{5.87}$$

where $b_i = X/Y_i$. The bit error probability when all interferers have the same mean power is given by

$$P_e = \frac{1}{2} \int_0^\infty \exp(-\beta r) p_r(r) dr$$

(5.88)

$$= \frac{I}{2b_i} \int_0^\infty \left(\frac{b_1}{r+b_1} \right)^{I+1} \exp\left[-\left(\beta r + \frac{b_1}{r+b_1} \right) \right] \sum_{i=0}^{I} C_I^{I-i} \frac{1}{i!} \left(\frac{ar}{r+b_i} \right)^i dr$$

where $\beta = \frac{1}{2}$ for NCFSK and $\beta = 1$ for DPSK.
When $a = 0$, the desired and cochannel interferers fade with Rayleigh statistics, (5.88) is simplified as

$$P_e = \frac{1}{2(I-1)!} \sum_{k=1}^{I} (k-1)!(-b_1\beta)^{I-k} - \frac{(-b_1\beta)^I}{2(I-1)!} \exp(b_1\beta) E_i(-b_1\beta)$$

(5.89)

where $Ei(.)$ is an exponential integral function defined as

$$E_i(z) = -\int_{-z}^\infty \frac{\exp(-t)}{t} dt$$

(5.90)

for $z<0$.
When all interferers have mutually different powers, the bit error probability for microcellular radio systems are found to be

$$P_e = \frac{1}{2} \sum_{i=1}^{I} \int_0^\infty \frac{b_i}{(r+b_i)^2} \left(1 + \frac{ar}{r+b_i} \right) \exp\left[-\left(\beta r + \frac{ab_i}{r+b_i} \right) \right] dr \prod_{\substack{j=1 \\ j \neq i}}^{I} \frac{b_j}{b_j - b_i}$$

(5.91)

In a special case with $a=0$, the error probabilities are given by

$$P_e = \frac{1}{2} \sum_{i=1}^{I} b_i\beta \exp(-b_i\beta) E_i(-b_i\beta) \prod_{\substack{j=1 \\ j \neq i}}^{I} \frac{b_j}{b_j - b_i}$$

(5.92)

5.8.2 GMSK Performance in Interference and Fading

GMSK modulated signals are extensively used in mobile radio systems. The performance of GMSK has been analyzed by a number of researchers under different channel conditions. Murota and Hirade [34] evaluated the performance of GMSK experimentally. Elnoubi [35] analyzed GMSK with discriminator detection in presence of fast fading. Varshney and Kumar analyzed GMSK in the presence of channel impairments like fast fading, co-channel interference, and frequency selective fading. The authors claimed to have developed closed form expression for BER, but the BER expression is not easy to apply [36].

The GMSK modulator output is given by

$$s_d(t) = \cos(2\pi f_c t + \phi_s(t)) \tag{5.93}$$

where

$$\phi_s(t) = \int_{-\infty}^{t} 2\pi f_d \sum_{n=-\infty}^{\infty} a_n g(v - nT) dv \tag{5.94}$$

where f_d is the phase deviation constant. The phase derivative, $\phi'_s(t)$ is given by

$$\phi'_s(t) = 2\pi f_d \sum_{n=-\infty}^{\infty} a_n g(t - nT) \tag{5.95}$$

For bit interval centered at $t = 0$ and when mark or space is transmitted in this interval, $\phi'_s(0)$ is written as

$$\phi'_s(0) = 2\pi f_d \left[\pm g(0) + \sum_{\substack{n=-\infty \\ n \neq 0}}^{\infty} a_n g(t - nT) \right] \tag{5.96}$$

The term under the summation represents ISI, and only few of these terms may be taken into consideration. The signal is contaminated by co-channel interferer and is represented similar to (5.93) with d replaced by c. The input to the discriminator is given by

$$e(t) = s_d(t) + s_d(t - \tau) + s_c(t) + n(t) \tag{5.97}$$

where

$$n(t) = x_n(t)\cos\omega_c t - y_n(t)\sin\omega_c t \tag{5.98}$$

and $x_n(t)$ and $y_n(t)$ are low pass independent zero mean Gaussian processes. Equation (5.97) can also be written in quadrature form as

$$e(t) = X(t)\cos\omega_c t - Y(t)\sin\omega_c t$$
$$= R(t)\cos(\omega_c t + \Psi(t)) \tag{5.99}$$

The probability density function of the limiter-discriminator output, $\Psi''(t)$ can be written as [36]

$$P(\Psi'') = \frac{1}{2}\left(\frac{\sigma_2}{\sigma_1}\right)^2 (1-\rho^2)\left[\Psi''^{-2} - \frac{2\sigma_2\rho\Psi''}{\sigma_1} + \left(\frac{\sigma_2}{\sigma_1}\right)^2\right]^{-3/2} \tag{5.100}$$

where

$$\sigma_1^2 = \sigma_s^2 + \sigma_s^2 + \sigma_n^2 + \sigma_{s(t-\tau)}^2$$

$$\sigma_2^2 = \sigma_s^2(\sqrt{2}\pi f_D)^2 + \sigma_c^2(\sqrt{2}\pi f_D)^2 + \sigma_{s(t-\tau)}^2(\sqrt{2}\pi f_D)^2 +$$
$$\sigma_s^2\phi_s'^2(t) + \sigma_c^2\phi_c'^2(t) + \sigma_{s(t-\tau)}^2\phi_{s(t-\tau)}'^2(t) + \frac{\sigma_n^2\pi^2 B^2}{3} \tag{5.101}$$

$$\rho = \frac{\sigma_s^2\phi_s'^2(0) + \sigma_c^2\phi_c'^2(0) + \sigma_{s(t-\tau)}^2\phi_{s(t-\tau)}'^2(0)}{\sigma_1\sigma_2}$$

The probability of error depends on the data sequence of the desired signal and that of interferer. The probability of error for the zeroth signalling interval $(a_0 = +1$ or $-1)$ is given by [36]

$$P_{e,1}(a_1, ..., b_0, b_1, b_{-1}, ..., t_c) = \tag{5.102}$$

$$\int_{-\infty}^{0} P(\Psi''|a_0 = 1, a_1, a_{-1}, a_2, ..., b_0, ..., c_t)d\Psi''|$$

$$P_{e,-1}(a_1, ..., b_0, b_1, b_{-1}, ..., t_c) = \qquad (5.103)$$

$$0$$
$$\int_{-\infty} P(\Psi'|a_0 = -1, a_1, a_{-1}, a_2, ..., b_0, ..., c_t)d\Psi'|$$

The average probability or error is

$$P_e - \frac{1}{2}\overline{P_{e,1}(a_1, ..., b_0, b_1, b_{-1}, ..., t_c)} + \qquad (5.104)$$
$$\frac{1}{2}\overline{P_{e,-1}(a_1, ..., b_0, b_1, b_{-1}, ..., t_c)}$$

Thus, the probability of error depends on the evaluation of the integrals in (5.102) and (5.103). For more details, the reader is referred to [35] and [36].

5.8.3 π/4-QDPSK Performance in Interference and Fading

In this section, we analyze π/4-QDPSK modulation, which was introduced in Chapter 4, and evaluate its performance in the presence of typical mobile channel impairment i.e. fading, shadowing and co-channel interference. Most of this material is taken from [37]. This modulation scheme is chosen since it is the most promising among multi-level phase modulation schemes on the basis of SNR and spectrum efficiency [20][21]. In coherent QPSK, each symbol conveys two binary bits and is encoded into four possible phases of the modulated carrier, located at equidistant from each other in the complex plane. In differential QPSK, the information symbol resides in the phase difference $\Delta\phi_n$ of two consecutive modulated carrier phases. In QDPSK, the set of phase difference is

$$\phi = \frac{\pi(2n-1)}{4}; \qquad n = -1, 0, 1, 2 \qquad (5.105)$$

A possible mapping between the set of symbols and the set of phase differences for the four symbols is π/4, 3π/4, -3π/4, -π/4 for symbols 0, 1, 2, and 3 respectively. Assuming zero initial carrier phase, the subsequent phases will take their values from the set {0, ±π/4, ±π/2, ±3π/4, π}. Figure 5.14 shows such phase transitions.

5.8.4 Error Bounds for π/4-QDPSK

Figure 5.15 shows a block diagram of QDPSK system operating in the presence of noise, fading and interference. The transmitted phase $\phi_s(t)$ has the form

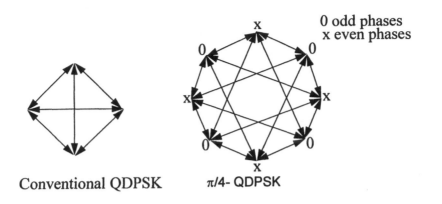

Conventional QDPSK   π/4- QDPSK

Figure 5.14 Signal Space diagram of conventional and π/4-QDPSK

$$\phi_s(t) = \sum_{i = -\infty}^{\infty} \frac{n\pi}{4} p(t - iT) \qquad (5.106)$$

with $n \in \{0,1,...,7\}$ and $p(t)$ a rectangular unit pulse of duration T.

The signal $s(t)$ is distorted by the channel fading, interference and additive noise. The input $e(t)$ to the BPF at the receiver is given by

$$e(t) = s(t) + i(t) + n(t) \qquad (5.107)$$

We analyze the system by detecting signals on the in-phase and quadrature branches and then combining the outputs into symbols. The detection block diagram of QDPSK is shown in Figure 5.16. Let E_I and E_Q be the events that the samplers make erroneous decisions on channel I and Q respectively. The probability of symbol error, denoted P_s, can be expressed as $P_s = Prob(E_I \cup E_Q)$. For any two events E_I and E_Q, the following inequalities hold

$$max(P(E_I), P(E_Q)) \le P_s = P(E_I \cup E_Q) \le P(E_I) + P(E_Q) \qquad (5.108)$$

Assume that the probabilities of the bit error on each channel I and Q are the same, i.e. $P(E_I) = P(E_Q)$, then (5.108) becomes

$$P_L = P(E_I) = P(E_Q) \le P_s = P(E_I \cup E_Q) \le 2P_L = P_U \qquad (5.109)$$

P_L is thus the lower bound bit error rate and the upper bound P_u is twice of P_L. Since the two bounds are related, only one bound, for instance the lower bound P_L needs to be evaluated. Referring to Figure 5.17, P_L is given by

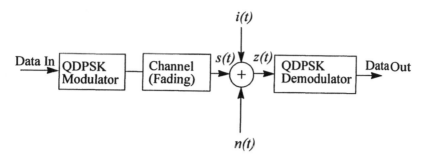

Figure 5.15 Block diagram of a QDPSK system in AWGN

$$P_L = \frac{1}{2}Prob\left[P_L| \pm \frac{\pi}{4}\right] + \frac{1}{2}Prob\left[P_L| \pm \frac{3\pi}{4}\right]$$

$$= \frac{1}{2}Prob[q_I < 0,\ \cos\Delta\phi_s > 0] + \frac{1}{2}Prob[q_I > 0,\ \cos\Delta\phi_s < 0]$$

(5.110)

where q_I is the output of channel I. Now we consider transmission over a Rayleigh fading channel.

Using the lowpass complex envelope representation, the components in (5.107) can be written as:

$$s(t) = Re\{z_s(t)exp(j(\omega_c t + \phi_s(t)))\}$$

$$i(t) = Re\{z_i(t)exp(j(\omega_c t + \phi_i(t)))\}$$

$$n(t) = Re\{z_n(t)exp(j\omega_c t)\}$$

(5.111)

Here, $\phi_i(t)$ is the phase determined by the data sequence of the interference signal $i(t)$ while $\phi_s(t)$ is for the desired signal $s(t)$. $z_s(t)$, $z_i(t)$ and $n(t)$ are zero mean Gaussian processes. The output of the BPF is given by

$$e(t) = Re[z_1(t)exp(j\omega_c t)]$$

(5.112)

where

$$z_1(t) = z_s(t)e^{j\phi_s(t)} + z_i(t)e^{j\phi_i(t)} + z_n(t)$$

(5.113)

If ω_c and T are selected so as to satisfy $\omega_c T = 2n\pi$ (n an integer), the one symbol time delayed signal is written as

$$e(t-T) = Re[z_1(t-T)exp(j\omega_c(t-T))]$$

$$= Re[z_1(t-T)exp(j\omega_c t)] \tag{5.114}$$

$$= Re[z_2(t)exp(j\omega_c t)]$$

where $z_2(t) = z_1(t-T)$. The LPF output of *I*-channel detector $q_I(t)$ is expressed as a function of $z_1(t)$ and $z_2(t)$[22]

$$q_I(t) = \frac{1}{2}Re[z_1(t)z_2{}^*(t)]$$

$$= \frac{1}{4}Re[z_1{}^*(t)z_2(t) + z_1(t)z_2{}^*(t)] \tag{5.115}$$

At sampling instants $t = nT$, (5.115) becomes

$$q_I(t) = \frac{1}{4}Re[z_1{}^*(nT)z_2(nT) + z_1(nT)z_2{}^*(nT)] \tag{5.116}$$

We define two Gaussian vectors *A* and *Z* as

$$A' = \begin{bmatrix} z_s & z_i & z_n \end{bmatrix} \tag{5.117}$$

and

$$Z' = \begin{bmatrix} z_1 & z_2 \end{bmatrix} \tag{5.118}$$

Their covariance matrices K_A and K_Z are computed as:

$$K_A = \frac{1}{2}\langle (A - \langle A \rangle)^*(A - \langle A \rangle) \rangle$$

$$K_A = \begin{bmatrix} \sigma_s^2 & 0 & 0 \\ 0 & \sigma_i^2 & 0 \\ 0 & 0 & \sigma_n^2 \end{bmatrix} \tag{5.119}$$

$$K_z = \begin{bmatrix} \sigma^2 & \sigma^2\rho^* \\ \sigma^2\rho & \sigma^2 \end{bmatrix}$$

where $\sigma_s^2 = 1/2 \langle z_s.z_s{}^* \rangle$, $\sigma_i^2 = 1/2 \langle z_i.z_i{}^* \rangle$, $\sigma_n^2 = 1/2 \langle z_n.z_n{}^* \rangle$, $\sigma^2 = \sigma_s^2 + \sigma_i^2 + \sigma_n^2$, and $\sigma^2\rho = \sigma_s^2\rho_s(T) \ e^{j\Delta\phi_s} + \sigma_i^2\rho_i(T) \langle e^{j\Delta i} \rangle + \sigma_n^2\rho_n(T)$. Here

$\rho_s(T), \rho_i(T), \rho_n(T)$ are the correlation coefficient functions of the processes $z_s(t)$, $z_i(t)$, $z_n(t)$ respectively. $\Delta\phi_s$, $\Delta\phi_i$ represent the phase differences between the two consecutive samples, i.e. $\Delta\phi_d = \phi_d(nT) - \phi_d((n-1)T)$; $(d=s,i)$.

Assume the interference signal is another $\pi/4$-QDPSK modulated signal with equally probable symbols, then $<\exp(j\Delta\phi_i(T))> = 0$ holds. Therefore

$$\sigma^2 \rho = \sigma_s^2 \rho_s(T) exp(j\Delta\phi_s) + \sigma_n^2 \rho_n(T) \tag{5.120}$$

We transform Z into $U' = [U_1 \; U_2]$ by the transformation $U = LZ$, where

$$L = \begin{bmatrix} 1 & 1 \\ 1 & -1 \end{bmatrix} \tag{5.121}$$

The covariance matrix U is

$$K_U = L K_z L^T$$

$$= \begin{bmatrix} 2\sigma^2 + \sigma^2 \rho + \sigma^2 \rho* & \sigma^2 \rho - \sigma^2 \rho* \\ \sigma^2 \rho* - \sigma^2 \rho & 2\sigma^2 - \sigma^2 \rho - \sigma^2 \rho* \end{bmatrix} \tag{5.122}$$

We can write (5.116) as

$$q_I = \frac{1}{8}(U_1 U_1* - U_2 U_2*) \tag{5.123}$$

The joint pdf $p(U_1, U_2)$ can be obtained as [23]

$$p(U_1, U_2) = \frac{1}{(2\pi)^2 4\sigma^4 (1 - |\rho|^2)} \exp\left\{ -\frac{1}{8\sigma^4 (1 - |\rho|^2)} \right. \tag{5.124}$$

$$\left. \frac{1}{4\sigma^4 (1 - |\rho|^2)}[K_{11} U_1 U_1* + K_{22} U_2 U_2* + K_{21}(U_1* U_2 - U_1 U_2*)] \right\}$$

where

$$K_{11} = 2\sigma^2 (1 + r_{re})$$

$$K_{22} = 2\sigma^2 (1 - r_{re}) \tag{5.125}$$

$$K_{21} = -2j\sigma^2 \rho_{im}$$

Equation (5.124) can transformed in polar coordinates, using $U_1 = R_1 \exp(j\theta_1)$ and $U_2 = \exp(j\theta_2)$, and is written as

$$p(R_1, \theta_1, R_2, \theta_2) = \frac{R_1 R_2}{(2\pi)^2 4\sigma^4 (1-|\rho|^2)} \exp\left\{-\frac{1}{8\sigma^4(1-|\rho|^2)}\right. \tag{5.126}$$

$$\left. \frac{1}{4\sigma^4(1-|\rho|^2)}\left[K_{11}R_1^2 + K_{22}R_2^2 - 2jK_{21}\sin(\theta_1 - \theta_2)\right]\right\}$$

The joint pdf of R_1 and R_2, is thus

$$p(R_1, R_2) = \frac{R_1 R_2}{4\sigma^4(1-|\rho|^2)} I_0\left(\frac{2\rho_{im}R_1 R_2}{4\sigma^4(1-|\rho|^2)}\right) \tag{5.127}$$

$$\exp\left(-\frac{1}{4\sigma^4(1-|\rho|^2)}\left[(1-\rho_{re})R_1^2 + (1+\rho_{re})R_2^2\right]\right); (R_1 \le \infty, R_2 \le \infty)$$

where ρ_{re} and ρ_{im} are real and imaginary parts of ρ. We can write q_l in terms of R_1 and R_2 as $q_l = (1/8)(R_1^2 + R_2^2)$, and $\text{Prob}[q_l < 0]$ and $\text{Prob}[q_l > 0]$ as

$$\text{Prob}[q_l < 0] = \text{Prob}[R_1 < R_2] = \frac{1}{2}\left(1 - \frac{\rho_{re}}{\sqrt{1-\rho_{im}^2}}\right)$$

$$\tag{5.128}$$

$$\text{Prob}[q_l > 0] = \frac{1}{2}\left(1 + \frac{\rho_{re}}{\sqrt{1-\rho_{im}^2}}\right)$$

The lower bound on probability of error is given by

$$P_L = \frac{1}{4}\left\{2 - \frac{1}{2}\left[\frac{\rho_{re}(\pi/4)}{\sqrt{1-\rho_{im}^2(\pi/4)}} + \frac{\rho_{re}(\pi/4)}{\sqrt{1+\rho_{im}^2(\pi/4)}} - \right.\right. \tag{5.129}$$

$$\left.\left. \frac{\rho_{re}(3\pi/4)}{\sqrt{1-\rho_{im}^2(3\pi/4)}} - \frac{\rho_{re}(3\pi/4)}{\sqrt{1+\rho_{im}^2(3\pi/4)}}\right]\right\}$$

In multipath fading environment $\rho_s(\tau)$ is

$$\rho_s(\tau) = J_0(2\pi f_m T) \tag{5.130}$$

and for uncorrelated noise samples using $\rho_n(T) = 0$, we have

$$P_L = \frac{1}{2}\left[1 - \frac{\sigma_s^2 J_0(2\pi f_m T)}{\sqrt{2\sigma^4 - \sigma_s^4 J_0^2(2\pi f_m T)}}\right] \qquad (5.131)$$

where $\sigma^2 = \sigma_s^2 + \sigma_i^2 + \sigma_n^2$. Defining γ the average CNR, and Λ the average CIR

$$\Gamma = \frac{\sigma_s^2}{\sigma_n^2}, \qquad \Lambda = \frac{\sigma_s^2}{\sigma_i^2} \qquad (5.132)$$

then (5.131) can be written as

$$P_L = \frac{1}{2}(1 - \Gamma\Phi^{-1/2}J_0(2\pi f_m T)) \qquad (5.133)$$

where

$$\Phi = 2\left(\Gamma + \frac{\Gamma}{\Lambda} + 1\right)^2 - \Gamma^2 J_0^2(2\pi f_m T) \qquad (5.134)$$

Equation (5.17) is a general result which is applicable to fading channels subjected to interference. When co-channel interference is not present, i.e. $\Lambda \rightarrow \infty$, (5.134) may be rewritten as

$$P_L = \frac{1}{2}(1 - \Gamma\tilde{\Phi}^{-1/2}J_0(2\pi f_m T)) \qquad (5.135)$$

where

$$\tilde{\Phi} = 2(\Gamma + 1)^2 - \Gamma^2 J_0^2(2\pi f_m T) \qquad (5.136)$$

P_L from (5.136), in the absence interference, are plotted for two values of $f_m T = 0.01$ and 0.001 and shown in Figure 5.16. It is shown that the system degrades considerably under fading conditions. For example, an additional 12 dB is needed to maintain SER at 10^{-2} at a fading rate of 0.001. The degradation is more at higher carrier to noise ratios. The presence of fast fading results in an irreducible error rate, i.e. the lowest error rate achievable at infinite CNR. This error rate worsens as the fade rate increases. For an order of magnitude increase in the fade rate, approximately two order of magnitude increase in the irreducible error rate of is observed.

When faster varying signal envelope is Rayleigh faded with a mean having lognormal distribution, the mean power of $z_s(t)$ (measured over relatively short distances) is no longer constant over greater distances. The quantity σ_s^2 thus becomes a random variable having lognormal distribution. To

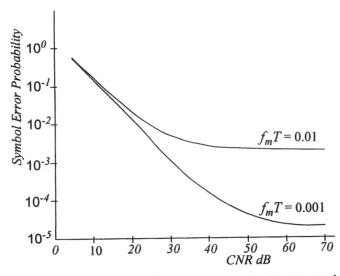

Figure 5.16 Theoretical bounds for symbol error probability of π/4-QDPSK in a Rayleigh fading environment

evaluate performance in the presence of shadowing pdfs of $\gamma(t)=\sigma_s^2/\sigma_n^2$ and $\delta(t) = \sigma_i^2/\sigma_n^2$ are required. These are derived to be

$$p(\gamma) = \frac{\log e}{\gamma\sqrt{2\pi}\sigma}\exp\left(-\frac{(10\log\gamma - 10\log\Gamma)^2}{2\sigma^2}\right)$$

$$p(\delta) = \frac{\log e}{\delta\sqrt{2\pi}\sigma}\exp\left(-\frac{(10\log\delta - 10\log\Delta)^2}{2\sigma^2}\right)$$

(5.137)

where $\Gamma = <\gamma(t)>$ and $\Delta = <\delta(t)>$.

Assuming the rate of change of the signal envelope due to shadowing is slow compared with fast fading such that the signal envelope remains constant over a symbol period, P_L can be obtained as

$$P_L = \int_0^\infty \int_0^\infty P_L(\gamma, \delta)P(\gamma)p(\delta)d\gamma d\delta$$

(5.138)

where $P_L = (\gamma, \delta) = 1/2 \ (1- \gamma\Phi^{-1/2} \ J_o \ (2\pi f_m T))$, and $\Phi = 2(\gamma+\delta+1)^2 - \gamma^2 J_o^2(2\pi f_m T)$. The above integral is difficult to solve analytically, thus a numerical solution is sought. In the absence of interference equation (5.138) reduces to (5.136).

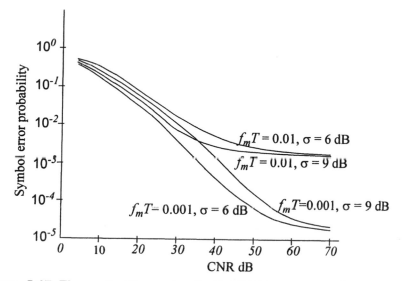

Figure 5.17 Theoretical bounds of π/4-QDPSK symbol error probability in a Rayleigh-lognormal fading

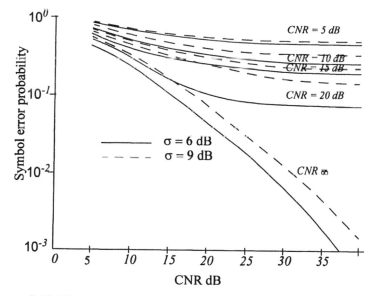

Figure 5.18 Theoretical error bounds of π/4-QDPSK in a Rayleigh-lognormal channel subjected to cochannel interference (f_mT = 0.01)

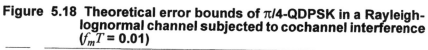

In Figure 5.17 plots produced from [37] when the signal is subjected to lognormal shadowing and interference are given. The results for $f_mT = 0.01$ and 0.001 for two degrees of shadowing severity (6 and 9 dB) are shown.

The degradation due to shadowing is small at low CNR (or high error rate), and increases with increasing CNR. For example, at SER of 10^{-1}, an additional 2 dB is required when shadowing depth increases from 0 to 6 dB, but this difference increases to 5 dB for SER of 10^{-2}. It seems that the degradation due to noise at low CNR is replaced by the degradation due to shadowing at high CNR.

The presence of shadowing has considerable influence on the SER performance. This is shown as shift in the SER curve to the right by 4 dB for $\sigma = 6$ dB and 8 dB at $\sigma = 9$ dB at a moderate CNR of 15 to 35 dB.

At very high CNR (>50 dB), an irreducible error rate equal to that when only fast fading was present. This indicates that the system performance at high CNR's is limited by fast fading only and shadowing has little effect.

Now, we turn our attention to the cases when the interference is also present. Equation (5.138) will be used to get the numerical results under several operating conditions. Figure 5.18 shows theoretical bounds of $\pi/4$-QDPSK symbol error probability in a Rayleigh-lognormal channel subjected to cochannel interference for the fade rate of $f_mT = 0.01$. Results for fade rate of $f_mT = 0.001$ are shown in Figure 5.19.

It is evident from the figures that the SER is strongly sensitive to interference. For example, when CIR is reduced from infinity to 20 dB in the presence of shadowing of standard deviation $\sigma = 6$ dB, the SER degrades from 1.5×10^{-5} to 5×10^{-2} which is approximately three orders of magnitude worse.

The presence of interference removes the large difference between the irreducible SERs for the two fading rates. For example, in the absence of interference, the irreducible SERs for $f_mT = 0.01$, and 0.001 are 2×10^{-3}, and 1.5×10^{-5}, respectively. When CIR=15 dB, the irreducible SERs for $f_mT = 0.01$, and 0.001 are 1.2×10^{-1}, and 1.1×10^{-1} respectively which are more or less indistinguishable. This may suggest that in the presence of interference the fade rate has very little impact on the irreducible error rate.

It should be noted that while different fade rates have very little effect on the system performance in the presence of interference, different standard deviations of shadowing produce quite divergent outcomes. In other words, the system is more susceptible to shadowing than fading when interference is present.

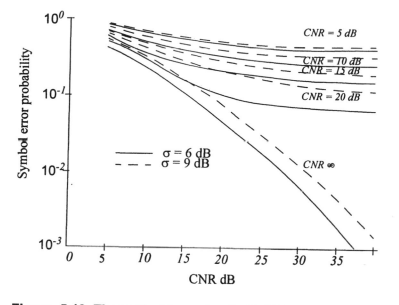

Figure 5.19 Theoretical bounds of π/4-QDPSK symbol error probability in a lognormal-Rayleigh channel subjected to interference, $f_m T = 0.001$

5.9 Environmental Noise

Chapter 4 examined the effects of thermal noise on the performance of mobile communication systems. In the preceding sections the effect of interfering signals was considered. A third form of interference is electromagnetic pollution. This arises from numerous electrical devices operating in the neighborhood of the receiver. Examples include electric motors, fluorescent lights, electrical switches and power line transients. In the mobile environment a major source of this nature is electromagnetic interference from motor vehicles ignition.

Operators of mobile radio systems have no control over such emissions. Fortunately, noise intensity decreases with increasing frequency. At 900 MHz its effect is small. However, at VHF (150 MHz band) and UHF (450 MHz) frequencies, it constitutes a serious problem which must be included in system analysis.

Characterization of environmental noise is difficult because of its statistical nonstationarity. Several types of noise models have been proposed

[24]-[28]. Some models are empirical, based on measured results. Others are physically motivated and include parameters such as antenna patterns and the location, distribution and nature of the emitters. Of the latter type, Middleton's work [27][28] is particularly important.

In practice, a number of parameters are used to measure noise in the context of electromagnetic compatibility. These include average-to-rms ratio, peak signal and average signal. Impulsive noise has been characterized statistically by its Amplitude Probability Distribution (APD), Noise Amplitude Distribution (NAD) and ratio of r.m.s. to average levels. To a considerable extent, the use of these measures has been limited by the type of application in mind. For example, noise amplitude distribution has been used in the prediction of FM receiver signal degradation, whereas amplitude probability distribution has been used in digital communications performance prediction. The following sections discuss some commonly used noise characterization parameters, and examine their usefulness in the context of mobile communication systems.

5.9.1 Amplitude Probability Distribution

The envelope of the noise at the receiver output is often represented by the amplitude probability distribution. It is normally plotted on Rayleigh paper, and shows the overall percentage of time for which the noise envelope exceeds any particular level; there is no information how the time is made up. The ordinate is expressed in dB relative to kT_oB. The appropriate point on the curve can then be plotted from

$$Prob(a \geq L) = P(L) = \frac{1}{T}\sum_{k=1}^{n} \Delta t_k \qquad (5.139)$$

5.9.2 Average Crossing Rates

The APD curve can be supplemented by a graph showing the number of times any particular level is crossed in a specified time interval. If, in a record of length T seconds, a given level y is crossed $n(y)$ times (in positive going direction) then the average crossing rate (ACR) is $n(y)/T$. This is usually expressed in the form of a graph of amplitude against crossing rate, on a logarithmic scale. ACR in conjunction with APD curves produce information about the average pulse duration and the average pulse interval, but not their distributions.

5.9.3 Pulse Width and Pulse Interval Distributions

Pulse duration distribution $P_d(\Delta t)$ and pulse interval distribution $P_i(\Delta t)$ respectively are given by:

$$P_d(\Delta t) = \frac{\text{numer of pulses with duration } \Delta t}{\text{total number of pulses at level L}}$$

$$P_i(\Delta t) = \frac{\text{number of intervals of length } > \Delta t}{\text{total number of intervals at level L}} \qquad (5.140)$$

The pulse duration distribution provides the results on the duration of the pulse at the output of the receiver filter. The pulse duration depends on the receiver bandwidth, the input impulse strength, and the rate at which the noise impulses arrive. Thus, the pulse duration distribution provides a measure of the cumulative effect of other three parameters.

The pulse interval distribution is the distribution of the pulse arrivals. The arrival rate coupled with amplitude probability distribution may be used to evaluate digital transmission system in the presence of impulsive interference.

5.9.4 Impulsive Noise Measurements

Measurements on the foregoing parameters are reported in the litera-ture [29]-[31]. In [30][31] the measurements and analyses of impulse noise in mobile environment is reported with considerable detail. The noise statistics depend on the frequency of operation and on the surroundings of the "victim" mobile, whether in a suburban, urban, or city core setting, the results on the amplitude and the number of pulses/sec statistics are summarized in Table 5.2 to Table 5.4.

Table 5.2		Amplitude and pulses/sec statistics in suburban areas			
Frequency (MHz)	Envelope amplitude dBkT$_o$B			Number of pulses/sec	
	$P=10^{-4}$	$P=10^{-2}$	$P=0.5$	45dBkT$_o$B	25dBkT$_o$B
40	56	36	16	10	1000
80	50	31	12	8	360
150	45	22	8	5	300
408	36	19	<receiver	<1	90
900	30	18	noise	<1	12

In the above table, probability that the noise envelope in dB relative to kT_oB is exceeded. For example, at 408 MHz the probability that the noise envelope exceeded 36 dB relative to kT_oB is 10^{-4}, for probability of 10^{-2}, it is decreased to 19 dB. It is also noted that as the frequency goes higher, the noise envelope is decreasing.

The last two columns give the number of pulses/sec at two reference levels - 45 and 25 kT_oB. Again the number of pulses that exceed the two level decrease with an increase in frequency. 40 MHz it is 10 dB/decade. The pulse duration distribution and pulse interval distribution also indicate high noise at 40 MHz, since the pulse duration frequently exceeds the reciprocal of the post detection bandwidth.

In Table 5.3, the results of measurements in urban areas are summarized. The trend is similar to that given in Table 5.2, but the noise in urban areas is more intense.

Specifically, the level exceeded with 10^{-4} probability is 58 dB kT_oB at 40 MHz, 55 dB at 150 MHz, and 32 dB at 900 MHz, an increase of 2 to 4 dB over the suburban areas. At higher frequencies (greater than 450 MHz) the receiver noise level becomes dominant.

Table 5.3   ***Amplitude and pulses/sec statistics in urban areas***

Frequency (MHz)	Envelope amplitude dBkT$_o$B			Number of pulses/sec	
	$P=10^{-4}$	$P=10^{-2}$	P=0.5	45dBkT$_o$B	25dBkT$_o$B
40	58	37	17	18	1100
80	55	32	14	12	600
150	48	25	10	8	300
408	38	22	<receiver	1	200
900	32	20	noise	<1	70

Table 5.4 shows the results in city core areas. The noise intensity is even higher than that in urban area. The additional intensity is coming from sources other than the motor vehicles. A similar trend is also observed in the number of pulses measured in city core area.

Table 5.4   ***Amplitude and pulses/sec statistics in city core areas***

Frequency (MHz)	Envelope amplitude dBkT$_o$B			Number of pulses/sec	
	$P=10^{-4}$	$P=10^{-2}$	P=0.5	45dBkT$_o$B	25dBkT$_o$B
40	61	43	22	50	3500
80	56	36	14	32	2000
150	48	30	10	17	1100
408	39	22	<receiver	~1	600
900	34	21	noise	<1	100

The pulse interval distribution (PID) and pulse duration distribution (PDD) curves at both 408 and 900 MHz were also measured. It is reported that probability of measuring very long pulses, in excess of one millisecond, was not feasible with the length of data available. The pulse interval distribution indicates no crossovers at 900 MHz, and the curves are more widely spread than those at 408 MHz. It is seen that higher probabilities of longer intervals and a major noise contribution comes from low level pulses at 900 MHz.

5.10 Summary

In this chapter, analysis of co-channel adjacent channel interference has been studied in some detail. The first part of the chapter addresses the general problem of interference in the presence and the absence of fading and other channel vagaries. The second part of the chapter looked at the interference problem with regard to analog FM communications. In the third part of the chapter, effect of interference on digital communications was considered. In addition to fading, it has been seen that interference from other signal sources is a severe problem. For a particular user, these sources include other mobile transmitters which may be operating on the same frequency, an adjacent band, or one which is harmonically related to that of the user. In addition, environmental noise is usually present in the form of electromagnetic pollution, especially in the urban areas favored by the mobile user.

In the face of this hostile environment, the designer is required to implement a system which combines efficiency, reliability and reasonable cost. One of the tools at the designer's disposal, that of modulation, has already been examined in Chapter 4. This chapter concludes the analysis of the mobile radio channel and signalling environment. The next two chapters will consider two more important areas, those of antennas and of signal processing.

Problems

Problem 5.1

(a) Verify equation (5.15) for a cellular system operating in Ricean-Rayleigh-lognormal propagation environment.

(b) State under what conditions, the simplified equations (5.17) and (5.18) are valid.

Problem 5.2

(a) What is the difference between coherent and non-coherent addition of interfering signals?

(b) Give at least two examples where coherent and non-coherent interference additions can be found in practical mobile environment.

Problem 5.3

For a macrocellular system operating with a cell radii of 5 km, the desired power to interference ratio, P_d/P_i, of 16 dB is required. If the propagation attenuation rate is $\alpha = 3$, from the base to the breakpoint and $\beta = 4.7$ beyond the break point. Calculate the reuse distance and the cluster size.

Problem 5.4

(a) In a cellular environment, the $(S/I)_{min}$ is found to be 8 dB. What should be the input carrier to interference ratio, if both the desired and interference signals are modulated with index, $\Phi = 2$?

(b) Using the results in Part (a), find the minimum reuse distance and the cluster size. Assume that propagation attenuation constant is 3.5.

Problem 5.5

Explain why in a fading environment, the output signal to adjacent channel interference can be poorer than the signal to co-channel interference despite heavy attenuation of adjacent signal by the IF filter.

Problem 5.6

Consider a situation where base station is receiving signals from different mobiles at two adjacent channels. The base receiver provides adjacent channel isolation of 70 dB. The desired signal source is located 4 Km away. What is the minimum distance from which the second mobile can transmit its signal without causing interference to the desired signal? Assume path loss model (i) loss varies as $d^{3.5}$ (ii) loss varies as $d^{4.2}$, where d is the distance between the transmitter and the receiver.

Problem 5.7

Drive from basic principles equation (5.91).

Problem 5.8

(a) At a certain location in a cellular environment, the mobile is experiencing co-channel interference. The input carrier to interference power ratio is found to be $R^2 = 3$. Calculate worst baseband *SIR* when both the interference and desired signal are modulated at $\Phi = 2.7$ radians.

(b) To reduce the interference effect the modulation index of the of the desired signal is increased to $\Phi = 3$. What improvement do we get in the worst case *SIR*.

(c) The desired signal modulation index is retained but the modulation index of the interference signal is reduced by 20%. What improvement do we get in the worst case *SIR*.

(c) Given the parameters as in (a) but in a fading environment, what will be *SIR* if mean carrier to noise ratio is given to be 18 dB?

Problem 5.9

Compare the performance of $\pi/4$-QDPSK with that of GMSK in the presence of co-channel interference and Rayleigh fading when the mean signal to noise ratio is 12 dB and mean signal to interference ratio is found to be in the range of 6 dB to 18 dB.

References

[1] W.L. Firestone, "A review of communication interference problems," *IEEE Trans. Veh. Tech.* vol. VT-14, No. 3, pp. 57-66, March 1965.

[2] W.C. Lee, *Mobile Communications Engineering*, pp. 374, McGraw Hill, 1982.

[3] Y-D. Yao and A.U.H. Sheikh, "Co-channel interference modelling and performance analysis of microcell systems for wireless personal communications," *Can. J. Elect. and Comp. Eng.* vol. 19, no. 1, pp. 27-35, 1994.

[4] P. Hartley, "Short distance attenuation measurements at 900 mhz and 1.8 ghz using low antenna heights for microcells," *IEEE-J.Select. Areas Comm.*, vol. SAC-7, no. 1, Jan. 1989, pp. 5-11.

[5] *Report of the Advisory Committee for the Land Mobile Radio Services*, Federal Communication Commission, Superintendent of Documents, U. S. Government Printing Office, Washington D.C.

[6] D.C. Cox, "Linear amplification with nonlinear components," *IEEE Trans. on Commun.* vol. COM-22, pp. 1942-1945, Dec.1974.

[7] D.C. Cox and R.P. Leck, "Component signal separation and recombination with nonlinear components," *IEEE Trans. on Commun.* Vol.COM-23,pp. 1281-1287, Nov.1975.

[8] A.J. Rustako, Jr. and Y.S. Yeh, "A wide-band phase-feedback inverse-sine phase modulator with application towards a LINC amplifier," *IEEE Trans. on Commun.* vol.COM-24, pp.1139-1143, Oct. 1976.

[9] V.K. Prabhu and L.H. Enloe, "Interchannel interference considerations in angle modulated systems," *Bell System Technical Journal*, vol. 48,pp. 2333-2358, September 1969.

[10] V.K. Prabhu and H.E. Rowe, "Spectral bounds of a PM wave," *Bell System Technical Journal*, vol. 48, pp. 769-811, March 1969.

[11] W.C. Jakes (Ed.), *Microwave Mobile Communications*, J. Wiley, 1974.

[12] V.K. Prabhu, "Error rate considerations for coherent phase shift keyed systems with co-channel interference," *Bell System Technical Journal*, vol., pp. 743-767, March 1969.

[13] J. Goldman, "Multiple error performance of PSK with co-channel interference and noise," *IEEE Trans. on Commun.*, vol. COM-19, pp.420-430, August, 1971.

[14] C. Colavito and M. Sant'agostino, "Binary and quaternary PSK radio systems in a multiple-interference environment," *IEEE Trans. on Commun.*, vol. COM-21, pp. 1056-1067, Sept. 1973.

[15] J.S. Engel, "The effects of co-channel Interference on the Parameters of a Small-Cell Mobile Telephone System," *IEEE Trans. on Veh. Tech.*, vol. VT-18, pp.110-116, November 1969.

[16] G. A. Arredondo, "Analysis of radio paging error in multi-transmitter mobile systems," *IEEE Trans. on Commun.* vol. COM-21, No.11, pp. 1310, November 1973.

[17] R.C. French, "Common channel multi-transmitter data systems," *The Radio and Electronic Engineer*, vol. 50, No.9, pp. 439-446, September 1980.

[18] A.S. Rosenbaum, "PSK Error Performance with Gaussian Noise and Interference," *Bell System Technical Journal*, vol.48, pp. 413-442, Feb. 1969.

[19] A.S. Rosenbaum, "Binary PSK error probabilities with multiple co-channel interferences," *IEEE Trans. Comm.*, vol. COM-18, pp.241-253, June 1970.

[20] Y. Akaiwa and Y. Nagata, "Highly efficient digital mobile communication with a linear modulation method," *IEEE Select. Areas., Comm.*, vol. SAC-5, no. 5, pp. 890-895, June 1987.

[21] J.A. Tarallo and G.I. Zysman, "Modulation techniques for digital cellular systems," *IEEE Veh. Tech. Conf. Record*, pp. 245-248, Sept. 1988.

[22] J. Horikoshi and T. Morinaga, "Multipath interference characteristics of dpsk in land mobile digital communications," *Trans. of IECE of Japan*, vol. J65-B,no.12, pp.1559-1560, Dec. 1982.

[23] M. Schwartz, W.R. Bennet, and S. Stein, *Communication Systems and Techniques*, McGraw Hill Inc., New York, 1966.

[24] A.U.H Sheikh, "Ignition interference in land mobile environment: measurements, characterization and models," *Proc. IEEE National Symposium on EMC*, April, 1984.

[25] B.I. Kuz'min, "Impulsive interference and noise immunity analysis (a review)," *Radio Electronics and Communication Systems*, vol. 24, pp. 1-11, 1981.

[26] E.C. Field Jr. and M. Lewinstein, "Amplitude probability distribution model for VLF/ELF atmospheric noise," *IEEE Trans. on Comm.*, vol. COM-26, pp. 83-87, Jan. 1978.

[27] D. Middleton, "Man-made noise in urban environments and transportation systems," *IEEE Trans. on Commun.*, vol. COM-21, pp. 1232-1241, 1973.

[28] D. Middleton, "Statistical-physical models of electromagnetic interference," *IEEE Trans. on Electromagnetic Compatibility*, vol. EMC-19, pp. 106-127, August, 1977.

[29] R.T. Disney and A.D. Spaulding, *Amplitude and Time Statistics of Atmospheric and Man-made Radio Noise*, Environmental Science Services Administration Tech. Report ERL ISO-ITS, Feb. 1970.

[30] A.U. Sheikh and J.D. Parsons, "The frequency dependence of urban man-made radio noise," *The Radio and Electronic Engineer(JIERE),* vol. 53, pp. 92-98, March 1983.

[31] J.D. Parsons, and A.U. Sheikh, "Statistical characterization of vhf man-made radio noise," *The Radio and Electronic Engineer(JIERE)*, vol. 53, pp. 99-106, March 1983.

[32] Y. -D. Yao and A.U. H. Sheikh, "Outage probability analysis for microcell mobile radio systems subject to fading and shadowing," *Electronics Letters,* vol. 26, pp. 864-866, 1990.

[33] Y. -D. Yao and A.U. H. Sheikh, "Bit error probabilities of NCFSK and DPSK signals in microcellular mobile radio systems," *Electronics Letters,* vol. 28, pp. 363-364, 1992.

[34] K. Murota and K. Hirade, "GMSK modulation for digital mobile telephony," *IEEE Trans. Commun.,* vol. COM-29, pp. 1044-1050, July 1981.

[35] S.M. Elnoubi, "Analysis of GMSK with discriminator detection in mobile radio channels," *IEEE Trans. Veh. Tech.,* vol. VT-35, pp. 71-76, May 1986.

[36] P. Varshney and S. Kumar, "Performance of GMSK in a land mobile radio channel," *IEEE Trans. Veh. Tech.,* vol. VT-40, no. 3, pp. 607-614, 1991.

[37] M.T. Le, *Performance of π/4-QDPSK Over Mobile Fading and Shadowed Mobile Radio Channel*, M.Eng. Thesis, Carleton University, Ottawa, Canada, 1990.

Chapter 6

ANTENNAS FOR MOBILE RADIO SYSTEMS

Transmission and reception of radio waves require antennas. In providing gateways to transmission and reception of radio signals, antennas play important roles both at the base station and mobile terminal. Though, higher transmit power and greater receiver sensitivity can improve radio link quality and reliability, the regulatory authorities place a ceiling to the transmit power in order to limit interference to other users. The receiver sensitivity is determined by manufacturing costs; receivers with greater sensitivity are much more expensive. Thus, good antenna design can potentially deliver substantial improvement in the system performance.

The gain and directivity are the two important parameters that define the antenna quality. For fixed link parameters like transmit power and receiver sensitivity, the signal to noise ratio determines the link quality, which in this case depends on the antenna gain. Furthermore, the antenna directivity can discriminate between the signal and interference thereby improving the signal to interference ratio. The directions from which signal and interference arrive continuously change due to the movement of at least one terminal forming the link, antennas that adapt to the direction of the signal have been more desirable.

The impact of the propagation environment on radio signals was studied in some detail in Chapters 2 and 3. In earlier studies, to simplify analysis we assumed omnidirectional transmitting and receiving antennas. In the case of fixed communication systems, the practice is to use high gain directive antennas. In Chapter 3, it was seen that the Doppler spread in mobile environment could be considerably reduced if we use directive antennas. In future, adaptive antennas or smart antennas having beam steering capability are expected to play greater role in wireless communication systems. Besides antenna directivity, polarization could be used to our advantage. In some applications, signals on the horizontal and vertical polarization can be used to provide substantial gain through diversity. In this chapter we focus on radiating characteristics and physical dimensions of the base station and mobile antennas. We shall examine methods in which antennas could be used to modify the received signal behavior so that the link performance could be improved. In particular, we evaluate the influence of radiation pattern and polarization on the characteristics of the received signal and the interference.

We shall also discuss the recent trends in antenna design for mobile radio systems.

6.1 Base Station Antennas

The primary function of the cell site or base station antenna is to provide adequate radio coverage over the entire cell. Even if we install an omnidirectional antenna, the resulting antenna pattern may not remain uniform over 2π radians in the azimuth plane, because the environmental irregularities surrounding the antenna site usually distort the coverage pattern. The needed radio coverage pattern in a certain environment depends on several factors e.g. tall buildings, hills, tunnels, interference from other neighboring antennas, and the amount of coverage overlap needed to reduce call drops and smoother operation of handoffs are some of the factors influencing the coverage pattern. Thus, often it becomes necessary to engineer the radiation pattern so that the coverage pattern is uniform regardless of the presence of environmental irregularities.

At the base station more than one antenna may be needed to obtain a satisfactory propagation pattern. Multiple antennas and their physical sizes raise several technical and environmental concerns. Cross-coupling between multiple antennas at the same physical location can result in intermodulation distortion and interference [1]. The large size of base antenna make them more conspicuous thereby altering the sight of surroundings. The base station antennas are designed for radiation pattern, gain and its physical shape. A good antenna design and proper installation can rid of several possible degradation effects on the radio link.

As compared to antenna used in mobile terminals, there is more freedom in the case of base station antennas to create desired radiation characteristics. Since, interfering signals place an upper limit to the performance of mobile wireless links, antennas can play a significant role in controlling interference. For example, base station antennas may be designed to produce a weaker propagation/reception characteristic in the direction of high interference. A placement of a null in that direction will probably eliminate interference. To fill the coverage gap thus created, another base station at a more suitable location may be used. Intercell interference is also a significant problem in cellular systems. Tilting the base station antenna downwards results in a significant reduction in the intercell interference thereby reducing the frequency reuse distance and increasing the spectrum efficiency.

At base stations, two types of antenna are commonly used: omni-directional and directional. The omni directional antennas are either dipole or folded dipole. Yagi or corner reflectors are used as directional antennas. Yagi arrays have been used whenever a good directivity in the antenna pattern is desired. Corner reflectors are specifically used when the cell is to be split into sectors.

The collinear antennas give uniform coverage in azimuth but have approximately 20° beamwidth in elevation with sidelobes having gains of approximately 12 dB below the main lobe. Antennas with gain between 5 and 9 dB having a bandwidth between 35 and 65 MHz are available. These antennas have power rating up to 500 Watts.

As noted in Chapter 4, Doppler spread due to mobility and multipath propagation places an upper bound on the performance in the form of irreducible error rate. The relatively narrow beamwidth of directive antennas lowers the fade rate as observed in Chapter 3. It was also observed that lowering random frequency modulation and large phase deviations associated with it improves the system performance dramatically. Under special circumstances, blocking signal propagation in one or more directions may be needed when interference from more or more directions is of concern. A desired overall radiation pattern can be obtained by controlling the feed gains and phases of antenna array consisting of two or more antennas. The radiation pattern in the elevation can be altered by rotating the array.

6.2 Advanced Base Station Antennas

Multiple narrow beam antennas, adaptive beam formers, and steerable beam smart antenna are alternatives to conventional omnidirectional and broad sector-beam antennas. A cylindrical array antenna of 3.5 meters diameter was investigated in [2] with an aim to increase CDMA system capacity. It has been shown that the overall system capacity improves from 33 users for conventional omni-directional antenna to 1107 for an antenna consisting of 36 beams, each beam having a width of 10°. The capacity estimates are based on systems paramcters of 20 MHz bandwidth, 32 kbits/sec bit data rate, and 25 dB signal to noise ratio. The desired bit error rate was set at 10^{-3} for fully loaded CDMA system. The results show that a proper antenna design could pay a handsome dividend in the form of increased capacity and revenues. The importance of antenna design to cellular mobile systems is gaining recognition and advanced design antennas are expected to play even more important role in enhancing spectrum efficiency than was previously recognized.

6.3 Installation of Base Station Antennas

Before a base station antenna is installed, many technical and environmental problems need to be surmounted. In recent years, environmental awareness is on the increase and the location and shape of base station antennas have become an environmental concern. Installation of environmentally insensitive antenna design or mounting is hotly contested. Also, not in my backyard phenomenon is also hitting the mobile operators hard because of health scare due to radiation. These concerns are influencing base station antenna design, location, and type of installation. Communities particularly in the developed countries are demanding that installed antennas should not be distinguishable from the surrounding landscape but these should be merged into it. The concerns include failure of the antenna mounting to conform to the surrounding building architectures and environmental landscaping. Details like its size and shape of the antenna, its location and type of installation envisaged determine whether such installation is suited to the given environment. Furthermore, in some applications, it may be desirable to use multiple antennas at a certain site to engineer a desirable propagation pattern. Thus, antenna mounting design has now developed into an art form.

Several architectural companies now specialize in antenna mounting designs, which make the antennas either practically invisible or merge into surrounding landscape. For example, many cell site antennas towers in New Jersey, USA emulate trees and are hardly distinguishable from the background forest. Since base station antennas are significantly larger than those used for portables, their mergence into the environment may be sometimes difficult. Obviously, these constraints accruing from environment leave little room to maneouver and it is possible that the environmental regulations in existence require compromises in the antenna design. In these circumstances, it may become difficult to get a suitably smoothed radiation patterns. These compromises in design usually result in loss in coverage and may be in the system capacity.

In order to provide in excess of 99% coverage, dead spots in the radio coverage are not desirable as these often lead to call drops. Excessive number of call drops will result in a reduction in the customer base. The dead spot problem arises due to the presence of nulls in the antenna radiation pattern. These nulls become more significant when beamwidth in the elevation is compressed to extract higher gain in the azimuth. This is illustrated in Figure 6.1.

Similarly, coupling between transmit and receive antennas resulting intermodulation products and interference is an important technical consideration when deciding on the spacing between antennas. The location of angular propagation nulls within the coverage area depends on the antenna height and

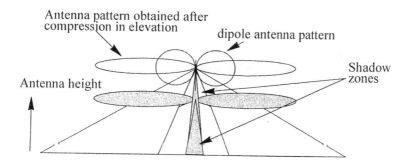

Figure 6.1 Impact of antenna height and pattern on coverage dead spots

beamwidth. As the antenna height is increased the dead spots move outwards away from the antenna location. Thus, reduction in dead spots can be achieved by reducing the antenna height. In this case, transmit power need not increase. When the gain of the antenna is lowered, its radiation pattern become more uniform and nulls become less defined or deep. Similar effect can be obtained by adjusting the tilt angle. However, note that tilting the antenna may adversely affect the gain and the beamwidth of the antenna. Also, a possibility exists that antenna tilt will produce deeper shadows of permanent nature due to nearby objects. The option of increasing the power to reduce dead zones is not desirable because it contributes to an increased interference in other cells.

The second problem of coupling between the transmit and the receive antenna arises because of insufficient physical isolation between them. The coupling between the transmit and receive antennas can be reduced by increasing the physical spacing between them either horizontally or vertically or both. A stronger coupling increases the receiver inband noise and reduces the receiver sensitivity, thereby reducing the receiver gain. Coupling between two or more transmitting antennas may also arise.

The signal isolation between antennas increases with an increase in the distance between them. In case of horizontal spacing one can use free space propagation loss formula given in (2.4). The isolation is given by:

$$H_{iso} = 22.0 + 20\log\left(\frac{d}{\lambda}\right) - (G_T + G_R) \qquad (6.1)$$

where λ and d are in meters. For vertical spacing, isolation in dB is given by:

$$V_{iso} = 28.0 + 40\log\left(\frac{d}{\lambda}\right) \tag{6.2}$$

It is seen that the vertical isolation depends on only frequency and the distance and is independent of the antennae gains. Since transmitter and receiving antennas have finite gains, therefore vertical separation is much better than horizontal separation. For example, at 900 MHz, isolation H_{iso} and V_{iso} for 5m antenna separation in horizontal and vertical plane are $45-(G_T+G_R)$ dB and 75 dB respectively. It is worth mentioning here, that the position and spacing between the antennas also determine the number of maxima and minima in the radiation patterns.

It is seen from (6.2), that spacing could be large depending on how much isolation is required. A larger isolation in vertical installation would require higher base station transmit tower. This may not be desirable from installation or cost concerns. The other alternative to increase isolation is to use widely separated transmit and receiver frequencies. For example, in DAMPS and GSM the transmit and receive frequencies are separated by 45 MHz. The out of band noise could then be filtered out with the use of multi-couplers, duplexers, and isolators.

To summarize, dead spots in the coverage, coupling between different antennas, and fast rate of signal fades could result in a substantial degradation in the system capacity.

6.4 Antenna Combiners and Multicouplers

Antenna combiners, multi-couplers and duplexers are essential devices to connect antenna systems to RF power amplifiers in case of transmitters and to front ends in the case of receivers. A duplexer is a standard mobile terminal device. It provide sufficiently high isolation between the transmitter and the receiver so that simultaneous transmission and reception of signals can take place. As mentioned earlier, a large separation between the transmit and receive frequencies increases isolation. For example, separation of 45 MHz between the transmit and receive frequencies has been standardized in mobile communications.

Antenna combiners and receiver multicouplers are used at cell sites. For a base station that has only one transmit and one receive antenna, multi-coupler and transmitter combiner are not needed; in this case a duplexer forms an essential part of the system.

In a cellular system operating in a large city, several channels are required to meet the traffic demand in each cell; the number of them usually vary between 20 and 30. The exact number of channels depends on traffic intensity, transmission format and the access method. When several RF signals are combined on a single antenna, potential of signal distortion exists. The number of signals that can be combined at each antenna are therefore decided on the basis of their frequencies and power so that distortion is negligible. Combining TDMA signals produce more distortion as compared to combining FDMA channels. Thus, the number of FDMA channels that can be combined is larger than the number of TDMA channels for the same level of distortion.

A good multiplexer is characterized by low insertion loss, high power handling capability, low intermodulation products, simplicity of maintenance and high operational stability. In practice, two types of multiplexers are commonly used: one type consists of hybrid circuits and circulator without frequency selective resonators, and the other type consists of circulators, cavity resonators and a junction box. The two types of transmitter multiplexer systems are shown in Figure 6.2. A particular multiplexer type is chosen on the basis of cost and total signal attenuation loss.

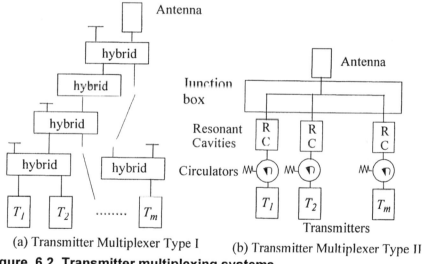

(a) Transmitter Multiplexer Type I   (b) Transmitter Multiplexer Type II

Figure 6.2 Transmitter multiplexing systems

6.5 Antennas for Mobile Terminals

A mobile antenna is an essential part of a mobile radio terminal. The mobile antennas are generally omnidirectional because the signal can arrive in a continuously changing random direction. The continuous terminal mobility and changes in the vehicle orientation does not allow fast and continuous adjustment to keep the antenna orientated towards the arrival direction of the strongest signal; the adaptive antennas with electronic beam steering are the exceptions. It is obvious that the portable terminals are unlikely to use multi-element adaptive arrays because of their small size. Thus, small in size, having reasonable gain and preferably omnidirectional radiating pattern are the desirable characteristics of mobile antennas.

The basic vehicular antenna is a unity gain quarter wave (whip) antenna which has a broad radiation pattern. An elevated feed 3 dB and 5dB collinear, and collinear on glass are improvements over the basic antenna type. The basic antenna works well in densely populated urban areas where signal may be expected to arrive equally from all directions.

The mobile antenna performance can be quite variable because of terminal mobility, its orientation with respect to the base station electric field polarization, proximity to body and changes in the surrounding environment. The coverage in the azimuth and elevation set the requirements on radiation pattern, antenna gain and cross polarization. The radiation pattern is affected by the physical dimensions of the antenna, the size of the ground plane, its proximity to human body or any other conducting object, and the position and type of the antenna mount. Here, the antenna mount relates to the position of the antenna on a certain part of a vehicle, i.e. roof, wind-screen, or window.

The antenna standing wave ratio (SWR), or related reflection coefficient, are important in determining transmitter power output; ideally the antenna SWR should be unity; a high SWR causes antenna inefficiencies. These factors also affect the power handling or voltage rating of the antenna [3]. The size of the antenna ground plane is important as it determines the radiation pattern. For vehicular antenna, the ground plane radius has been recommended to be equal to one half wavelength at 30 MHz and one wavelength at frequencies between 100 and 1250 MHz [4]. In the case of small portable terminals, limited terminal size has insufficient ground plane size, which makes the antenna performance frequency dependent. To reduce the dependency of the antenna on frequency, a larger ground planes up to 5λ is desirable. The shape of the ground plane also alters the radiation pattern of the antenna.

The antenna aperture, consequently the gain, can be increased by stacking two or more antenna sections. The first application of stacking results in the 3 dB collinear design, which has two radiators, upper $5/8 \lambda$ long and lower $\lambda/4$, separated by an air-wound coil. Further increase in gain is obtained by stacking three elements separated by two air coils. In the elevated feed type the antenna feed is elevated over a quarter wave length polycarbonate tube, a further improvement is the on-glass type where the elevating section is made even longer. Practical considerations related to antenna gain determine the number elements to be stacked.

The choice on antenna type depends on the environment where mobile is most likely to operate. In urban areas where signal arrives from all directions with nearly equal strength, unity gain antenna is perhaps the best choice. In sub-urban area where cell separation is larger and the signal is expected to come from lower angles, 3 dB collinear may be a better choice. Higher gain antennas are good in rural areas because they help increase the coverage area.

Recent research indicates that for extremely high frequency (>20 GHz) adaptive antenna array with fast beam steering may be the answer to large variability in the received signal to interference levels. The adaptive algorithms in general are unable to track the swift changes in the direction of signal arrival. This is the major limitation of the adaptive antennas. It is obvious that a large number of antenna elements forming an array can be accommodated in a relatively small physical size only at very high frequencies (>20 GHz). It is shown in Section 6.6 that directional mobile antenna are more useful in a dispersive environment. The coverage requirements depends on the type of application; for vehicle-to-vehicle communications coverage in the azimuth plane suffices whereas a wide angle coverage both in elevation and in azimuth are required for systems that use higher base station antennas.

6.5.1 Mounting Considerations for Mobile Antennas

The dependency of the antenna radiation pattern on the dimensions of the ground plane requires that the antenna mounting position be chosen care fully; in this regard, it is believed that the roof of the vehicle is the most suitable location for the antenna. It is possible to derive an expression for the antenna radiation pattern, however because of the complexity of the automobile structure, a simplified model of the vehicle as a ground plane becomes necessary [5][6][7][8]. For example in reference [6], the geometric theory of diffraction has been used to reduce vehicular body effects to a wedge and curved surface diffraction problem (Figure 6.3(a)). The car is represented in the form of three perfectly conducting parallel plates displaced relative to each other as shown in Figure 6.3 (b). The effects of the windshield and the internal structure of the automobile is ignored.

A Piecewise Sinusoidal Model (PWS) is used in which the entire length of the plate is divided into a large number of elementary lengths of dimension $l \ll \lambda$. The current distribution in an elementary length is given by

$$I(z) = I_n \frac{\sin(k(l - |z - z_n|))}{\sin(kl)} \qquad (6.3)$$

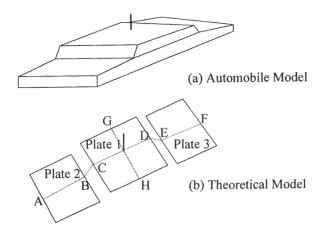

(a) Automobile Model

(b) Theoretical Model

Figure 6.3 Vehicle body model

where $k = 2\pi/\lambda$ and I_z is the current coefficient associated with the nth PWS dipole.

The electric field generated from this dipole is expressed as [8]:

$$E_n = \frac{30I_n}{j\sin(kl)}\left(\frac{exp(-jkR_{n+1})}{R_{n+1}\sin\theta_{n+1}}t_{n+1} - \right.$$
$$2\cos kl\frac{exp(-jkR_n)}{R_{n+1}\sin\theta_n}t_n + \left.\frac{exp(-jkR_{n-1})}{R_{n+1}\sin\theta_{n-1}}t_{n-1}\right) \quad (6.4)$$

The above equation indicates that the electric field is obtained by superposition of spherical wave components radiated from the top, bottom and centre of the PWS dipoles. The diffracted field is calculated using the points on the three parallel plates representing the automobile. In these calculations the complex ground dielectric constant is taken to be

$$\varepsilon_g = \varepsilon_r - j\left(\frac{\sigma}{\omega\varepsilon_o}\right) = \varepsilon_r - j\varepsilon_x \quad (6.5)$$

and the length of the dipole of $\lambda/4$ is used. In (6.5), ε_x is the imaginary part of ε_g, ε_o the free space dielectric constant, ω is the frequency, ε_r and σ is defined as relative dielectric constant and ground conductivity respectively. In this analysis, it is assumed that the antenna is located at the roof of the vehicle; such may not be the case in practice as this location is usually unacceptable to the user unless new types of antennas lesser visible become available. One possible type of antenna is a circular array consisting of several antenna elements.

The most common location of the antennas is either on the fender or the front or rear windshield. However, these locations seriously affect the resulting radiation pattern because of considerable reduction in the size of the ground plane, its angle relative to the ordinate, and asymmetrical location of the edges and curves of the vehicle body (diffraction points) relative to the antenna location. At 900 MHz, the side mirror, which now has the dimensions comparable to the wavelength, can be of considerable influence on the radiation pattern. An antenna located on the front fender will produce multiple nulls at both the rear and sides of the vehicle. The latter are much more pronounced on the side on which the antenna is located. In the next subsection, results of an experimental evaluation of mounting position are discussed.

6.5.2 Experimental Evaluation of Mounting Positions

Though the influence of the mounting position or the orientation of mobile antenna on the cellular system management is not available in any detail but its impact on the system capacity can be significant. An extended null in the antenna radiation pattern implies none or inadequate coverage in that direction. Such an absence in the coverage is likely to impact the system performance in several ways. First, because of non-uniform coverage, a loss in the system capacity could result. Second, when the terminal moves slowly, the number of handovers is likely to increase. Such an increase results from failure of the mobile antenna to find adequate signal strength in radio blind spots. A large number of handovers can lead to an overload of system management system [9]. Thirdly, because of the desirability of uniform antenna radiation pattern, some design complexities will result in an increase in antenna system design and installation costs.

In the following we summarize the results of an experimental measurements carried out at CRC on the distortion of antenna patterns when the antenna is mounted at different locations [10]. In these experiments several antenna locations were tested and resulting antenna directional patterns examined. A motor vehicle was mounted on a rotating platform and the far field radiation pattern of a vehicle mounted antenna measured. Different antenna types and locations on a vehicle body were investigated. Several interesting observations were made. First, the roof of the vehicle is far the best position for antenna mount. It provides relatively more uniform radiation pattern. Second, the antenna mounted on a glass surface always results in lower antenna gain. Third, when the antenna is mounted on windows using clips, it provides the worst directional pattern leaving large number of dead spots. Fourth, the dead spots in the antenna pattern can result in an increased number of handovers may ultimately result in a reduction in the system capacity.

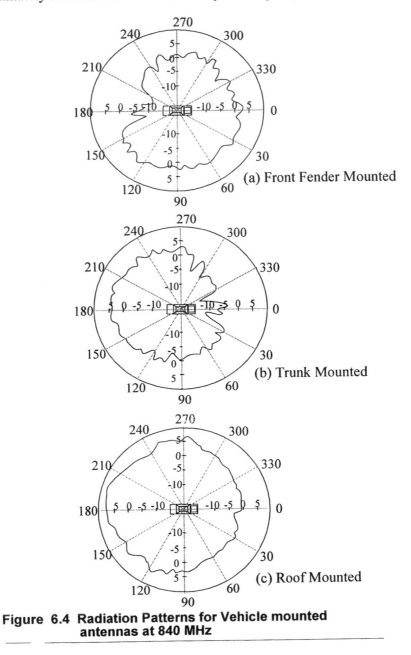

(a) Front Fender Mounted

(b) Trunk Mounted

(c) Roof Mounted

Figure 6.4 Radiation Patterns for Vehicle mounted antennas at 840 MHz

The results obtained are shown in Figure 6.4. It is seen that when the antenna is mounted on the front fender the radiation pattern shows approximately 9 dB reduction in the gain in the direction in which the car trunk is pointing. Similar observation is made when the antenna is mounted on the trunk of the car. In this case a sharp drop in the antenna gain towards the front of the car is seen. As mentioned earlier the roof mounting gives uniform gain in all azimuth directions.

In another study, the radiation pattern of a vertical (*E*-plane) $\lambda/4$ wave whip mounted on wind-screen was measured at 160 and 460 MHz. The results were are compared with those of $\lambda/2$ wave dipole (0 dB) standard antenna [11]. The radiation pattern suffers distortion as the operating frequency is increased.

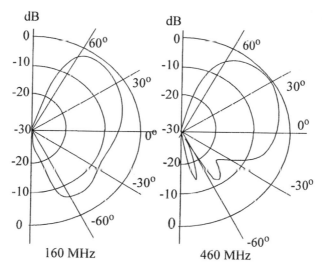

Figure 6.5 Vertical E(Plane) radiation pattern of $\lambda/4$ wave whip mounted on screen of a motor vehicle

6.6 Reception with Directive Antennas

The directional characteristics of base station and mobile antennas could play modify the mobile channel characteristics. A change in the directivity of the base station antenna towards the direction of signal arrival increases the signal power and limits interference. The directivity confines the coverage to the antenna footprint. Directive antennas are also useful in creating sectors, which may be used to increase frequency reuse. In this regard, corner antennas are used to split the cells into sectors. For example, a ing sectors, which may be used to increase frequency reuse. In this regard, corner antennas are used to split the cells into sectors. For example, a cell split into six sectors will reduce the number of co-channel interfering cells by approximately 1/6. Multi-element antenna arrays can be implemented to generate multiple beams. The beams can be used to achieve radio coverage in discrete azimuth angles to suit particular geographical and traffic conditions. Even radio frequencies can be reused simultaneously in different beams. This technique of reusing frequencies in different beams is called space division multiplexing (SDM). Spatial diversity may also be implemented at the base station using multi-element antennas, where individual antennas may or may not be directional in character. In a situation where propagation is to be confined to smaller zones, tilting of antennas towards the ground is beneficial [12]. Lee has shown that such tilting produces a notch in the centre of the horizontal radiation pattern which increases with the increase in the tilt angle. This property is quite effective in reducing co-channel interference in small cell cellular systems [13].

Mobile antenna directivity has been studied and reported in the literature [14]-[19]. Directional mobile antennas may be used to lessen the effects of fading by reducing the antenna gain in the direction of the incoming interference. Consider an ideal antenna with a uniform gain over the beam width β degrees and zero gain over all angles outside of the sector. Under the assumption of uniform scattering, the spectral effect of the antenna is to reduce the Doppler spectrum of an omnidirectional antenna in to a narrow slice. Thus, if the antenna is directed broadside to the direction of travel, the slice is taken out of the middle of the omnidirectional spectrum and its width is given by $2f_m \sin(\beta/2)$. Similarly when the narrowbeam antenna is directed along the direction of the vehicle travel, the slice of the spectrum is located at the edge of the omnidirectional spectrum and is of width $f_m[1-\cos(\beta/2)]$ [7]. The received power spectra in these cases are given below.

For a broadside direction:

$$S(f) = \frac{G_o}{f_m \sqrt{1 - \left(\frac{f}{f_m}\right)^2}}; \qquad \text{if } \left|\frac{f}{f_m}\right| < \sin\frac{\beta}{2}$$

$$= 0; \qquad \text{otherwise} \tag{6.6}$$

and along the direction of the vehicle travel:

$$S(f) = \cfrac{G_o}{f_m \sqrt{1 - \left(\cfrac{f}{f_m}\right)^2}}; \qquad \text{if} \quad 1 > \frac{f}{f_m} > 1 - \cos\frac{\beta}{2}$$

$$= 0; \qquad \text{otherwise} \tag{6.7}$$

where

$$G(a - \phi) = G_o; \quad |a - \phi| < \frac{\beta}{2}$$

$$= 0; \qquad \text{otherwise} \tag{6.8}$$

Lee [14] used an antenna array to confirm this phenomenon experimentally. It was shown that the power received by antennas having different beam widths and omnidirectional antennas is the same. Thus, in a uniform scattering environment, the gain of the antenna in a particular direction plays very little role in improving the link. This is only reasonable if the signal power is uniformly distributed over 2π radians in the azimuth. The obvious advantage in using directive antennas is to limit the angular range from where the interference could arrive and reduce the spectral width of the received signal and therefore its fade rate. This is shown in Figure 6.6. This technique has application in highly built up areas, such as a city core. Outside of such an area an omnidirectional antenna is usually most effective since the received signal is usually directional in nature, and the vehicle bearing is a random process.

6.7 Antenna Polarization and Signal Characteristics

When randomly oriented buildings and other obstacles reflect and diffract the radio waves, the polarization of the radio waves may change depending on the surface irregularities and orientations. The alternative way of using polarization is to transmit the same signal in both horizontal and vertical polarizations. The cross-polarized signals could be combined to reduce the effects of fading due to multipath propagation [15][16]. In particular, polarization diversity can reduce the effects of signal dropouts in personal communications where a hand held transceiver experiences changes in orientation. The effectiveness of polarization diversity depends on cross polarization coupling, which is defined as the ratio, expressed in decibels, of the median magnitude of the polarized transmitted signal to that of the orthogonally polarized one.

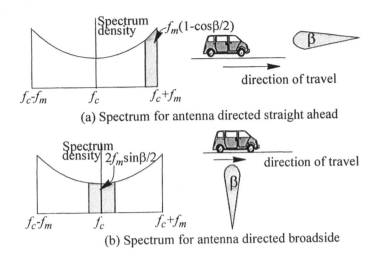

(a) Spectrum for antenna directed straight ahead

(b) Spectrum for antenna directed broadside

Figure 6.6 Receiver input spectra for broadside and straight directed antennas

In [16], field measurements of the cross coupling between vertical and horizontal polarized signals is reported. The measurements were conducted within buildings and houses and around them, where major application of personal communications is envisaged. Measurements show that significant cross coupling of orthogonally polarized signals ranging from -1.8 dB to +3.5 dB exists. The larger values occur in small buildings. In large buildings, and when the signal penetrates through the walls, the cross coupling approaches 0 dB, which indicates that equal power is distributed in the two polarizations examined.

From the above discussion it is seen that strong cross coupling will help portable units to receive adequate signal strength in any orientation, at least in terms of the median or mean signal magnitude. It has been suggested [17] that the field distributions in the horizontal and the vertical planes have little correlation. Thus, a fade in one plane does not necessarily imply a simultaneous fade in the other. This clearly suggests that polarization diversity can be used in portable communications.

6.8 Effect of Antenna Pattern on Handoffs

In the preceding section discussions centered on the impact of the antenna radiation pattern on the signal strength and Doppler spectrum. The importance of the antenna pattern on the overall system performance may not be immediately apparent. However, directional pattern of the antenna resulting from its mounting and structure of the ground plane could have important bearing on the system handover rate. A secondary effect of the nulls in the antenna directional pattern may be in the form of an increase in the number of handovers and the resulting increased management overhead of the system. If the mobile antenna exhibits a large null in the directional pattern due to its mounting position, a premature handover may be affected. In a densely populated environment where several tens of vehicles may be present an increased in the handover frequency may result in congestion of Mobile Switching Centre (MSC).

Losee and Shoshtek discovered that in a slowly changing RF environment, a null in the mobile antenna pattern may find itself pointing towards the direction of the signal arrival long enough to trigger handover as a result of sudden loss in the signal level. This effect, known as Losee-Shoshtek effect, results in an increase in the number of premature handovers [9].

6.9 Adaptive Antennas

The explosive growth of mobile communication systems during the past few years have resulted in saturation of cellular systems. Not only the subscriber base has increased beyond 70% of the population in some areas, a number of new applications have resulted in more pressure towards creation of additional capacity. Thus, the future generation of mobile systems are expected to meet increasing demands for services in the presence of several constraints like limited spectrum, increased interference because of a larger number of customers, fading and shadowing. On top of these requirements, the new systems are expected to surpass the service quality offered by the fixed network. The most plausible solution to the problem of high demand is to assign more spectrum as the mobile phones are quickly replacing the fixed telephones. However, this option is not likely to materialize because of limited available spectrum. The other alternative is to increase frequency reuse coupled with deployment of interference minimizing methods.

Several solutions to overcome the capacity problem have been proposed. For example, a reduction in the cell size has a positive impact on the capacity but this solution increases interference significantly. It is recognized that smaller cell sizes can not be avoided in a quest for higher capacity, therefore current research has been directed towards methods to control interference and to create additional capacity without needing additional spectrum. The methods to control interference include adaptive equalizers to cancel interference, multiuser detectors to make the receivers more immune to interference, interference cancellers - parallel, serial or hybrid, and antenna designs that identify the directions of interference arrival and create nulls in those directions.

Creation of new capacity without needing additional spectrum is an interesting challenge. This challenge relates to increasing supportable data rate from 1 bit/Hz to several tens of bits/Hz. To achieve this objective, information theoretic solution was proposed in the form of Bell Labs Layered Array for Space and Time (BLAST). The BLAST approach exploits multipath, that is, it uses the scattering characteristics of the propagation environment to enhance reception reliability by treating the multiplicity of scattering paths as separate parallel subchannels [20],[21].

In BLAST, the single user data stream is split into multiple substreams, which are transmitted using an antenna array. Since parallel streams are created, the transmission rate is increased in proportion to the number of antennas. At the receiver an array of antenna is again used to pick the multiple transmitted substreams and their scattered images. Each receiving antenna sees all of the transmitted substreams but superimposed. If the multipath scattering is sufficiently uncorrelated, the constituent streams are scattered differently. Using sophisticated signal processing, slight difference in scattering allows the substreams to be identified.

The BLAST signal processing algorithm used at the receiver is the heart of the technique. The signal processing first extracts the strongest substream followed by others in the order of their strengths. Under the widely used theoretical assumptions of independent Rayleigh scattering, the theoretical capacity of BLAST architecture grows roughly linearly with the number of antennas, even when the total transmit power is held constant.

6.10 Summary

This chapter has briefly introduced the subject of antennas within the context of base station and mobile antenna characteristics. No attempt was made to go into electromagnetic theory to derive characteristics of antennas. It is clear that at the base station, directional antennas may be used to suit special features of the terrain and avoid undue interference from other transmitters. Moreover, it is desirable that the radio coverage provided by a base station should be limited to its cell and any transmission beyond the boundaries of the cell be avoided. In this regard adaptive antennas that control the direction of radiation can be designed.

At the mobile unit, the picture is substantially different, because a mobile that continuously change its orientation can not maintain a fixed direction. In general, an omnidirectional antenna is therefore preferred. Only in the presence of a highly scattered signal, a directional antenna may prove to be more useful.

The limitation of spectrum can be overcome by increasing spectrum utilization. In this respect space division multiplexing or BLAST systems have been suggested. One way in which antennas can be effectively used to combat signal fading is by implementing diversity. Chapter 7 is devoted to signal processing where role of diversity, multiuser detection, and adaptive equalization has been discussed in some detail.

References

[1] J. Gardiner and S.A. Kotsopoulos, "Relationship between base station transmitter multi-couling requirements and frequency planning strategies for cellular mobile radio," *Proc. IEE*, vol.132, Part F, pp.384-387, August 1985.

[2] W.S. Davies, R.J. Lang and E. Vinnal, "The challenge of advanced base station antennas for future cellular mobile radio systems," *Proc. IEEE International Workshop on Digital Mobile Radio*, Melborne, March 1987.

[3] R.A. Burberry, "The specification of methods of measurement of antennas in the mobile service, 30-1250," *IEE Colloquium on Antenna Systems for Mobile Communications*, March 1983.

[4] A.L. Davidson, "Mobile antenna gain at 900 MHz," *IEEE Trans. on Veh. Tech.* VT-24, pp. 54-58, 1975.

[5] K. Nishikawa and Y. Asano, "Vertical patterns of mobile antenna at UHF frequencies," *IEEE Veh.Tech Cof. Record*, pp. 44-49, 1985.

[6] K. Nishikawa, "Effect of automobile body and earth on radiation pattern," *Trans. IECE (Japan)*, vol. E67, pp. 555-562, 1984.

[7] D.W. Horn, "Vehicle-caused pattern distortion at 800 MHz," *IEEE Veh. Tech. Conf. Record*, pp. 197-204, 1983.

[8] D.W. Horn, "Selection of vehicular antenna configuration and location through use of radiation pattern," *IEEE Veh. Tech. Conf. Record*, 1973.

[9] S. Shostek, "Can cellular be sold," *Telephone Engineer & Management*, July 15, 1987, pp. 72-77, July 1987.

[10] A.U.H. Sheikh, *Report on the Testing of Cellular Telephones and Antennas, Part I*, Lapp-Hancock Associates Ltd., Ottawa, Canada, 1990.

[11] T. J. McMullin, "Gain antennas for mobile applications," *IEEE Trans. on Vehicular Communications*, vol. VC-14, pp. 141-148, 1965.

[12] Lee, W.C.Y, *Cellular mobile radio telephone system using tilted antenna radiation patterns*, (U.S. Patent 4,249,181, Feb 3, 1981).

[13] Lee, W.C.Y., *Mobile Communications Design Fundamentals*, Sams 1986.

[14] Lee, W.C-Y., "Preliminary investigation of mobile radio signal fading using directional antennas on the mobile unit," *IEEE Trans. Veh. Communications*, vol. VC-15, pp. 8-15, Oct. 1966.

[15] W.C.Y. and Y.S. Yeh, "Polarization diversity system for mobile radio," *IEEE Trans. on Communications,* vol. COM-20,pp. 912-913, October 1972.

[16] D.C. Cox, "Universal portable radio communications," *National Communications Forum,* Chicago, Il., September 24, 1984.

[17] R.H. Clarke, "A statistical theory of mobile radio reception," *Bell System Technical Journal,* vol. 47,pp. 957-1000, July-August 1968.

[18] Gans, M.J., "A power spectral theory of propagation in the mobile radio environment," *IEEE Trans. on Veh. Tech.,* vol., VT-21, No. 1, pp. 27-37, Feb. 1972.

[19] Jakes, W. C., *Microwave Mobile Communications,* J. Wiley, 1974.

[20] G. J. Foschini, "Layered space-time architecture for wireless communications in a fading environment when using multi-element antennas," Bell Labs Technical Journal, Autumn, pp. 41-59, 1996.

[21] G. J. Foschini and M. J. Gans, "On limits of wireless communication in a fading environment when using multiple antennas," Wireless Personal Communications, vol. 6, no. 3, pp. 311-335, 1998.

Chapter 7

SIGNAL PROCESSING IN WIRELESS COMMUNICATIONS

Signal to noise ratio, channel outage, error rate, average error burst length are some of the measures that are used to determine quality of service (QoS). The task of a communication engineer is to achieve high signal to noise ratio, low signal outage probability and minimum error rate. However, meeting all these criteria may not be easy in mobile radio environments. The channel shadowing degrades the received signal to noise ratios by tens of decibels [1], thereby could produce long signal drop outs. The channel fading also results in signal dropouts whenever the interference or noise overwhelms the desired signal. The reduction of signal to noise ratio and the signal drop-outs distort the received signal to the extent that the detected signal becomes inaudible if it is voice or full of error bursts if the information is digital.

To reduce the severity of the signal impairments two solutions are possible. The first is that of increasing the transmit power; this approach is usually not acceptable due to several reasons. For example, an increase in the transmit power results in a proportional increase in the interference level, this necessitates an increase in the frequency reuse distance, which results in a degradation in the system capacity. The second solution, generally preferable, uses signal processing to reduce the outage probability, enhance signal to noise ratio, and improve error performance.

In mobile communications, several signal processing techniques are used to restore the signal quality. Diversity is the best known among them and it is widely used in systems that operate over fading channels. Applicable to both digital and analog systems, it generally results in a significant reduction in signal dropout periods or fade durations, thus reducing the distortion in the received signal. The underlying principle of the diversity is to change the statistical description of the received signal such that the resulting signal to noise has a lower probability of falling below the reception threshold.

Equalization, thought to be difficult to implement in the user terminal few years ago, is now an integral part of many digital communications systems; its use is deemed necessary over frequency selective channels. The adaptive equalizers are now used in the user terminals as well as in the sub-systems at base stations. The underlined principle of equalizer is to make the magnitude of the channel gain constant and phase linear over the signal bandwidth so that the resulting channel transfer function is as close as possible to that of an ideal bandpass channel characteristics.

The channel coding is the third type of signal processing technique. Its use is as essential as the radio itself. Error control coding is used to improve the reliability of signal reception; as without channel coding, the mobile systems are incapable of delivering required reliability. It has been observed that the raw error rate over the mobile channel varies over a wide range (from 10^{-1} to 10^{-4}) and to achieve a BER performance within a narrow range, the use of all the three techniques mentioned above may be necessary.

In this relatively long chapter, Sections 7.1 through 7.7 deal with several aspects of diversity reception systems. The impact of diversity on the channels impairments like noise, interference, random FM, Average crossing rates, and average fade duration are discussed in a comprehensive manner. This is followed by discussion on coding for mobile radio in Section 7.8. Section 7.9 is devoted to the subject of equalization. In this section, a case is made for a very powerful combination of diversity, equalization and channel coding is argued. Finally, the chapter summary is provided in Section 7.10.

7.1 Basic Diversity Classification

Diversity means the state of being composed of different parts, elements, or individuals. In the context of mobile communications, it implies the use of more than one copies of the signal. Such a use results in more robust signal as it results in a reduction of the effect of short term fading. This effect was first discovered by A. de Haas while experimenting with two spaced antennas at HF [2]. It was observed that the short term fluctuations of the signal received at one antenna tend to be independent of the signal fluctuations at the other antenna. Thus, switching between antennas can be arranged to receive strongest of the copies of the signal. Thus, the average signal power improves. As indicated in Chapter 3 {Section 3.8}, if two channels are sufficiently separated in time, frequency, or space, the short term signal fluctuations on the two channels are more or less independent. The probability that the signals on both antennas will fade simultaneously is therefore much smaller relative to the probability of fading on either of channels alone. For example, if probability of signal fading below a certain threshold at each antenna is, say 0.1, the probability that signals on two antenna will fall simultaneously below this threshold will be 0.01. This result is based on the assumption that the signals on the two branches fade independently. It is possible that signal fluctuations on two antennas may have some correlation, even then, the output signal will have reduced fading. Obviously, signals transmitted at different frequencies, polarizations, times or received at different spatial locations or polarization may be combined to achieve a similar reduction in the fading severity of received signal. Several diversity techniques are introduced below.

7.1.1 Space Diversity

Space diversity uses two or more antennas spaced at a sufficient distance "d" from each other such that fading of the signals on the antennas are mutually independent. Following the discussion in Section 3.8, it will be assumed that this condition is met for most practical purposes even if the cross-correlation coefficient between two signals is small.

Of the mathematical models presented in Chapter 3 to define the multipath field, the most complete is probably that of Clarke [3]. The normalized spatial autocorrelation function of the electric field is given by a Bessel function of the first kind order zero (A.26, Appendix A),

$$L_e(\zeta) = J_0^2\left(\frac{2\pi\zeta}{\lambda}\right) \qquad (7.1)$$

Similar autocorrelation functions exist for the two orthogonal magnetic components. It is observed that $L_e(\zeta)$ decreases rapidly with the spatial displacement, and reaches a value of 0.3 at approximately $\lambda/4$ and is zero at 0.378λ. Thus, signals received by the electric field sensing antennas displaced by a distance of not less than $\lambda/4$ should be sufficiently uncorrelated.

7.1.2 Frequency Diversity

The discussion in Chapter 3 showed that the cross-correlation function [Appendix A] between the envelopes of tones received at two frequencies can be used to estimate the coherence bandwidth of the radio channel. In other words, if the same signal is transmitted at two frequencies separated by more than the channel coherence bandwidth, the received signals at these frequencies will be sufficiently uncorrelated. Thus, these signals can be used to reduce degradation due to fading. Since frequency diversity uses more than one carrier to transmit the same signal, it is usually not recommended for use in mobile communication systems because of spectrum scarcity. However, indoor wireless communication, operating in frequency range 20-60 GHz where bandwidth is in abundance, may be an exception.

7.1.3 Time Diversity

In time diversity, a signal is transmitted at two different times, which are sufficiently apart such that the received signals are sufficiently uncorrelated. The assumption is that during the elapsed time between transmissions, the channel has changed sufficiently to warrant decorrelation. The time between the transmissions is usually taken to be the reciprocal of the fading bandwidth, and is inversely proportional to the vehicle velocity. Thus, time diversity does not work for stationary vehicle operating in a static environment. Note that time diversity may work where the environment surrounding the stationary terminal change rapidly. An example of such a scenario is fixed wireless link operating in a public place. If there are M branches, a total delay of $M/2f_m$ is required in transmitting the message. This kind of delay between diversity branches is not desirable for speech communications. As in the case of frequency diversity, time diversity is spectrum inefficient and therefore is rarely used in most mobile applications except where the channel is expected to change very fast.

7.1.4 Polarization Diversity

Polarization diversity is extensively used in UHF/microwave communications. In this technique the signal is transmitted at the same frequency but with different polarizations. A single receiving antenna with two feed horns or dipoles is used to receive cross polarized signals. With cross linearly polarized feeds, the two received signals fade in an uncorrelated manner [4]. In mobile communication systems, however, some correlation between differently polarized signals is expected, so achievable diversity gain is relatively small. It should also be noted that polarization diversity restricts the number of diversity branches to two. Notwithstanding these factors, polarization diversity is actively being considered for implementation in some future personal communications systems [5].

7.1.5 Angle Diversity

In the discussion on space diversity it was assumed that components of the signal arrive at the receiving antennas from all directions with equal probability. Since signal is reflected from scatterers of different type and characteristics, it is reasonable to assume its components are independent. It suggests that antennas with narrow beams could be employed to devise an angle diversity system. Angle diversity has been extensively used in the troposcatter channels in the past [6]. Advanced or smart antennas with steerable or fixed multiple narrow beams are now used in indoor and outdoor wireless systems.

7.2 Methods of Signal Recombination

The availability of two or more independent fading signals raises the important question of how to use these signals in order to mitigate the effect of fading. The method of using these signals is termed as diversity combining or signal recombining. In the following we discuss several diversity combining techniques. However, first we review the characteristics of the signal that are important to diversity. Mobile communication channels have two types of variation:

(a) Long term variations:

The movement of a mobile in shadowed regions, which are caused by hills and other large irregularities in the terrain bestow long term variations in the transmit signal. All diversity techniques meant for combating short term variations are not effective against long term variations of the signal. However, use of two or more transmitting antennas, a technique which is called macro diversity can overcome the long term variations. It is discussed in Section 7.7.

(b) Short term variations:

It was noted in Chapter 3 that relatively short term fluctuations in the signal magnitude, observed when the signal propagates over the mobile channel, are due to the presence of local scatterers. The resultant signal fluctuations are, in general, modeled to follow Rayleigh statistics. In the analyses that follow, only Rayleigh fading is considered. The techniques developed can be used for other statistical descriptions of signals. Furthermore, short term fading appears to behave phenomenologically as if the total signal were the sum of many individual signal components arriving over a multiplicity of radio paths with infinitesimally small mutual delays. Given M such equivalent and statistically independent signal transmission paths, the diversity receiver operates on signals on these paths in order to improve signal legibility over that if the signal were received over a single radio path. It may do this by either choosing at any instant of time the best of the M signals (selection or switching diversity), or combining them in a linear manner (combining diversity).

In general, the diversity receiver works on M available signals to form the resultant signal. i.e.

$$s(t) = a_1 s_1(t) + a_2 s_2(t) + \ldots + a_j s_j(t) + \ldots + a_M s_M(t)$$

$$= \sum_{j=1}^{M} a_j s_j(t) \tag{7.2}$$

where $s(t)$ is the output of diversity combiner, a_j is the weight assigned to the signal $s_j(t)$, $j = 1,2,\ldots M$. The signals may be combined either at IF or at post-detection stage. These combining techniques are known as pre-detection and post detection combing respectively. In the pre-detection technique, signals are combined after they are co-phased but prior to detection while in the post-detection the signals are combined after detection. It is evident that in the post-detection combining, the phase information is lost. The consequence of this is a loss in the diversity gain of approximately 3 dB.

In the following subsections, four principal types of combining techniques are discussed. These are scanning, selection, maximal ratio and equal gain combining. Most diversity systems make use of any of these basic techniques or their modified versions. For example, the switched diversity method is a degenerate form of the dual selection diversity technique.

7.2.1 Scanning Diversity

The scanning diversity is a form of switched diversity with two or more antennas. A selector scans the antennas in a predetermined sequence until an antenna with a signal level above the threshold level is found. This antenna is connected to the receiver input. When the signal level in the selected branch falls below the threshold level, the channels are scanned afresh in the same sequence. Scanning diversity may also be considered within the class of selection diversity with a unique selection procedure. This form of diversity will not be considered further because it is hardly used in practice. Furthermore, the selection diversity described in the next section has performance superior to it.

7.2.2 Selection Diversity

The selection diversity is an improved form of scanning diversity. Instead of scanning all available antennas or signal branches $s_1, s_2, ... s_m$, until an acceptable one is found, the system monitors (simultaneously or not necessarily in a particular sequence) all of them and selects the strongest; the remaining $M-1$ signals are discarded. Equation (7.2) may be rewritten as

$$s(t) = a_j s_j(t); \quad a_j \geq a_k, k = 1, 2, ..., M, k \neq j \tag{7.3}$$

The jth channel is selected such that $a_j \geq a_k$, $k=1, 2,..., M, k \neq j$. The design criterion used in (7.2) is

$$\begin{aligned} a_j &= 1 \quad\quad j = k \\ a_j &= 0 \quad\quad j \neq k \end{aligned} \tag{7.4}$$

The performance of any diversity scheme is measured in terms of the outage probability, which is defined as the probability of the received signal falling below a threshold level. Alternatively, the term diversity gain is used, which is defined as the reduction in dB the transmission power needed to yield the signal outage probability identical to that achieved in the absence of diversity. The distribution function of signal to noise ratio for an M-branch diversity system is simple to obtain, provided the branch noise powers $E[n_j^2]$ are equal. This is generally a reasonable assumption since all the receivers used in the diversity branches have similar noise figures. If the branch noise power $E[n_j^2] = N$, then the instantaneous signal to noise ratio in the jth branch, γ_j, is defined as

$$\gamma_j = \frac{\text{local mean signal power per branch}}{\text{mean noise power per branch}} = \frac{\langle r_j^2 \rangle}{2N} \tag{7.5}$$

and the mean signal to noise ratio as

$$\Gamma_j = \frac{\text{Mean signal power per branch}}{\text{Mean noise power per branch}} = \langle \gamma_j \rangle \tag{7.6}$$

The probability density function of an individual signal envelope r_j is Rayleigh with mean power $2\sigma^2$ and is given by:

$$p(r_j) = \frac{r_j}{\sigma^2} exp\left(-\frac{r_j^2}{2\sigma^2}\right) \tag{7.7}$$

Therefore,

$$\gamma_j = \frac{\langle r_j^2 \rangle}{2N}$$

$$\Gamma_j = \frac{\int_0^\infty \frac{r_j^3}{\sigma^2} exp\left(-\frac{r_j^2}{2\sigma^2}\right) dr_j}{2N} = \frac{\sigma^2}{N} \tag{7.8}$$

Using $p(\gamma_j) \, d\gamma_j = p(r_j) \, dr_j$, we get,

$$p(\gamma_j) = \frac{1}{\Gamma_j} exp\left(-\frac{\gamma_j}{\Gamma_j}\right) \tag{7.9}$$

Assume γ_s to be the reception threshold or the minimum signal to noise ratio required for acceptable reception. The probability that the carrier-to-noise ratio γ_j in any one branch is less than or equal to γ_s is:

$$P(\gamma_j \leq \gamma_s) = \int_0^{\gamma_s} p(\gamma_j) d\gamma_j = 1 - exp\left(-\frac{\gamma_s}{\Gamma_j}\right) \tag{7.10}$$

Thus, the probability of signal outage that γ_j in all the M branches are simultaneously less than or equal to γ_s; this is given by:

$$P(\gamma_1, ..., \gamma_M \leq \gamma_s) = \prod_{j=1}^{M} Prob(\gamma_s \leq \gamma_s) = \prod_{j=1}^{M} \left(1 - exp\left(-\frac{\gamma_s}{\gamma_j}\right)\right) \quad (7.11)$$

The following four cases are of practical interest. These are analyzed for diversity gain and outage probability under the following operating conditions:

(i) the mean signal to mean noise ratio in all branches is equal;

(ii) the instantaneous signal-to-noise ratio, averaged over one RF cycle, at the output of the combiner is lower than the branch mean signal to noise level; this condition occurs when the instantaneous signal-to-noise ratio on every branch is below its mean value;

(iii) the mean signal-to-noise ratio in one of the M branches is lower than the resulting signal-to-noise ratio at the output of the combiner;

(iv) the signals on all branches have some degree of correlation between them.

Case (a) $\gamma_1 = \gamma_2 .... = \gamma_M = \gamma$

This case is typical of mobile communication system operations. Here, the mean[†] signal strength to mean noise ratio for each of the diversity branches is equal. The probability of the signal fading below a given level γ_s is then given by:

$$P(\gamma \leq \gamma_s) = P(\gamma_1, \gamma_2, ..., \gamma_M \leq \gamma_s) = \left(1 - exp\left(-\frac{\gamma_s}{\Gamma}\right)\right)^M \quad (7.12)$$

In Figure 7.1, plots of the outage for systems with M-branch diversity are presented. Note that for γ_s/Γ equal to -10 dB, the 2-branch diversity brings the outage from 10^{-1} to 10^{-2}. In other words, for 1% outage, implementation of two and four branch diversity systems will result in savings in transmitting power of 10 and 16 dB respectively. This is a substantial improvement over the no diversity system. Note, however, that when the number of branches are increased beyond 2, the realizable gain is gradually decreasing with an increase in the diversity branches.

---

† *Usually median signal power is measured in the mobile environments. If we know the statistics of the signal, the mean signal power can be derived from the median signal power.*

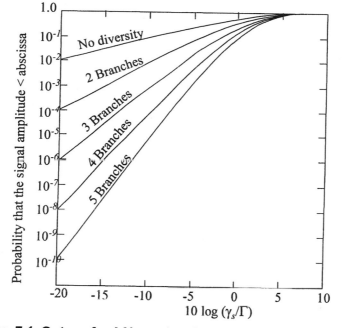

Figure 7.1 Outage for *M*-branch selection diversity combiner

Case (b) $\gamma_j \ll \Gamma_j, j = 1, 2, \cdots\cdots, M$

This case suggests that the instantaneous signal-to-noise ratio on all branches is less than their respective mean signal to mean noise ratio. The mean signal to mean noise ratio in each branch is assumed to be identical. Since the signal-to-noise ratio at the output of the combiner is identical to the signal on one of *M* branches, therefore the outage is given by:

$$P(\gamma \le \gamma_s) = \prod_{j=1}^{M} \left(1 - exp\left(-\frac{\gamma_s}{\Gamma_j}\right)\right)$$

$$P(\gamma \le \gamma_s) \cong \prod_{j=1}^{M} \frac{\gamma_s}{\Gamma_j} = \frac{\gamma_s^{M}}{\prod_{j=1}^{M} \Gamma_j}, \qquad \text{for } \frac{\gamma_s}{\Gamma_j} \ll 1 \quad \text{all } j \tag{7.13}$$

Case (c) $\Gamma_M < \gamma \ll (\Gamma_1, \Gamma_2, \cdots\cdots \Gamma_{M-1})$

Here, the combiner output signal-to-noise ratio is greater than the mean signal to noise ratio of the branch *M* but is smaller than on each of the remaining *M-1* branches. The outage is given by:

$$Prob(\gamma_1, \gamma_2, ..., \gamma_M) = \left[1 - exp\left(-\frac{\gamma_s}{\Gamma_M}\right)\right]\left(\frac{\gamma_s^{M-1}}{\prod_{1=1}^{M}\Gamma_j}\right) \qquad (7.14)$$

If $\gamma_s \gg \Gamma_M$, the first term of the product on the right hand side of equation (7.14) will be very close to unity, and the outage may be approximated by:

$$Prob(\gamma_1, \gamma_2, ..., \gamma_M \leq \gamma_s) = \frac{\gamma_s^{M-1}}{\prod_{j=1}^{M-1}\Gamma_j} \qquad (7.15)$$

It should be noted that if $\Gamma_M \gg \gamma_s$, the diversity is ineffective since the instantaneous S/N is adequate in the chosen branch and change over to other branches does not take place. Here we have shown that if, out of M, K branches have short term average S/N lower than the other M-K branches, then only M-K branch diversity will be effective, i.e.

$$P(\gamma \leq \gamma_s) = \frac{\gamma_s^{M-K}}{\prod_{j=1}^{M-K}\Gamma_j}; \Gamma_{M-K+1}, \Gamma_{M-K+2}, ..., \Gamma_M \leq \gamma_s \qquad (7.16)$$

Case (d) Correlation between branch signals

In this case either the antennas are not sufficiently spaced so that the signals on them are mutually correlated to some degree or the signal reaches the antennas via paths having with some correlation[†]. In practice, this means that the signals on different antennas go through fades at the same time thereby reducing the effectiveness of diversity. The probability that the combiner output S/N is below some selected level of reception is given by:

$$P(\gamma \leq \gamma_s) = P(\gamma_1, \gamma_2, ..., \gamma_M \leq \gamma_s)$$
$$= \int_0^{\gamma_1} d\gamma_1 \int_0^{\gamma_2} d\gamma_2 \cdots \int_0^{\gamma_M} P(\gamma_1, \gamma_2, ..., \gamma_M) d\gamma_M \qquad (7.17)$$

where $p(\gamma_1, \gamma_2, \gamma_3, \cdots, \gamma_M)$ is the joint probability density function of signal-to-noise ratios at M branches. From Section 3.8, we have the joint probability density function of the envelopes as:

[†]. *This statement is applicable to space diversity. For other types of diversity, similar results are obtained if branch signals are uncorrelated. For example, in the case of frequency diversity the frequencies are spaced more than the channel coherence bandwidth.*

$$P(r_1, r_2, ..., r_m) = \int_{-\pi}^{\pi} d\theta_1 \int_{-\pi}^{\pi} d\theta_2 \cdots \int_{-\pi}^{\pi} P(r_1, \theta_1, r_2, \theta_2, ..., r_M, \theta_M) d\theta_M \quad (7.18)$$

Here a two branch diversity system with correlated signals is considered. The correlated multiplicative fading processes on the two branches are assumed to be jointly Gaussian, with complex cross-covariance ρ. The normalized covariance between branches is closely approximated by $|\rho^2|$.

If R_1 and R_2 are normalized relative to $r/(2N_k)^{1/2}$, the envelope magnitudes at the two branches, and Γ_1, Γ_2 represent the mean SNR's on the two branches ($\Gamma_k = <R_k^2>$), then the joint *p.d.f.* of R_1, and R_2 at any instant is (see appendix A).

$$P(R_1, R_2) = \frac{4R_1 R_2}{\Gamma_1 \Gamma_2 1 - |\rho|^2} I_0 \left(\frac{2|\rho| R_1 R_2}{1 - |\rho|^2 \sqrt{\Gamma_1 \Gamma_2}} \right) \cdot$$
$$exp \left(-\frac{1}{(1-|\rho|^2)} \left(\frac{R_1^2}{\Gamma_1} + \frac{R_2^2}{\Gamma_2} \right) \right) \quad (7.19)$$

For a threshold level, x, the outage is given by:

$$P(\gamma < x) = 1 - exp\left(-\frac{x}{\Gamma_1}\right) [Q(b, |\rho|a) + Q(a, |\rho|b)] + \quad (7.20)$$
$$exp\left(-\frac{x}{1-|\rho|^2}\left(\frac{1}{\Gamma_1} + \frac{1}{\Gamma_2}\right)\right) I_0 \left(\frac{2|\rho| x}{1 - |\rho|^2 \sqrt{\Gamma_1 \Gamma_2}} \right)$$

where

$$a = \sqrt{\frac{2x}{\Gamma_1(1-|\rho|^2)}} \quad \text{and} \quad b = \sqrt{\frac{2x}{\Gamma_2(1-|\rho|^2)}} \quad (7.21)$$

and $Q(x,y)$, is defined as

$$Q(x, y) = \int_0^\infty exp\left(-\frac{x^2 + y^2}{2}\right) I_0(a\lambda)\lambda d\lambda \quad (7.22)$$

Under normal operating conditions the mean signal-to-noise ratios on the two branches are equal. It is possible, however, that the mean signal to noise ratios on these branches do not remain equal. The effects of unequal branch signals and the correlation between the signals may be demonstrated by considering the following cases.

(i) $\Gamma_1 = \Gamma_2$

It ha been shown [7] that as correlation between the branch signals increases, the diversity gain reduces. Even at a high correlation figure of $\rho = 0.95$ a diversity gain of 4.2 dB is realized at 1% outage.

(ii) $\Gamma_1 \neq \Gamma_2$

When the mean signal to noise ratio on two branches is not identical, the combiner output will remain connected to the stronger branch for longer period of time that is the presence of the weaker branch will not be felt unless the instantaneous *SNR* on the stronger branch drops below that on the weaker branch. Thus, as the imbalance between the branch signal increases, the diversity gain decreases. For example, for $\Gamma_1 = 2\Gamma_2$, and $\Gamma_1 = 10\,\Gamma_2$ (for ρ as high as 0.95), the diversity gain degrades by 1.5 and 5 dB respectively relative to the case of equal strength branch diversity [7]. In practical systems, to achieve a meaningful diversity gain, the correlation coefficient between the branch signals should be kept below 0.8 [8]. At low outage rates, *Prob* $(\gamma < \gamma_s)$ is approximately given by:

$$P(\gamma < \gamma_s) \approx \gamma_s^2 \left[\frac{1}{\Gamma_1 \Gamma_2} \frac{1}{(1 - |\rho|^2)} \right] \qquad \gamma_s \ll \Gamma_1, \Gamma_2 \qquad (7.23)$$

7.2.3 Maximal Ratio Combining (MRC)

In maximal ratio combining (MRC) the branch signals are combined so that the combiner output signal to noise ratio is maximized. The combiner output may be written as:

$$v(t) = \sum_{j=1}^{M} a_j v_j(t) \qquad (7.24)$$

If P denotes the local power ratio of *s(t)*, then the maximal ratio combiner realizes

$$P = \sum_{j=1}^{M} P_j \qquad (7.25)$$

For purposes of analysis the following assumptions are made:

(v) signal and noise in each branch are mutually independent and the noise is additive. Thus

$$v_j(t) = g_j(t)u(t) + n_j(t) \qquad (7.26)$$

(vi) Here $u(t)$ represents the transmitted signal and $g_j(t)$ is the time varying gain of the signal on the jth branch. It is also assumed that $g_j(t)$ varies slowly relative to the instantaneous variation of $u(t)$.

(vii) the signals $v_j(t)$ are coherent at the input of combiner;

(viii) the noise components are incoherent (uncorrelated) and have zero mean, i.e.

$$\langle n_i(t)n_j(t)\rangle = \langle n_i(t)\rangle \langle n_j(t)\rangle \tag{7.27}$$

The signals may be combined at the IF stage (pre-detection combining) or following the detection stage (post detection combining). Since detection of FM signals is non-linear operation therefore pre-detection combining is preferred for mobile radio communication. In the following we discuss pre-detection combining. The complex envelope of the IF signal of kth branch is given by:

$$v_k(t) = g_k(t)u(t) + n_k(t) \tag{7.28}$$

Then the complex envelope output at the combiner output is given by:

$$v(t) = \sum_{k=1}^{M} a_k(t)v_k(t) \tag{7.29}$$

The gains $a_k(t)$ are complex weighting factors, changing with time as the signal of the branch changes due to fading.

The instantaneous combiner output is given by:

$$
\begin{aligned}
v(t) &= \sum_{k=1}^{M} a_k[g_k(t)u(t) + n_k(t)] \\
&= u(t)\sum_{k=1}^{M} a_k(t)g_k(t) + \sum_{k=1}^{M} a_k(t)n_k(t) \\
&= s(t) + n(t)
\end{aligned}
\tag{7.30}
$$

The instantaneous output noise power is

$$N = \frac{1}{2}\langle |n(t)|^2\rangle = \sum_{k=1}^{M} |a_k|^2 N_k \tag{7.31}$$

Assuming the normalization

$$\frac{1}{2T}\int_{-T}^{T} |u(t)|^2 dt = 1 \tag{7.32}$$

instantaneous signal-to-noise ratio at the output of the combiner is

$$\gamma = \frac{1}{2}\frac{\left|\sum_{k=1}^{M} a_k g_k(t)\right|^2}{\sum_{k=1}^{M} \left|a_k(t)\right|^2 N_k} \qquad (7.33)$$

Rather than using the standard techniques of differentiation, Brennan [9] used Schwartz's inequality to find the condition which maximizes the combiner output signal-to-noise ratio. From Schwartz's inequality, if the parameters a_k and b_k are complex numbers, then

$$\left|\sum_{k=1}^{M} a_k^* b_k\right|^2 \leq \left(\sum_{k=1}^{M} \left|a_k\right|^2\right)\left(\sum_{k=1}^{M} \left|b_k\right|^2\right) \qquad (7.34)$$

This can be interpreted as an inequality on the scalar product of two complex valued vectors relative to their individual magnitudes. A necessary and sufficient condition for equality to hold is that the two vectors be parallel and were in phase. Thus, application of this result to equation (7.33) results in a condition for maximum SNR at the output of the combiner. Rewriting equation (7.34) as

$$\left|\sum_{k=1}^{M} \frac{g_k(t)}{\sqrt{N_k}} a_k(t) \sqrt{N_k}\right|^2 \leq \left(\sum_{k=1}^{M} \frac{\left|g_k(t)\right|^2}{N_k}\right)\left(\sum_{k=1}^{M} \left|a_k(t)\right|^2 N_k\right) \qquad (7.35)$$

$$\frac{1}{2}\frac{\left|g_k(t)a_k(t)\right|^2}{\left|\sum_{k=1}^{M} a_k(t)\right|^2 N_k} \leq \frac{1}{2}\sum_{k=1}^{M} \frac{\left|g_k(t)\right|^2}{N_k} \qquad (7.36)$$

$$\gamma \leq \frac{1}{2}\sum_{k=1}^{M} \frac{\left|g_k(t)\right|^2}{N_k}$$

For maximum γ the equality holds and the condition on gains α_k is obtained as:

$$a_k(t) = \frac{K g_k(t)}{N_k}, k = 1, 2, ..., M \qquad (7.37)$$

where K is a constant.

Equation (7.37) implies that the optimum weighting for each branch is proportional to the branch channel gain and inversely proportional to the branch noise power. The phase of the signals on all branches should be aligned to some common reference. Since each term in the right hand side of equation (7.36) represents the instantaneous signal-to-noise ratio of the diversity branches, therefore the combiner instantaneous S/N is given by:

$$\gamma = \sum_{k=1}^{M} \gamma_k \qquad (7.38)$$

In order to align the branch signals, the gain and phase of each branch must be estimated to determine the weighting factors used in the combiner. Obviously, any error in the estimate of the phase or the magnitude will result in some degradation in the diversity gain. It has been shown that an overall phase error of 36.5° will cause a loss of not more than 1 dB and 60° will cause a loss of not more than 3 dB. Similarly, errors in the relative magnitudes of 0.5 and 1 dB will result in losses of up to 1 dB and 3 dB respectively [7].

7.2.3.1 Distribution with Equal SNR in all Branches

To evaluate the performance of MRC diversity, we again find the outage at the combiner output. We assume that the variables γ_k are independently distributed. The *pdf* of γ, $p(\gamma)$, can then be determined conveniently by first determining its characteristic function. The evaluation of this is simplified since γ and γ_k are positive real, and the Laplace transform can be used [7].

$$F(s) = \int_0^\infty exp(-\gamma s) p(\gamma) d\gamma = \langle exp(-\gamma s) \rangle$$

$$= \left\langle exp\left(-s \sum_{k=1}^{M} \gamma_k\right) \right\rangle = \prod_{k=1}^{M} \langle exp(-\gamma_k s) \rangle \qquad (7.39)$$

Substituting for $p(\gamma_k)$ from equation (7.9), to get distribution of γ as

$$p(\gamma) = \frac{1}{(M-1)!} \frac{\gamma^{M-1}}{\Gamma^M} exp\left(-\frac{\gamma}{\Gamma}\right), \text{ when } \Gamma_j = \Gamma \text{ all } j \qquad (7.40)$$

$$Prob(\gamma < \gamma_s) = \int_0^{\gamma_s} p(\gamma) d\gamma$$

$$= \frac{1}{(M-1)! \Gamma^M} \int_0^{\gamma_s} \gamma^{M-1} exp\left(-\frac{\gamma}{\Gamma}\right) d\gamma \qquad (7.41)$$

which is the outage probability, shown in Figure 7.2 for various values of M. Comparing Figure 7.2 with Figure 7.1 shows that MRC has lower outage probability compared with the selection diversity. For two branch diversity the outage is given by

---

$$Prob(\gamma < \gamma_s) = 1 - \left(1 + \frac{\gamma_s}{\Gamma}\right)\exp\left(-\frac{\gamma}{\Gamma}\right), \quad \text{for} \quad M = 2 \tag{7.42}$$

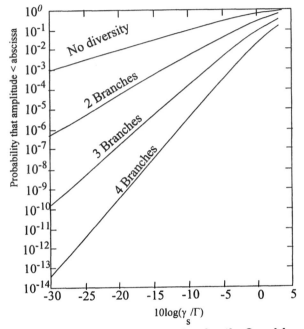

Figure 7.2 Outage for *M*-branch maximal ratio Combiner

When correlation between signal branches exists, the outage probability for two branch MRC diversity with correlation is given by [11]

$$Prob(\gamma < \gamma_s) = \int_0^{\gamma_s} p(\gamma) d\gamma$$
$$\tag{7.43}$$
$$= 1 - \frac{1}{2|\rho|}\left\{(1 + |\rho|)\exp\left[-\frac{\gamma_s}{\Gamma(1 + |\rho|)}\right] - (1 - |\rho|)\exp\left[-\frac{\gamma_s}{\Gamma(1 - |\rho|)}\right]\right\}$$

7.2.4 Equal Gain Combining (EGC)

Maximal ration combining, though highly effective, is a sophisticated technique which requires substantial circuitry. Equal gain combining (EGC) is

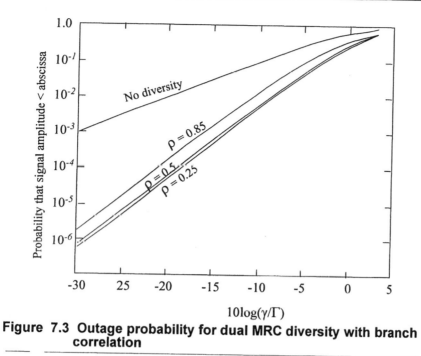

Figure 7.3 Outage probability for dual MRC diversity with branch correlation

a simpler and less expensive alternative achieved at the cost of only a small loss in performance. In equal gain combining, the weighting factor α_k in (7.33) and (7.37) have equal magnitude but their phases are aligned to a single reference. Two cases are considered: first, in which signals on the two branches are uncorrelated, and the second when the signals have some correlation between them.

Case (a) Uncorrelated branch signals
Equation (7.36) is rewritten as:

$$\gamma = \frac{1}{2}\frac{\left[\sum_{k=1}^{M}|g_k|\right]^2}{\sum_{k=1}^{M}N_k} \tag{7.44}$$

where parameters $|g_k|$ are the individual signal envelope levels.

It should be noted that the assumption of equal noise power levels in all branches is crucial to the effectiveness of equal gain combining, otherwise branches with high noise and low signal levels could dominate the *SNR* at the combiner output, since in practical systems it is difficult to measure *S/N* ratio instead we measure (*S+N*) rather than signal strength only.

When the signal envelopes g_k are Rayleigh distributed, the short term distribution at the output of the combiner involves a normalized sum of *M* such signals, we define a variable *y* as

$$y = \frac{1}{\sqrt{2N_T}}\sum_{k=1}^{M} |g_k| \tag{7.45}$$

A simple transformation $\gamma = y^2$ will yield the distribution of signal-to-noise ratio. The combiner output is given by the sum of the outputs of the signals from the individual branches after the signals have been co-phased. Determination of the probability distribution of the sum of squares of Rayleigh distributed signals is possible only through numerical methods. The plots of outage probability is similar to that of MRC but is slightly worse than that for MRC. The distribution of the signal to noise ratio of the combiner output at low values of γ is given by [10]:

$$P(\gamma) = \frac{2^{M-1}}{(2M-1)!}\left(\frac{1}{\prod_{k=1}^{M}\frac{N_k}{N_T}}\right)\left(\frac{\gamma^{M-1}}{\prod_{k=1}^{M}\Gamma_k}\right) \tag{7.46}$$

The mean value of the combined signal to noise ratio may be found quite easily, since the signal on the various branches is uncorrelated [10]

$$\langle\gamma_s\rangle = \Gamma\left[1 + \frac{(M-1)\pi}{4}\right] \tag{7.47}$$

It can be seen that for an infinite number of diversity branches, the signal to noise ratio at the equal gain combiner is only 1.05 dB lower than the output of the maximal ratio combiner.

To maximize the Equal Gain combining performance, we need to minimize $p(\gamma)$ for small values of γ. This leads to the condition [7],

$$\prod_{k=1}^{M}\frac{N_k}{N_T} = \left(\frac{1}{M}\right)^M \tag{7.48}$$

With this condition, (7.46) becomes

$$p(\gamma) = \frac{2^{M-1} M^M \gamma^{M-1}}{(2M-1)! \dfrac{M}{\displaystyle\prod_{k=1}^{M} \Gamma_k}} \tag{7.49}$$

and the outage probability is written as [7]

$$p(\gamma < \gamma_s) = \int_0^{\gamma_s} p(\gamma)d\gamma = \frac{(2M)^M \gamma_s^M}{(2M)! \dfrac{M}{\displaystyle\prod_{k=1}^{M} \Gamma_k}} \tag{7.50}$$

It has been shown that for very large number of branches, the Equal Gain combiner requires 1.34 dB more power than the maximal ratio combining for low outage.

<u>Case (b) Branch signal correlation</u>

When cross correlation coefficient between the envelopes of the branch signals, ρ_r, is not negligible and is closely approximated by $|\rho|^2$ where $|\rho|$ is the magnitude of normalized complex cross-variance of complex Gaussian process [7]. The cumulative probability distribution for the instantaneous carrier-to-noise ratio of two branch diversity below a certain threshold value, γ_s, is approximated by [12]:

$$Prob(\gamma \le \gamma_s) = 1 + \frac{(1 - \sqrt{\rho_r})exp\left(-\dfrac{2a\gamma_s}{\Gamma}\right) - (1 + \sqrt{\rho_r})exp\left(-\dfrac{2b\gamma_s}{\Gamma}\right)}{2\sqrt{\rho_r}} \tag{7.51}$$

where

$$a = \frac{g_2}{2(1 - \sqrt{\rho_r})}, \quad b = \frac{g_2}{2(1 + \sqrt{\rho_r})}$$

$$g_2 = 1.16; \quad \rho_r < 1 \tag{7.52}$$

$$g_2 = 1; \quad \rho_r = 1$$

We use (7.51) and (7.52) to find that the outage probability drops from 0.048 for no diversity to 0.011 and 0.023 for $\rho_r = 0.5$ and 0.8 respectively when $10 \log 10(\gamma_s/\Gamma) = -10$ dB. This shows that despite high branch correlation, improvement in outage can still be obtained. These results are of particular interest in the case of base station diversity where spacing between the antennas is usually limited. As indicated in Chapter 3, the distance between the base station antennas for complete decorrelation or branch independence may be of the order of several tens of meters.

EXAMPLE E-7.2.1

Find the correlation coefficient required between signals on two branch diversity system so as to obtain a gain of 8 dB over the average carrier to noise ratio of a single channel for at least 99.7% of the time.

Solution

The type of diversity combining is not specified. We shall solve the problem for selection and maximal ratio combining.

1. Selection Diversity

We use (7.20) and its approximation in (7.23) to calculate $Prob(\gamma < x)$. The respective mean carrier to noise (CNR) for the two branches may be written as

$$\Gamma_k = \frac{\text{Mean signal power in branch k}}{\text{Mean noise power in branch k}} \tag{7.53}$$

(7.23) is reproduced hers, as

$$P(\gamma < \gamma_s) \approx \gamma_s^2 \left[\frac{1}{\Gamma_1 \Gamma_2} \frac{1}{\left(1 - |\rho|^2\right)} \right] \qquad \gamma_s \ll \Gamma_1, \Gamma_2 \tag{7.54}$$

For a single branch, the probability that the instantaneous CNR, γ, is below a level γ_s, is given in (7.10), which we can use to find threshold γ_s relative to Γ. Thus, $\gamma_s/\Gamma = 10 \log 0.003 = -25.22$ dB.

In order to obtain an 8 dB gain in the two branch system, the signal level must remain at -17.7 dB or higher. Now assuming, $\Gamma_1 = \Gamma_2$, we have

$$0.003 = Prob(\gamma < \gamma_s) \approx \frac{(\gamma_s/\Gamma)^2}{1 - |\rho|^2} = \frac{(0.01897)^2}{1 - |\rho|^2} \tag{7.55}$$

or $|\rho| = 0.9381$.

2. Maximal Ratio Combining

We use (7.43) to find the correlation coefficient as $|\rho| = 0.9762$.

This result suggests that maximal ratio combining is more susceptible than the selection diversity to the correlation between the branch signals.

7.2.5 Switched Diversity Scheme (SDS)

It is difficult to implement selection diversity in practice since it uses variable threshold, which is provided by one of the diversity branches. In the operation of the selection diversity, all branches are examined, compared with each other and the branch with the largest signal is selected. Selection diversity has, thus, an adaptive infinitely variable threshold equal to the instantaneous value of the signal plus noise on the branch not currently under use. This is the major drawback in its implementation.

Switched diversity differs in that a fixed threshold is used. Consider the common case of a two-branch system. Whenever the signal on the branch in use (Branch 1) goes below the selected threshold, switching to the other branch (Branch 2) takes place. If the signal on Branch 2 is also below the threshold, the antenna switches back to Branch 1. Switching will continue until an acceptable signal level on one of the two branches is found. This switching technique is known as a 'switch and examine' strategy. The rapid switching between branches is obviously not desirable, since it is a source of noise; it can be avoided by following a so called 'switch and stay' strategy. These strategies are shown in Figure 7.4.

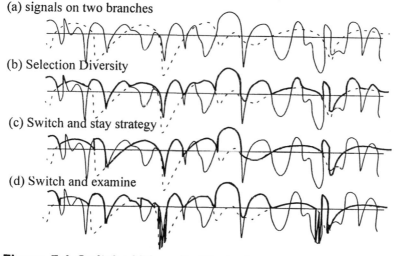

Figure 7.4 Switched Diversity Strategies

In the switch and stay strategy, switching to Branch 2 takes place whenever the signal on Branch 1 falls below a selected threshold regardless of the signal level on Branch 2. The noise introduced due to rapid switching is thus greatly reduced. Figure 7.5 shows the switched diversity receiver block diagram.

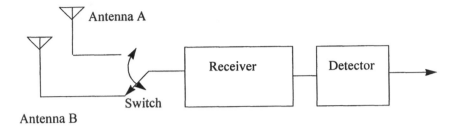

Figure 7.5 A schematic of switched diversity receiver

An analysis of the switch and stay strategy requires determination of the resultant cumulative probability distribution of the output signal. This derivation is presented in the following subsection.

7.2.6 Cumulative Probability Distribution of the Switched Diversity Output

Several authors have derived the expression for the probability distribution of the output of the switched diversity system [13]-[15]. Here, we shall follow the method used by Parsons [15]. As mentioned earlier, the switched diversity is a degenerate form of two branch selection diversity. The two branches carry signals $r_1(t)$ and $r_2(t)$, which are independent and identically distributed Rayleigh processes, with identical means and autocorrelation functions. The envelope, $R(t)$, of the output signal is also a stochastic process. $R(t)$ consists of portions of the envelopes of signals $r_1(t)$ and $r_2(t)$. This is shown in Figure 7.7.

For a threshold of A, we define

$$q = Prob(r \le A) = 1 - exp\left(-\frac{A^2}{2\langle r^2 \rangle}\right) \tag{7.56}$$

and

$$p = Prob(r > A) = exp\left(-\frac{A^2}{2\langle r^2 \rangle}\right) \tag{7.57}$$

The average duration that r is above A is

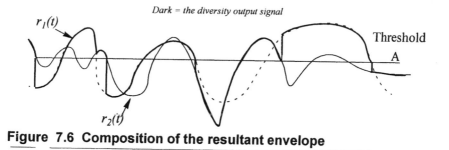

Figure 7.6 Composition of the resultant envelope

$$\tau_p = \frac{p}{q}\tau_q \quad \text{and} \quad \tau_q = \frac{\sigma}{\sqrt{\pi}}\frac{exp\left(-\frac{A^2}{2\sigma^2}\right) - 1}{A f_m} \tag{7.58}$$

τ_q is the average fade duration (see (3.59)) and $<r^2> = \sigma^2$.

We observe that the signal switches from r_1 to r_2 at $t = t_1$ and returns to r_1 at $t = t_0'$. The average duration between t_1 and t_0' is $\tau_p/2$. The normalized autocorrelation coefficient of $r_1(t+\tau_p/2)$ is

$$\rho = \frac{\langle r_1(t)r_2(t)\rangle - \langle r_1(t)\rangle^2}{\langle r_1(t)^2\rangle - \langle r_1(t)\rangle^2} = \frac{R_r\left(\frac{\tau_p}{2}\right) - \langle r_1(t)\rangle^2}{\langle r_2(t)^2\rangle - \langle r_1(t)\rangle^2} = J_0\left(\sqrt{\pi}\frac{\sigma}{A}\right) \tag{7.59}$$

where $J_0(.)$ is the zero order Bessel function, and

$$R_r(\tau) \approx \frac{\pi\sigma^2}{2}\left(1 + \frac{J_0^2(\tau)}{4}\right) \tag{7.60}$$

In most cases the switching will be performed for small values of ρ, implying independence between $r_1(t)$ and $r_2(t)$. It is seen from Figure 7.7 that two types of transitions exist, successful and unsuccessful. The type is determined by whether or not the signal increases after switching. The signal level following a successful transition remains above the threshold, whereas after an unsuccessful transition the signal remains below the threshold, at least initially.

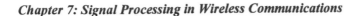

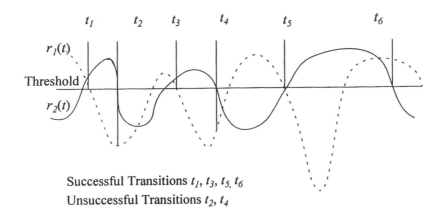

Successful Transitions t_1, t_3, t_5, t_6
Unsuccessful Transitions t_2, t_4

Figure 7.7 Successful and unsuccessful transitions

First, consider segments that follow a successful transition. Assume that we require the statistics of the combiner output about an arbitrary level A. In periods that follow successful transitions, the probability that $R(t) \leq A$ is conditioned on the signal on the branch in use is above the threshold T. We also note that $r_1(t)$ and $r_2(t)$ are statistically indistinguishable. Thus, we can write

$$
\begin{aligned}
P[R \leq A] &= P[R \leq A | (r \geq T)] \\
&= \frac{P_r(A) - q}{1 - q}, \quad A \geq T \\
&= 0, \quad A \leq T
\end{aligned}
\tag{7.61}
$$

Consider now the second type, the unsuccessful transition. The periods following an unsuccessful transition can be split into two parts, corresponding to times spent below and above the threshold level T. For the first of these the probability that $R(t) < A$, when the signal on the branch in use is below the threshold T, is given by

$$
\begin{aligned}
P[R \leq A] &= P[R \leq A | r \leq T] \\
&= \frac{P_r(A) - q}{1 - q}, \quad A \leq T \\
&= 1, \quad A \geq T
\end{aligned}
\tag{7.62}
$$

For the second part of the period, the situation is identical to that which follows a successful transition and (7.61) applies.

To obtain the overall cumulative distribution, it is necessary to combine (7.61) and (7.62) with appropriate weighting to account for the relative times over which these apply. Let τ_p represents the average duration when $r(t)$ is above the threshold. Since transitions take place at random times, the average duration of the period following a successful transition is $\tau_p/2$. Similarly, if τ_q represents the average duration when $r(t)$ is below threshold level, then the average duration of segments which follow an unsuccessful transition is $\tau_q/2 + \tau_p$. The probabilities of successful and unsuccessful transitions are *(1 q)* and *q*, respectively, and the weighting appropriate to each distribution is proportional to the probability of occurrence multiplied by the average duration. Equation (7.61) applies following successful transition for $\tau_p/2$ and following unsuccessful transition for a time τ_p. Thus, the appropriate weighting is $(1-q)\,\tau_p/2 + q\,\tau_p$.

Similarly, equation (7.62) applies following unsuccessful transitions for $\tau_q/2$ and the weighting is

$$\frac{q\tau_q}{2} = \frac{q^2\tau_p}{2(1-q)} \tag{7.63}$$

since

$$\tau_q = \frac{q\tau_p}{1-q} \tag{7.64}$$

The relative normalized weighting is therefore, *1 - q²: q²*.

The probability density function obtained by differentiating $P[R < A]$ with respect to A is given by:

$$\begin{aligned} P_R(A) &= (1+q)p_r(A), && A \geq T \\ &= qp_r(A), && A < T \end{aligned} \tag{7.65}$$

Figure 7.8 shows theoretical cumulative distribution based on the above equation. It is observed that distributions for the switched diversity with different threshold levels touch the ideal selection diversity curve at the selected threshold level. Well above this threshold the distributions become Rayleigh. This should not be of any concern since there is very little to be gained at these high levels. Well below the threshold level, the curves are displaced but closely approximate a Rayleigh slope. Switched diversity offers most of the gain in the region immediately above threshold level, and it is evident that the optimum threshold level should be set slightly above the level corresponding to the lowest acceptable SNR. There is no point in a lower setting because the signal would become unusable before switching took place. A setting near this level minimizes the number of switching and maximizes the probability of successful transitions.

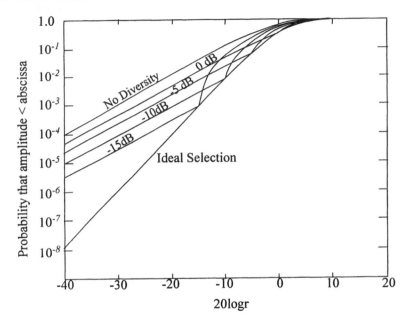

Figure 7.8 Probability of outage for switched diversity

From Figure 7.8 it can be seen that for signal level of -15 dB, and a threshold of -10 dB the switched diversity decreases outage from 0.012 to 0.0008 which is equivalent to diversity gain of more than 13 dB.

7.3 Impact of Diversity on Signal Impairments

The preceding sections examined several diversity techniques that are used to reduce the outage probability, improve signal to noise ratio, or reduce the transmit power. These performance measures show the extent to which diversity is effective against signal fading. This section considers the impact of diversity on other sources of signal degradation introduced by the transmission medium, notably noise and interference. In particular, the improvement in SNR in analog communication and bit error rate in digital communication will be analyzed. Most of material in this section is taken from [17].

7.3.1 Impact of Diversity on Noise

Chapter 4 examined the signal-to-noise ratio at the output of an FM discriminator in the absence of any pre-emphasis and de-emphasis at the transmitter and receiver respectively. It was shown in Section 4.2.2 that the power spectral densities of the three noise components at the discriminator output were [16]:

Quadratic Noise

$$\varpi_q(f) = 4\pi \frac{2(1 - exp(-\gamma^2))^2}{B\rho} f^2 exp\left(-\frac{\pi f^2}{B^2}\right) \tag{7.66}$$

Click Noise

$$\varpi_c(f) = 8\pi B \frac{exp(-\gamma)}{\sqrt{2}(\gamma + 2.35)} \tag{7.67}$$

Signal Suppression Noise

$$\overline{N_s} = \overline{n_s^2(t)} = \overline{(f(t) - \overline{f(t)})^2} \overline{S_0} \tag{7.68}$$

where

$$\overline{S_0} = \overline{s_0^2(t)} \tag{7.69}$$

and $f(t)$ is defined as

$$f(t) = 1 - exp(-\gamma(t)) \tag{7.70}$$

As indicated in Section 7.2.2, the input carrier-to-noise ratio in each branch is exponentially distributed as

$$p(\gamma_i) = \frac{1}{\Gamma} exp\left(-\frac{\gamma_i}{\Gamma}\right) \tag{7.71}$$

a) Maximal ratio Combining:

For the M branch maximal ratio diversity, the carrier-to-noise ratio is distributed as given by equation (7.31):

$$p_M(\gamma) = \frac{\gamma^{M-1}}{(M-1)! \Gamma^M} exp\left(-\frac{\gamma}{\Gamma}\right) \tag{7.72}$$

The noise powers for the three components are [17]:

$$N_q(\gamma) = \int_{-W}^{W} \varpi_q(f)df = \frac{a(1 - exp(-\gamma))^2}{\gamma} \tag{7.73}$$

$$\overline{N_q} = \int_0^{\infty} N_q(\gamma)p(\gamma)d\gamma = \frac{a}{(M-1)\Gamma}\left[1 - \frac{2}{1+\Gamma^{M-1}} + \frac{1}{1+2\Gamma^{M-1}}\right], \text{ when } M \geq 2 \tag{7.74}$$

where

$$a = \frac{4\pi^2 W^3}{3B}\left[1 - \frac{6\pi}{10}\left(\frac{W}{B}\right)^2 + \frac{12\pi^2}{56}\left(\frac{W}{B}\right)^4 + \ldots\right] \tag{7.75}$$

and

$$N_c(\gamma) = \int_{-W}^{W} \varpi_c(f)df = bexp(-\gamma)[\pi(\gamma + \zeta)]^{-\frac{1}{2}} \tag{7.76}$$

where

$$b = 8\pi^2 \sigma W = \frac{8\pi^2 W^2}{\sqrt{ln4}}\alpha \tag{7.77}$$

and $\alpha = (B/2W)$ and $\zeta = 2.35$

$$\overline{N_c} = \int_0^{\infty} N_c(\gamma)p_m(\gamma)d\gamma \tag{7.78}$$

This integral solved by Davis [16] gives the click noise power at the output of the discriminator as

$$\overline{N_c} = \frac{b}{\Gamma^M(M-1)!\sqrt{\pi}}I_M \tag{7.79}$$

with I_M given by recursive expression as

$$I_M = \frac{2M-3}{3\beta}I_{M-1} + \frac{\zeta}{2\beta}J_{M-1}, \text{ when } M \geq 2 \tag{7.80}$$

For $M \geq 2$, $J_M = I_{M-1} - \zeta J_{M-1}$; we can calculate I_M recursively if I_1 and J_1 are known. These are given by:

$$I_1 = \sqrt{\frac{\pi}{\beta}} exp(\zeta\beta) erfc(\sqrt{\zeta\beta}) \qquad J_1 = \frac{2}{\sqrt{\zeta}} - 2\sqrt{\beta\pi} exp(\zeta\beta) erfc(\sqrt{\beta\zeta})$$

$$\beta = 1 + \frac{1}{\Gamma}$$

(7.81)

For an M branch maximal ratio combiner the output signal power is obtained by averaging the discriminator output over the distribution of γ, i.e.

$$\bar{S} = S_0 \int_0^\infty (1 - exp(-\gamma))^2 p(\gamma) d\gamma = S_0 \left(1 - \frac{1}{(1+\Gamma)^M} \right)^2$$

(7.82)

The signal suppression noise is given by

$$\bar{N}_s = S_0 \left[\int_0^\infty (1 - exp(-\gamma))^2 p_M(\gamma) d\gamma - \left(1 - \int_0^\infty p_M(\gamma) exp(-\gamma) \right)^2 \right]$$

$$\bar{N}_s = S_0 \left[\frac{1}{(1+2\Gamma)^{2M}} - \frac{1}{(1+\Gamma)^{2M}} \right]$$

(7.83)

The baseband signal to noise ratio is defined as:

$$SNR - \frac{\bar{S}}{\bar{N}_q + \bar{N}_c + \bar{N}_s}$$

(7.84)

Figure 7.9 shows plots of *SNR* against *CNR* for various B/2W and diversities of orders 2, 3 and 4. It is seen that with the increase in the diversity order, the threshold effect becomes more visible. The limiting value of *SNR* occurs when the sum of the quadrature noise and the click noise is much smaller than the signal suppression noise.

b) <u>Selection Diversity:</u>
　　　A similar procedure yields an expression for baseband SNR for M branch selection diversity. The distribution of γ is given from equation (7.11) as:

$$p_M(\gamma) = \frac{M}{\Gamma} \left(1 - exp\left(-\frac{\gamma}{\Gamma} \right) \right)^{M-1} exp\left(-\frac{\gamma}{\Gamma} \right)$$

(7.85)

The quadratic noise and the click noise components are given by:

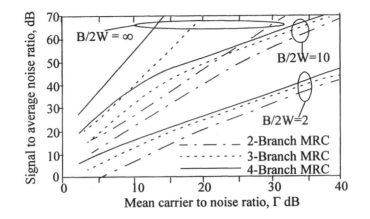

Figure 7.9 Output signal to noise ratio for number of diversity branches

$$\overline{N}_q = \frac{aM}{\Gamma}\int_0^\infty \frac{(1 - exp(-\gamma))^2}{\gamma}\left[1 - exp\left(-\frac{\gamma}{\Gamma}\right)\right]^{M-1} exp\left(-\frac{\gamma}{\Gamma}\right)d\gamma$$

$$\overline{N}_q = \frac{aM}{\Gamma}\sum_{j=0}^{M-1}\binom{M-1}{j}(-1)^j log\left[\frac{(j+1+\Gamma)^2}{(j+1)(j+1+2\Gamma)}\right] \qquad (7.86)$$

$$\overline{N}_c = \frac{8\pi MbW}{\Gamma\sqrt{2}}\int_0^\infty \frac{exp(-\gamma)}{\sqrt{\gamma + 2.35}}\left(1 - exp\left(-\frac{\gamma}{\Gamma}\right)\right)^{M-1} exp\left(-\frac{\gamma}{\Gamma}\right)d\gamma$$

Again the output signal and signal suppression noise are given by

$$\bar{S} = S_0\left[1 - M\sum_{j=1}^{M-1}\frac{(-1)^j}{j+1+\Gamma}\right]^2$$

$$N_s = S_0\left\{M\sum_{j=0}^{M-1}\binom{j}{M-1}\frac{(-1)^j}{1+j+2\Gamma} + \left[M\sum_{j=0}^{M-1}\frac{(-1)^j}{1+j+\Gamma}\right]^2\right\} \qquad (7.87)$$

Davis [16] showed that the output signal-to-noise ratio gives results almost identical to that for maximal ratio combining except for an offset in the abscissa of approximately $(M!)^{1/M}$.

c) Equal Gain Combining:

In this case the distribution of the resultant instantaneous carrier-to-noise ratio is not known, therefore the approximation given in [17] is used

$$p_M(\gamma) = \frac{\gamma^{M-1}}{(M-1)!\gamma_x^M}exp\left(-\frac{\gamma}{\Gamma}\right) \qquad (7.88)$$

where

$$\gamma_x = \frac{2\Gamma}{M}\left(\frac{\Gamma\left(M+\frac{1}{2}\right)}{\frac{1}{2}\Gamma}\right)^{1/M} \qquad (7.89)$$

and $\Gamma(.)$ is the Gamma function.

It is seen that this distribution is similar to that for maximal ratio combining, except that term γ_x instead of Γ is used. The relation between Γ and γ_x is given by equation (7.85). Note that the plots of C/N and signal to average noise ratio for equal gain and for maximal ratio combining differ only by a shift in the abscissa. For the same output signal to average noise ratio equal gain combining requires a larger C/N.

7.3.1.1 Impact of Diversity on Random FM

Earlier we showed that diversity techniques are effective in reducing signal outage that are caused by channel fading. In Chapter 3, we also saw that when the signal goes through deep fade, random FM increases. Since diversity changes the statistics of the signal envelope such that the probability of the signal envelope fading below a threshold is significantly reduced, therefore the impact of random FM will be reduced. The result is derived in this section.

Section 3.5 showed that the probability distribution of random FM for a given envelope r is Gaussian with a standard deviation proportional to maximum Doppler frequency. For selection diversity, the combiner produces an output $\theta'(t)$, where

$$\theta'(t) = \frac{d\theta(t)}{dt} = \frac{d\theta_i(t)}{dt}, r_i(t) \geq r_j(t) \quad \text{all } j \qquad (7.90)$$

The probability that at some time t the value of $\theta'(t)$ is less than α is given by

$$P(\theta'(t) < \alpha) = \sum_{j=1}^{M} P(\theta'_i < \alpha \cup r_i \geq r_j); \quad \text{all } j \tag{7.91}$$

The probability density of θ' is therefore

$$p(\theta') = \int_0^\infty dr_1 \int_0^\infty dr_2 \cdots \int_0^\infty dr_M p(r_1, \theta') p(r_2) p(r_3) \cdots p(r_M) \tag{7.92}$$

Using results in Appendices A, B and C, we can write

$$p(\theta') = \frac{M}{2} \sqrt{\frac{\sigma}{\mu_2}} \sum_{j=0}^{M-1} \binom{M-1}{j} (-1)^j \left(k+1+\frac{\Gamma}{\mu_2}\theta'^2\right)^{-\frac{3}{2}} \tag{7.93}$$

The asymptotic form as θ' tends to infinity is

$$p_M(\theta') \approx \frac{2M!}{(M-1)!2^{2M}\sqrt{\frac{\mu_2}{\sigma}}}\theta'^{2M+1} \tag{7.94}$$

Davis [18] has shown that the mean square random FM is given by

$$
\begin{aligned}
\langle\theta'^2\rangle &= \int_{-\infty}^\infty \theta'^2 p_M(\theta') d\theta' \\
&= \frac{M\mu_2}{2\sigma} \sum_{j=1}^{M-1} \binom{M-1}{j} (-1)^{j+1} \log(j+1); \quad M \geq 2
\end{aligned}
\tag{7.95}
$$

The power spectrum of random FM is obtained from its autocorrelation function. Davis [18] uses the method of first evaluating the autocorrelation function for two branches and extending the result for M branches. If $r_2(t)$, $\theta_1(t)$ are the envelope and the phase of the one diversity branch and $r_2(t)$ and $\theta_2(t)$are those for the second branch, then the following variables are defined:

$$
\begin{aligned}
r_1 &= r_1(t) & s_1 &= r_1(t+\tau) \\
r_2 &= r_2(t) & s_2 &= r_2(t+\tau) \\
\theta_1 &= \theta_1(t) & \phi_1 &= \theta_1(t+\tau) \\
\theta_2 &= \theta_2(t) & \phi_2 &= \theta_2(t+\tau)
\end{aligned}
\tag{7.96}
$$

The output of the selection diversity can be written as:

$$\theta(t) = \begin{cases} \theta_1, r_1 \ge r_2 \\ \theta_2, r_1 \le r_2 \end{cases}$$

$$\theta(t+\tau) = \begin{cases} \phi_1, s_1 \ge s_2 \\ \phi_2, s_1 \le s_2 \end{cases} \tag{7.97}$$

The autocorrelation function for two branches is then given by

$$\Re_2(\tau) = A + B + C + D \tag{7.98}$$

where

$$A = \int_{S_1} \theta_1 \phi_1 P(r_1, s_1, \theta_1, \phi_1) P(r_2, s_2) dr_1 dr_2 ds_1 ds_2 d\theta_1 d\phi_1$$

$$B = \int_{S_2} \theta_1 \phi_1 P(r_1, s_1, \theta_1) P(r_2, s_2, \phi_2) dr_1 dr_2 ds_1 ds_2 d\theta_1 d\phi_1$$

$$C = \int_{S_3} \theta_2 \phi_1 P(r_1, s_1, \theta_1) P(r_2, s_2, \phi_2) dr_1 dr_2 ds_1 ds_2 d\theta_2 d\phi_1 \tag{7.99}$$

$$D = \int_{S_4} \theta_2 \phi_2 P(r_1, s_1) P(r_2, s_2, \theta_2, \phi_2) dr_1 dr_2 ds_1 ds_2 d\theta_2 d\phi_2$$

and $S_1 = \{r_1 \ge r_2$ and $s_1 \ge s_2\}$, $S_2 = \{r_1 \ge r_2$ and $s_1 < s_2\}$, $S_3 = \{r_1 < r_2$ and $s_1 \ge s_2\}$, $S_4 = \{r_1 < r_2$ and $s_1 < s_2\}$.

Davis [18] solved these integrals and expressed $\Re_2(\tau)$ as

$$\Re_2(\tau) = 2\left(\frac{b^2 + ad}{2a^2}\right) ln\left(\frac{2}{1 + \sqrt{1 - a^2}}\right) \tag{7.100}$$

where $a = \psi(\tau) = g(\tau)/b_o$, $b = \psi'(\tau) = g(\tau)'/b_o$, $d = -\psi'(\tau) = g(\tau)'/b_o$.
The general result for M branches is

$$\Re_M(\tau) = M\frac{(b^2 + ad)}{2a^2} f_M(a) \tag{7.101}$$

where

$$f_1(a) = -ln(1 + a^2)$$

$$f_2(a) = ln\left(\frac{2}{1 + \sqrt{1 - a^2}}\right)$$

(7.102)

For large M, only the series expansions are known. For small values of a, $f_M(a)$ is approximated by $f_M(a) \cong a^2/M^2$ and

$$f_M(1) = \sum_{k=0}^{M-1} \binom{M-1}{k}(-1)^k ln(k+1), M \neq 1$$

(7.103)

To find an expression for the asymptotic spectrum for random FM, the Fourier transform of the autocorrelation function is required. Davis [18] has obtained the approximate spectra for $N = 2$ and 3 and are given below:

$$S_2(f) \cong \frac{1}{2}\frac{\left(2\pi^2 f_m^2\right)^{3/2}}{(2\pi f)^2}; \qquad S_3(f) \cong \frac{3K\left(2\pi^2 f_m^2\right)^{3/2}}{(2\pi f)^2}$$

(7.104)

where the value of constant K equal to 0.375 is obtained through empirical evaluation. These results indicate a reduction of 13.6 and 16.2 dB in the random FM for two and three branch diversity system respectively. It is clear that, while selection diversity does not eliminate random FM completely, it does reduce it substantially.

For equal gain and maximal ratio combining, two distinct cases exist, i.e., pre-detection and post-detection combining. In pre-detection combining the signals for both the techniques are co-phased before combining. Co-phasing can be accomplished with reference to the phase of either a pilot [19] or to one of branches [20]. In the case of pilot transmission the phase of the pilot is subtracted from the phase of the signal, thus eliminating the random FM if the pilot and the signal are significantly correlated. When one of the branches is used as a reference, the phase reference is lost whenever the signal on the reference branch experiences a deep fade, and co-phasing does not remain satisfactory. Another possibility is to use the phase reference derived from the output of the combiner. This is a form of feedback system. Whatever the method we use to derive the phase, the random FM can be reduced.

In the case of post-detection combining, the random FM is the weighted sum of random FM on all of the branches. Thus, for M branch diversity, the spectrum of the combined signal will have random FM bandwidth of Mf_m. The signal power at the output of the combiner will increase as M^2 whereas the random FM power will increase as M. The result is that the diversity gain is equal to 10 log M, much lower than that obtained for selection diversity.

Adachi and Parsons [21] extended the results of [17] to the case when signals on the two diversity branches are correlated. It has been shown that when d/λ becomes nearly equal to $f_D \times$ time duration of the random FM noise, selection diversity can remove almost all segments of large random FM noise. In other words, even with a smaller antenna spacing, it is possible to reduce random FM noise.

7.3.2 Impact of Diversity on Interference

In cellular mobile systems frequencies must be reused to achieve high capacity. When small size cells are used signal to interference ratio becomes more important than signal to noise ratio. Because of the presence of shadowing and fading the interfering signals could propagate beyond the boundaries of the cell. Under these circumstances, the system performance may be limited by the co-channel interference. It has been shown in Chapter 5 that in the presence of fading the signal-to-interference ratio no longer improves as the cube of the r.m.s. deviation ratio. The use of diversity can help to restore this improvement. Reduction of interference through diversity has been extensively studied and reported in the literature [22][23][24][25][26].

In this subsection, the aim is to determine the effect of diversity on the probability that the signal-to-interference ratio exceeds a predetermined level, receiver threshold or the receiver capture ratio. Suppose that the desired signal, $s_d(t)$, is received in the presence of an interfering signal, $s_i(t)$. Each of these signals is transmitted with a pilot tone and both the signal and the pilot are subject to Rayleigh fading. Only selection diversity and maximal ratio diversity will be considered; similar analysis can be used to get results for equal gain combining.

a) Selection Diversity

Consider an M branch diversity system. The signal and interference magnitude on each branch is R_i and I_i respectively, $i=1,2,...,M$. All $2M$ random variables are independent and have distributions:

$$p(R_i) = \frac{2R_i}{\sigma_R^2} exp\left(-\frac{R_i^2}{\sigma_R^2}\right); \quad \text{when } R_i \geq 0$$

$$= 0; \quad \text{when } R_i < 0$$

(7.105)

$$p(I_i) = \frac{2I_i}{\sigma_I^2} exp\left(-\frac{I_i^2}{\sigma_I^2}\right); \quad \text{when } I_i \geq 0 \tag{7.106}$$

$$= 0; \quad \text{when } I_i < 0$$

Thus, both the signal and interference are Rayleigh distributed with average power σ_R^2 and σ_I^2 respectively.

The short term output of each antenna is sensed and the antenna with the largest composite signal (signal plus interference) is connected to the front end of the receiver. The probability that the signal-to-interference ratio exceeds a given level K^2 is given by [24]:

$$P_r\left(\frac{S^2}{I^2} \geq K^2\right) = \sum_{i=1}^{M} P_r\left(\frac{R_i^2}{I_i^2} \geq K^2 \middle| \left(R_i^2 + I_i^2 \geq R_k^2 + I_k^2\right)\right) \tag{7.107}$$

$$P_r[[R_i^2 + I_i^2 \geq R_k^2 + I_k^2]; \quad k = 1, 2, ..., i-1, i+1]$$

Sciff [24] has shown that the above probability can be written as:

$$P_r\left(\frac{S^2}{I^2} \geq K^2\right) = \sum_{j=1}^{M} P_r\left[\frac{R_1^2}{I_1^2} \geq K^2 \middle| R_1^2 + I_1^2 \geq R_j^2 + I_j^2\right]; \quad j = 2, 3, ..., M] \tag{7.108}$$

$$= MP_r\left[\frac{R_1^2}{I_1^2} \geq K^2, R_1^2 + I_1^2 > R_j^2 + I_j^2\right]; \quad j = 2, 3, ..., M$$

or

$$P_r\left[\frac{S^2}{I^2} \geq K^2\right] = MR^2 \sum_{p=0}^{M-1} \sum_{q=0}^{i} \binom{M-1}{p}\binom{p}{q}(-1)^q R^{-2q}\left(\frac{R^2}{1-R^2}\right)^p g_{pq} \tag{7.109}$$

where

$$g_{pq} = [qR^2 + p + 1 - q]\left[(1+q)R^2 + p - q + K^2(qR^2 + p + 1 - q)\right]^{-1} \tag{7.110}$$

and $R^2 = \sigma_R^2/\sigma_I^2$, the local average-signal to average interference ratio. For various values of M, Figure 7.10 shows the probability of signal-to-interference ratio falling below a level K^2 (the capture ratio of the receiver) as a func-

---

tion of the difference between the mean signal to noise ratio and the capture ratio.

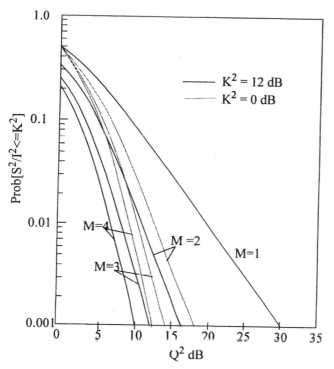

Figure **7.10 Probability of signal to interference ratio falling below a threshold level K^2**

We define $20 \log_{10} Q$ as the amount R^2 exceeds K^2. In other words it is receiver capture ratio expressed in dB. For a fixed probability of outage, say 0.01, we observe that the receiver capture ratio improves approximately 20 dB to 11 dB for two branch selection diversity for $K = 12$. Thus, diversity results in an improvement in the capture ratio of the receiver. A receiver with a poorer capture ratio experiences greater improvement in performance.

b) Maximal Ratio

The performance of diversity system in countering interference when both the desired signal and interferer are Rayleigh distributed has been reported in [17]. In [26] the work has been extended to include Ricean distribution. Also, some results have been produced on the impact of diversity on BER of BPSK signalling. In here, we follow the development reported in [26]. In analyzing the system, quasi static approach is assumed, that is the rate at which the channel changes is much slower than the transmission rate. The total received signal at the antenna consists of the desired signal, contributions from N co-channel interferers and additive white Gaussian noise. We can write the received signal as

$$R(t) = S_d(t) + \sum_{i=1}^{N} S_i(t) \tag{7.111}$$

where the desired signal is given by

$$S_d(t) = \sqrt{P_d}\,\alpha_d s(t)\cos(\omega_o t + \phi_d) \tag{7.112}$$

$s(t)$ is given by

$$s(t) = \sum_{m=-\infty}^{\infty} a_d(m)g_T(t - mT - \tau_d) = v_d(t) \tag{7.113}$$

The expressions for the interfering signals are obtained by replacing d with i. τ_d for the desired signal is taken to be 0. The random phases ϕ_i of the interfering signals are uniformly distributed over $[0,2\pi]$ and α_i are fading channel gains. It is also assumed that all the interfering signals have same average power. The random variables α_i, ϕ_i, and τ_i for $1 \leq i \leq N$ are assumed independent and identically distributed. Zero ISI is assumed. The filtered received signal is then

$$r(t) = \sqrt{P_d}\,\alpha_d v_d(t) + \sqrt{P_I}\sum_{i=1}^{N}\alpha_i v_i(t)\cos\phi_i \tag{7.114}$$

where $v_d(t)$ and $v_i(t)$ are obtained from (7.113). The sampled output is given by

$$r[k] = \sqrt{P_d}\,\alpha_d a_d[k] + \sqrt{P_I}\sum_{i=1}^{N}\alpha_i z_i[k]\cos\phi_i \tag{7.115}$$

and

$$z_i[k] = \sum_{m=-\infty}^{\infty} a_i[m]g(kT - mT - \tau_i) \tag{7.116}$$

which is a random time series with the following properties: $E[z_i[k]] = 0$, $E[z_i[k]z_j[m]] = (1-\beta/4)\delta_{ij}\delta_{km}$, and $E[z_i[k]^2] = 1-\beta/4$ where β is the roll-off factor of the combined transmit, channel and receive filters. If the receive station has M antenna elements, the array output at the sampling instant k, is given by

$$\mathbf{r}[k] = \sqrt{P_d}\mathbf{c}_d a_d[k] + \sqrt{P_I}\sum_{i=1}^{N} \mathbf{c}_i z_i[k] \tag{7.117}$$

where

$$\mathbf{c}_d^T = \left[\alpha_{d,1}\ \alpha_{d,2}e^{j\phi_{d,2}}\ \dots\ \alpha_{d,M}e^{j\phi_{d,M}}\right] \tag{7.118}$$

$$\mathbf{c}_i^T = \left[\alpha_{i,1}\ \alpha_{i,2}e^{j\phi_{d,2}}\ \dots\ \alpha_{i,M}e^{j\phi_{d,M}}\right] \tag{7.119}$$

represent the channel complex gains for the desired and co-channel interference. If the weight vector of the antennas is $\mathbf{w}$, the output of the combiner is then

$$y[k] = \mathbf{w}^H\mathbf{r}[k] = y_d[k] + y_I[k] \tag{7.120}$$

The *SIR* over the observation time over which the channel does not change significantly is given by [26]

$$\mu = \gamma\frac{\left|\mathbf{w}^H\mathbf{c}_d\right|^2}{\sum_{i=1}^{L}\left|\mathbf{w}^H\mathbf{c}_i\right|^2} \tag{7.121}$$

where γ is defined as

$$\gamma = \frac{P_s}{P_I(1-\beta/4)} \tag{7.122}$$

Equal Gain Combining

The mean signal power to mean interference power for Rayleigh-Rayleigh case is given by [26]

$$\mu_r = \gamma \frac{1 + (M-1)(\pi/4)}{N} \qquad (7.123)$$

EXAMPLE E-7.3.1

Compare the gain for equal gain combining for the following cases: (i) two branch three interferers (ii) three branch two interferers. The channels are assumed to fade with i.d.d. Rayleigh distribution.

<u>Solution</u>

Using directly (7.123), we have for the case (i) the output SIR is 2.25 dB below γ, and case (ii) the output is 1.09 dB above γ. These results indicate that improvement is obtained only if the degrees of freedom in antenna are more than the number of diversity branches.

When the desired signal is Rice distributed and the interfering signals are Rayleigh distributed, the mean signal to mean interference power for large Ricean factor K is approximated by

$$\mu \cong \gamma \frac{M}{N} \qquad (7.124)$$

The real advantage of diversity is to alter the statistics of the output signal so that the outage probability is lowered. To find the outage probability, we need to derive the probability density function of the output signal. It has been shown that the density function of the SIR, μ, for the Rayleigh-Rayleigh case is given by [26]

$$f_\mu(\mu) = \frac{\Gamma(N+M)}{\Gamma(N)\Gamma(M)} \gamma^L \frac{\mu^{M-1}}{(\gamma+\mu)^{M+N}} \qquad (7.125)$$

When the desired signal is Ricean distributed, the density function of SIR is [26]

$$f_\mu(\mu) = e^{-MK} {}_1F_1\left(N+M, M; MK\frac{\mu}{\gamma+\mu}\right) \gamma^N \frac{\Gamma(N+M)}{\Gamma(N)\Gamma(M)} \frac{\mu^{M-1}}{(\gamma+\mu)^{M+N}} \qquad (7.126)$$

The outage probability is then given by

$$P_{out} = Pr[\mu < \mu_p] = \int_0^{\mu_p} f_\mu(\mu)d\mu \qquad (7.127)$$

The outage probability is plotted in Figure 7.11 for several values of M, the number of diversity branches. It is seen that the outage decreases with increase in the number of branches.

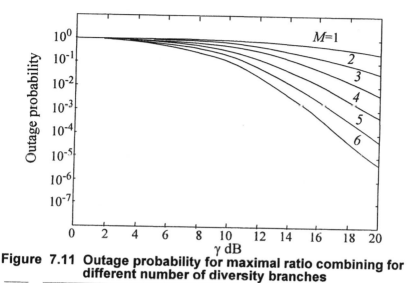

Figure 7.11 Outage probability for maximal ratio combining for different number of diversity branches

7.3.3 Impact of Diversity on Average Fade Duration and Crossing Rate

Adachi *et.al.*[27] derived general expressions for the level crossing rates (LCR) and average fade duration for diversity combining schemes. The average crossing rate, N_R, for selection, equal gain and maximal ratio combining are given to be:

<u>Selection Diversity</u>

$$N_R = 2\sqrt{\frac{-\rho_{11}''}{\pi}} \frac{R}{\sigma\sqrt{2}} \exp\left(-\frac{R^2}{2\sigma^2}\right)\left\{1 - \exp\left(-\frac{R^2}{2\sigma^2}\right)\right\} \qquad (7.128)$$

<u>Equal Gain Combining</u>

$$N_R = \sqrt{\frac{-\rho_{11}''}{\pi}} \exp\left(-\frac{R^2}{2\sigma^2}\right)\left[\frac{R}{\sigma\sqrt{2}} \exp\left(-\frac{R^2}{2\sigma^2}\right) + \left(\frac{R^2}{\sigma^2} - 1\right)\frac{\sqrt{\pi}}{2}\operatorname{erf}\left(\frac{R}{\sigma\sqrt{2}}\right)\right] \qquad (7.129)$$

<u>Maximal Ratio Combining</u>

$$N_R = \sqrt{\frac{-\rho_{11}''}{\pi}} \left(\frac{R}{\sigma\sqrt{2}}\right)^3 \exp\left(-\frac{R^2}{2\sigma^2}\right) \qquad (7.130)$$

where ρ_{11} is the complex correlation function, and " denote its second differential.

The normalized LCR (N_R/f_D) is shown in Figure 7.12. It is seen that diversity reduces LCR substantially. For example, for a normalized signal level of - 20 dB, approximately two order of magnitude reduction is achieved.

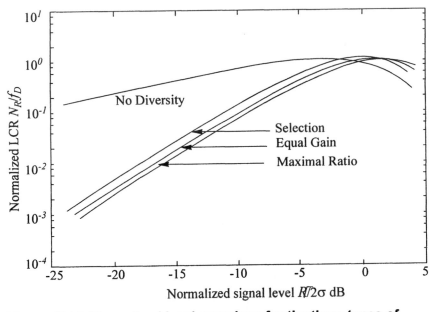

Figure 7.12 Normalized level crossings for the three types of diversity combining

The general expression for the average fade duration, τ, for the three diversity schemes considered above are as follows:

Selection Diversity

$$\overline{\tau}_R = \frac{1}{2}\sqrt{\frac{\pi}{-\rho_{11}''}} \frac{\exp\left(\frac{R^2}{2\sigma^2}\right) - 1}{R/(\sigma\sqrt{2})} \qquad (7.131)$$

Equal Gain Combining

$$\overline{\tau}_R = \sqrt{\frac{\pi}{-\rho_{11}''}} \frac{1 - \exp\left(-\dfrac{R^2}{2\sigma^2}\right) - \dfrac{R}{\sigma\sqrt{2}}\exp\left(-\dfrac{R^2}{2\sigma^2}\right)\sqrt{\pi}\,\mathrm{erf}\left(\dfrac{R}{\sigma\sqrt{2}}\right)}{\exp\left(-\dfrac{R^2}{2\sigma^2}\right)\left[\dfrac{R}{\sigma\sqrt{2}}\exp\left(-\dfrac{R^2}{2\sigma^2}\right) + \left\{\dfrac{R^2}{\sigma^2} - 1\right\}\sqrt{\pi}\,\mathrm{erf}\left(\dfrac{R}{\sigma\sqrt{2}}\right)\right]}$$ (7.132)

Maximal Ratio Combining

$$\overline{\tau}_R = \sqrt{\frac{\pi}{-\rho_{11}''}} \frac{\exp\left(\dfrac{R^2}{2\sigma^2}\right) - \left(1 + \dfrac{R^2}{\sigma^2}\right)}{\dfrac{R}{\sigma\sqrt{2}}}$$ (7.133)

AFD curves are shown in Figure 7.13. The curves show that diversity reception can halve the AFD irrespective of combining technique used when the diversity branches are independent.

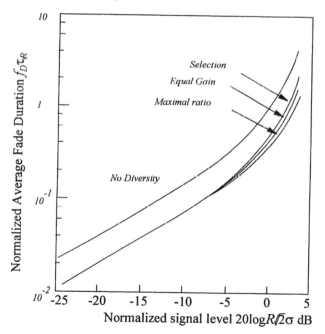

Figure 7.13 Normalized average fade duration for the three types of diversity combining

7.4 Diversity and Digital Transmission

Earlier it was shown that diversity is very effective in mitigating the signal degradation caused by multipath fading and interference. All of the diversity techniques discussed can be also implemented in digital systems. Switched diversity has been implemented for digital PSK, DPSK and FSK [28][29]. The difference in SNR's required by various modulation methods to achieve equal error rates remain relatively similar over a wide range of error rates [29].

Frequency Shift Keying is the predominant form of modulation used over mobile channels. Coherent FSK, non-coherent FSK, and FSK with a discriminator detection are other possibilities. Coherent FSK is not recommended for use over fading channels because phase recovery is difficult. It has been shown that discriminator detection provides error rate performance, for high carrier-to-noise ratio, similar to that of CFSK when $f_d = 0.35 f_s$ [31]. For analysis purposes, therefore, the results for coherent FSK can be used for discriminator detection in a Rayleigh fading environment; for large carrier-to-noise ratio the error rate is given by [11]:

$$P_e = \frac{1}{2}\left(1 - \frac{1}{\sqrt{1 + \frac{2}{\Gamma}}}\right)$$

(7.134)

where Γ is the mean SNR. The mean probability of error is obtained by averaging the conditional error probability over the density function of the signal to noise ratio, i.e.

$$P_e = \int_0^\infty p(\gamma) P_e(\gamma) d\gamma$$

(7.135)

Substituting from equations (7.66), (7.78), and (7.85) for $p(\gamma)$ we obtain the probability of error using, CFSK with various forms of diversity [33], [34],

For Maximal ratio Combining:

For coherent FSK, the probability of error is given by

$$P_{mrc} = \frac{1}{2}\prod_{k=1}^{M} \frac{1}{(1 + 0.5\Gamma_k)}$$

(7.136)

For DPSK, the error probability is given by

$$P_{mrc} = \frac{1}{2} \prod_{k=1}^{M} \frac{1}{(1+\Gamma_k)}$$
(7.137)

For Selection Diversity:

$$P_{sel} = \frac{1}{2} \frac{M!}{\displaystyle\prod_{k=1}^{M} (k+0.5\Gamma)}$$
(7.138)

For Equal Gain Combining:
For coherent FSK, the probability of error (at high signal to noise ratio) is given by

$$P_{eq} = \frac{\left(\frac{M}{2}\right)^{M} \sqrt{\pi}}{2\left(M-\frac{1}{2}\right)!} \prod_{k=1}^{M} \frac{1}{(1+\Gamma_k)}$$
(7.139)

For non-coherent detected FSK, the error probability is given by:

$$P_{eq} = \frac{\left(\frac{M}{2}\right)^{M} \sqrt{\pi}}{2\left(M-\frac{1}{2}\right)!} \prod_{k=1}^{M} \frac{1}{(1+0.5\Gamma_k)}$$
(7.140)

In [35] Arnold and Bodtsmann presented an expression for discriminator detected FSK system that uses two branch maximal ratio combining. The expression is based on the assumption that the performance of discriminator detection is very similar to that of coherent detection. The probability of error is

$$P_{mrc} \approx \frac{1}{2} \left[1 - \sqrt{\frac{\Gamma}{(2+\Gamma)}} - \frac{1}{\Gamma} \left(\frac{\Gamma}{(2+\Gamma)}\right)^{\frac{3}{2}} \right]$$
(7.141)

For selection diversity when discriminator detection is used, the error probability is given by:

$$P_{sel} \approx \frac{4}{(2 + \Gamma')(4 + \Gamma')} \tag{7.142}$$

where Γ' is $= 1.633 \Gamma$ (i.e. mean SNR dB + 2.13 dB).

Selection diversity has been shown to combat frequency selectivity and the optimum threshold level is found to be insensitive to vehicle speed.

EXAMPLE E-7.4.1

Compare the error rates delivered by two branch FSK discriminator detected maximal ratio and selection diversity systems for the input mean signal to noise power ratios of 5, 10, and 15 dB.

Solution

We use (7.141) and (7.142) and calculate the error rates as follows:

Diversity Type	Signal to noise ratio (dB)		
	5	10	15
MRC	0.0329	5.56×10^{-3}	6.79×10^{-4}
Selection	0.061	0.0107	1.342×10^{-3}

7.5 Optimum Diversity Combining

The diversity combining schemes discussed in the preceding sections have been found to be effective in improving system performance in the presence of channel impairments like fading and interference. In all analyses, independence of interference and signals on diversity branches is a critical assumption. This may not be realistic in cases where the same interference signal may be present on all branches. The branch weights of a diversity system may be adjusted to maximize SINR. The weights can be adjusted with the use of one of several available algorithms. Figure 7.14 shows a typical M-branch space diversity combiner. We follow the analysis given in [36].

The signal received by ith branch is $s_i(t)$ is split into quadrature components, x_{Ii} and x_{Qi}. These are then multiplied by branch weights which are generated by a weight generator algorithm. The weighted signals are them summed to form the array output $s_o(t)$. The weight vector is written as

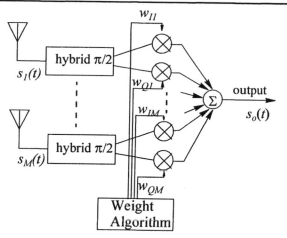

Figure 7.14 M branch space diversity system

$$
\mathbf{w} = \begin{bmatrix} w_{I1} \\ \cdot \\ \cdot \\ \cdot \\ w_{IM} \end{bmatrix} - j \begin{bmatrix} w_{Q1} \\ \cdot \\ \cdot \\ \cdot \\ w_{QM} \end{bmatrix} \tag{7.143}
$$

and the received signal vector is given by

$$
\mathbf{x} = \begin{bmatrix} x_{I1} \\ \cdot \\ \cdot \\ \cdot \\ x_{IM} \end{bmatrix} - j \begin{bmatrix} x_{Q1} \\ \cdot \\ \cdot \\ \cdot \\ x_{QM} \end{bmatrix} \tag{7.144}
$$

The received signal consists of the desired signal, noise and interference can be written as

$$
\mathbf{x} = \mathbf{x}_d + \mathbf{x}_n + \sum_{j=1}^{L} \mathbf{x}_j \tag{7.145}
$$

where $\mathbf{x}_d, \mathbf{x}_n$, and $\mathbf{x}_j$ are the received signal, noise, and jth interference signal vectors. L is the number of interferers. If the transmitted average powers of the desired and the interference signals are normalized to unity, i.e.

$$E[s_d^2(t)] = 1$$

$$E[s_d^2(t)] = 1 \qquad \text{for } 1 \le j \le L \tag{7.146}$$

we can write

$$\mathbf{x} = \mathbf{u}_d s_d(t) + \mathbf{x}_n + \sum_{j=1}^{L} \mathbf{u}_j s_j(t) \tag{7.147}$$

where $\mathbf{u}_d$ and $\mathbf{u}_j$ are the desired and jth interfering signal propagation vectors.

The received interference-plus-noise correlation matrix is given by

$$\mathbf{R}_{nn} = E\left[\left(\mathbf{x}_n + \sum_{j=1}^{L} \mathbf{x}_j \right)^* \left(\mathbf{x}_n + \sum_{j=1}^{L} \mathbf{x}_j \right)^T \right] \tag{7.148}$$

where the superscripts * and T denote conjugate and transpose respectively. when interfering signals and noise are uncorrelated, it can be shown that

$$\mathbf{R}_{nn} = \sigma^2 \mathbf{I} + \sum_{j=1}^{L} E[\mathbf{u}_j^* \mathbf{u}_j^T] \tag{7.149}$$

where σ^2 is the noise power and $\mathbf{I}$ is the identity matrix. The weight vector that maximizes the output SINR is [36]

$$\mathbf{w} = \alpha \mathbf{R}_{nn}^{-1} \mathbf{u}_d^* \tag{7.150}$$

where

$$\mathbf{R}_{nn} = \sigma^2 \mathbf{I} + E[\mathbf{u}_1^* \mathbf{u}_1^T] \tag{7.151}$$

Defining mean signal power to mean noise plus interference ratio as

$$\Gamma = \frac{\Gamma_d}{1 + \sum_{j=1}^{L} \Gamma_j} \tag{7.152}$$

we determine the probability density function of the output of the combiner as [37]

$$p(\gamma) = \frac{e^{-\gamma/\gamma_d}\left(\frac{\gamma}{\gamma_d}\right)^{M-1}(1+M\Gamma_1)}{\Gamma_d(M-2)!}\int_0^1 (1-t)^{M-2}e^{-(\gamma/\gamma_d)M\Gamma_1 t}dt \quad (7.153)$$

and the outage probability is

$$p\left(\gamma < \frac{\gamma_s}{\Gamma_d}\right) = \int_0^{\gamma_s}\frac{e^{-x}(x)^{M-1}(1+M\Gamma_1)}{\Gamma_d(M-2)!}\left\{\int_0^1 (1-t)^{M-2}e^{-(x)M\Gamma_1 t}dt\right\}dx \quad (7.154)$$

In [38] simplified closed form expressions for outage probability for maximal ratio combining in the presence of multiple interferers are given. We follow the development in [38]. The input to an *M*-element interferers is written as

$$X(n) = s_o(n)\mathbf{A}_o + \sum_{i=1}^{N} s_i(n)\mathbf{A}_i + \mathbf{n}(n) \quad (7.155)$$

and its covariance matrix is given by

$$\mathbf{R}_{xx} = \mathbf{A}_o\mathbf{A}_o^H + \sum_{i=1}^{N}\mathbf{A}_i\mathbf{A}_i^H + \sigma^2\mathbf{I} \quad (7.156)$$

where $s_o(n)$, $s_i(n)$, $i = 1,2,..,N$ are the transmitted data sequence, $\mathbf{A}_o$ and $\mathbf{A}_i$, $i = 1,2,...,N$ the M-dimensional directional of arrivals of the desired and N co-channel interfering users respectively. Each component of vectors $\mathbf{A}_o$ and $\mathbf{A}_i$, $i = 1,2,...., N$ is assumed to follow a complex Gaussian distribution with mean power P_o and P_i, $i = 1,2,..., N$, respectively. The MRC weighting vector in (7.155), therefore is

$$\omega_{MRC} = \mathbf{A}_o \quad (7.157)$$

The *SINR* of MRC in the presence of N co-channel interferers is given by

$$\text{SINR} = \frac{\omega_{MRC}^H\mathbf{A}_o\mathbf{A}_o^H\omega_{MRC}}{\omega_{MRC}^H\mathbf{R}_{nn}\omega_{MRC}} \quad (7.158)$$

where

$$\mathbf{R}_{nn} = \sum_{i=1}^{N} \mathbf{A}_i \mathbf{A}_i^H + \sigma^2 I \qquad (7.159)$$

We define $x_i = \mathbf{A}_i^H \mathbf{A}_i$ and $\rho_i = |\mathbf{A}_i^H \mathbf{A}_i|^2 / (x_o x_i)$, $i = 1,2,..,N$. When N CCI's are present, the SINR is given by

$$\text{SINR} = \frac{x_0}{\sum_{i=1}^{N} \rho_i x_i + \sigma^2} \qquad (7.160)$$

Define $z_i = \rho_i x_i$, its pdf is given by

$$f(z_i) = \frac{1}{P_i} \exp\left(-\frac{z_i}{P_i}\right) \qquad (7.161)$$

and is $z = \sum_{i=1}^{N} z_i$, which has pdf

$$f(z) = \frac{z^{N-1} e^{-\frac{z}{P}}}{P^N \Gamma(n)} \qquad (0 \le z_i \le \infty) \qquad (7.162)$$

where x_0 is the desired signal, x_i weighted by ρ_i is the ith interference, N number of co-channel interference and σ^2 is the noise variance. The pdf of a new random variable $w = z + \sigma^2$ is given by

$$f(w) = \frac{(w - \sigma^2)^{N-1} e^{-(w - \sigma^2 / P)}}{P^N \Gamma(N)}; \qquad (w \ge \sigma^2) \qquad (7.163)$$

The SINR $y = x/w$ has a pdf

$$f_Y(y) = \int_{\sigma^2}^{\infty} w f_{x,w}(yw, y)\, dw \qquad (7.164)$$

which can be rewritten as

$$f_Y(y) = ke^{-y\sigma^2} y^{m-1} \sum_{i=1}^{M} (\sigma^2)^{M-i} C_M^i \frac{(N(M-1)+i-1)!}{\left(y+\frac{1}{P}\right)^{N(M-1)+i}} \tag{7.165}$$

where $k = 1/P^N \Gamma(M)\Gamma(N)$.

The outage probability for dual diversity and single interferer is given by

$$P_Y(y < \beta) = \sum_{i=0}^{2} (\sigma^2)^{2-i} C_{2(1+i-1)!}^{i} \int_0^{\beta} \frac{e^{-y\sigma^2}}{\left(y+\frac{1}{P}\right)^{1+i}} dy \tag{7.166}$$

The pdf of SINR for M-branch MRC diversity with two interferers with unequal powers is given by

$$f_2(y) = \frac{e^{-y\sigma^2} y^{M-1}}{\Gamma(M)(P_1 - P_2)} \sum_{i=0}^{M} (\sigma^2)^{M-i} C_{M^i}^i \left(\frac{1}{\left(y+\frac{1}{P_1}\right)^{i+1}} - \frac{1}{\left(y+\frac{1}{P_2}\right)^{i+1}} \right) \tag{7.167}$$

Similarly, for three CCI's pdf of $z = z_1 + z_2 + z_3$ is found as

$$f_3(z) = \frac{e^{-y\sigma^2} y^{M-1}}{\Gamma(M)} \sum_{i=0}^{M} (\sigma^2)^{M-i} C_{M^{i!}}^i \tag{7.168}$$

$$\left(\frac{1}{\left(y+\frac{1}{P_1}\right)^{i+1}} \frac{P_1}{(P_1 - P_2)(P_1 - P_3)} + \frac{1}{\left(y+\frac{1}{P_2}\right)^{i+1}} \frac{P_2}{(P_2 - P_1)(P_2 - P_3)} + \right.$$

$$\left. \frac{1}{\left(y+\frac{1}{P_3}\right)^{i+1}} \frac{P_1}{(P_3 - P_1)(P_3 - P_2)} \right)$$

We can find the outage probability using (7.167) and (7.168).

7.6 Diversity Reception Systems

A number of diversity combining systems have been described in the foregoing sections. The simplest of them, in concept, is the post detection combining receiver which consists of separate RF, IF and detector modules for each branch. The detected outputs are added to form a final output. Several IF combining techniques have been proposed in an attempt to improve performance and to reduce the equipment complexity. Two well known phasing techniques are used extensively. One uses a multiple heterodyne receiver [39], and the other a separate pilot just outside the modulation band [40]. All of the IF combining systems require an entirely new receiver design and are not suitable with existing receivers. Since a large capital investment has already been made by the users, an add-on system greatly facilitates the implementation of diversity.

With this incentive, single receiver pre-RF combining systems have been proposed [41][42] that use the techniques of Lewin [40]. These are described in the following section.

7.6.1 Phase Perturbation Methods

For any pre-detection combining technique, the phases of the signals on diversity branches are usually aligned. For example, for a two branch combiner, an adjustable phase shifter is placed in one of the antenna leads as shown in Figure 7.15. The phase shifter adjustment operates in a hill climbing mode using AGC line voltage as an input signal. The AGC time constant is critical to the operation; it must be large enough to average out the faster random variations but not so large that the combiner is unable to follow the phase variations.

In a system described by Lewin, the signal on a certain diversity branch is aligned to a reference branch [40]. The system schematic is shown in Figure 7.16. The signal on Channel B is passed through a phase shifter, which is controlled by the output of a phase detector. The Channel B signal is phase perturbed by a small deviation phase perturbation circuit driven by a perturbation low frequency sinusoidal oscillator. The signal from the phase perturbation network is added to the reference branch signal. The summed signal experiences a 180 degree phase shift when the perturbing signal is changed from lagging to leading or vice-versa. The output of the summation network is input to a receiver whose output is proportional to perturbed phase. The output of the phase detector is then used to adjust the phase shifter in Channel B. A dual of this technique has also been proposed in a which a small amplitude perturbation is applied to the RF signal. The resulting phase perturbation is detected by the receiver and can be used to provide control information [43].

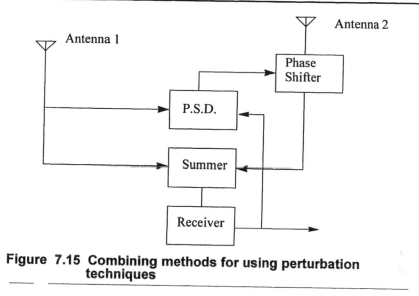

Figure 7.15 Combining methods for using perturbation techniques

This scheme may cause oscillation of the control voltage about the optimum value because of quantization effects in the phase shifter. A modified version which eliminates the problem is to use single sided RF modulation without carrier suppression as shown in Figure 7.17. This system can be used for both AM and FM receivers.

7.6.2 Phase Sweeping Method

Another technique employing a single receiver uses a method of continuously sweeping the phase in one of the signal paths. This has been studied by several workers [44][45]. The basic phase-sweeping diversity receiver is shown in Figure 7.18. If the phase shifter continuously changes its value between 0 and 2π radians, there are instances when the two signals are in phase and a peak output appears.

Rapid phase fluctuations, resulting in interference to other channels, must be avoided for proper operation of the system. A phase change which is linear with time is suitable, and can be implemented by using a repetitive saw-tooth phase modulation. This is equivalent to frequency translation of the whole spectrum by the inverse of the repetition period of the sawtooth wave. The majority of the systems reported in literature applied this technique to AM systems, but it is equally applicable to FM systems.

The use of multiple front end receivers becomes attractive when receiver noise figure and intermodulation performance are the major considerations. For example, the perturbation can be introduced either after the RF amplifier or in the local oscillator. Similarly, since the phase sweeping

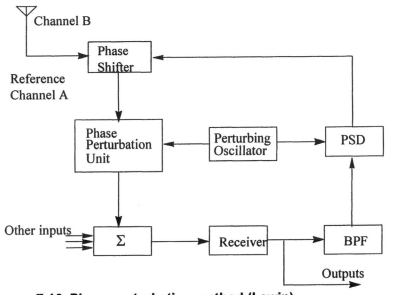

Figure 7.16 Phase perturbation method (Lewin)

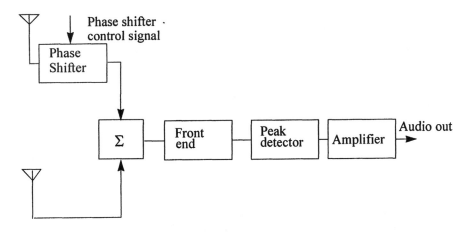

Figure 7.17 Phase sweeping diversity receiver

method is equivalent to frequency shift, the desired result could be obtained by using local oscillators at different frequencies. This would require high-Q

IF amplifiers to avoid adjacent channel rejection. By the use of AGC for individual branches, a maximal ratio combining system can be realized.

7.6.3 Special Receivers

Predetection combiners known as heterodyne phase strippers are used to co-phase the branch signals. A block diagram of such a system is presented in Figure 7.18. This circuit has the property that the output phase is independent of the input. The circuit action is described as follows.

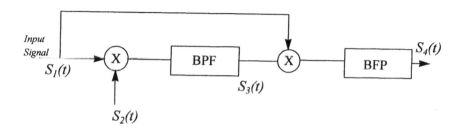

Figure 7.18 Heterodyne phase stripper

Let the input signal be given by

$$S_1(t) = A(t)\cos[\omega_c(t) + \phi(t) + \phi_0] + A_p(t)\cos[\omega_p t + \phi_0] \qquad (7.169)$$

where the first term on the right hand side represents the modulated carrier and the second term is a pilot tone. The frequencies, ω_c and ω_p, are sufficiently close to warrant high correlation between the pilot and the signal. The signal is multiplied by the output of a local oscillator

$$S_2(t) = B\cos(\omega_0 t + \theta) \qquad (7.170)$$

where $\omega_o < \omega_c$ or ω_p.

The output of the filter F_1 is centered at the difference frequency $\omega_d = \omega_c - \omega_0$ or $\omega_d = \omega_p - \omega_0$; the filter bandwidth being narrow will reject the modulation sidebands.

The filter output is multiplied by input signal $S_1(t)$. The resulting signal, filtered by filter F_2 having a bandwidth wide enough to pass the modulation sidebands, is given by

$$A_2\cos(\omega_d - \theta + \phi_0) \qquad (7.171)$$

The filter F_2 is centered either at $\omega_p\text{-}\omega_d$ or $\omega_c\text{-}\omega_d$. If the pilot tone frequency is the centre frequency of the filter, the result is given by:

$$S_4(t) = k[A(t)\cos((\omega_c-\omega_p+\omega_0)t+\theta+\phi(t))+A_p\cos(\omega_0 t+\theta)] \quad (7.172)$$

and if the pilot tone is at the carrier frequency, the output is

$$S_4(t) = k[A(t)\cos(\omega_0 t+\phi(t)+\theta)+A_p\cos((\omega_p+-\omega_c+\omega_0)t+\theta)] \quad (7.173)$$

It is seen that the output $S_4(t)$ has acquired the phase of the local oscillator and the static phase of the input signal has been cancelled. With several branches a single local oscillator will co-phase the branch signals prior to combining. A pre-detection combiner using the phase stripper principle is shown in Figure 7.18.

7.6.4 Granlund Receiver

In the previous subsection, we observed that all input signals are reduced to a certain common frequency and their phases are synchronized to those of the local oscillator. Granlund [46] proposed a receiver, its schematic shown in Figure 7.19, which is in the form of a regenerative closed loop. The shown receiver configuration has two main advantages over the conventional approach that uses local oscillators. First, it does not require high isolation between the local oscillator and the output; second, for FM signals the filter F_1, although narrow-band, need not be very sharp. The operation of the Granlund receiver is described below.

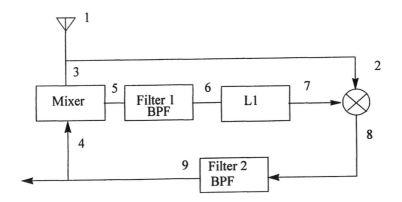

Figure 7.19 Gruland receiver block diagram

The received signal at Point 1 is given by

$$S_1(t) = A(t)\cos(\omega_c t + \phi(t) + \phi_r(t) + \phi_n(t)) \tag{7.174}$$

This signal is divided into two equal parts at Point 2 and Point 3. Thus $S_2(t)$ and $S_3(t)$ are given by

$$S_2(t) = S_3(t) = \frac{1}{2}S_1(t) \tag{7.175}$$

The output of the filter F_2, at frequency ω_2 with a phase α is mixed with the signal $S_3(t)$ to get $S_5(t)$

$$S_5(t) = \frac{A(t)}{2}\cos(\omega_c t + \phi(t) + \phi_r(t) + \phi_n(t))k\cos(\omega_2 t + \alpha')$$

$$S_5(t) = \frac{kA(t)}{4}\cos[(\omega_2 - \omega_c)t - \phi_r(t) - \phi_n(t) + \alpha'] + \tag{7.176}$$

$$\cos((\omega_2 + \omega_c)t + \phi(t) + \phi_n(t))$$

Here α' is the phase shift experienced by the signal when it is passed through filter F_2. $S_5(t)$ is passed through a narrow-band limiter-filter amplifier to give an output

$$S_7(t) = K\cos[(\omega_2 - \omega_c)t + \alpha' - \phi_r(t) - \phi_n(t)] \tag{7.177}$$

where K is the limiter gain, and at Point 8 the output is written as,

$$S_8(t) = S_2(t)S_7(t)$$

$$= K\cos[(\omega_2 - \omega_c)t + \alpha' - \omega_r(t) - \omega_n(t)]\frac{A(t)}{2}\cos(\omega_c t + \phi(t) + \phi_r + \phi_n) \tag{7.178}$$

When filtered by F_2, signal $S_8(t)$ will be identical to one assumed earlier, provided the phase shift contributed by the filter F_2 is negligible. In this case the output of this filter is the signal stripped of its random phase; a new phase α' is now acquired. Signals from several such branches can be combined to achieve advantage due to diversity.

7.7 Macro-Diversity

Rapid Rayleigh fluctuations are mitigated significantly with the use of diversity techniques, which are described in the preceding sections. These techniques assume that the mean signal strength remains constant over fast fluctuating envelope. In the presence of shadowing, the signal envelope undergoes a slow variation, described statistically by lognormal distribution, as the terminal maneuvers through the shadow zone, with the result that the performance deteriorates. It is clear that shadowing creates gentle ups and downs in the coverage. This effect can be reduced through the use of macro diversity that uses supporting base stations [47] or multiple antennas. Shadowing is particularly serious in the case of cellular mobile communications operating in heavily built areas which are sources of major traffic. In a cellular system (see Chapters 11 and 12), the initial system will contain only one base station located at the centre of the cell, resulting in extensive shadow effects. As the system grows towards maturity, increasing demands render the single antenna system inadequate in any case, regardless of shadow effects.

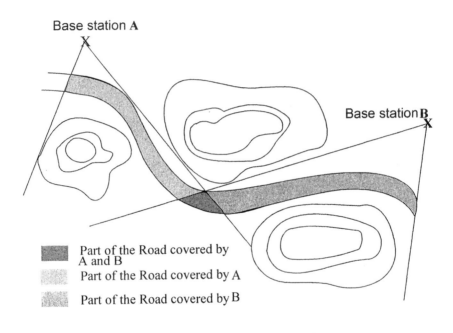

Figure 7.20 Macro diversity

Systems with sectorized antennas (three 120° sector or six 60° sectors) can be used to alleviate the above mentioned problems. The sectored antenna not only allows a more efficient frequency re-use pattern, but also counters congestion. In addition, signals on two or more antennas widely spaced within a cell can be combined to give signal improvement through diversity. The desired signal as well as the interference signal (if present) will follow a lognormal distribution. In this section mathematical expressions are presented to quantify the effect of diversity against signal coverage holes and interference.

Figure 7.20 shows a typical cell served with three corner antennas A, B and C. The mobile unit served within the cell selects the base station that provides highest of the signals at the mobile.

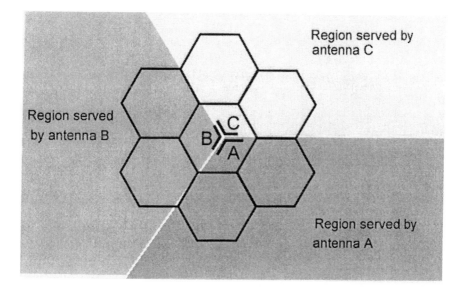

Figure 7.21 A cell served by three corner antennas

The signal selected by the mobile is given by:

$$S = Max(S_A, S_B, S_C) \qquad (7.179)$$

where S_A, S_B, and S_C are the signal strengths at the mobile from the three base stations. All the three signals in dB are normally distributed. This is given by:

$$F_i(\beta) = \int_{-\infty}^{\beta} \frac{1}{\sqrt{2\pi}\sigma} \exp\left\{-\frac{(x-m_i)^2}{2\sigma^2}\right\} dx; \quad i=A,B,C \qquad (7.180)$$

where σ is the standard deviation of the shadow fading and is assumed to be the same for all the base stations: m_i is the mean (or median) signal strength in decibels and is related to the distance from base to mobile by $10\,\alpha\log_{10}(1/r_i)$; and α is the propagation constant, typically between 3 and 4 (see section 2.4).

The distribution function of S is given by:

$$F_s(\beta) = Prob[S \le \beta]$$

$$= F_A(\beta)F_B(\beta)F_C(\beta) \qquad (7.181)$$

7.8 Error Control in Mobile Communications

The presence of channel impairments produce errors in a digital transmission system. To make the system reliable, we discussed channel conditioning techniques such as diversity and equalization. These techniques do not lead to desired reliability as fading and ISI are not completely mitigated. Thus, we seek more techniques to enhance reliability. The first of these technique is Automatic Repeat reQuest (ARQ). A brief discussion on ARQ follows.

7.8.1 Automatic Repeat reQuest (ARQ)

ARQ depends on the principal of repeating transmission of the message until reception of error free message is acknowledged. The simplest of these techniques is Stop and Go ARQ. In this method, the transmitter waits for an acknowledgment from the receiver. If the acknowledgment is positive, then the next message is transmitted. However, if the acknowledgment is negative, the message is retransmitted. Under adverse channel conditions, a message is retransmitted several times. The Stop and Go ARQ mechanism is shown in Figure 7.22 (a). Here 4th and 6th messages receive negative acknowledgment (NACK), so these are repeated[†]. Note that the transmission becomes active only after receiving an acknowledgment. It may be observed that such methods is not efficient since the channel remains idle between transmissions. Depending on the system architecture, the acknowledgments can be transmitted on either a separate paging channel or the message channel. Packet retransmissions continue until an acknowledgment is received from the destination. The stop-and-go ARQ is particularly effective in the presence of high error rates. However, over these unreliable channels throughput figures can be very poor. The throughput is a function of network and propagation delays. The throughput for the wait and Go ARQ scheme is given by

$$\eta = \frac{K(1 - P_B)}{N + R_S \Delta T}; \quad \text{bits per symbol} \qquad (10.182)$$

where K is the number of information bits/block, N block length in symbols, P_B block error probability, R_s signalling speed over the channel in symbols/sec and ΔT is overall round trip delay.

---

†. No acknowledgment means negative acknowledgment.

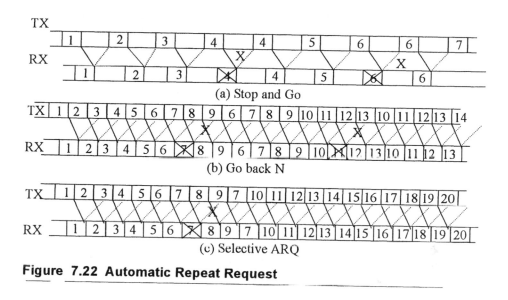

(a) Stop and Go

(b) Go back N

(c) Selective ARQ

Figure 7.22 Automatic Repeat Request

When the error rate is low, however, it is often more efficient to send packets in groups of N packets and require an acknowledgment only after transmission of every group. Such an approach leads to two variations of ARQ: Go-Back-N and Selective ARQ. In the Go-Back-N ARQ, the last N packets are retransmitted when one or more packets of the group have been received incorrectly. In Figure 7.22 (b), the transmitter receives negative acknowledgment for message #7. By this time, the transmitter was in the process of sending the message #9. The transmitter goes back to message # 6 and retransmits messages #6,7,8, and 9 before sending message #10. It can be seen that this method also has a drawback of repeating some messages unnecessarily. The throughput of Go-Back-N is an improvement over wait and go ARQ because the waiting period to receive acknowledgment for every packet is eliminated. The throughput for Go-Back-N ARQ is given by:

$$\eta = \frac{K(1-P_B)}{N+DP_B} \qquad (10.183)$$

where $D = R_s\,\Delta T$ is the overall delay. The other parameters are as defined in (10.182).

The efficiency of Go-Back-N can be improved further, since some packets, out of N transmitted, are received correctly and these are retransmitted unnecessarily. In the Selective ARQ scheme, each packet is numbered;

---

after a sequence of packets is transmitted, the destination sends a NACK (not acknowledged) signal for the packets which have not been received correctly. The transmitter then retransmits the unacknowledged packets. The throughput of Selective ARQ is given by

$$\eta = \frac{K}{N}(1 - P_B) \tag{10.184}$$

It is not difficult to see that the use of ARQ provides a very reliable method of communications. However, in these cases the channel utilization is not very good. The channel throughput can be increased tremendously if ARQ is accompanied by a minimal of channel coding.

7.8.2 Channel Coding

Channel coding is a process whereby some redundancy is added to the messages so that the receiver can detect and correct errors induced by impaired channels. From the signal space point of view, coding may be viewed as the process of increasing the Euclidean distances between the signal points thereby increasing the detection reliability of the messages. The most primitive error detection code is called single parity check code. This code could be even or odd parity check code. In even check code, the number of ones in every code word is arranged to be even. In odd parity the number of ones is odd. It is obvious that single parity codes fail to detect two or more errors in a code word.

The structure of the code word may be changed in several ways. For example, a data block of length, k, may be transformed into another block of longer length, n, to result in so called (n,k) block code. The ratio k/n defines the code rate. When the original message is retained and is appended by the code redundancy bits, then it is called a systematic code. Convolutional codes fall in another class of codes, where the data is passed through a finite state linear shift register. The output of the shift register being the convolution of the input data and the shift register connections is convolutional encoded data - hence the name convolutional codes.

7.8.2.1 Block Codes

A block of k symbols is transformed into a block of n symbols to form (n,k) code. For binary bits k with a redundancy of $n-k$ bits, 2^k tuples is mapped into 2^n, thus increasing the distance between them. This means more reliable detection and correction. Hamming codes are both binary and non-binary codes. These fall in a class of codes where

$$(n, k) = (2^m - 1, 2^m - 1 - m) \tag{7.185}$$

where m is any positive integer. The minimum distance for any (n,k) Hamming code is $d_{min} = 3$.

Extended Golay Codes is obtained by adding a parity bit to the perfect (23,12) Golay codes. A t-error correcting code is called a perfect code if its standard array has all the error pattern of t or fewer errors and no others as coset leaders. Extended Golay (24,12) code has a minimum distance $d_{min} = 8$.

A subclass of linear codes comprises cyclic codes. These codes have property that all cyclic shifts of a code word are also code words. This property adds a structure to the code, which can be exploited to simplify encoding and decoding operations. The encoding and decoding of these codes can be accomplished with the use of shift registers representing coding or decoding polynomials.

BCH Codes is generalization of Hamming Codes. These form a large class of cyclic codes that include both binary and non-binary alphabets. BCH codes can be constructed with parameters

$$n = 2^m - 1$$
$$n - k \le mt \tag{7.186}$$
$$d_{min} = 2t + 1$$

where m (>3) and t are arbitrary positive integers.

For a hard decision decoding, the probabilities of error for different modulation formats are given as [34]

For Coherent PSK, the error probability is given by

$$p = Q(\sqrt{2\gamma_b R_c}) \tag{7.187}$$

where R_c is the code rate, k/n. For transmission of code word bits using FSK, the probability of error is

$$p = Q(\sqrt{\gamma_b R_c}) \tag{7.188}$$

for coherent detection and for non-coherent detection it is given by

$$p = \frac{1}{2}\exp\left(-\frac{1}{2}\gamma_b R_c\right) \tag{7.189}$$

In soft decision decoding, the block error probability depends on the cross-correlation coefficients. The maximum correlation coefficient between a pair of coded waveform is [34]

$$\rho_{max} = 1 - \frac{2}{n} d_{min} \qquad (7.190)$$

Under the assumption that all M code words have the maximum cross-correlation between them, the upper bound on the probability of word error is given by [34]

$$P_M \le 1 - \frac{1}{\sqrt{2\pi}} \int_{-\infty}^{\infty} e^{-v^2/2} \left(\frac{1}{\sqrt{2\pi}} \int_{-\infty}^{y + \sqrt{4\gamma_b R_c d_{min}}} e^{-x^2/2} dx \right)^{M-1} dv \qquad (7.191)$$

Reed-Solomon Codes fall in the class of non-binary block codes. The elements of the code word are selected from an alphabet of q symbols, where $q = 2^k$ is usually selected. A systematic (N,K) non-binary code consists of K information symbols and $N-K$ code redundancy symbols. Reed-Solomon codes have good minimum distance property and can be decoded by efficient hard decision decoding algorithm.

7.8.2.2 Convolutional Codes

Figure 7.23 shows a very simple example of a convolutional encoder. It has three stages, in general encoder with K stages is called encoder with constraint length of K. k information bits are shifted into the shift register at a time. For each k bits shifted, n output bits are produced. Thus, the code rate is k/n. The code shown in the figure is 1/2 rate code with constraint length of 3.

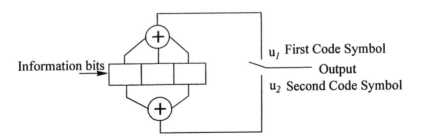

Figure 7.23 Convolutional Encoding Concept

The convolutional encoder is a finite state machine, therefore Maximum Likelihood Sequence Estimation (MLSE) is an optimum decoder for convolutional codes. For this purpose, Viterbi decoding algorithm can reduce the computational burden of (MLSE).

7.8.2.3 Interleaving

The channel multipath propagation and fading cause random and burst errors. Pure burst correction codes like Fire Code can not be used. Reed Solomon codes that can correct Burst and Random error correction codes for reasonable sized blocks are expensive to correct. We can use relatively simple random error correcting codes if data streams are manipulated prior to transmission. Interleaving is such manipulation. There are several methods by which data is manipulated, use of block interleaver is the simplest method. In this the data at the transmitter is written in columns and is read out in rows before transmission. Figure 7.24 shows an example of a block interleaver. The data stream consisting of symbols 1 to 24 is written in columns as shown in the left hand side matrix. Because of channel impairments the received data experiences an error burst of five symbols i.e. symbols 6, 10, 14, 18 and 22 are in error.

$$\begin{bmatrix} 1 & 5 & 9 & 13 & 17 & 21 \\ 2 & 6 & 10 & 14 & 18 & 22 \\ 3 & 7 & 11 & 15 & 19 & 23 \\ 4 & 8 & 12 & 16 & 20 & 24 \end{bmatrix} \qquad \begin{bmatrix} 1 & 5 & 9 & 13 & 17 & 21 \\ 2 & ⑥ & ⑩ & ⑭ & ⑱ & ㉒ \\ 3 & 7 & 11 & 15 & 19 & 23 \\ 4 & 8 & 12 & 16 & 20 & 24 \end{bmatrix}$$

Figure 7.24 Example of a block interleaver

Before the received data is decoded, the data written in rows is read in columns. The received sequence is then

1, 2, 3, 4, 5, ⑥, 7, 8, 9, ⑩, 11, 12, 13, ⑭, 15, 16, 17, ⑱, 19, 20, 21, ㉒, 23, 24

It is seen that the error burst has been broken into random errors. The gap between errors is called interleaving depth. It is also obvious that the interleaving and deinterleaving process results in delays. Thus, for real time applications like digital speech, long interleaving delays are not tolerated. Also, too shallow an interleaving depth will not be very effective against burst errors. A compromise between delay and interleaving depth is usually sought.

Other interleaving techniques are also used, convolutional interleaving is used in conjunction with convolutional coding. An example of convolutional interleaving is shown in Figure 7.25.

In the interleaving shown, the code symbols are sequentially shifted into the bank of N registers. Each register stores J symbols more than the preceding stage. As a symbol enters a shift register, the oldest symbol is pushed

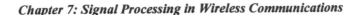

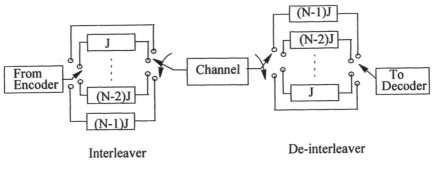

Interleaver   De-interleaver

Figure 7.25 Convolutional Interleaver

out, which is collected by the commutator. A similar de-commutation process is used at the receiver.

7.9 Equalization in Mobile Communications

The passage of signals through multipath mobile radio fading channels produces inter-symbol-interference (ISI) which causes severe signal distortion. Figure 7.26 shows the response of a distorting channel to discrete input. The familiar solution of increasing signal power to counter fading does not work as it fails to unravel the ISI. With an increase in power, the ISI also increases proportionally. Thus, to counter ISI, some corrective action to unravel ISI is needed. Such corrective action is provided by a filter called equalizer.

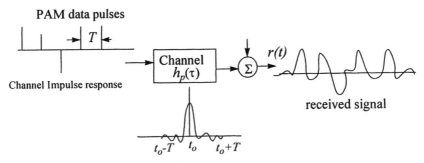

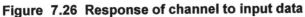

Figure 7.26 Response of channel to input data

The earliest applications of channel equalizations are found in the tele-communications networks where echoes from hybrids are suppressed [32]. In these networks once a link is established, the channel remains practically unchanged and the corrective filter or equalizer does not need any readjustment during the call. When the channel impulse response changes, the equalizer must adapt itself to the new reality. Such is the case of mobile communications where channel varies continuously. In practice transversal or lattice filters are configured as equalizers. A schematic of a typical adaptive equalizer is shown in Figure 7.27.

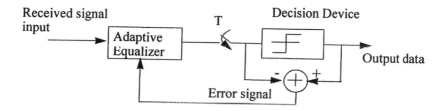

Figure 7.27 A typical adaptive equalizer

In the figure it is seen that the output of the equalizer filter is sampled every T seconds and is compared with the output decision. It is assumed that the decisions made on the incoming data are in general correct. The mean square of the error thus produced is minimized by adjusting the equalizer coefficients using one of several algorithms. The update signal adjusts the filter or equalizer coefficients.

The steepest descent (SD), the least mean square (LMS), the normalized least mean square (NLMS) and recursive least square (RLS) are commonly used algorithms. Each of these algorithm uses a certain minimization criterion. As the channel impulse response changes, the algorithm tries to track the channel variations. It is obvious that if the channel changes faster than the time between updates, the equalizer is unable to track the channel variations and errors propagate. Since the adaptation procedures depends on the quality of decisions, bad decisions can lead the equalizer to instabilities. Thus, relative rate of fading and update speed are important factors in choosing an algorithm. Furthermore, the round-off errors in mathematical operations are also important because these errors could buildup to a stage leading to instability. Some algorithms e.g. RLS are susceptible to round off errors.

Adaptive equalizers are extensively used for elimination of ISI. The basis of elimination of ISI lies in choosing the strongest path among several available. This results in gains similar or greater than those for selection diversity because the equalizer uses a certain optimum criterion for the selection process.

Several equalizer structures are possible; transversal filters structure is the most commonly used. A brief theory of adaptive equalization is presented in the next section. For more details on this subject, the reader is referred to [34].

7.9.1 Theory of Equalization

To minimize distortion, Minimum Peak Distortion and Minimum Mean Square Error are the two most used criteria. The problem of equalization may be formulated as finding coefficients of a digital filter or equalizer according to one of the two above mentioned criteria. The location of the equalizer in a communication link is shown in Figure 7.28.

The data symbols, $\{x_n\}$, are passed through a communication link comprising of a transmit filter, a dispersive channel, AWGN, and a receive filter. The sampled output of the receive filter consists of the desired signal, distortion due to channel and the noise. Due to band limitation by the receive filter, the noise at the output of this filter is no longer white and should be whitened so that available results on the performance of digital systems in additive white Gaussian noise could be used. Figure 7.28 (b) shows an equivalent communication system that consists of a composite filter, a noise whitening filter and equalizer. The theory of equalization will leads us towards finding the best receive filter according to a certain performance criterion e.g. minimum mean square error.

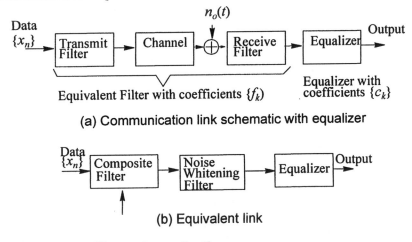

(a) Communication link schematic with equalizer

(b) Equivalent link

Figure 7.28 Channel equalization

The passband transmitted signal is

$$s(t) = Re\left[\sum_n x_n a(t-nT)\exp(j2\pi f_c t)\right]$$ (7.192)

where $a(t)$ is the transmit pulse shape. The received passband signal is

$$r(t) = s(t)*h_p(t) + n_o(t)$$ (7.193)

The output of the equalizer is,

$$y(t) = c(t)*[r(t)\exp(-j2\pi f_c t)] = \sum_n x_n h(t-nT) + n(t)$$ (7.194)

which may be represented in complex notation as $y(t) = y_r(t) + j\, y_i(t)$.

The question is: "Given the received signal $y(t)$, what is the best receive filter?" The answer may be found by the following considerations. First, we need to state a criterion by which we can say that a best filter has been achieved. As stated earlier, minimum peak distortion and minimum mean square error are the two most used criteria.

7.9.1.1 Minimum Peak Distortion Criterion

The equivalent impulse response of the discrete linear filter model of the channel and the equalizer is given by

$$q_n = \sum_{j=-\infty}^{\infty} c_j f_{n-j}$$ (7.195)

The peak distortion at the output is

$$D(c) = \sum_{\substack{n \neq 0 \\ n = -\infty}}^{\infty} |q_n| = \sum_{\substack{n \neq 0 \\ n = -\infty}}^{\infty} \left| \sum_{j=-\infty}^{\infty} c_j f_{n-j} \right|$$ (7.196)

For an infinite tap equalizer, it is possible to select tap weights such that $D(c)$ $=0$, $q_n = 0$ for $n = -\infty$ to $+\infty$ except for $n = 0$, when $q_o = 1$. This condition means that

$$Q(z) = C(z)F(z) = 1$$ (7.197)

Thus,

$$C(z) = \frac{1}{F(z)} \tag{7.198}$$

where $Q(z)$, $F(z)$ and $C(z)$ are the z-transforms of the equivalent filter, the discrete impulse response of the channel, and the equalizer respectively. The equalizer is an inverse filter of the channel. Since such a filter completely eliminates ISI, it is called a zero forcing equalizer. The concept of zero forcing equalizer is shown inFigure 7.29.

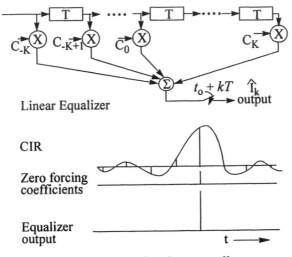

Figure 7.29 Concept of zero forcing equalizer

When the equalizer length is not infinite that is it has $2K + 1$ taps, and the channel impulse response is of length L, then there are $2K + L$ non-zero values in the response $\{q_n\}$. However, we can adjust only $2K+1$ taps, thus in the case of finite length equalizer, the distortion can not be completely eliminated.

EXAMPLE E-7.9.1

A passage of a raised cosine pulse through a noise free channel with ISI results in the following sampled output sequence,

$$x_k = \begin{bmatrix} \dots & 0 & -0.4 & 0.2 & 1 & -0.3 & 0.05 & 0 & \dots \end{bmatrix} \tag{E-7.9.1}$$

(a) Determine the coefficients of a three tap linear equalizer based on zero-forcing criterion.
(b) For the coefficients determined in (a), determine the output of the equalizer when the input is an isolated pulse.
(c) Determine the residual ISI.

Solution

(a) For zero forcing criterion

$$\begin{bmatrix} 1 & 0.2 & -0.4 \\ -0.3 & 1 & 0.2 \\ 0.05 & -0.3 & 1 \end{bmatrix} \begin{bmatrix} c_{-1} \\ c_0 \\ c_1 \end{bmatrix} = \begin{bmatrix} 0 \\ 1 \\ 0 \end{bmatrix}$$

(E-7.9.1)

$$\begin{bmatrix} c_{-1} & c_0 & c_1 \end{bmatrix}^T = \begin{bmatrix} 1.0395 & 1.2474 & 0.3222 \end{bmatrix}^T$$

(b) The output of the equalizer is given by

$$q_m = \sum_{n=-1}^{1} c_n x_{m-n}$$

(E-7.9.2)

Using the above equation, we get the output sequence as

$$\left\{ 0, 0.12888, 0.5634, 0.15588, 0.9999, -4.5 \times 10^{-5}, 0.05274, 0 \right\}$$

(E-7.9.3)

The residual ISI is found to be

$$\left\{ 0, -0.12888, -0.5634, -0.15588, 0.00001, 4.5 \times 10^{-5}, -0.05274, 0 \right\}$$

(E-7.9.4)

7.9.1.2 Minimum Mean Square Criterion

When using the MSE criterion, the tap weight coefficients $\{c_j\}$ are adjusted to minimize the mean square value of the error, $\varepsilon_k = I_k - \hat{I}_k$, I_k is the information symbol at kth signalling interval and $\hat{I}_k$ is its estimate. The cost function to be minimized is

$$J = E|I_k - \hat{I}_k|^2$$

(7.199)

The MSE criterion is satisfied, when the equalizer transfer function is given by

$$C(z) = \frac{F^*(z^{-1})}{F(z)F^*(z^{-1}) + N_o} \qquad (7.200)$$

When we use noise whitening filter with the equalizer, the equivalent equalizer filter transfer function is obtained as

$$C'(z) = \frac{1}{F(z)F^*(z^{-1}) + N_o} \qquad (7.201)$$

This is the case of infinitely long equalizer. For a finite length equalizer, the output of the equalizer is in the kth signalling interval is

$$\hat{I}_k = \sum_{j=-K}^{K} c_j v_{k-j} \qquad (7.202)$$

where v_k is the input to the equalizer at kth signalling interval. The cost function is

$$J(K) = E|I_k - \hat{I}_k|^2 = E\left| I_k - \sum_{j=-K}^{K} c_j v_{k-j} \right|^2 \qquad (7.203)$$

Minimization of the cost function is equivalent to forcing the error signal to be orthogonal to the signal samples. This condition results in $2K+1$ simultaneously equations expressed as

$$\Gamma C = \xi \qquad (7.204)$$

where C denotes the column vector of the equalizer tap gain coefficients and Γ is Hermitian covariance matrix with elements Γ_{ij}, and ξ is $2K+1$ dimensional column vector. The optimum solution for the equalizer taps is

$$C_{opt} = \Gamma^{-1}\xi \qquad (7.205)$$

The fractionally spaced equalizer (FSE) is based on sampling of the incoming signal at least as fast as the Nyquist rate. Figure 7.30 shows a schematic of a fractionally spaced equalizer. The FSE removes the difficulty with the symbol rate equalizer which compensates for the aliased received signal but it can not compensate for the channel distortion inherent in the incoming spectrum. The optimum solution for the tap functions is [34],

---

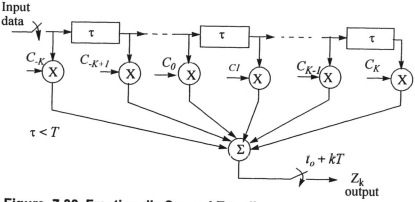

Figure 7.30 Fractionally Spaced Equalizer

$$C_{opt} = A^{-1}\alpha \qquad (7.206)$$

where **A** is the covariance matrix of the input data and α is the vector of cross-correlations.

7.9.1.3 Decision Feedback Equalizers

Figure 7.31 shows the schematic of a decision feedback equalizer. The equalizer consists of two filters, a feed forward filter and a feedback filter.

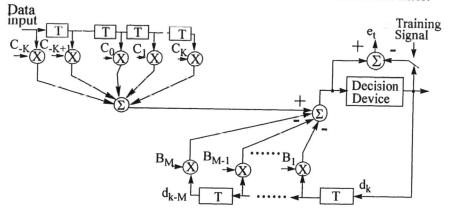

Figure 7.31 Decision Feedback Equalizer

The feed forward filter is similar to that used in the linear transversal equalizer. The previously detected symbols are fed to the feedback filter to remove the precursor from the previously detected symbols. The forward filter removes the post cursor. In [34], it shown that if the previously detected symbols are correct, ISI is completely eliminated. In the absence of decision errors, the minimum MSE is given by

$$J(2K) = 1 - \sum_{-2K}^{0} c_j f_{-j} \qquad (7.207)$$

When ISI is zero, the equivalent spectrum, $X(e^{j\omega T}) = 1$, and $J_{min} = N_o/(1+N_o)$. The equivalent output signal to noise ratio is $\gamma_\infty = 1/N_o$. The inequality $e_{min}(MF) \leq \varepsilon_{min}$ (DFE) $\leq \varepsilon_{zf}$ (DFE) holds for the three types of equalizers.

7.9.2 Adaptation Algorithms

The linear equalizers discussed in the foregoing subsections are used when the channel is time invariant. In actual practice, the impulse response of communication channels continuously vary with time and the equalizer coefficients must adapt to channel variations automatically. The minimum peak distortion criterion and minimum mean square error criterion can be used in devising algorithms to adjust the equalizer taps automatically.

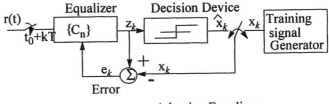

Adaptive Equalizer

Figure 7.32 Adaptive Equalizer

The simple recursive algorithm to adjust the coefficients of zero forcing algorithm is

$$c_j^{(k+1)} = c_j^{(k)} + \Delta \varepsilon_k I_{k-j}^*, \qquad j = -K, ..., -1, 0, 1, ...K \qquad (7.208)$$

where Δ is the adjustment step size, ε_k is the error signal at time $t = kT$, the term $\varepsilon_k I_{k-j}^*$ is the estimate of the cross correlation between the error and data signal. Since the channel is generally unknown, a training sequence is used in the beginning. At the end of the training sequence, the equalizer coefficients are converged to optimum values and the data decisions are sufficiently reliable to enable the equalizer to continue adaptation.

The basic LMS algorithm is given by

$$\hat{C}_{k+1} = \hat{C}_k + \Delta\varepsilon_k V_k^* \tag{7.209}$$

where V_k is the vector of the received signal samples, Δ is a positive number chosen small enough to ensure convergence, and ε_k is the error at time kT.

7.9.3 Diversity and Equalization

In the foregoing sections, the use of diversity in combating fading and equalization in countering ISI was studied. The use of diversity is mainly used when the transmission channel had flat fading characteristics. The use of equalization is more suited to the cases whereby the signal bandwidth is wider than the channel coherent bandwidth and ISI is produced as a consequence. An equalizer faces difficulties when multipath channel fade rate is too fast for the equalizer to track it. In this regard, a combination of diversity and equalization proves to be a powerful technique. The diversity changes the fade statistics and the equalizer mitigates inter-symbol interference. A combination of coding, diversity and equalization produces even a powerful technique to combat wideband channel impairment. For more detail the reader is referred to [48] - [51].

7.10 Summary

In a mobile system, the received signal level varies because of type of environment around the mobile terminal, its mobility, polarization and direction of travel. All of these variations can be regarded as annoyances; paradoxically, they also represent a method of enhancing reliability. If signal variations at two locations, for instance, are statistically uncorrelated, then combining them after some processing reduces the effects of severe fading. Using this approach it has been seen that substantial improvements are possible in reception. Detailed discussion was presented on the influence of diversity on the noise, amplitude and other signal parameters. Coding for channel is another important signal processing needed to make the mobile radio channel more reliable. Coding is presented as an essential requirement in any reliable communications. Equalization has been found to be very effective weapon against inter-symbol-interference. Several types of equalizers were discussed. In the presence of channel variations, adaptation algorithms were presented. A combination of diversity, channel coding, and equalization is identified as a powerful technique to combat most of mobile channel impairments.

Chapter 7 concludes the second part of this text, dealing with signal characterization and processing techniques. We have shown that the poor signal environment associated with mobile radio can be countered by the use of advanced signal processing methods. In particular, diversity combining is seen to be a powerful technique in alleviating the influence of channel impairments. In subsequent chapters we begin an examination of systems that combine techniques described in the preceding chapters to produce reliable end-to-end transmission.

Problems

Problem 7.1

(a) Explain how more than one independently fading channels can be created for a vehicular speed of 80 km/hr at 930 MHz.

(b) In time diversity, a signal is transmitted with a certain delay over which the channel is assumed to have changed sufficiently to create two uncorrelated copies of the signal. Find time delays between transmissions for vehicular speeds of 100, 60, 30 and 0 km/hr. Assume operating frequency of 890 MHz.

(c) State the condition under which time diversity does not work.

Problem 7.2

(a) In a 4-branch selection diversity, the mean signal to noise of each branch is randomly varying with a uniform distribution over 6 dB. Derive an expression for the outage probability.

(b) Draw conclusions on the diversity gain from the results obtained in part (a).

Problem 7.3

(a) The probability of fading 10 dB below the mean signal to noise ratio of a Rayleigh faded signal has been found to be 0.12. Determine the mean SNR in dB relative to the threshold.

(b) What will be the probability of signal fading to 10 dB below the mean SNR when a two branch selection diversity is used?

(c) Repeat (b) when the cross-correlation coefficient between the branch signals is 0.6.

Problem 7.4

(a) Compare the improvement in the signal to noise ratio (dB) over a single branch when the number of diversity branches, M, equals 2, 3, 4, 8, 16 for Selection, MRC, and Equal Gain combining.

(b) Plot the results obtained in (a). What inference do you draw from the graph?

Problem 7.5

Compare achievable S/N ratios for Maximal Ratio and Selection Diversities of order 3. The following parameters are given.

     i) Input Mean signal to Noise Ratios are equal in all branches and is equal to 13 dB.

     ii) Signal bandwidth is 3 KHz

     iii) Transmission bandwidth is 26 KHz

Problem 7.6

Compare the performance of equal gain combining with switched diversity when the threshold A in the case of switched diversity is varied from -20 dB to +10 dB in steps of 5 dB.

Problem 7.7

A convolutional code is described by polynomials

$$g_1 = \begin{bmatrix} 1 & 1 & 0 \end{bmatrix} \qquad g_2 = \begin{bmatrix} 1 & 0 & 1 \end{bmatrix} \qquad g_3 = \begin{bmatrix} 0 & 1 & 1 \end{bmatrix} \qquad \text{(P-7.8.1)}$$

(a) Draw schematic of the encoder.
(b) Draw the state-diagram of the code.
(c) Draw the trellis diagram of the code.

Problem 7.8

A three tap zero-forcing equalizer is needed for a dispersive channel having ISI spanning three symbols and characterized by the values $x(0) = 1.01$, $x(1) = 0.35$ and $x(-1) = 0.15$. Also find the residual ISI when the taps are set optimally.

References

[1] S.H. Lin, "Statistical behavior of a fading signal," *Bell System Technical Journal*, vol. 50, pp. 3211-3270, 1971.

[2] S.H. Van Wambeck and A.H. Ross, "Performance of diversity receiving systems," *Proc. IRE* vol. 39, pp. 256-264, 1951,

[3] R.H. Clarke, "A statistical theory of mobile radio reception," *Bell System Technical Journal*, vol. 47,pp. 957-1000, July-August 1968.

[4] G.L. Grisdale, J.B. Morris, and D.S. Palmer, "Fading of long-distance radio signals and a comparison of space and polarization diversity in the 6-18 Mc/s range," *Proc. IEE*, vol. 104, pp. 39-51, Jan. 1957

[5] W.C.-Y. Lee and Y.S. Yeh, "Polarization diversity system for mobile radio," *IEEE Trans. on Comm. Tech.*, vol. COM-20, pp. 912-923, October 1972.

[6] J.H. Vogelman, J.L. Ryerson and M.H. Bickelhaupt, "Tropospheric scatter system using angle diversity," *Proc. IRE*, vol. 47, pp. 688-696, May 1959.

[7] M. Schwartz, W.R. Bennet and S. Stein, *Communication Systems and Techniques*, McGraw Hill, 1966.

[8] H. Staras, "Diversity *Reception With Correlated Signals*," *J. Applied Physics*, vol. 27, pp. 93-94, January 1956.

[9] D.G. Brennan, "Linear *Diversity Combining Techniques*," *Proc. IRE*, vol. 47, pp. 1075-1101, June 1959.

[10] B.B. Barrow, "Diversity combination of fading signals with unequal mean strengths," *IRE trans. on Comm. Syst.*, vol. CS-11, pp. 73-78, March 1963.

[11] W.C.Y. Lee, *Mobile Communication Engineering*, McGraw Hill Book Company, 1983.

[12] W.C.Y. Lee, "Mobile radio performance for a two-branch equal-gain combining receiver with correlated signals at the land site," *IEEE Trans. on Veh. Tech.*, vol. VT-27, pp. 239-243, 1978.

[13] W.E. Shortall, "A switched diversity receiving system for mobile radio," *IEEE Trans. on Comm. Tech.*,vol. COM-21, pp. 1269-1275, November 1973.

[14] A.J. Rustako, Y.S. Yeh, and R.R. Murray, "Performance of feedback and switched space diversity 900 mhz fm mobile radio systems with rayleigh fading," *IEEE Trans. on Comm. Tech.*, vol. COM-21, pp. 1257-68, November 1973.

[15] J.D. Parsons, "Experimental switched diversity system for v.h.f. mobile radio," *Proc. IEE*, vol. 122, pp. 780-784, August 1975.

[16] B.R.Davis, "FM noise with fading channels and diversity," *IEEE Trans. on Comm. Tech.*, vol. COM-19, pp. 1189-1200, December 1971.

[17] W.C. Jakes Jr. (Ed.), *Microwave Mobile Communications*, J. Wiley, 1974.

[18] B.R. Davis, "Random FM in mobile radio with diversity," *IEEE Trans. on Comm. Tech.*, vol.COM-19, pp. 1259-1267.

[19] C.C. Cutler, R. Kompfener and L.C. Tillotson, "A self steering array repeater," *Bell System Technical Journal*, vol.42, pp. 2013-2032, 1963.

[20] C.W. Earp, *Radio Diversity Systems*, U.S. Patent No. 2,683,213, July 6, 1954.

[21] F. Adachi and J.D. Parsons, "Random FM noise with selection diversity combining," *IEEE Trans. on Communications*, vol. 36, no. 6, June 1988, pp. 752-754.

[22] F. Adachi, "Periodic switching diversity effect on co-channel interference performance of a digital fm land mobile radio," *IEEE Trans. on Veh. Tech.*, vol. VT-27, pp. 220-223, 1978.

[23] L. Lundquist and M.M. Peritsky, "Cochannel interference rejection in a mobile radio space diversity system," *IEEE Trans. on Veh. Tech*, vol. VT-20, pp. 68-75, 1971.

[24] L.Schiff, "Statistical suppression of interference with diversity in a mobile radio environment," *IEEE Trans. on Veh. Tech.*, VT-21, pp. 121-128, November 1972.

[25] W.L. Aranguran and R. E. Langseth, Baseband performance of a pilot diversity system with simulated Rayleigh fading signals and co-channel interference," *IEEE Trans. Veh. Tech.*, vol. VT-22, pp. 164-172, November 1978.

[26] A. Shah and A. M. Haimovich, "Performance analysis of maximal ratio combining and comparison with optimum combining for mobile radio communications with cochannel interference," *IEEE Trans. on Veh. Tech*, vol. VT-49, pp. 1454-1463, July 2000.

[27] F. Adachi, M.T. Feeney and J.D. Parsons, "Effect of correlated fading on level crossing rates and average fade duration with pre-detection diversity reception," *Proc. IEE*, vol. 135, Pt. F, no.1, Feb 1988, pp. 11-17.

[28] J.H. Winters, "Switched diversity with feedback for DPSK mobile radio systems," *IEEE Trans. Veh. Tech.*, vol. VT-32, pp. 134-150, February 1983.

[29] M.A. Blanco and K.J. Zdunek, "Performance and optimization of switched diversity systems for the detection of signals with rayleigh fading," *IEEE Trans. on Comm.*, vol. COM-27, pp. 1887-1895, Nov. 1979.

[30] J.D. Parsons and A. Pongsupaht, "Error rate reduction in Vah mobile radio data systems using specific diversity reception techniques," *Proc. IEE* Pt. F., vol. 127,pp. 475-484, Dec. 1980.

[31] T.T. Tongue and P.H. Witty, "Carrier transmission of binary data in a restricted band," *IEEE Trans. on Comm.*, vol. COM-18, pp. 295-304, August 1970.

[32] A. Gersho, "Adaptive equalization of highly dispersive channels," *Bell System Technical Journal*, vol. 48, pp. 55-70, January 1969.

[33] M. Schwartz, W. Bennet, and S. Stein, *Communication Systems and Techniques*, McGraw Hill Book Company, 1966.

[34] J.G. Porkies, *Digital Communications*, McGraw Hill Book Company, 1995.

[35] H. W. Arnold and W.F. Boatman, "Switched-diversity FSK in frequency selective rayleigh fading," *IEEE Trans. on Veh. Tech.*, VT-33, pp 156-163, 1984.

[36] J. H. Winters, "Optimum combining in digital mobile radio with cochannel interference," *IEEE Journal on Selected Areas in Communications*, vol. SAC-2, no. 4, pp. 528-539, July 1984.

[37] V.M. Bogachev and I. G. Kiselev, "Optimum combing of signals in space diversity reception," *Telecommunication and Radio Engineering*, vol. 34/35, pp 83, October 1980.

[38] J. Cui and A. U. H. Sheikh, "Outage probability of cellular radio systems using maximal ratio combining in the presence of multiple interferers," *IEEE Trans. on Communications*, vol. 47, no. 8, pp. 1121-1124, August 1999.

[39] J.D. Parsons, P.A. Ratliff, M. Henze and M.J. Withers, "Single receiver diversity systems," *IEEE Trans. on Communications*, vol. COM-21, pp. 1276-1280, November 1973.

[40] L. Lewin, "Diversity reception and automatic phase correction," *Proc. IEE*, vol. 109B,pp. 295-304, November 1973.

[41] J.D. Parsons, M. Henze, P.A.Ratliff and M.J. Withers, "Diversity techniques for radio reception," *Radio and Electronic Engineer*, vol. 45, pp. 357-367, July 1975.

[42] J.D. Parsons and P.A. Ratliff, "Self-phasing aerial array for F.M. communication links," *Electronics Letters*, vol. 7, No. 13, pp. 380-381, July 1971.

[43] S. Kazel, "Antenna pattern smoothing by phase modulation," *Proc. IRE*, vol. 52, No. 4, pp.435, April 1964.

[44] M.J. Withers, "A diversity technique for reducing fast-fading," *IERE Conference Proceeding, Radio Receivers and Associated Systems*, 1972.

[45] O.G. Villard Jr., J.M. Lomasney, and N.M. Kawachika, "A mode-averaging diversity combiner," *IEEE Trans. on Antennas and Propagation*, vol. AP-20, No. 4, pp. 463-469, July 1972.

[46] J. Granlund, *Topics in the Design of Antennas for Scatter*, MIT Lincoln Laboratory Tech. Report No. 135, November 1956.

[47] S.W. Halpern, "The theory of operation of an equal gain pre-detection regenerative diversity combiner with rayleigh fading channels," *IEEE Trans. On Communication Technology*, vol. COM-22, Nov.8, pp.1099-1106, 1974.

[48] S-C Lin and V. K. Prabhu, "Diversity combining and equalization of frequency - selective fading signals with unequal mean strengths," IEEE Veh. Tech. Conference Record, pp. 753-757, 1998.

[49] D. D. Falconer, M. Abdulrahman, N.W.K. Lo, B.R. Petersen, and A. U. H. Sheikh, "Advances in equalization and diversity for portable wireless systems," Digital Signals Processing, vol. 3, pp. 148-163, 1993.

[50] C.L.B. Despin, D.D. Falconer, and S. A. Mahmoud, "Compound strategies of coding, equalization, and space diversity for wideband TDMA indoor wireless channels," *IEEE Trans. Veh. Tech.*, vol. 41, no. 4, pp. 369-379, 1992.

[51] S. Tantikovit and Asrar U. H. Sheikh, "Joint multipath diversity combining and MLSE equalization (Rake-MLSE receiver) for WCDMA," *IEEE-VTC'2000-Spring Record*, pp. 435-439, Tokyo, Japan, May 2000.

Chapter 8

MULTIPLE ACCESS COMMUNICATIONS

The unabated growth in the use of wireless communications is continuing and new services are being introduced continuously. The phenomenal growth has led to spectrum congestion even though additional spectrum has been added several times over the past few years. The spectrum shortage still remains a major concern as a double digit growth rate per year put more pressure on the spectrum. In order to meet the growing demand, several solutions are possible.

The first is obvious - seek new spectrum in the higher frequency bands, say in the microwave frequencies. Some progress in this direction has already been made. In an earlier meeting of WARC, 1850 to 1930 MHz band was assigned to Personal Communications Systems. In the WRC-2000, additional spectrum in the 2.5 GHz and 5 GHz bands was assigned. Furthermore, 20-60 GHz band is being investigated for use in the indoors mobile communication systems as well as for wireless local loop (WLL) applications. Additional allocations have been made for satellite communications in the L band.

The second option is to migrate non-mobile radio services to cable or optical fiber and assign released blocks of spectrum to mobile communications. A close observation suggests that the TV broadcasters and armed forces are largest users of the spectrum which can be classified as more suitable for mobile communications. It is possible to use wireless local loop (WLL) in the 28 to 38 GHz band to distribute television programs. Furthermore, it is not even necessary to use broadcast mode to distribute TV programs; the cable network has been found to be more appropriate. In a single six megahertz wide channel used for TV broadcast, several thousand users could be provided mobile communication services. However, this issue is so controversial and politically sensitive that it is unlikely to be resolved in the foreseeable future.

The third solution, which is being pursued with vigour, is to use the assigned spectrum more efficiently. In any case, greater attention is being given to increase the spectrum efficiency as much as possible so that greater number of users are provided with mobile telecommunications services.

In the case of mobile communications, efficiency may be defined in two ways; the channel efficiency and spectrum efficiency. The channel efficiency is related to signal and modulation format. For example, the channel efficiency in the case of voice communications can be increased by reducing the transmission bandwidth with the use of low bit voice encoders. The term spectrum efficiency is related to the number of channels that can be carved out of a given bandwidth and should be distinguished from channel efficiency. The spectrum efficiency is defined as the number of users that can share a given bandwidth in a unit area i.e. users/MHz/km^2 while meeting the grade of service (GoS) specifications. The inclusion of area in the definition of the spectrum efficiency is used to indicate frequency reuse. It is easy to see that when frequency is not reused, the spectrum efficiency is proportional to channel efficiency. The overall system efficiency is obtained by combining the channel and the spectrum efficiencies.

Consider for example, the Advanced Mobile Phone System (AMPS) where channel bandwidth of 30 kHz is used for each call in one direction. The spectrum efficiency is increased more than three folds when low rate voice encoding is combined with spectrum efficient QPSK modulation to create three TDMA channels as is the case in IS-54 Digital AMPS (DAMPS) or IS-136. This increases the spectrum efficiency by more than three folds[†]. The proposed half rate vocoder will allow six TDMA channel further doubling the spectrum efficiency. The number of channels reserved to gain access to the system relative to the number of message channels determines the signalling overhead. These channels known as overhead channels do not contribute to spectrum efficiency. The channel efficiency may be increased by reducing signaling and coding overheads. For example, in AMPS, IS-54, and IS-136 out of 820 channels 21 are reserved for paging and access. Similar overhead is used in IS-95, the Qualcom (QCDMA) system. In the 3G systems, paging and access overhead can be reduced even further.

Further gains in efficiency may be achieved by employing Discontinuous Transmission Principle. The discontinuous transmission reduces the average interference, which allows a reduction in the cluster size thereby increasing the system capacity. A reduction in the cell size also increases the capacity. However, in this case the base as well as mobile transmit power must also be lowered to maintain acceptable signal to cochannel and adjacent channel interference at the cell edges. Thus, cell size, low rate voice encoder, multiple access, channel bandwidth and frequency reuse factor all play distinct roles in determining overall spectrum efficiency. In this chapter, we shall concentrate on the capacity of the channels used for random access, paging, and message transmission.

---

†. *The increase in capacity includes trunking gain.*

8.1 Multiple Access Methods

A system where many users share a common spectrum is called multi-user communication system. Obviously, a user intending to share the spectrum has to get an access to the system prior to making use of the common resource for information transmission. The process of access to a common resource takes place in two stages. In the first stage, a terminal lets the system know its intention to use common resource. In the second stage, the shared resource is actually used. To accomplish these tasks, three types of channels, access, paging and message, are used.

The access channels are used to make request to use a common resource consisting of message channels. An access channel, also known as random access channel is always used for mobile to base or reverse link communications. The paging channel is used in the forward or base to mobile direction to transmit several types of messages including grant of permission to use common resource, specific channel allocation, to send periodical updates on the system status, signals to alert the users for incoming calls, and to mark the beginning and the end of a communication sessions. A mobile user intending to gain access to the system, transmits on one of several random access channels; the actual number of access channels depends on many factors including the total number of subscribers, transmission rate and the average number of messages per unit time to be transmitted. The number of random access channels available and their status are continuously updated on a designated forward link paging channel. To provide a particular grade of service (GoS) to the customers, adequate access and message capacity is needed to avoid congestion.

On reception of the access request, the system control centre uses a paging channel to assign a free pair of channels, one each for forward link and the reverse link. On reception of the channel allocation message, the user terminal tunes to the assigned channels, sends acknowledgment to the system control centre to the effect that the terminal has been tuned to the assigned frequencies. For this purpose the terminal uses the assigned reverse link. This procedure is followed by exchange of security and authentication information. The message channels are arranged in a certain manner called multiple access method. Four types of multiple access methods are discussed in this chapter. These are FDMA, TDMA, CDMA and OFDMA.

8.2 Capacity of Forward Paging Channel

The capacity of a communication channel is defined in terms of the average number of seconds it remains busy in one hour. The maximum capacity of a communication channel is 3600 seconds, which is defined as equivalent to 1 Erlang. The total capacity of a forward paging channel is 1 Erlang. The average number of users that can share the channel is determined from the volume of the average traffic generated by a single user/hour and the total number of users. This is equivalent to the length of an average call in seconds. Dividing one Erlang by the average traffic single user gives the total number of users that can be supported by the channel. This is of course the theoretically maximum number of users that can be supported by a single channel.

Suppose the average forward message is N bits long, and the channel transmission rate is R bits/sec. The average traffic generated by a single user is $m \times N/(3600\ R)$ Erlangs assuming that each user on the average generates m messages per busy hour[†]. Thus, the maximum number of users that can be supported per paging channel is given by

$$N_p = \frac{1}{mN/(3600R)} = \frac{3600R}{mN} \qquad (8.1)$$

For example, if the channel transmission rate is 4800 bits/sec and each user generates on the average one message of length 240 bits per busy hour, then a single paging channel will support 72,000 mobiles. It should be noted that the actual number of users may be substantially smaller since it is difficult to achieve full utilization the channel.

EXAMPLE E-8.2.1

In a certain cellular system, three channels are assigned for paging. Each cluster of three cells is assigned one paging channel. The data rate on each channel is limited to 2.4 kbits/sec. Each forward paging message is 18 byte long, which protected by 1/2 rate code before transmission. If the paging channel can achieve maximum efficiency of 67%, determine the number of users each cell can support. Assume that each user generates on the average 1.2 messages per hour.

Solution

We shall estimate the number of supportable users in a cluster of three cells and divide it by 3 to get the desired answer. We use modified form of (8.1) to estimate the number of users per cell as:

$$N_p = \frac{3600R}{mN} \times \frac{0.67}{3} = \frac{3600(2400)}{1.2(15 \times 8 \times 2)} \times \frac{0.67}{3} = 6700 \qquad (E-8.2.1)$$

[†]. *Busy hour is defined as the hour during which the traffic peaks.*

Thus, each cell has a paging capacity of 6,700 users.

8.2.1 Cell Traffic and Channel Requirements

The number of traffic or message channels required to meet demand in a cell is determined by the number of users within that cell, call attempt rate, holding time, and the desired grade of service. The call attempt rate is the average number of calls a user originates within a unit time. The unit time is taken to be the busiest hour of the day. The holding time is the average call duration. Usually the grade of service is expressed in terms of blocking probability, although mean waiting time may also be used. The probability that no channel is available to service the call is called the blocking probability.

In addition to average call duration, call attempt distribution is also an important factor in determining the number of channels needed in a particular area to provide the required grade of service. In estimation of the number of channels, it is generally assumed that each customer during its idle time originates calls randomly and independent of the status of all other users. The assumption that all users act independently may not be strictly valid since calling a subscriber located within the service area introduces dependency. However, the probability of a call being originated and terminated within the same cell is so insignificant that a small degree of dependency can be ignored. The probability that n calls arrive in a time duration of t seconds is modeled as Poisson distribution. This is given by:

$$P_n(t) = \frac{(\lambda t)^n}{n!} exp(-\lambda t) \tag{8.2}$$

Here $P_n(t)$ is the probability that n calls occur in time interval $(0, t)$ and λ is the average call arrival rate. The Poisson distribution model of the number of arriving calls leads to an exponential distribution of call duration or holding time [1][2]. The probability that any randomly selected call will last for T second or greater is given by:

$$P(T) = 1 - exp(-\lambda T) \tag{8.3}$$

The average holding time of all customer calls is equal to $1/\lambda$.

The number of channels required to meet the user demand satisfying service type and quality depends on the call handling discipline, which may be one of three types. The most commonly used call handling method in telecommunications is called Blocked Call Cleared. It is also known as Erlang B. In this discipline, if a subscriber requests for a channel and if the channel is not available, the call is cleared from the system that is the subscriber has to retry to get access to the system. Most of the telephony calls belong to this discipline. In the second discipline called, Blocked Call Delayed, a blocked call is held until a free channel becomes available. This discipline requires queueing of incoming calls particularly during the times when traffic demand becomes high. In this case the queue length is theoretically infinite which means that the blocked call may be delayed for an infinite time. In the third strategy, called Blocked Call Held or Erlang C, a call that fails to find a free channel is held for the average call holding time, which is a system design parameter. If a channel becomes available during the holding time, the call is served for the remainder of the average call duration, otherwise it is cleared from the system. After the call is cleared, the user must retry. The use of call discipline depends on the application. For example, Erlang B is used in telephony, and Erlang C is used in the private fleet control networks.

For the Blocked Call Cleared concept or Erlang B discipline, the probability of blocking is well known. For n available channels and Poisson call arrival of rate of λ, and call holding time, H, of $1/\mu$, the probability of blocking is given by [3]:

$$B(n, a) = E_{1, n}(a) = \frac{a^n / n!}{\sum_{m = 0}^{n} a^m / m!}; \quad a = \frac{\lambda}{\mu} \quad (8.4)$$

where m is the number of busy channels. The above expression is also known as the First Erlang Loss Function. Figure 8.1 shows the number of channels required as a function of both offered traffic and probability of blocking.

For the Blocked Calls Held discipline, it is clear that each user remains in the system for one holding time. With assumptions similar to those for the Erlang B formula, Jakes [3] derived the blocking probability as:

$$P\left(n, \frac{\lambda}{\mu}\right) = \sum_{j = c}^{\infty} \frac{(\lambda / \mu)}{j!} \exp(-\lambda / \mu); \quad n \geq 0 \quad (8.5)$$

Here, n is the number of channels.

In the Blocked Calls Delayed or Erlang C strategy the blocked calls are allowed to queue up and wait for service in the order of arrival. In other words the system allows an infinitely long queue. The probability that the call will be delayed is given by [3]:

$$C(n, a) = p_0 \frac{a^n}{n!} \frac{n}{n - a} \quad (8.6)$$

where p_o is the number of busy channels and is given by

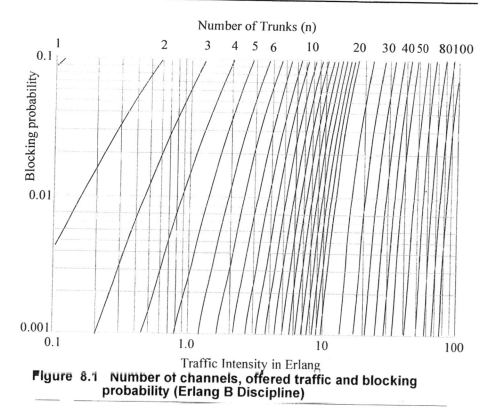

Figure 8.1 Number of channels, offered traffic and blocking probability (Erlang B Discipline)

$$P_0 = \sum_{m=0}^{n-1} \frac{a^m}{m!} + \frac{a^n}{n!} \frac{n}{n-a}$$
(8.7)

Figure 8.2 shows the probability of delay against offered traffic for Erlang C discipline. In the Erlang B strategy the blocked calls leave the system and do not occupy a channel, resulting in high channel utilization and minimal delay. This strategy is commonly used in switched telephone systems.

EXAMPLE E-8.2.2

A bus management system has five channels and operates on the principle of blocked calls delayed. If in the 10% of the calls are delayed, the load per user is 0.045 Erlang and λ is 1.2 calls/hour,

    (i) determine the number of users that can be supported by this system,

Number of Trunked Channels

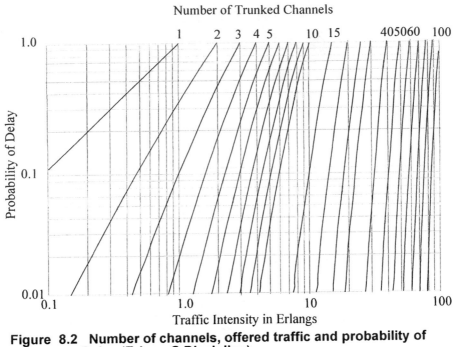

Figure 8.2 **Number of channels, offered traffic and probability of Delay (Erlang C Discipline)**

   (ii) determine the probability that a delayed call will have to wait for more than 10 seconds, if the probability of delay is given by

$$P_r[\text{delay} > t | \text{delay} > 0] = \exp\left[-\frac{(n-A)t}{H}\right] \qquad \text{(E-8.2.2)}$$

where H is the holding time, n is the number of channels, and A the traffic intensity.

Solution

   (i) Using Erlang C Chart, for 10% probability of delay with $n = 5$, traffic intensity is approximately equal to 2.15. Thus, the number of users is 2.15/0.45 = 47.8 say 48 users.
   (ii) Given $\lambda = 1.25$, holding time is = (0.045/1.25)×3600 = 129.6 seconds. The probability that the delayed call will have to wait before being served is

$$P_r[\text{delay} > t | \text{delay} > 0] = \exp\left[-\frac{(n-A)t}{H}\right] = \exp\left[-\frac{(5-2.15)}{129.6} \times 10\right] \quad \text{(E-8.2.3)}$$
$$= 0.8025$$

8.3 Random Channel Access

In switched telephone systems, the subscribers are permanently connected to the local switching office (SO) or a center via fixed communication lines. When a subscriber lifts the telephone handset, the connection to the local switching office becomes alive and the current flowing through the circuit loop indicates to the SO that a request for a message channel is made. On detecting the loop current, the switching office signals the subscriber on the availability of a message channel with a tone. Unfortunately, in wireless communications, such permanent connection is not provided and a connection with the switching office must be established to send a request for a message channel. Such requests are sent over what we call random access channels. The name random access is given to the procedure since the control centre is unaware of the time when a request for a channel will be made.

Since requests from mobile users arrive randomly and independently, collisions between requests from several users are likely. The access request message is usually kept short in order not to occupy the channel a user for unacceptably long time and block the others from using it. The message is assembled in the form of a package (known as packet) with destination address, originating party address, error protection and some control information. The request messages may be transmitted with or without knowing the channel state i.e. whether occupied or free. For a small number of users, access and paging functions may be combined in one channel but this is exception and the not the rule. As the number of subscribers grows, access and paging functions are moved to separate channels. The number of channels needed for access is determined by the number of users, rate of requests from each user, length of request message and the maximum allowable collision probability. The random access to system is achieved by using one of several random access methods like ALOHA, Slotted ALOHA, CSMA, Polling and their variations.

8.3.1 ALOHA Random Access Method

In 1970, a random access method for packet radio was developed at the University of Hawaii and was named ALOHA after a Hawaiian word for *hello*. The ALOHA system (also known as Pure ALOHA) [4] is the first example of channel sharing in this context. The system considers a large number of widely dispersed users, each wanting to transmit a packet over a shared high-speed channel. It is assumed that the users generate packets randomly in an independent manner, and the aggregate traffic is low. The packets are transmitted immediately upon generation. Transmitted packets may be lost either because two or more simultaneously occurring requests collide or number of errors are too numerous to be correctable. In most of the analyses, collision is considered to be the only cause of packet loss. In the analysis that follows, it is assumed that the latter cause predominates and the effect of noise is neglected. If a request originating from one mobile collides with a request from another mobile access failure occurs and access to channel is retried after a random delay. Random and independent delay minimizes the chance of repeated packet collisions between the same terminals. It is possible that randomly rescheduled transmissions collide with requests from other users. In the analyses, a large (infinite) population and long rescheduling delays are assumed. Thus, for a system operating under steady state conditions, the new request rate, Λ_{new}, and total request rate, Λ, are related as,

$$\Lambda_{new} = \Lambda(1 - P_r).$$

(8.8)

where P_r is the probability that the user's attempted request experiences a collision. The probability of collision is given by

$$P_c = \left[1 - exp\left(-\frac{2\Lambda T\tau_a}{n}C_a\right)\right]$$

(8.9)

where τ_a is the length of request message in seconds, n takes a value of 1 and 2 for pure and slotted ALOHA respectively, and C_a is the number of access channels. In Pure ALOHA, partial as well as full collisions occur whereas in the slotted ALOHA only full collisions occur because packets are transmitted within specified packet time slots. The operational difference between pure and slotted ALOHA is illustrated in Figure 8.3.

The phenomenon of collision is shown in Figure 8.3(a) where packets from user (a) to (h) arrive randomly at the base station and because these packets overlap (collide) each other, none of the packets survives collision. To quantize this effect, we examine network performance in terms of channel throughput and the average delay experienced in delivering the packet to destination [5]. The delivery of the packet is delayed because successful transmission happens after multiple tries and collisions. The probability of collision between packets is small when channel traffic is low. When the offered traffic increases because of more users or longer messages or more frequent transmissions, the collision probability increases significantly.

Assume that the start times of packets on the channel is a Poisson point process with parameter λ packets per second; λ is equal to the average pack-

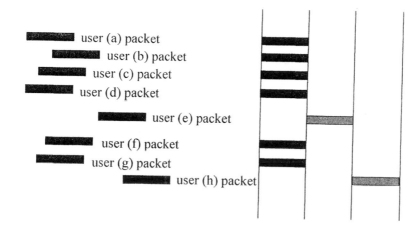

(a) Pure ALOHA Transmissions   (b) Slotted ALOHA Transmission

user (a) packet
user (b) packet
user (c) packet
user (d) packet
user (e) packet
user (f) packet
user (g) packet
user (h) packet

Figure 8.3 Transmission strategies in ALOHA and Slotted ALOHA

ets/sec/user multiplied by the number of the users. If the packet length is τ seconds, the normalized offered traffic G is defined as:

$$G = \lambda\tau \tag{8.10}$$

If we assume that only those packets which do not overlap with any other packet are received correctly (effect of noise is neglected), we may define λ' < λ as the rate of occurrence of those packets which are delivered to the destination successfully. Then, we define the normalized channel throughput S as:

$$S = \lambda'\tau \tag{8.11}$$

The probability that k packets are generated during a packet time is given by Poisson distribution

$$P[k] = \frac{G^k e^{-G}}{k!} \tag{8.12}$$

The probability that no packet is generated in this time is e^{-G}. Since the vulnerable period or the collision window is twice the packet length, the average number of packets during this time is equal to 2G. The probability that a packet will not experience collision with any other packet is therefore, *exp(-2G)*. This gives the channel throughput as

$$S = G exp(-2G) \hspace{4cm} (8.13)$$

This relationship is plotted in Figure 8.4.

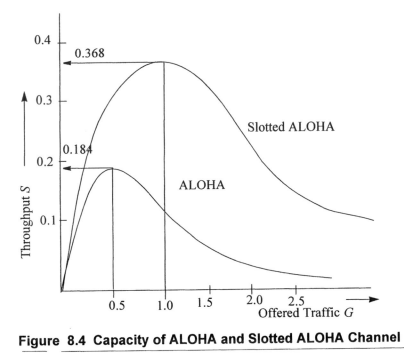

Figure 8.4 Capacity of ALOHA and Slotted ALOHA Channel

The capacity of the Pure ALOHA system (the maximum value of *S*) is seen to be 1/2*e*, or 0.184; it occurs when the normalized offered traffic is 0.5. Beyond this point, increasing the offered traffic results in greater number of collisions, which leads to a decrease in the channel throughput. As the offered traffic further increases, a greater number of collisions leaves very little channel time for successful transmission. The low throughput of Pure ALOHA is mainly due to unscheduled transmissions of packets, which results in partial collisions of packets. If some discipline is imposed unto packet transmissions such that only full packet collisions are allowed, the throughput can be improved.

The Slotted ALOHA system is a disciplined version of Pure ALOHA. In Slotted ALOHA, the base station divides the channel times into time slots equal in length to the packet duration. All users transmit packets only within the time slot boundaries. Since the packets are generated at random times, the generated packets anywhere within the time slot are held till the start of the next slot. This is shown in Figure 8.3. It is seen that Slotted ALOHA does not allow partial overlap of packets; packets are either spared from collision or are destroyed in full. In the figure, it is seen that none of the packets is spared in Pure ALOHA but two packets are collision free (successfully delivered) in the case of Slotted ALOHA. This simple illustration shows that the capacity of Slotted ALOHA is expected to be higher than that of Pure ALOHA. This is indeed the case. The capacity of the Slotted ALOHA channel has been analyzed by several researchers [6][7]. Since, in the case of Slotted ALOHA, the collision window is equal to one packet length, therefore the probability that a collision does not take place in this window is $exp(-G)$. The throughput, S, is therefore given by

$$S = Gexp(-G) \tag{8.14}$$

This is plotted in Figure 8.4. The maximum system capacity is $1/e$, or 0.368, which is double that of Pure ALOHA.

The delay in delivering a packet depends on the number of tries before the packet is successfully delivered. Consider transmission of a packet. The probability that transmission of the current packet is unsuccessfully

$$P_F = 1 - exp(-G) \tag{8.15}$$

The probability that K attempts are needed to deliver the packet successfully is

$$P_K = exp(-G)(1 - exp(-G))^{K-1} \tag{8.16}$$

The expected number of transmissions before a packet is successfully transmitted is

$$E - \sum_{k=1}^{\infty} kP_k = \sum_{k=1}^{\infty} ke^{-G}(1-e^{-G})^{k-1} \tag{8.17}$$
$$= e^G$$

It is seen that a small increase in traffic beyond $G = 0.5$ for Pure ALOHA and $G = 1$ for Slotted ALOHA, decreases the throughput sharply. When the offered traffic increases significantly, the throughput tend to go to zero because the expected number of tries before a packet is successfully delivered grows exponentially. Note that the above result does not consider other channel impairments like noise, fading, shadowing and interference.

8.3.2 Carrier Sense Multiple Access

CSMA stands for carrier sense multiple access. The system requires in addition to random access or reverse channels at least one forward or Paging channel. The forward or paging channel is used for transmission of the reverse or access channel status, whether busy or idle. The presence of a carrier on the forward channel indicates that the access channel is busy. When a packet becomes available for transmission at the mobile terminal, the forward channel is sensed (examined) and its status whether idle or busy is determined. If the channel status is found to be idle, the packet is transmitted otherwise it is delayed till the access channel status changes to idle.

Several variations of CSMA are possible; these include persistent, *p*-Persistent, and Digital Sense Multiple Access (DSMA). The objective behind all these variations is to reduce the length of the collision window. The collision window is the duration of the time in which collision is possible. The collision window is made of propagation time, channel switching time, channel access time, transmitter attack time, and base hangtime. The signal propagation time is the round trip propagation time between the mobile and base station. The channel switching time is defined as the time the mobile terminal takes to switch itself from listening mode to transmission mode. The channel access time is the time that mobile takes to get access to the channel. The transmitter attack time is defined as the time that the transmitter takes to stabilize after the terminal changed its mode from listen to transmit. After the elapsed of this time, the transmitter becomes ready to transmit.

Since the channel is sensed prior to an access is attempt, collisions are controlled. Thus, the CSMA method is an efficient access procedure. The channel throughput is substantially increased over that of ALOHA protocol. The base station provides the status of the reverse channel (mobile to base) on the forward channel. Since the mobile has information on the status of the random access channel, the probability of packets colliding is significantly reduced. The collisions may still occur under two conditions. The first is related to the absence of complete simultaneity between the activities of the users. The time over which collisions can take place is the sum of the round trip propagation time and the time required to switch the device (e.g. base station switch over time) from listening to transmit mode. For different users, the round trip propagation times are different because of spatial dispersion of their locations. In the second situation, a terminal fails to detect the carrier because of some obstruction between the base station and the mobile. Having erroneously detected the channel status to be idle, the terminal proceeds with transmission of a packet, which obviously collides with a packet from another user. This situation is known as the hidden terminal problem. More generally, a hidden terminal is the one, which despite being present within the coverage area fails to ascertain the current channel status.

A variation of CSMA in which the system status is obtained by detecting the status of a particular bit rather than a carrier. In this variation, instead of detecting in the absence of a carrier, the mobile unit detects the status of a designated bit within a packet to ascertain channel status. The important difference is that the default condition (desired signal not detected) denies transmission rather than allowing it. This variation addresses the hidden terminal problem.

CSMA is usually used in terrestrial packet broadcast networks because the propagation delay between the nodes is much shorter than the duration of the packet. This is an important assumption in the CSMA system design. This assumption is usually satisfied in practical land based systems. Long propagation delays in satellite communications renders this method impractical. The achievable throughput of the CSMA access channel is much higher than that for slotted ALOHA and it could be as high as 90% under ideal conditions.

Non-persistent and *p*-persistent are the two distinct classes of CSMA system. In the non-persistent CSMA, if the channel is indicated to be busy, a remote terminal unit (RTU) always reschedules the packet transmission according to some retransmission rules. After the elapse of a random delay, the terminal senses the channel and decides whether to transmit or reschedule the packet. If the channel is sensed idle, a packet is transmitted. Slotted CSMA is also possible for fixed length packets. The algorithm works as follows:

Packet ready

1) If channel status clear

Transmit

Else reschedule sense

Go to 1.

The throughput, *S*, of non-persistent CSMA is given by:

$$S = \frac{Ge^{-aG}}{G(1 + a + b) + e^{-aG}} \tag{8.18}$$

where G is offered traffic and parameters, a and b are normalized collision interval and hangtime respectively. The maximum throughput, S_{max}, is given by:

$$S_{max} = \frac{(\sqrt{(3a + 4b + 4ab)} - a)}{(a + 2a + 2ab)} \tag{8.19}$$

The collision interval consists of remote transmitter's attack time, mobile to base and base to mobile propagation times, decision time, set busy time, and busy detect time. The busy hangtime is the delay in detecting that the channel is no longer busy.

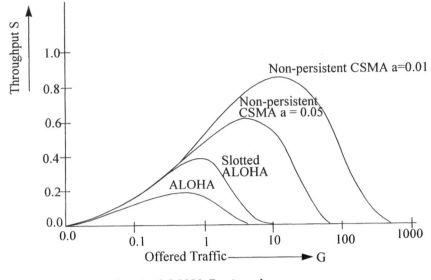

Figure 8.5 Throughput of CSMA Protocol

The p-persistent protocol is based on the desire that the channel should not be let go idle if a packet is available at the terminal to transmit. The value of p is a design parameter, which depends on the application. When $p=1$, 1-persistent protocol emerges which lets the terminal transmit with a probability of one when the channel is sensed idle. If the channel is sensed busy it waits till the channel is sensed idle and only then the terminal transmits the packet. The system algorithm is listed in the following:

> Packet ready
>
> 1) If channel status clear
>
> transmit
>
> else Go to 1.

In case of p-persistent, the packet is transmitted with a probability of p when the channel is sensed idle and is rescheduled with a probability of 1-p. The value of p is chosen to reduce the probability of collision and channel idle time. The access algorithm is listed below.

> Packet ready
>
> 1)Channel status
>
> Case of channel status: idle

Draw random number

If random number < 1-p, delay by τ seconds

Go to 1

Else transmit packet

Case of Channel status: busy

Repeat 1.

The transmitter attack time is perhaps the most critical parameter that adversely affect the throughput. It is less than 70 msec for FM voice and could be reduced to less than one milliseconds by careful design of transceiver synthesizer. The present day inexpensive digital transceiver possesses 5-10 msec attack time. The propagation time is 0.1 msec/30 km path. The base busy detect time depends on modem design and ranges between 1 and 20 msecs. The busy hangtime should be longer than the average signal fade duration but it should not be too long to adversely affect the throughput. Figure 8.5 shows a comparison of throughputs between non-persistent CSMA, slotted and pure ALOHA protocols. It is seen that the CSMA throughput could be much larger than what is possible with ALOHA.

8.3.3 Polling

Polling is another method to share a common communication channel for packet transmission. In polling, channel time is divided into time slots, which are assigned to the active users. A user transmits a packet only in the time slot assigned to it. The central node keeps the control over the network; it schedules reception and transmission from each active user once or more in a defined cycle time. It is clear that this strategy completely avoids contention between the users.

Consider a set of N mobile units. Let M be the average time between message origination for the busiest mobile unit. Suppose that the base station polls each mobile every NT seconds so that there are N time slots, one available to each mobile every NT seconds. In order that the time slot occurrences keep up with the most rapid message origination rate it is necessary that each time slot has a duration T, which is given by

$$T \leq \frac{M}{N} \tag{8.20}$$

where N is the number of mobile units, NT the time between a particular unit's time slots and M the minimum average time between message origination for any one mobile.

Furthermore, since each mobile unit may originate a message with uniform probability over the interval NT, average delay time D for each unit will be:

$$D = \frac{NT}{2} \tag{8.21}$$

Though at first glance an efficient transmission method, polling does suffer from a number of drawbacks. One is added complexity in maintaining the slot cycle. The allocation of slots to a previously inactive user requires some additional protocol, such as registration of new users to the list of existing users; this procedure requires the use of one or more registration slots in the polling cycle. Evidently the use of the registration slot can result in contention, with a possible system bottleneck. Alternatively, all free slots could also allow random access. Also, during its time slot each set has to switch from receive to transmit mode with the result that the slots have to be lengthened to accommodate transmit attack times. Further, in mobile environment relative propagation delays require guard time between the slots. This time is clearly wasted. Furthermore, the channel bandwidth for polling will be N times that of ALOHA and CSMA protocols. The higher bandwidth could result in intersymbol-interference (ISI), which may lead to degradation in the system capacity. The length of the polling cycle leads to random delays in transmitting a packet. This is analogous to the delay in accessing a specific sector of a hard disk memory. Despite these drawbacks, polling is an efficient access strategy, and throughput in excess of 90% can be achieved over channels that are void of impairments.

8.4 Performance of ALOHA and Slotted ALOHA Over Impaired Channels

The mobile radio channel is beset with a number of impairments; fading, shadowing and interference are the most important ones. These impairments may affect the channel throughput in a negative or positive ways. For example, the desired packets and the interfering packets reach the base station via channels that fade independently. In these operating conditions, it is possible that some of the interference packets experience fades while the desired packet is unaffected. In this case, the channel throughput is likely to improve. However, when the traffic is high, the number of colliding packets is expected to rise with the result the throughput will actually degrade. Since it is relatively rare that the desired packet is hit by more than one interfering packet, it is possible that the presence of fading and shadowing improves the channel throughput. Similar argument could be made for the case when channel shadowing is present. In the following section, we examine the throughput when channel impairments are present. We also discuss the role played by the receiver capture in aiding reception of packets in the presence of weaker interfering packets.

8.4.1 Impact of Fading and Shadowing on ALOHA

In the mobile environment, the power level difference between packets may arise due to channel fading, shadowing, and spatial distribution of the users (near-far effect). We follow the development given in [12] The mobile channel is characterized by noise, Rayleigh fading, and lognormal shadowing. It is also assumed that as the result of spatial distribution of the users near-far effect is present. The near-far effect arises whenever received signal power from an interfering user located nearer to the base station swamps the signal from the desired located far from the base.

The channel throughput is defined as the average number of packets successfully transmitted in a packet length duration†. Obviously, the maximum throughput is one. It is assumed that each user generates traffic that follows Poisson distribution. The total traffic arriving at a receiver from several sources has the distribution,

$$p(\lambda) = G^k \exp(-G) / k! \qquad k = 0, 1, 2, ... \qquad (8.22)$$

with a mean channel traffic G. The channel throughput is,

$$S = GP \qquad (8.23)$$

A packet is transmitted successfully if and only if the following conditions are met.

(i) No error occurs due to channel impairments. We may assume that a packet is detected correctly if the received power from it is above some threshold, W_t, and corresponding probability is denoted as,

$$P_t = Pr(w_c > w_t). \qquad (8.24)$$

(ii) Acknowledgment is received correctly, acknowledgment failure results in packet retransmission.

(iii) The desired packet captures the receiver.

Thus,

---

†. *In this section, we analyze the slotted ALOHA case, the analysis of pure ALOHA follows a similar procedure.*

$$P = p(0)P_t P_a + p(1)P_t P_a P_{capt}(1) + p(2)P_t P_a P_{capt}(2) + \ldots \tag{8.25}$$

$$P = P_t P_a \exp(-G) \sum_{i=0}^{\infty} [GP_{capt}(1)]^i / i!$$
$$= P_t P_a \exp(-G)\exp[GP_{capt}(1)] \tag{8.26}$$

where P_a is probability that the acknowledgment is received correctly. When both P_t and P_a equal 1 and $P_{capt}(1)$ approaches 0, (8.23) and (8.26) reduce to the case of standard slotted ALOHA in the absence of noise and the receiver capture effect.

The probability density function of shadowing is given by

$$P_{wc}(wc) = \frac{C}{w_c}\exp\left[-\frac{(\log w_c + A)^2}{B}\right], \qquad w_c > 0 \tag{8.27}$$

where

$$C = \frac{\log e}{\sqrt{2\pi}\sigma}$$

$$A = -\log W_o + \frac{\sigma^2}{2\log e} \tag{8.28}$$

$$B = 2a^2$$

and W_o is the mean value of w_c and σ is its standard deviation in Bels.

When shadowing and channel noise are present, the packet may not be detected correctly. The probability that the packet is transmitted successfully is written as,

$$P_t = \frac{1}{\sqrt{\pi}} \int_{L_1}^{\infty} \exp(-u^2)du \tag{8.29}$$

where

$$L_1 = \frac{\log R_1 + \frac{a^2}{2\log e}}{\sqrt{2}a} \tag{8.30}$$

and $W_I = R_I W_o$ $(R_I \geq 0)$.

When both lognormal shadowing and Rayleigh fading are present, the w_c is exponentially distributed while its mean is lognormal distributed with an area mean of W_o. The conditional probability of w_c is given by,

$$p_{w_c | W_c}(w_c | W_c) = \frac{1}{W_c} \exp\left(-\frac{w_c}{W_c}\right) \tag{8.31}$$

and the pdf of W_c $(W_c > 0)$ is

$$p_{W_c}(W_c) = \frac{C}{W_c} \exp\left(-\frac{(\log W_c + A)^2}{B}\right) \tag{8.32}$$

Combining (8.31) and (8.32), we have

$$p_{w_c}(w_c) - C\int_0^\infty \exp(-w_c u) \exp\left(-\frac{(A - \log u)^2}{B}\right) du \tag{8.33}$$

8.4.1.1 Capture Effect

The transmission of a packet fails on two accounts, when the packet collides with other packets or the channel impairments induce more errors than can be corrected. If the received power of the desired packet is higher than the interfering power by a certain margin, known as the capture ratio, it is possible that a desired packet captures the receiver even in the presence of other overlapping or interfering packets. The capture probability $P_{capt}(1)$ is given by

$$P_{capt}(1) = \Pr\left(\frac{w_c}{w_{u1}} > R\right) \tag{8.34}$$

where w_c is the power of the packet concerned, w_{ui} is the power of the interfering packet, and R is the capture ratio. A receiver with $R = 1$ is said to have a perfect capture, since it can recover the desired packet even if the interference power approaches the power of the desired packet. A receiver with $R = \infty$ cannot capture the desired packet even if the interference power approaches zero.

8.5 Access Protocol Analyses for Finite Population-Finite Buffer Systems

In the previous section, several random access methods were analyzed. In the analyses, it was assumed that the user population is infinite and each user generates a very small traffic. It was also assumed that each user does not have more than one packet ready for transmission at a time. However, in practical systems these assumptions may not be realistic particularly for high speed multimedia communication systems, where each user may maintain a finite length queue of packets for transmission. It is also more realistic to assume finite user population.

The analysis of finite-population finite buffer packet networks is very tedious. Markovian method is known to be the exact technique, but this technique is very cumbersome because even a modest sized system has very large number of states. Thus, the analysis involves calculation of very large number of state transition probabilities. A widely used approximate approach is to describe the overall system using two Markov chains. One is used to describe the number of busy users in the system since only busy users may contend for the channel; the other describes the number of packets in an arbitrary user called the tagged user, since all users are assumed to have identical statistical behaviour because of channel symmetry. In this case state transmission probability matrices have to be calculated. For larger number of users, the complexity due to large number of states remains [8].

The equilibrium point analysis (EPA) is an approach in which transition probabilities are not calculated [9]. In EPA the number of calculations is small, and it gives fairly accurate results. However, because of modeling complexity, it is difficult to extend the model to more complicated systems such as heterogeneous systems or homogeneous systems with bulk packet arrivals.

In [10], [11] a new method called Tagged User Analysis (TUA) was proposed for analysis of finite-user finite-buffer systems (FUFB) that use different random access protocols including ALOHA, S-ALOHA, CSMA, etc. In this method it is assumed that each user operates at its equilibrium probability. The packet service point is determined with the help of state flow diagram.

8.6 Multiuser Transmission Techniques

A successful access to the system is followed by a grant of a message channel out of many available. The method the message channels are structured within the assigned spectrum is known as multi-user transmission strategy. The communication resource dedicated to transmission of messages can be considered to be made of time-frequency space. In some circumstances, the third dimension of power is also added to define the communication source. The messages of a finite time duration, occupying bandwidth are transmitted. For example, consider a TDMA system in which a signal occupying a bandwidth B Hz is transmitted in a burst T seconds long.

The channels may be organized in the time-bandwidth space in three ways: time continuous-frequency slots, frequency continuous-time slots, and both time and frequency continuous. Each of these structures is classified as a multiplexing technique.

The multiplex strategy, which uses time continuous - frequency slots is called Frequency Division Multiplexing (FDM). Frequency division multiple access/frequency division multiplex, as the name suggests, divides the available bandwidth, B_w, into a set of a number of orthogonal channels each of bandwidth B_c. A channel is continuously used over the duration of the message. Every link in wireless system is full duplex; the duplex link requires one channel in the base to mobile (forward channel) direction and another in the mobile to base. (reverse channel). Since each carrier frequency can carry one channel, the technique is called single carrier per channel (SCPC). FDM is essentially a narrowband system. Figure 8.6 shows the channel organization in FDMA.

It should be noted that within each channel the transmit power may be allowed to vary and such variation does not affect the other channels. A minimum carrier to interference is maintained in order to get uniform BER performance. This means that each user can choose to transmit power according to the need and without any regard to powers in the other channels. It is worth noting however, that if the same channel is reused at another physically separate location, an increase in transmit power will adversely affect the carrier to interference ratio at that location. The difficulty with FDMA channel organization lies in its limited transmission rate, therefore it is generally used for narrowband applications.

Wideband mobile communications systems in various forms have been proposed [13][14][15] and studied [16][17]. These include time division multiple access systems (TDMA) and code division multiplexed systems. Time Division Multiple Access (TDMA), relatively new to wireless mobile comminutions, has been extensively used for several decades in the switched telephone network and satellite communications. Digital signal processing techniques and their very cost effective VLSI implementation are the primary reason behind its choice. The channels arranged in time slots - frequency continuous manner in the time-frequency space are said to be time division multiplexed. The channel time is divided into slots which are arranged into frames. Each active user is allocated a unique slot within a frame, therefore several

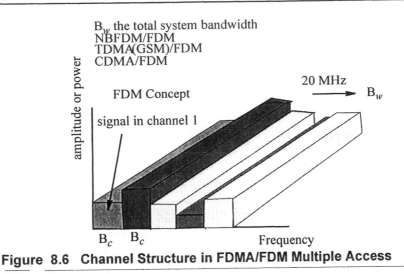

Figure 8.6   Channel Structure in FDMA/FDM Multiple Access

channels are supported per carrier. During each slot, the entire channel bandwidth is used. Thus, in contrast to FDMA, the transmission in TDMA is discontinuous, that is the users transmit in bursts which are confined to slots specifically assigned to them. Each slot has a preamble (frame sync), word sync, control and supervisory bits, and user data. The number of slots per frame depends on several factors including RF bandwidth, modulation format and transmission rate. A set of time slots are assembled into a frame. A typical frame structure is shown in Figure 8.10 (b). The frame length and the number of slots per frame vary from system to system. For example, North American TDMA system supports six slots (three users)/channel, GSM sixteen slots (eight users)/channel and DECT 24 slots/channel. As mentioned earlier, TDMA can be implemented either as a narrowband or a wideband system. North American systems, IS54 and IS136 are narrowband TDMA systems, GSM uses moderately wide bandwidth while DECT is a wideband implementation. In TDMA, the transmission and recovery of the signal is attained by opening a time gate over the duration of slot, i.e. τ seconds.

In TDMA, the message for transmission is broken into time bursts of length equal to the time slot. At the receiver, these bursts are collected to assemble the message. Note that to assemble the contents of the slots into message, precise track of the slot boundaries is kept. TDMA is a choice multiple access method for several important second generation digital land mobile systems such as GSM, DECT, DAMPS and PDC in Japan. The use of non-overlapping frequencies in FDMA and time slots in TDMA effectively create channels that are orthogonal in one of the dimension of the time-frequency space. The orthogonality between the channels effectively controls inter-channel interference. Figure 8.7 illustrates the principle behind TDMA.

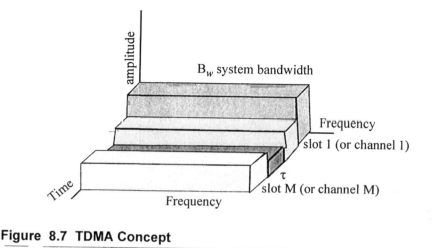

Figure 8.7 TDMA Concept

Code Division Multiple Access/Frequency Division Multiple Access is a radically different multiple access concept. The total system bandwidth may be used as a single block or is divided into spectral blocks having band-width much wider than that of the signal bandwidth. In this method, the time-frequency-power space is used simultaneously by all users. If each user message is encoded with a unique signature code which is orthogonal to all other signature codes used in the systems then all multiplexed signals can be separated. This multiplex strategy, known as Code Division Multiple Access (CDMA), can be used by many users to share time-frequency-power resource. The signature codes usually occupy much wider bandwidth than the message signal - a reason CDMA signals are also known as spread spectrum systems. The spread spectrum technology is well suited to mobile communi-cations because it provides considerable protection against impairments of the mobile radio communication channels [18][19][20][21].

Thus, the signals from the active users intermingles with each other. At the receiver, the individual signals are separated by correlating with the uniquely assigned codes assigned to the individual users. Although, the avail-able time-frequency-power space can be used fully in CDMA, difficulties can arise due to lack of orthogonality between codes. Moreover, any change in the power by one user directly affects all other users sharing the bandwidth. To optimize the use of spectral resource a power balance between the users is usually maintained. This is known as power control; it is deemed necessary in direct sequence CDMA systems.

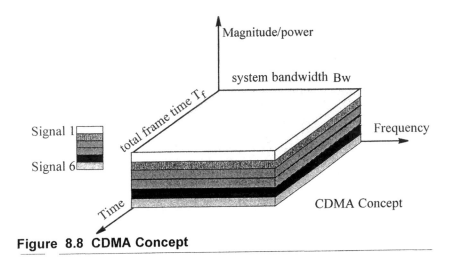

Figure 8.8 CDMA Concept

In frequency modulated SCPC systems, when the interference level increases beyond a specified protection ratio, the receiver is captured by the interference and intelligent communication becomes difficult. Similar effect is produced in the TDMA when the interference from the users in the neighbouring cells occupying the same time slot renders the channel unsuitable for communication. However, in spread spectrum transmission as the number of users increases, the resulting rise in the cumulative interference is spread over the entire set of users thereby degrading the quality of each user sharing the frequency. Thus, increasing number of users produce gradual degradation in the quality of all users and in this case a hard limits to the number of users does not exist. Several spectrum spreading techniques are possible, but in interference limited systems such as mobile radio, direct sequence and frequency hopped systems are considered to be the most suitable [22][23][24][25]. Hybrid forms such as time hopping and frequency chirp also exist and share the attributes of these systems. In the following, we evaluate the characteristics of the three above-mentioned multiplex strategies when applied to wireless mobile communications. In particular we establish a comparison between the capacities offered by these multiple access methods.

8.6.1 Frequency Division Multiple Access (FDMA)

First, we consider a cellular mobile communication system that uses frequency division multiple access. We consider North American advanced mobile phone systems (AMPS) as an example. The total allocated bandwidth of 50 MHz is divided equally in two 25 MHz blocks, one each for use in the forward and reverse direction. Each block is equally divided into two sub-blocks of 12.5 MHz which are allocated to two competing operators. Thus, each operator is allocated 12.5 MHz in each direction. 45 MHz separates the forward and reverse direction channel blocks. In AMPS channel bandwidth of 30 kHz is chosen, thus a total a of 832 channels are available. The analysis presented here can be applied to other systems in a straight forward manner. In European and Japanese analog systems 25 kHz channelization is used. To reduce interchannel interference, a guardband between the channels is used. Therefore, the message spectrum width is narrower than the channel spacing. To reduce inter-system interference, a wider guardband is allowed at the edge of the allocated sub-block. For example, a guard band of 10 kHz is allowed at the edges in AMPS.

A typical FDMA system architecture is shown in Figure 8.9. Each base station in a cellular system is assigned a set of RF carriers. The modulated carriers are multiplexed and transmitted using a single antenna. On demand, each user is assigned a duplex pair (a forward and reverse channels) centered at RF carriers 45 MHz apart. The FDMA architecture though has many advantages over TDMA and CDMA but it also suffers from several drawbacks. The main advantages are: low ISI, lower overhead, and simple hardware. The Inter-Symbol Interference (ISI) arises in digital communications because of significant channel delay spread relative to the symbol period. This happens when the transmission bandwidth is wider than the channel coherence bandwidth. The digital FDMA systems are traditionally narrowband systems where the symbol period is much longer than the average channel delay spread, therefore ISI is either very small or insignificant. The FDMA system uses channels on a continuous basis thereby periodic timing and synchronization controls are not needed and only fewer bits are required for signalling and control. Thus, FDMA hardware is much simpler than those for other multiplexing technique as it does not require adaptive equalization and slot timing adjustment control; framing and synchronization in FDMA is also simpler. In FDMA, the multiplexed signals are separated with bandpass filters.

The major drawbacks of FDMA systems include higher base station cost, requirement of a duplexer in the mobile unit and perceptible degradation of link quality during handoffs. Duplexer, a high cost component (approximately 10% of the total cost of a mobile terminal), is essential in FDMA mobile unit for simultaneous reception and transmission. At the base station, FDMA system requires individual transmitter and receiver for each channel, a requirement that augments the base station cost as well as the real estate to house it. In FDMA systems intermodulation products at the base station is of major concern. Finally, the degradation in the link quality happens because interruption in the continuous time link is needed to switch over to a newly assigned channel.

In Figure 8.9, it is shown that the individual signal, s_j is modulated, translated at frequency f_j, combined with other modulated and translated signals before being transmitted using a single antenna. The receiver separates the desired signal, s_j, with a BPF centered at f_j, down converts it using a carrier and demodulates the down converted signal. Finally, a lowpass filter is used to recover the message.

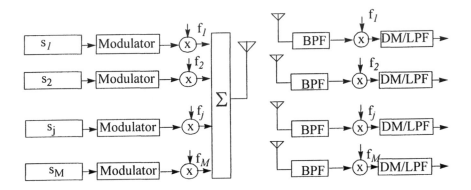

Figure 8.9 FDMA Receiver Architecture

8.6.2 Time Division Multiple Access (TDMA)

In conventional TDMA system, frequency duplex channelization, forward and reverse link channels having large frequency offset (45 MHz in North American IS-54 System) is used. Figure 8.10(a) shows architecture of TDMA system. The frame structure is shown in Figure 8.10(b). A frame hierarchy where frames are arranged in multiframes and superframes are arranged in superframes. A variation to TDMA is Time Division Duplex (TDD) where a single RF channel time is used for two way communications. The channel time is divided into frames, half of the frame divided into m forward link slots while the other half is divided into m reverse link slots. A TDD frame structure is shown in Figure 8.10(c).

The discontinuous mode of transmission and reception in TDMA requires sizable number of overhead bits. Similarly, inclusion of guard time between the slots reduces the usable channel time. Synchronization, control bit overhead and slot guard time could use up to 30% of the channel time; the

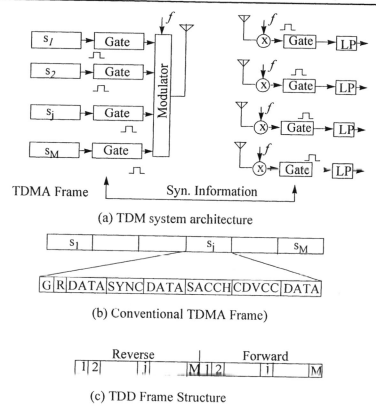

(a) TDM system architecture

(b) Conventional TDMA Frame)

(c) TDD Frame Structure

Figure 8.10   TDMA and TDD system structures

actual overhead varies from system to system and type of application. Since a single transmitter and receiver unit can be shared among m TDMA channels, the cost of a TDMA base station could be $1/m$ of an FDMA base station. It is also smaller in physical dimensions. Duplexer is not needed in TDMA since the time between the assigned slots is sufficient for switch over from one frequency to another. TDMA base station costs are lower relative to a same size FDMA base station. This advantage is crucial in recommending TDMA system for micro- and pico-cellular systems. In the pico-cellular systems, base stations on lamp posts are envisioned. The intermodulation products being of major concern in FDMA, are of minimal worry in TDMA provided adequate guard time between slots is provided. Similarly interchannel interference can be dealt with ease. The main disadvantage of TDMA lies in its complexity related to synchronization and dynamic slot alignment subsystem, equalization to mitigate ISI resulting from channel delay spread, interchannel interference, and sometimes echoes. However, solutions to these complexities contribute to an improvement in the system performance.

Figure 8.7 shows the TDMA concept. The total bandwidth is divided into a number of RF channels called carriers. The RF channel bandwidth is several times wider than the information signal bandwidth. This bandwidth is known as carrier spacing. The time on each carrier is divided into frames; each frame consists of a number of time slots of width, τ seconds. Each user is assigned a time slot on the forward and a corresponding slot on the reverse channel. The number of slots depends on the RF bandwidth. The transmission and reception therefore take place during these time slots, which are time orthogonal to each other. As in the case of FDMA, each user is free to transmit power according to the needs and such transmission does not affect the other time slots.

8.6.3 Direct Sequence Spread Spectrum

The use of FDMA and TDMA in cellular environment requires substantial real time coordination in order to use system resources efficiently. For example, assignment of channels to users while keeping intercell interference within an acceptable limit could be formidable task in FDMA when the number of frequencies is in hundreds. Similarly, slot-frequency assignment and management could become quite complex in TDMA system. One of the main attributes of CDMA is that it requires very little channel coordination. Code division has been used extensively in military communication networks for its resistance to jamming and channel Multipath, robustness against eavesdropping, and difficult unauthorized detection. Application of CDMA to wireless mobile communications is relatively new, although a proposal to use it for mobile communications in 1979 raised a considerable controversy. In CDMA, a large number of users share the same nominal frequency and are permitted to transmit simultaneously each using a unique code. This creates an intense mutual interuser interference. At a receiving terminal, the desired signal is recovered from the composite signal using the spreading code used at the transmitter to spread the signal. It is clear that the receiver must possess the knowledge of that code. It has been shown that if universal frequency reuse is employed in a sectored cellular system, capacity can be improved many times over that of FDMA or TDMA systems [13].

CDMA though having good attributes of robustness and high capacity, it suffers from two major problems; synchronization and power control. Although a central timing control is not necessary as in TDMA, the transmitter and the receiver forming a CDMA link must work in synchronism. A single user direct sequence system is illustrated in Figure 8.11. The energy in the baseband signal or modulated RF carrier is spread in bandwidth by multiplica-

---

tion the signal with a broadband spreading sequence. Spreading factors between few tens to about one thousand are typically used [22]. Typically, a transmission bandwidth of several MHz is required.

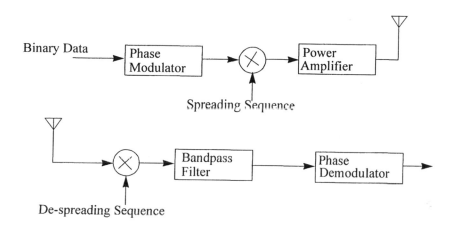

Figure 8.11 A schematic of DS SS system

The spreading function can occur at baseband or at RF. In the former case, the resulting signal is transmitted either by continuous phase shift modulation or by multiphase modulation. The spreading sequence allocated to each user is chosen from a set of mutually orthogonal sequences [22]. The sequence generator is clocked at a rate much higher than the baseband signal data rate. The inverse of the clock rate determines the spreading sequence bandwidth.

The first function of the receiver is to extract the desired signal from the multiplexed signal. The received signal, after translation to either baseband or an appropriate IF, is multiplied by a replica of the spreading sequence used at the transmitter. If the received signal spreading code is perfectly synchronized with the locally generated spreading code, the spreading code is removed from the received signal by coherent demodulation, and the energy spread over the transmission bandwidth will be collected into its original relatively narrow information signal bandwidth. The interfering signal power, however, is spread over a wideband by this process and the multi-user interference is completely eliminated if mutual orthogonality exists between the spreading codes for the desired and the interfering signals. If the spreading codes are not mutually orthogonal, the signals .from other users appear as interference called the multi-user interference (MUI) or multiple access interference (MAI) in the demultiplexing process. This process results in an improvement in the signal-to-noise ratio. This is shown in Fig. 11.6. The output of the correlator is conventionally demodulated to recover the signal. In practice, however, it is difficult to construct a large set of completely orthogonal codes and some residual multi-user interference remains. The presence of multipath or channel delay spread may also result in self-interference. Under certain channel conditions self interference sets an upper limit to the performance.

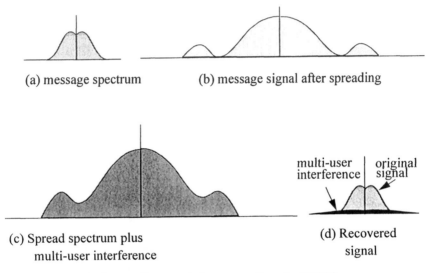

(a) message spectrum   (b) message signal after spreading

(c) Spread spectrum plus multi-user interference

(d) Recovered signal

Figure 8.12 Spreading and de-spreading in CDMA

The spread spectrum process gain G_p is defined as the ratio of transmission bandwidth to the bandwidth of the information bearing signal or post-decorrelation bandwidth,

$$G_p(f) = \frac{\text{transmission bandwidth}}{\text{post-correlation bandwidth}} \qquad (8.35)$$

This definition assumes that the signal power is spread uniformly over the spread bandwidth. The processing gain is a measure of the ability of the receiver to discriminate between the desired signal and others in the transmission band. In direct sequence spectrum spreading, the signal power spectrum has components at frequencies which are integer multiples of $1/M\Delta$, i.e.:

$$G_s(f) = \frac{1}{M\Delta} G(f) \sum_{k=-\infty}^{\infty} \delta\left(f - \frac{k}{M\Delta}\right) \qquad (8.36)$$

where M is the length of the sequence, Δ the chip duration, $\delta(f\text{-}f_0)$ is the Dirac delta function and $G(f)$ represents the envelope of the signal spectrum. For a maximum length sequence, $G(f)$ takes the form:

$$G(f) = \Delta \left[\frac{\sin \pi \Delta f}{\pi \Delta f} \right]^2 \qquad (8.37)$$

for $f \neq 0$ and for $f = 0$

$$G(f) = \frac{\Delta}{M} \qquad (8.38)$$

The ability to realize the full potential of processing gain depends on the cross-correlation between the spreading code of the received signal and the locally generated despreading code. A typical M-channel CDMA base station architecture is shown in Figure 8.13. Each signal is first modulated using a suitable modulation technique. Each modulated signal is then spread in bandwidth by multiplying it with a unique spreading sequence. The spread spectrum signals are added and frequency translated prior to transmission.

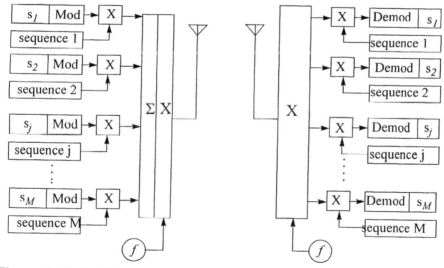

Figure 8.13 Architecture of DS-CDMA Multiple Access System

At the receiver the composite signal is down converted and is correlated with locally generated sequences. Since members of the code sequences may have imperfect cross correlation properties, the signals from all other users after despreading process appear as correlation noise. The output signal-to-noise ratio is then given by:

$$SNR_{output} = \frac{\int_{-\infty}^{\infty} H(f)R_{y1}(f)df}{\int_{-\infty}^{\infty} H(f)R_{yI}(f)df + N} \tag{8.39}$$

where $H(f)$ is the transfer function of the post-correlation filter or integrate-and-dump circuit.

$R_{y1}(f)$ is the Fourier transform of the correlation function of the received code sequence and the locally generated sequence (cross power spectrum) and $R_{yI}(f)$ is the sum of the Fourier transforms of the correlation functions of the locally generated code with that of the other interferers i.e.

$$R_{YI}(f) = \sum_{i=2}^{k} R_{yi}(f) \tag{8.40}$$

and N is the output noise power. For a noise free receiver, the processing gain definition reduces to the ratio of the output SNR to the input SNR. In practical systems, the achievable processing gain is usually lower than the designed value because of receiver nonlinearities, finite receiver noise figure and limited receiver dynamic range. Direct sequence spread spectrum also suffers from the near/far problem. This is discussed in more detail in Chapters 5 and 11. Probably its most serious limitation accrues from the loss of synchronization as a result of deep fading that is encountered in the mobile radio environment. Moreover, a slow recovery of a conventional receiver from the loss of synchronism results in additional degradation.

8.6.3.1 Frequency Hopped Spread Spectrum

Frequency hopping is another form of code division multiplexing. In this system, the carrier frequency of the narrow band modulated signal "hops" over a set of RF carriers [26][27]. Basically there are two types of hopping, slow and fast. In slow hopping, several symbols are transmitted during each hop. A variation proposed by Viterbi [28], and more suitable to mobile communications, requires transmission of discrete frequencies at a chip rate. Here a message symbol is divided into a number of chips and frequency is hopped every chip. This is known as a frequency hopped multi-level FSK (FH-MFSK) system. Since, frequency hops several times during a symbol duration, this is known as fast frequency hopping system.

Frequency hopping provides inherent diversity against frequency selective fading. The requirement of synchronization is waived in this architecture since the signal can be detected non-coherently. Figure 8.14 shows usage of time frequency space in Fast Frequency Hopped system. The figure shows m possible frequencies for hopping. The time is divided into chips of duration τ seconds. The figure shows hopping patterns of three signals. Each signal has a unique hopping pattern, which is generated by a pseudorandom sequence generator. At the receiver, same sequence is used to recover the baseband signals from the hopping frequencies. It is possible that two signals may hop to a single frequency at the same time. This is known as a hit and it causes degradation in the performance of the FH systems.

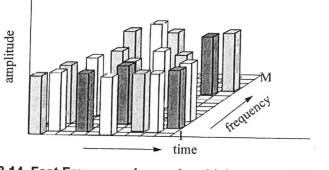

Figure 8.14 Fast Frequency hopped multiple access (FFHMA)

To achieve this form of coding, a PN code generator controls a variable frequency synthesizer which in turn hops frequency over a wide band in a specific order. The hop rate is dependent on the rate of information transmission or the chip rate. Figure 8.15 shows a typical transmitter and receiver for multi-level FSK-FH. The input signal is modulated to a suitable format and the narrowband signal thus generated is shifted in frequency with the frequency translator every chip time.

Over mobile channels, this system is most effective with noncoherent detection. For multiple access systems, a message symbol consists of k bits is modulo-2 added to one of L k-bit words. This means that each message symbol is divided into L k-bit words, each of which identifies a user address. The modulo-2 sum is used to generate a tone at intervals equal to T/L second where T is the basic signalling interval. At the receiver, the baseband signal is modulo-2 added with the address code at each T/L second interval and the output is passed through code word detector which delivers k bits of data. For example, a data symbol represented by a frequency f_2 is transmitted as a sequence of frequencies f_{17}, f_{15}, f_{16}, f_{13}, f_{14}, f_{12}, and f_{11}. At the receiver the received frequencies follow the same sequence except in the seventh chip time f_{r3} instead of f_{r1} is received. This error occurs due to noise or interference. The transmitted and received signals are shown is Figure 8.16.

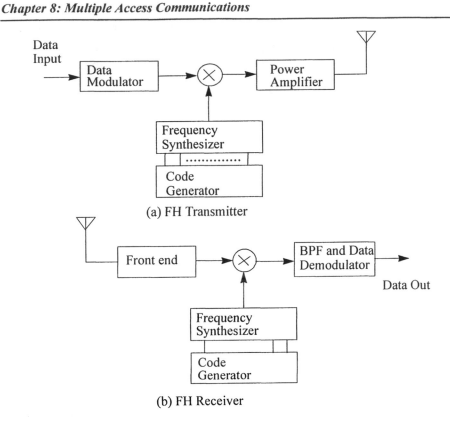

(a) FH Transmitter

(b) FH Receiver

Figure 8.15 FH Transmitter and Receiver

8.6.3.2 Comparison of Spread Spectrum Techniques

Frequency hopped and direct sequence systems provide spectrum spreading to combat channel impairments. A major disadvantage of direct sequence spread spectrum system is the requirement of synchronization and tracking of the phase of the received signal. This task is usually difficult in the presence of fast fading. This difficulty does not exist, however, in an environment which does not exhibit fast fading characteristics, e.g., an indoor communication system where multipath propagation is the major channel impairment. This type of impairment results in intersymbol interference (ISI). MFSK-FH overcomes dynamic range and nonlinearity problems that arise in direct sequence spread spectrum systems. However, for fast hopping systems the technological constraints in producing fast frequency synthesizers are major obstacles.

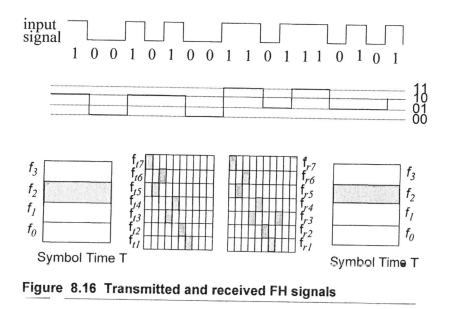

Figure 8.16 Transmitted and received FH signals

8.7 Orthogonal Frequency Division Multiple Access (OFDMA)

The M-ary FSK is the basis for OFDM. The problem of ISI encountered in transmission of high speed data over channels with significant channel delay spread can be reduced if data rate is slowed or pulse length is increased. The original serial data stream at R bits/sec is multiplexed into N parallel data streams of rate R/N. Thus, the receiver has to deal with N receive paths. In doing so the pulse length is increased proportionally and the ratio of channel delay spread to symbol time on each path becomes small. Of course the larger the number of parallel paths, the better is the immunity to the channel delay spread. The major challenge is to decide the number of parallel paths and find a method to generate a large number of orthogonal carriers. This challenge is met by choosing the transmission pulse to be rectangular since this choice has the advantage of forming the pulse and modulation by using simple Inverse Discrete Fourier Transform (IDFT), which can be implemented efficiently as an Inverse Fast Fourier Transform. The Fourier transform of a rectangular pulse produced subcarrier spectrum of shape $\sin(x)/x$, which has infinite bandwidth. Obviously, these spectra overlap. However, these can be separated because of their mutual orthogonality - a reason why this method is known as Orthogonal Frequency Division Multiplex (OFDM). The spacing between the carriers is chosen in such a way that only the desired signal is available at carriers and all other signals pass through zeros. To preserve orthogonality the receiver must be perfectly synchronized to the receiver both in terms of time and frequency. Furthermore, the presence of multipath may result in the selection of a path that could not maintain orthogonality between the signal carriers. In this regard, the OFDM symbol is prolonged by addition of a tail preceding the symbol prior to transmitting it. At the receiver, the added tail it is removed to restore orthogonality. Thus, the transmitted symbol in slot s and carrier l are disturbed by the Fourier transform of the channel impulse response, H_{sl}. The received signal may be written as

$$z_{sl} = a_{sl} H_{sl} + n \tag{8.41}$$

As observed from preceding discussion, OFDM system essentially requires synchronization and channel estimation blocks. With regard to synchronization, we require both frequency and timing estimation and their compensations. For details on synchronization reader is directed to [35]. Channel estimation is also needed in other mobile radio systems, for example implementation of Rake receiver. This will be briefly discussed in Chapter 9.

8.8 Efficiencies and Capacities of Multiple Access Techniques

In previous sections, several multi-user transmission techniques were introduced. Many salient features of each were discussed. Usage of a particular transmission method depends on application at hand (narrow or wideband) and the environment (urban, suburban, rural) in which the system is to operate. For example, for high data rate applications, FDMA is not suitable. In an environment, which exhibits large delay spreads, the use of CDMA may be more appropriate because of its ability to exploit path diversity. The use of TDMA for high speed transmission has disadvantage of complexities resulting from the use of channel equalizers to counter ISI. In mobile wireless communication, overall efficiency is chosen as the basis on which a choice on a particular multiuser transmission technique is made. The overall efficiency is made up of several efficiencies. Since both the time and the bandwidth may be used as a measure of spectrum usage, overall efficiency is determined by the efficiency of using time-frequency space.

8.8.1 Spectral Efficiency Due to Modulation

The spectrum allocated to mobile communications is very limited, therefore spectrum efficiency is an important aspect of mobile radio systems. The objective is to maximize the number of users that could share the given spectrum. The way how the resource in time and frequency is used determines the efficiency of RF spectrum. Thus the overall spectrum efficiency involves efficiencies due to modulation and multiple access. We consider the modulation efficiency first.

Consider a mobile radio system that provides service over an area of A Km^2. We can define the spectral efficiency due to modulation as the number of channels/unit bandwidth/unit area i.e. channels/MHz/km^2. The spectral efficiency due to modulation may be defined as:

$$\eta_m = \frac{(\text{Total number of channels available in the system})}{(\text{Bandwith})(\text{total coverage area})} \qquad (8.42)$$

We need to know the number of cells and the cluster size to calculate the total number of channels in the system. The total number of channels is given by the number of channels per cluster multiplied by the number of clusters. The spectral efficiency with respect to modulation is therefore:

$$\eta_m = \frac{(\text{Total number of channels/cluster})(\text{number of clusters})}{\text{Bandwith} \times A_c \times \text{number of clusters} \times \text{number of cells/cluster}}$$

$$\eta_m = \frac{\left(\dfrac{B_w}{B_c}\right) \times N_c}{B_w \times N_c \times A_c \times N} \qquad (8.43)$$

$$\eta_m = \frac{1}{B_c \times N \times A_c}$$

where η_m is modulation efficiency, B_w is the system bandwidth in MHz, B_c the channel spacing (MHz), N_c the number clusters, N number of cells per cluster i.e. frequency reuse factor of system, and A_c the cell area in km^2. Alternatively, the channel efficiency can also be measured in terms of the amount of traffic carried by the channel and the available capacity. The spectral efficiency may be defined in Erlangs/MHz/km^2. In each cell, a group of channels is used, therefore we should also include the trunking efficiency[†] in the definition of spectral efficiency.

$$\eta_m = \frac{(\text{Total traffic carried by the system})}{(\text{Bandwidth})(\text{Total coverage area})}$$

$$\eta_m = \frac{\text{Total traffic carried by}\left(\dfrac{B_w/B_c}{N}\right)\text{channels}}{B_w A_c} \qquad (8.44)$$

The trunking efficiency depends on blocking probability, which is a measure of quality of service (QoS) or grade of service (GoS). Including the trunking efficiency, η_t, in the definition of efficiency, we write,

$$\eta_m = \frac{\eta_t\left(\dfrac{B_w/B_c}{N}\right)}{B_w A_c} = \frac{\eta_t}{B_c A_c N} \qquad (8.45)$$

EXAMPLE E-8.8.1

North American Advanced Mobile Phone System (AMPS) is still in operation in several small markets of the United States. Each market is served by two competing service providers. Each provider is assigned half of the total cellular band of 50 MHz. The channel bandwidth is 30 kHz. The assigned spectrum to one of the operator is equally divided among channels in the forward and the reverse direction. The system covering a circular area of radius 31 km is divided into hexagonal cells of radius 2 km. Each user originates 1.25 calls during the busy hour and holds the line for an average duration of 150 seconds. The system specifies a call blocking probability of 2%. The frequency reuse factor is 12. Of the total number of channels, the top 21 channels are used for signalling. Calculate the spectral efficiency of the system.

Solution:

The number of channels = $12.5 \times 10^6/(30 \times 10^3) = 416$
Number of signalling channels = 21
Number of traffic carrying channels[†] = 416 - 21 = 395
Number of channels per cell = 395/12 = 33
Number of cells = $\pi \times 31^2/(\pi \times 2^2) = 240$

---

† *Trunking efficiency results from the use of one of several available channels by a user. These channels are shared among all users. Since the traffic originated from the users is random, therefore the channel time (total resource) is used more efficiently if the number of channels in a set is large. This concept allows higher subscriber loading.*

† *The signalling channels do not carry traffic generated by the users, therefore these are excluded from efficiency calculation.*

With 2% blocking, the total traffic carried by 33 channels (from tables or Erlang B formula) = 24.6 Erlang/cell or 1.958 Erlangs/km^2.
Number of call/hour = 24.6 x 3600/150 = 590.4 calls/hour/cell or 49.2 call/hour/km^2.
Number of users/cell = 590.4/1.25 = 472.3
Efficiency = η_m = 24.6 x 240/3016x12.5 = 0.157 Erlangs/MHz/km^2.

8.8.2 Multiple Access Spectral Efficiency

The multiple access efficiency is defined as the ratio of the time-frequency space dedicated for traffic to the total time-frequency space available to system. Thus, in the case of FDM, the multiple spectrum efficiency is equal to the ratio of the bandwidth used for messages to the total system bandwidth. In the case of TDMA, the multiplex spectrum efficiency is equal to the ratio of the time used for messages to the total time.

8.8.2.1 FDMA Spectral Efficiency

In FDMA, the multiple access spectral efficiency is given by:

$$\eta_a = \frac{N_T}{(B_w / B_c)} = \frac{B_c N_T}{B_w} \le 1 \tag{8.46}$$

where η_a is the multiple access spectral efficiency and N_T is the total number of voice channels in the service area. The numerator of (8.46) is the bandwidth used for useful information and the denominator is the total bandwidth.

EXAMPLE E-8.8.2

The multiple access efficiency for FDMA system of Example 8.8.1 is

$$\eta_a = \frac{30 \times 395}{12.5 \times 1000} = 0.948 \tag{E-8.8.1}$$

8.8.2.2 TDMA Spectral Efficiency

In TDMA, the channel resource is divided into slots, which are arranged in frames. The multiple access spectral efficiency of a TDMA system is given by the ratio of the time used for user information in a frame to the frame time, i.e.

$$\eta'_a = \frac{\tau M_t}{T_f} \tag{8.47}$$

where τ is the duration of a time used for useful information in each time slot, T_f the frame duration and M_t the number of time slots/frame. The total system bandwidth is divided into a number of RF channels, in each of these RF channel, TDMA frame structure is used. The efficiency is therefore:

$$\eta_a = \left(\frac{\tau M_t}{T_f}\right)\left(\frac{B_u N_u}{B_w}\right) \tag{8.48}$$

where B_u bandwidth of an individual user (the traffic channel bandwidth) during his or her time slot and N_u number of RF traffic channels (it is the bandwidth that users share while using different slots in a frame), and B_w is the system bandwidth.

The overall efficiency is given by

$$\eta = \eta_m \eta_a \tag{8.49}$$

EXAMPLE E-8.8.3

In IS-136, the D-AMPS, each operator is assigned 25 MHz bandwidth for duplex channels. The channel spacing is chosen to be 30 kHz. The channel time is structured in frames of 40 msec and each frame consists of six slots. The speech signal is encoded at 8 kbits/sec using VSELP. The digital speech is protected against errors with block encoding to produce encoded speech at 13 kb/sec. To this additional framing bits are added to produce an individual user data rate of 16.2 kb/s. Calculate the efficiency of TDMA system.

Solution:

Although the slot time is 6.67 milliseconds, but only a portion of the slot is used to for useful information; this portion is, $\tau = (13/16.2) \times (40/6) = 5.35$ msec. Given, $T_f = 40$ msec, $M_t = 6$, $N_u = 395$, $B_u = 30$ kHz, and $B_w = 12.5$ MHz.

$$\eta_a = \frac{5.35 \times 6}{40} \times \frac{30 \times 395}{12,500} = 0.76 \tag{E-8.8.2}$$

The spectral efficiency is 76% and remainder 24% is used to accommodate frame overhead.

8.8.3 Capacity and Frame Efficiency of TDMA System

8.8.3.1 Capacity

The capacity of a TDMA system is the number of RF channels per cell and it is given by:

$$N_u = \frac{\eta_b \mu}{v_f} \times \frac{B_w}{RN} \tag{8.50}$$

where N_u is the number of channels (RF carriers) per cell, η_b the bandwidth efficiency factor, μ bit efficiency (number of bits/symbol, 2 for QPSK), v_f voice activity factor (1 for TDMA), B_w forward or reverse direction bandwidth, R information rate plus overhead, and N the frequency reuse factor. The spectral efficiency η bits/sec/Hz is given by

$$\eta = \frac{N_u \times R}{B_w} \tag{8.51}$$

EXAMPLE E-8.8.4

Calculate the efficiency and the capacity of a certain TDMA system having the following parameters: bandwidth efficiency factor $\eta_b = 0.94$, bit efficiency (with QPSK) $\mu = 1.6$, voice activity factor $v_f = 1.0$, forward or reverse bandwidth $B_w = 12.5$ MHz, information bit rate $= 16.4$ kb/sec, and frequency reuse factor $N = 7$.

Solution

$$N_u = \frac{0.94 \times 1.6}{1.0} \times \frac{12.5 \times 10^6}{16.4 \times 10^3 \times 7} = 163.76 \tag{E-8.8.3}$$

N_u is the number of mobile users/cell.
Spectral efficiency

$$\eta = (163.76 \times 16.4)/(12.5 \times 1000) = 0.2148 \text{bits /sec/Hz} \tag{E-8.8.4}$$

8.8.3.2 TDMA Frame Efficiency

A TDMA frame consists of N_t slots for transmission of user bursts; each slot is separated from the adjacent slots by a guard time. The frame also has preamble bits which are used for synchronization. The frame efficiency is defined as the ratio of user information (data) bits to the total number of bits in the frame. The number of overhead bits per frame:

$$b_o = N_r b_r + N_t b_p + (N_t + N_r) b_g \tag{8.52}$$

where N_r is the number of reference bursts per frame, b_r number of overhead bits/reference burst, N_t number of traffic bursts (slots)/frame, b_p the number

---

of overhead bits per preamble per slot and b_g is the number of equivalent bits in each guard time interval.

The total number of bits per frame is

$$b_T = T_f \times R_{rf} \qquad (8.53)$$

where T_f is the frame duration, and R_{rf} bit rate of the radio frequency channel.

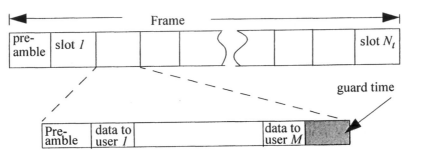

Figure 8.17 A typical TDMA frame

The frame efficiency is given by:

$$\text{Frame efficiency } \eta = (1 - b_o/b_T) \times 100\% \qquad (8.54)$$

The number of bits per data channel (user) per frame $b_c = RT_f$, where $R =$ bit rate of each channel.

$$\text{Number of channels per frame } N_{CF} = \frac{(\text{Total data bits})/(\text{frame})}{(\text{Bits per channel})/(\text{frame})}$$

$$N_{CF} = \frac{\eta R_{rf} T_f}{R T_f} = \frac{\eta R_{rf}}{R} \qquad (8.55)$$

8.8.4 Capacity of a DS-CDMA System

The capacity of DS-CDMA is a function of the processing gain, B_w/R, the energy per bit to interference power spectral density ratio, E_b/I_o, the voice activity factor, the DS-CDMA ideal frequency reuse efficiency, η_f, and the number of sectors in the cell antenna, G.

The received signal power at the cell site, from a mobile is $S = R \times Eb$. The signal to interference ratio is

$$\frac{S}{I} = \frac{(R \times E_b)/B_w}{I_o}$$

(8.56)

In a hexagonal cell structure, the interference accrues from users in the cell and users in tiers 1, 2, 3,.. cells. We assume that each user transmits the same power. The interference from the users in the outer tier cells experience greater attenuation. Thus, *S/I* ratio may be written as:

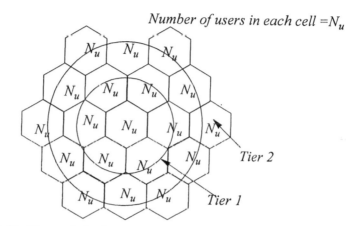

Figure 8.18 Multiuser interference scenario in CDMA

$$\frac{S}{I} = \frac{1}{(N_u - 1) \times [1 + 6 \times k_1 + 12 \times k_2 + 18 \times k_3 + \ldots]}$$

(8.57)

where N_u = number of mobiles in the band, Bw and k_i, $i = 1,2,3\ldots$ are the attenuation factors of the interference contributors from the terminals in the cells in tiers 1,2,3,..... If we define a frequency reuse efficiency, η_f, as

$$\eta_f = \frac{1}{[1 + 6 \times k_1 + 12 \times k_2 + 18 \times k_3 + \ldots]}$$

$$\frac{S}{I} = \frac{\eta_f}{(N_u - 1)}$$

(8.58)

and

$$\frac{E_b}{I_o} = \frac{B_w}{R} \times \frac{\eta_f}{(N_u - 1)}$$

(8.59)

We modify (8.59) to include the background thermal noise:

$$\frac{E_b}{I_o} = \frac{B_w}{R} \times \frac{\eta_f}{(N_u - 1) + \rho / S} \tag{8.60}$$

where ρ is signal to noise ratio, dividing it by S results in noise power. Since the DS-CDMA is time continuous and frequency continuous system, prohibition of transmission during the silent periods will reduce the multiuser interference. Thus, DS-CDMA can exploit voice activity factor to increase the number of simultaneous users while it is not feasible to do that in FDMA or TDMA. Equation (8.60) is modified to include voice activity factor, v_f, as:

$$\frac{E_b}{I_o} = \frac{\eta_f}{v_f} \times \frac{B_w}{R} \times \frac{1}{(N_u - 1) + \rho / S} \tag{8.61}$$

Accurate power control is an essential requirement for efficient operation of DS-CDMA; inaccuracies in power control degrades the system efficiency. Also, the bandwidth efficiency can be increased by reusing the frequency in cell sectors as well. We include bandwidth efficiency and capacity degradation due to imperfect power control, and number of sectors for completeness.

$$\frac{E_b}{I_o} = \frac{\eta_f}{v_f} \times \frac{B_w}{R} \times \frac{\eta_b C_d G}{(N_u - 1) + \rho / S} \tag{8.62}$$

$$N_u = \frac{\eta_f}{v_f} \times \frac{B_w}{R} \times \frac{\eta_b C_d G}{E_b / I_o} + 1 - \rho / S$$

$$N_u \cong \frac{\eta_f}{v_f} \times \frac{B_w}{R} \times \frac{\eta_b C_d G}{E_b / I_o} \tag{8.63}$$

where η_b, C_d, and G are bandwidth efficiency factor (< 1), capacity degradation factor to account for imperfect power control, and the number of sectors.

EXAMPLE E-8.8.5

It is desired to find the capacity and spectral efficiency of a single sector DS-CDMA system having the following parameters: total bandwidth = 12.5 MHz, bandwidth efficiency = 0.9, frequency reuse efficiency = 0.47, capacity degradation factor = 0.85, voice activity factor = 0.38, information bit rate = 9.6 kb/sec, E_b / N_o = 7.5 dB MHz. Neglect all other sources of interference.

Solution:

Substitution of these parameters in (8.63), results in $N_u = 112.89$ users (say 113 users). Spectral efficiency $\eta = (113 \times 16.2)/(12.5 \times 10^3) = 0.145285$ bits/sec/Hz.

The ratio of the capacity for DS-CDMA and TDMA system is given by:

$$\frac{N_{CDMA}}{N_{TDMA}} = \frac{c_d N \eta_f}{E_b / I_o} \times \frac{1}{v_f} \times \frac{G}{\mu} \qquad \text{(E-8.8.5)}$$

In the above consideration was not given to the loss of capacity due to the interchannel guard band. The presence of a guard band will reduce the number of traffic channels. This will result in capacity degradation. We now consider a case when F_e is lost on each side of the assigned band. The total number of channels is now given by

$$N_f = \frac{B_w - 2F_e}{W_c} \qquad \text{(E-8.8.6)}$$

As an example, for a bandwidth of 12.5 MHz, and top 21 channels assigned for control functions, the number of channels/cell for a cluster size of 4 is 98, if the guard band on each side of the spectrum is set to 10 kHz. The number of users that 98 channels per cell at a blocking rate of 2% is calculated to be 2064. We have assumed call origination rate of 1.25 calls per busy hour and call length of 120 seconds. If we consider interchannel guardband of say F_g, the total number of channels become

$$N_f = \frac{B - 2F_e}{W + F_g} \qquad \text{(E-8.8.7)}$$

For F_g of say 1 kHz, $F_e = 10$ kHz, the number of users/cell drops to 1994; the system experiences 3.4% loss in the capacity.

8.8.4.1 Spectrum Efficiency of CDMA

The spectrum efficiency of CDMA can be estimated with the use of an alternate methodology because the mostly understood concept of call blocking does not apply in this case. As the number of users per carrier increases, the signal to noise ratios of all users degrade till a stage is reached that if a user is added the link quality of all users degrades below an acceptable level. In other words, the capacity of DS-CDMA, the maximum number of users in CDMA, may be estimated from the quality threshold. The threshold may be defined in terms of either signal to noise ratio or achievable error rate. The signal to noise ratio at the input of the detector is given by

$$\frac{S}{I} = \frac{P}{(N_{cu} - 1)\frac{P}{G} + N_o B_w} \tag{8.64}$$

where N_{cu} is the number of users in a system that does not employ voice activity factor, P the transmit power, G the processing gain, B_w is the transmit bandwidth and N_o is the noise spectral density. Neglecting noise, we can rewrite (8.64) as

$$\frac{S}{I} \approx \frac{G}{(N_{cu} - 1)} \tag{8.65}$$

Assuming, interuser interference to be Gaussian distributed, the bit error probability for BPSK system is

$$BER = Q\left(\sqrt{\frac{S}{I}}\right) \tag{8.66}$$

where $Q(.)$ is Q-function defined as

$$Q(x) = \frac{1}{\sqrt{2\pi}} \int_0^x \exp\left(-\frac{\lambda^2}{2}\right) d\lambda \tag{8.67}$$

For other modulation methods, appropriate $Q(.)$ may be used. Table 8.1 shows the number of continuous transmissions (Erlangs) for several bit error rates. We have assumed a processing gain of 125 and BPSK modulation.

Table 8.1	*Single Cell Capacity of DS-CDMA System*

BER	Number of continuous users or channels/cell
10^{-2}	81
10^{-3}	25
10^{-4}	14

It is seen that lowering the BER requirement, more number of users can be accommodated. For 10^{-4}, only 14 Erlangs are possible in 1.25 MHz. Using average traffic per user of 0.0278 Erlang, neglecting the implicit trunking gain, we estimate the number of users which could be supported. These are shown in Table 8.2. In arriving at these figures, it is assumed that all channels are fully loaded and frequency is universally reused. It is also assumed that power control and synchronism are perfect. Imperfections in these will lead to some degradation in capacity.

| Table 8.2 | | Number of users/cell | | |

BER	Number of users per cell	C = 3	C = 4	C = 7
10^{-2}	2,916	8,748	11,664	20,412
10^{-3}	900	2,700	3,600	6,300
10^{-4}	504	1,512	2,016	3,528

8.8.4.2 Impact of Fading on FDMA, TDMA and CDMA Capacities

The wireless mobile channels experience fades. It is of interest to investigate the affect of fading of system capacity. Considering first FDMA, we see that the number of channels available for use are fixed and fading really does not rob any capacity on the primary basis that is the number of channels available are unaffected. The only degradation occurs in reduction of channel utilization; the average fade duration provides a measure of reduction in the channel utilization. Thus, the capacity is only affected on secondary basis because repeat transmissions are needed to overcome the effect of fading. Similar reasoning can be used to justify the loss in channel utilization in TDMA. Thus, the channel fading does not significantly affect the FDMA and TDMA capacities.

The case of CDMA is different. Here, the capacity is strongly dependent on the achievable signal to multi-user interference. The SINR under fading conditions is given by

$$SINR = \frac{E\{S^2\}}{Var(N) + Var(I)} \tag{8.68}$$

It has been shown that SNR under fading conditions can be written as [36]

$$SNR = \frac{\left[1 + \dfrac{2\sigma^2_f}{T_d \displaystyle\int_0^{T_d}\left(1 - \dfrac{\tau}{T_d}\right)\rho_f(\tau)d\tau}\right]}{SNR_G^{-1} + \displaystyle\sum_{i=2}^{K}\dfrac{P_i(1+\sigma^2_f)}{P_1 T_d}\displaystyle\int_0^{T_d}\hat\rho_{11}(\tau)\hat\rho_{ii}(\tau)\hat\rho_x(\tau)d\tau} \tag{8.69}$$

where σ^2_f is the fading envelope variance, T_d is the data bit duration which is equal to lT_c, P_i is the effective received power for the ith interferer, $\rho_{11}(t)$, $\rho_{ii}(t)$ and $\rho_x(t)$ are normalized autocorrelation functions for the signals of the desired user, ith interferer, and fading respectively. l is the length of the spreading code. The limiting value of probability of error is inversely proportional to the length of the code used.

$$\text{limit } G \to \infty P_r(E) = \frac{1}{4}\frac{P_i}{P_1}\frac{\langle C_i^2\rangle}{T_d^2} \tag{8.70}$$

$$\text{limit } G \to \infty P_r(E) = \frac{1}{6l} \tag{8.71}$$

where SNRG is the equivalent Gaussian SNR in the absence of fading and is given by

$$SNR_G = \frac{S^2}{Var(N)} = \frac{2P_1 T_d}{\eta_o} \tag{8.72}$$

Figure 8.19 shows the performance of a spread spectrum system that uses Gold codes of length 31 under fading and interference conditions.

We may summarize the above results that no impact of fading is found on the available channels in FDMA and TDMA in fixed channel assignment, however fading may have an impact on the capacity when dynamic channel assignment is used, and fading results in degraded performance (e.g. voice quality) and call drop probabilities (or outage probability). The CDMA system capacity is found to be susceptible to fading.

8.8.4.3 Multicell Capacity

The ideal multichannel capacity of a multicell system is given by the single cell capacity multiplied by number of cells, but the overall capacity may degrade due to interference from adjacent cells. The interference from

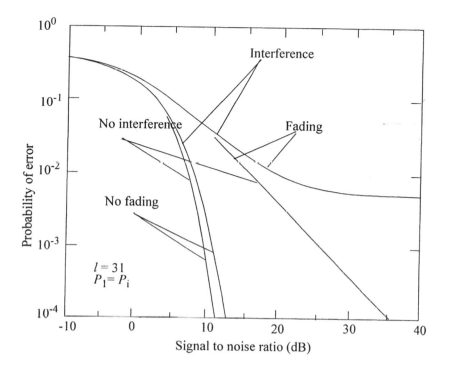

Figure 8.19 Performance of spread spectrum system under fading and interference conditions

adjacent cells may be strongly related to shadowing. This suggests dependency on the type of environment.

8.9 Summary

In this chapter, the concept of communication resource in terms of time-frequency space was introduced. The division of resource space in frequency and time resulted in the use of resource in frequency and time division multiplexed systems. The entirely different method of sharing the resource by devising multiplexing codes was introduced as code division multiplexed (CDMA) systems.

The expressions for the capacities of the three multiplexed systems were derived and the capacities of the three multiplexed systems were com-

---

pared. Finally, the effect of fading on the capacity degradation was investigated.

Problems

Problem 8.1:

At a certain base station, call arrivals is recorded every minute over the first half of the busy hour on a selected day. These are tabulated in the following.

Time	No of calls	Time	No of calls	Time	No of calls
1	2	11	3	21	7
2	3	12	6	22	8
3	1	13	5	23	5
4	6	14	1	24	3
5	4	15	2	25	1
6	7	16	4	26	3
7	9	17	5	27	6
8	8	18	7	28	2
9	4	19	8	29	7
10	7	20	3	30	5

   (i) Determine the mean arrival rate.

   (ii) Assume that the probability of arrival of a certain number of calls, n, is given by Poisson distribution

$$P[n] = \frac{r^n e^{-r}}{n!}.$$ (P-8.1.1)

where r is the mean arrival rate. Compare the observed results with that of measured. Comments on your result.

Problem 8.2

In a Pure ALOHA system, the packet arrival times are observed to follow Poisson distribution having an arrival rate of 200 packets/sec. If the transmission rate is 2 Mb/sec and the packets are 160 characters long, determine,

   (i) the packet duration,

   (ii) the traffic intensity,

   (iii) the normalized throughput of the system,

   (iv) the length of the packet that will maximize the throughput.

Problem 8.3

Repeat Problem 8.2, for Slotted ALOHA.

Problem 8.4

(a) Explain the difference between "blocked calls cleared" and blocked calls delayed disciplines.

(b) A particular cell in a blocked call delayed cellular system has 6 channels. If the system has a 6% probability of a delayed call, the load per user is 0.05 Erlang and $\lambda = 1.0$ calls/busy hour,

    (v) which discipline describes the system, Erlang-B or Erlang-C,

    (vi) determine the number of users that can be supported by this cell,

    (vii) determine the probability that a delayed call will have to wait for more than 10 seconds.

It is given that

$$P_r[delay > t | delay > 0] = \exp\left[-\frac{(n-A)t}{H}\right] \qquad \text{(P-8.5.1)}$$

where n is the number of channels, A is the traffic intensity, and H is the holding time.

Problem 8.5

A mobile communication System consultant is given a task of estimating the number of channels for the following cases:

(a) Mobile telephone service

    Mean holding time = 120 seconds

    Call originating rate = 1.2 calls/busy/hour/customer

    Projected user density 70 users/km^2.

    cell radius = 5km

    The blocking probability is required to be 0.02.

(b) Mobile dispatch system

    Number of taxis within the coverage area 750

    Average message length = 12 seconds

    Traffic origination rate = 0.2 calls/hr/customer

    Probability of delay = 0.65.

    What were his recommendations?

Problem 8.6

In a single cell system, each user generates on the average 3.5 calls per hour and each call lasts on the average 150 seconds. Determine,

   (i) the traffic intensity of each user

   (ii) the number of users that could use the system with 1% blocking if only channel were available

   (iii) the number of users that could use the system with 1% probability, if five trunks are available.

   (iv) the probability of blocking for five trunks, if a traffic hotspot is developed and the number of users is doubled from the number given in (iii).

Problem 8.7

A cellular mobile communication system for voice is allocated 25 MHz. The whole spectrum is divided into a number of 200 kHz channels. In each channel, 8 synchronous TDMA slots are implemented. Assume that no provision exists for guard band,

   (i) Determine the number of voice channels that can be supported by the system,

   (ii) sketch the frequency allocation of the voice channels and the channel frame structure.

Problem 8.8

A TDMA system transmits at 270.8 kb/sec to support 8 users per frame. Each user occupies one time slot per slot,

   (i) determine the raw data rate for each user,

   (ii) in each user time slot, guard and other overhead bits are inserted which effectively consume a rate of 10.1 kbps. What is the user traffic efficiency?

Problem 8.9

A TDMA based cellular system has a six slot frame structure and it transmits at 48.6 kbps to support 3 users per frame, that is each user occupies 2 slots. A slot on the reverse channel consists of 324 bits out of which 6 bits are used as guard bits. The other overhead bits are: 12 bits for supervisory control, 28 bits for synchronization, 12 bits for control channel use, and 6 bits are reserved for power ramp-up.

   (i) Sketch the structure of the time slot.

    (ii) Determine the raw data rate for each user.

    (iii) Determine the frame efficiency for the system.

    (iv) If a more efficient speech coder were available and each user is assigned one slot within a frame, determine the raw data and the frame efficiency.

References

[1] E.C. Molina, "Application of the theory of probability to telephone trunking problems," *Bell System Technical Journal*, vol. 6, pp. 461-494, 1927.

[2] A.K. Erlang, "Solution of some problems in the theory of probabilities of significance in automatic telephone exchanges," *Post Office Electrical Engineers' Journal*, vol. X, pp. 189-97, Jan. 1918.

[3] W.C. Jakes, Jr. *Microwave Mobile Communications*, John Wiley & Sons, 1975.

[4] N. Abramson, *The ALOHA System- Another Alternative for Computer Communications*, ALOHA System Technical Report B70-1, University of Hawaii, April 1970.

[5] A.B. Carleial and M.E. Hellman, "Bistable behaviour of ALOHA type systems," *IEEE Trans. on Comm.*, vol. COM-23, pp. 401-410, 1975.

[6] N. Abramson, "The throughput of packet broadcast channels," *IEEE Trans. on Communications*, vol. COM-25, pp. 117-128, 1977

[7] L.G. Roberts, "ALOHA packet systems with and without slots and capture," *Computer Communication Review*, vol. 5, pp.28-42, 1975.

[8] A.O. Fapojuwo, W.C. Chan and D. Irvine-Halliday, "An iterative approximation for modeling buffered CSMA-CD LAN's," *Journal of Electrical and Electronics Engineering*, vol. 8, no.2, pp. 102-111, 1988.

[9] S. Tasaka, "Dynamic behaviour of a CSMA-CD system with a finite population of buffered users," *IEEE Trans. on Comm.*, vol. COM-34, pp. 898-905, 1986.

[10] Tao Wan and Asrar U. H. Sheikh, "Performance and stability analysis of buffered slotted aloha protocol using tagged user approach," *IEEE Trans. on Veh. Tech.*, vol. 49, no. 2, pp. 582-593, 2000.

[11] Tao Wan and Asrar U. H. Sheikh, "Performance analysis of buffered CSMA/CD Systems," *Wireless Personal Communications*, vol. 18, no. 1, pp. 45-65, July 2001.

[12] A. U. H. Sheikh, Yu-Dong Yao, and Xiaoping Wu, "The ALOHA systems in shadowed mobile radio channels with slow or fast fading," *IEEE Trans. on Veh. Tech.*, vol 39, no. 4, pp. 289-298, November 1990.

[13] K.S. Gilhousen, I.M. Jacobs, R. Padovani, A.J. Viterbi and L.A. Weaver Jr. and C.E. Weatley, "On capacity of a cellular cdma system," *IEEE Trans. on Veh. Tech.*, vol. VT-40, pp. 303-312, May 1991.

[14] G.R.Cooper, and R. W. Nettleton, "A spread spectrum technique for high capacity mobile communications," *Proc IEEE Veh. Tech. Conference*, Orlando, pp. 98-103, 1977.

[15] P.A. Matthews, A.S. Bajwa, and D. Drury, "Spread spectrum signalling in land mobile radio systems," *Land Mobile Radio IERE Conference Proceeding*, No. 44, pp. 319-324, 1979.

[16] O. Yue, "Spread spectrum mobile radio, 1977-1982," *IEEE Trans. on Veh. Tech.*, vol. VT-32, pp. 98-105, 1983.

[17] U. Timor, "Improved decoding scheme for frequency hopped multi-level FSK system," *Bell System Technical Journal*, vol. 59, pp.471-484, 1980.

[18] G. Einarrson, "Address assignment for a time frequency coded spread spectrum system," *Bell System Technical Journal*, vol. 59, pp. 1241-1255, 1980.

[19] B.G. Haskell, "Computer simulation results on frequency hopped MSK radio-noiseless case," *IEEE Trans. on Communications*, vol. COM-29, pp. 125-132, 1981.

[20] D.J. Goodman and U. Timor, "Spread spectrum with variable-bit-rate speech transmission," *IEEE Trans. on Communications*, vol. COM-30, pp. 531-538, 1982.

[21] D.J. Goodman, P.S. Henry and V.K. Prabhu, "Frequency hopped multi-level fsk for mobile radio," *Bell System Technical Journal*, vol. 59, pp.1257-1275, 1980.

[22] G.R. Cooper and R.W. Nettleton, "A spread spectrum technique for high capacity mobile communications," *IEEE Trans. on Veh. Tech.*, vol. VT-27, pp. 264-275, 1978.

[23] P.S. Henry, "Spectral efficiency of a frequency hopped dpsk spread spectrum mobile radio system," *IEEE Trans. on Veh. Tech.*, vol.VT-28, pp.,November 1979.

[24] W.F. Utlaut, "Spread spectrum principles and possible applications," *ITU Telecommunications J.*, vol. 45,pp. 20-32, January 1978.

[25] G. Turin, "The effects of multipath and fading on the performance of direct sequence CDMA systems", *IEEE J.Selected Topics on Communications*, vol. SAC-2, pp.597-603, August 1984.

[26] Packets to the Editor, *IEEE Communication Magazine*, pp.2-3 September 1979.

[27] R.F.Ormondroyd, M.S. Shipton, "The feasibility of using spread spectrum communication systems for land mobile service on a non- interference basis with other users," *The Radio and Electronic Engineer (JIERE)*, vol. 50, pp. 407-418, 1980.

[28] R.C. Dixon, Spread Spectrum Systems. New York, Wiley, 1976.

[29] R. Eschenbach, "Application of spread spectrum radio to indoor data communications, *Proc. ICC*, 34.5.1-34.5-3, 1982.

[30] M. Khaverad, "Performance of non-diversity receivers for spread spectrum in indoor wireless communications," *AT&T Technical Journal*, vol. 64, pp. 1181-1210, July-August, 1985.

[31] M. Khaverad and P.J. McLane, "Performance of low-complexity channel coding and diversity for spread spectrum indoor, wireless communication," *AT&T Technical Journal*, vol.64, pp. 1927-1965, October 1985.

[32] O. Yue, "Frequency hopping, multiple-access, phase shift keying systems in a rayleigh fading environment," *Bell System Technical Journal*, pp. 861-879, July-August, 1980.

[33] R. Maummar and S.C. Gupta, "Performance of a frequency hopped multi-level FSK spread spectrum receiver in a rayleigh and lognormal shadow channel," *National Telecommunications Conference*, pp. 6.B.6.1-6.B.6.5, 1982.

[34] A.J. Viterbi, "A processing satellite transponder for multiple access by low rate mobile users," *Proc. Digital Satellite Communications Conference*," Montreal, 1978.

[35] J. G. Proakis, *Digital Communications*, Fourth Edition, McGraw-Hill, 2000

[36] C.S. Gardner and J. A. Orr, "Fading effects on the performance of a spread spectrum multiple access communication system," IEEE Trans. on Commu., vol. COM-27, no. 1, pp. 143-149, January 1979.

Chapter 9

MOBILE WIRELESS SYSTEMS AND SERVICES

The previous eight chapters laid the foundation of wireless communications by imparting a deeper understanding of the character of wireless channels and their impact on transmission of signals whether analog or digital. The concept of frequency reuse being necessary to increase the system capacity naturally results in interference scenarios, which when combined with other channel impairments infringes on the reliability of wireless systems. In Chapter 7, we discussed in considerable detail several signal processing techniques to make communications reliable in mobile environments. In this short chapter, mobile radio systems and services they provide are introduced. We defer details on several selected cellular systems to the ensuing chapters.

The wireless systems may be categorized in to two main classes - Public Land Mobile Systems (PLMS) and Private Mobile Radio Systems (PMRS). PLMS systems provide radio communication services to general public. These services include telephonic and paging. Cellular and paging systems are well known to the general public. It may come as a surprise to the reader that PMRS is a very large and prosperous sector of wireless communications. The services by PMRS are in general provided to several sectors of economy, which includes transportation, emergency, dispatch, law enforcement, armed forces and many others. Private mobile radio systems were growing at an annual rate of 15% since 1978 primarily due to emergence of new applications of mobile radio. The old systems were based on push to talk protocol and used voice as the mode of communications. However, lately PMRS have adopted new technologies related to cellular systems to create private data systems.

The PLMS provide mainly paging and telephonic services. Paging is the most basic form of mobile radio communications. Paging in fact falls in the category of radio broadcast system, which targets messages to selected subscribers. The intended subscriber upon receiving the message finds some other means of sending response to the message source. Paging services are provided in practically all bands assigned to mobile communications.

Prior to 1982, the year of cellular system inception in the United States, wireline telephone companies (Telcos) and Radio Common Carrier Operators (RCCO) offered mobile telephone services in North America on only 23 channels (11 VHF and 12 UHF) and 21 channels (7 VHF and 14

UHF) respectively. Thus, regardless of the population of any service area, at most 44 channels were available for mobile telephony. Because of meagre spectrum resource, the mobile radio services were limited to a small group of subscribers; only 44 simultaneous calls were possible in a given coverage. Obviously, this number of channels was insufficient to meet the demand for mobile telephone service particularly in large cities. In contrast, hundreds of frequencies in VHF and UHF bands were assigned to PMRS that provide public safety, emergency, ambulance, and other services. The demand for mobile telephony kept rising until the mobile radio systems became so congested particularly in major cities of North America that applicants had to wait for up to two years to get a license to operate a mobile telephone. It seemed that mobile radio communications was heading towards a bleak future.

The telecommunications regulatory authorities around the world remained under intense pressure to release new spectrum for mobile communications so that increasing demand for mobile services could be met. Besides telecommunications regulatory authorities, the mobile radio systems operators faced challenges on two accounts; to meet a substantial increase in the demand for wireless communications due to convenience it offered and to increase the radio coverage area, both requiring additional spectrum. The solution requiring an increase in the effective radiated power (ERP) and/or in increasing the height of the transmitter antenna was not satisfactory because it exacerbated the problem of interference and limited number of frequencies still remains a problem. For example, a system designed to cover an area within a 60 km radius might interfere with other stations within 160 km due to the presence of extended propagation. Clearly such a situation is not acceptable. Moreover, enlarged coverage area meant larger population and greater network congestion. In short, conventional solutions in mobile radio systems had the following severe limitations:

(a) inability to offer expanded services within the limits of channels allocated by the regulatory authorities;

(b) traffic congestion leading to impaired service quality;

(c) increased cochannel and inter-channel interference; this problem is exacerbated if higher powered transmitters are used in an effort to expand the coverage area over a wide area.

For many years, the regulatory authorities grappled with the problem of capacity shortage at local, national and international levels. The major obstacle was lack of agreements between various national regulatory authorities on the selection of the frequency bands for mobile telephone services. Beyond the deficiency of number of radio channels, however, was the realization that the conventional system was basically inefficient in its spectrum use and to increase capacity with more radio channels was not the only acceptable solution. The issue of increasing the capacity of mobile systems was considered by the regulatory authorities nationally and by the World Administrative Radio Congress[†] (WARC) internationally; it was decided that to increase the capacity of mobile radio systems, new systems with different attributes must be designed [1]. The foremost attributes of the new system were to have high spectrum efficiency and flexibility to accommodate future growth in traffic and coverage area without disruption. The other attributes included a provision of toll quality voice communication while keeping the service cost within the reach of the average consumer.

While the regulatory authorities were trying to finalize an agreement, developments of new systems remained stalled because no new spectrum was assigned. Moreover, in the absence of new developments in mobile telephone services the regulatory authorities were not convinced to open new frequency bands. The developers and the system operators were in an impasse and the future of mobile telephone communications started to look beak. One desirable solution to this dilemma was to find a technique that would accommodate a very large number of users by consuming a limited amount of spectrum.

The solution most suitable was to implement mobile communications based on cellular concept with allocation of some additional spectrum. The principal advantage of cellular communications lies in simultaneous reuse of frequencies at geographically disjoint areas[‡]. Theoretically, cellular radio can increase traffic capacity without limit [1]; in practice, the cochannel interference limits the reuse distance - the distance between the center of two areas where the same frequency is used.

The cellular solution was not new; D. H. Ring of Bell Laboratories informally proposed the concept in 1947 [2]. Several publications referred to the cellular concept in the late 60's and early 70's [3][4][5][6][7]. In 1971, in response to an invitation from FCC, AT&T filed a proposal for the development of a cellular mobile system. In 1974, the FCC allocated 40 MHz of spectrum on experimental basis to test the viability of the cellular concept, and invited proposals from the interested parties. In the same year, applications for two competing cellular systems were approved for field trials, one in Chicago and the other in Baltimore-Washington area.

The resounding success of field trials resulted in a further increase in pressure on WARC to assign additional spectrum to mobile radio communications. The landmark decision by WARC'79, to support new cellular technology and allocate the 800-900 MHz band on a global basis to mobile radio

---

†. *This organization is now known as World Radio Congress (WRC).*
‡. *In systems using DS-CDMA, frequency can be reused in the adjacent areas.*

communications, was a major step towards the development of commercial cellular telecommunication systems. Concurrent with the assignment to cellular systems, additional spectrum was allocated for private land mobile radio systems (PLMS). Forty MHz were allocated to high capacity cellular mobile telephone service, 30 MHz to dispatch services using current technologies, and an additional 45 MHz was kept in reserve. The national governments confirmed this allocation shortly thereafter. In 1986, the allocation to cellular systems was increased to 50 MHz because of congestion. With these decisions, the spectrum allocated to land mobile communication systems nearly quadrupled. The spectrum allocation stimulated intense interest mainly because of high potential financial returns. Mobile wireless communications, a field choked to near demise, due to inadequate spectrum, suddenly had a new lease on life.

In the following we shall introduce salient features of several mobile radio systems and the services these systems offer. In the ensuing chapters, where more details on the mobile radio systems that are currently in operation are given.

9.1 Mobile Radio Systems and Services

We shall first review the telecommunication services that are commonly offered in mobile communications. Our discussion starts with a brief review on paging systems that are currently in operation. It will be followed by brief descriptions of the conventional and Improved Mobile Telecommunication System (IMTS). We shall introduce working principles of cellular systems in this chapter; we defer details on several cellular systems are deferred till Chapters 11 and 12.

9.1.1 Paging Systems

In its basic form, this is the most elementary form of mobile communication. For the last several years, paging systems had been enjoying a superb growth. In recent years, this growth has somewhat stunted because of integration of many services including paging into cellular systems. In its most simplest form, a paging system works as follows.

An originating party requests the central paging control to page a message to the destination user. The central paging control receives the message and directs the base station to broadcast an alert signal over the network indicating that a message for the intended portable is waiting with the central control. The alerting message used was a beeping sound but recently the alerting message can be a combination of several tones. The destination user upon the receipt of the alert message, calls the central control via some other media, usually a nearest available public telephone, and receives the message. The destination user then calls the originating party using public telephone. Note that a paging system is necessarily a one-way, from base to mobile communication. The modern paging systems have developed into very sophisticated networks. The call to central control to retrieve the waiting message have been virtually eliminated by transmitting full message in the form of text or digitized voice instead of only an alert message. For more details on paging systems, see Chapter 10.

9.2 Conventional mobile systems

Conventional systems are two-way broadcast systems usually operated in simplex mode: only one communication path is available, which is used alternatively by the base station and the portable terminal, Thus, simultaneous two way or duplex mode of communication is not possible. This system is also known as "Push To Talk" (PTT) system. The major use of conventional systems is found in dispatch, as is the case in communications for police, ambulance, transport and taxi services. The old voice based conventional systems are still in use in some parts of the world. Because these systems lacked privacy and were prone to misuse by competitors. These are rapidly being replaced by digital systems, In most digital systems, the message transmitted in the form of alphanumeric data is also displayed on a screen as text.

The change over from voice to data has resulted in many fold improvement in the system capacity. This mode of communication is being taken over by a new system known as SMR (Special Mobile Radios) Services. The SMR Service is a more sophisticated form of conventional mobile system that works on shared basis using multichannel trunking systems. The older systems were operated independently by many small operators and most of the operations were confined to a single base station. Generally operated by public companies the new systems provide service to many fleet operators. The SMR's achieve high spectrum efficiency and provide transparency from eavesdropping by unauthorized users. A schematic of a typical conventional private land mobile system is shown in Figure 9.1.

9.3 Multichannel Trunked Systems

Multichannel trunked systems (MCTS) allow users to access one of several available duplex message channels. In the most primitive form, the mobile user searches a designated bank of channels and seizes the first free

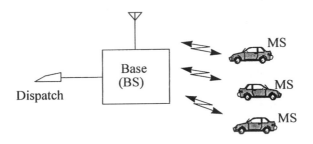

Figure 9.1 A typical configuration of PLMS

channel it finds to communicate. The MTC systems achieve higher spectral efficiency compared to the systems that use dedicated channels to their users. Multichannel systems were initially installed to provide telephones in vehicles. Business executives subscribing to this mobile telephone service, created a market which eventually caused very high congestion referred to earlier. In 1965, MTC systems allowed automatic dialling for the first time, this system known as Improved Mobile Telephone System (IMTS) may be considered as the forerunner of today's cellular mobile telephones. A multichannel system can be used to offer assortment of services including paging, although such provision is not allowed by the regulatory authorities in many countries.

The trunking principle is based upon the fact that each user requires a channel only for the duration of the message, which is typically between 10 to 20 seconds in short message services and could be as long as several minutes when communicating by voice. A typical trunking system is shown in Figure 9.2. In the system shown, the central switching matrix having access to all the three channels controls the activities on the channels and several centrally located dispatch positions. Three is the minimum number of channels because a trunked system with less than three channels does not provide enough trunking gain to justify the complexity needed in operating a trunking system. The mobile stations can access any of the trunking channels (three in the figure) and the consul point (CP). Thus, every mobile station must have a frequency synthesizer to tune to a particular channel in addition to capabilities to generate control signals for getting access to the system, automatically scan the message channels, select a channel for communication, and controlling the end to end link.

The grade of service in a trunking system is defined in terms of probability of call blocking and the wait time before access to the system is granted. A call is considered blocked, if a mobile had successfully transmitted a request for access, but a message channel could not be granted even after wait-

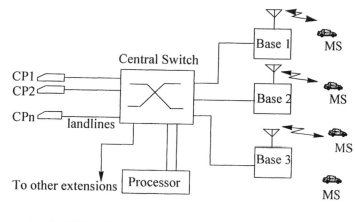

Figure 9.2 A typical Trunking system

ing for a certain specified time. The waiting time is usually defined in terms of average holding time. Thus, the system access process consists of three procedures: procedure to request for a message channel, procedure to hold calls in queues, and finally the process of actual granting a channel. The mobile terminal obtains access to the trunked systems on an access channel using one of several random access protocols. ALOHA, Slotted ALOHA, CSMA or CSMA-CD are notable examples of commonly used access proto cols. More details on random access procedures are given in Chapter 8.

Multichannel trunked systems are also being used in Private Mobile Radio Systems (PMRS) with a primary aim of providing voice and data services to business and other non-public users. Similarly, law enforcing agencies were pressed to use trunked systems because of greater emphasis placed by the regulatory bodies on achieving higher spectrum utilization. In this regard, MOBITEX, APCO-25, and TETRA are examples of systems used in PLMS. In the forthcoming sections, we briefly describe several multichannel trunked systems. Details on these systems are given in Chapter 10.

9.3.1 MOBITEX

MOBITEX is a multichannel wide area digital dispatch system. It provides textual, data and voice traffic for dispatch communications. It can also provide non-real time data services. Channels may be designated for voice or data. However, the voice channels may be opened up for data service if need arises. MOBITEX has a provision for group calls as well as for closed user groups. The MOBITEX system can be interconnected to public data and PSTN.

The architecture of MOBITEX is similar to that of a cellular system. The system consists of a MOBITEX Central Exchange, several base radio stations, area exchanges, main exchanges and mobile terminals. The channel is accessed via slotted ALOHA that uses a dynamic frame length control, that is the number of slots and slot lengths can be varied in order to optimize the channel utilization. The system transmission rate is 1200 bits/sec and uses FFSK modulation. The maximum packet length is 512 bytes or 4096 bits. The system uses (64,48) block codes for error detection and correction.

9.3.2 APCO-25

The Association of Public-Safety Communications Officials (APCO) developed APCO-25 standard seeking a high level of performance, efficiency, capabilities and quality in two way radio communications. The APCO-25 standard uses CQPSK modulation in 6.25 kHz channel bandwidth and C4FM in 12.5 kHz bandwidth. A large number of individual and talk-group addresses are provided. The aggregate bit rate is 9600 bits/sec. In the case of data transmission, the data packets consist of header, overhead information and data. The modulation uses frequency deviations of 1.8, 0.64, -0.64, and -1.8 kHz for dibits 01, 00, 10 and 11 respectively.

9.3.3 TETRA

Terrestrial Trunked Radio Access (TETRA) is an open digital trunked radio standard which resulted from the joint efforts of network operators, administrators, equipment manufacturers and users, under the auspicious of the European Telecommunications Standardisation Institute (ETSI). TETRA provides a range of bearer services, teleservices, and supplementary services that have joint voice plus data (V+D) capability. The direct mode operation (DMO) provides direct mobile to mobile communications. It also supports packet data services under the banner of what is called Packet Data Optimised (PDO). V+D and PDO specifications are based on the same physical radio platform i.e. modulation and operating frequencies.

TETRA finds many applications in transmission of secure speech and data, road transport, fleet management, vehicle location, and railways. More details on applications of TETRA are found in [9].

9.4 Cellular Mobile Telephone System

Research into methods to increase spectrum efficiency led to the concept of frequency reuse. In implementing cellular mobile telephone systems, large service areas are divided into hexagonal cells[†]. The assigned spectrum is divided into a number of channel sets, each set is used concurrently in cells

---

that are located at a distance called reuse distance. If the transmitted power is low and the reuse distance is sufficiently large, then the co-channel interference can be made small. A set of a certain number of neighboring cells form a cluster. The cell and cluster size are design parameters of a cellular system. These parameters are determined by the application in hand. For example, a low density rural area is divided into larger cells but relatively small size cells are used in urban areas. Radio coverage in each cell is provided by a base station, which uses a prescribed set of frequencies that nominally belong to that cell. Any mobile unit while within a certain cell must communicate on a pair of frequencies assigned to the cell. With frequency reuse, a much larger number of subscribers are accommodated than any other previously known system. It is of course understood that not all users are concentrated in one cell but are dispersed over the coverage area. Thus, in cellular systems, the capacity can be theoretically increased indefinitely by reusing the channels. However, cell size and system complexity put an upper limit to the achievable capacity.

The cellular mobile telephone service is designed to allow communications in several modes - fixed to mobile, mobile to mobile or mobile to fixed stations. A call from a mobile unit proceeds as follows. The subscriber pre-dials the desired number and picks up the hand set. As the hand set is lifted, the mobile transmitter sends a request message to the nearest base station over one of several dedicated access channels, identifying both the calling and called parties. On reception of such a request from the mobile, the base station passes the request to the control centre. The latter confirms the status of the caller as a legitimate user, determines the location of the called party and alerts the receiving station. At the same time an acknowledgment is sent to the calling party on a broadcast channel (also known as the forward paging channel) to the effect that such a request has been received and is being processed. The message also contains information on the frequencies to which the subscriber's equipment need to be tuned. If the called party happens to be another mobile, the broadcast message also contain instructions for the called party to tune to the specified frequencies. After tuning to the allocated frequencies, the mobile unit sends acknowledgment to the base station on the assigned message frequency. Upon reception of acknowledgments from each party, the call is set up and conversation begins. A supervisory signal indicating the end of the call is sent before the mobile transmitter is ordered for a shut down.

---

†. *The hexagon shape is for theoretical convenience. It is closest to a circular area matching the coverage area of an isotropic antenna. In practice, the shape of cells are in general very irregular.*

During the call, if the mobile unit crosses a cell boundary and moves into a neighboring cell, the successor cell base station takes control of the call, and another pair of channels is assigned so that the call could continue. Such handover of a call from one cell to another requires an exchange of considerable amount of control information between the mobile, base station and the switching station. Consider a mobile in Cell A being supervised by Base A. After a short time the mobile reaches a region where boundaries of three cells A, B, and C meet. When the signal strength from the mobile at the serving base station is indicated low, the neighboring base stations are directed to monitor the strength from the mobile[†]. The call is transferred to the base whose signal is judged to be the strongest; in this case Cell B is the declared to be the candidate for handover. To transfer the call, the base and the mobile are assigned a new pair of frequencies to which the mobile switches and the call continues. If a handover is indicated and no new frequencies are not available due to heavy traffic or some other reasons, the call is either prematurely disconnected or continued with the signal quality gradually degrading. The call disconnection phenomenon is called Call Drop. The handover mechanism is illustrated in Figure 9.3.

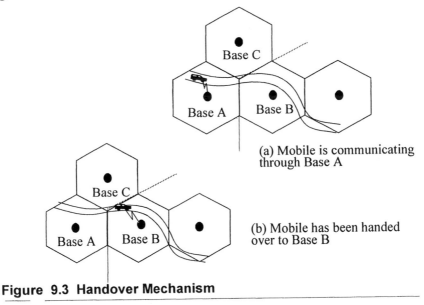

(a) Mobile is communicating through Base A

(b) Mobile has been handed over to Base B

Figure 9.3 Handover Mechanism

---

† *A system where base stations monitor the signal from the mobile is called base station assisted handover. Another system where handover decision is assisted by the mobile station is called mobile assisted handover (MOHO).*

A unique feature of cellular systems is that of wide area paging. When a cellular telephone number is dialled, the system automatically searches for the called party. The search proceeds on the basis that each user has a designated "home" base station (or operating area). If the target mobile is found, the system makes the connection with it regardless of the target mobile's location. The mobile control station equipment periodically monitors the control channels and reports to the serving base station, which identifies whether the user in its home area or not. This feature is used for automatic call forwarding since all the cellular phone numbers uniquely identify their home operating areas. Some users consider this feature to be an infringement of privacy, since users can be tracked without any difficulty. To overcome such concern, the subscriber can switch off his or her mobile telephone.

9.4.1 Cellular Systems in Operation

The concept of cellular mobile telephone system became a reality when in WARC assigned sufficient spectrum to implement cellular systems. The first cellular system started its operation in Nordic countries in 1981. In USA, the FCC assigned 40 MHz bandwidth (666 duplex channels of bandwidth 30 kHz each) to test the cellular concept. Subsequent to trials, the first cellular system was installed in North America in 1983. Since its inception in 1981, the cellular systems continue to grow and more spectrum was assigned This increased the total number of 30 kHz wide duplex channels to 832.

The first generation analog cellular systems, Nordic in Nordic countries, AMPS in North America, TAC in the UK, and NEC in Japan, were superseded by the second generation digital cellular systems. DAMPS replaced AMPS, GSM replaced other analog cellular systems, and PDC replaced NEC system in Japan. DAMPS retained the channel bandwidth of 30 kHz, GSM expanded the channel bandwidth to 200 kHz whereas PDC retained the 25 kHz bandwidth of NEC. A new cellular system based on code division multiple access was developed around 1994. This was a significant development as it provided the ground work for the third generation cellular systems. The bandwidth of 1.25 MHz was selected for IS-95.

Chapter 11, provides details on design fundamentals of mobile cellular system and Chapter 12 gives detailed description of several cellular systems.

9.5 Battery Life Considerations

Battery capacity and physical size play a crucial role in mobile communication system design. The intention is to have a pocket sized transceiver or even smaller; typical dimensions are 13 cm x 11 cm x 1.5 cm (215 cm^3).

This restricts power dissipation to 250 mW for 10°C and 500 mW for 15°C rise in the ambient temperature. At this power level, the efficiency of the electronic circuitry is considered to be about 50%. In addition, the gain of the portable transceiver antenna is affected by the proximity of the operator's body [10]. A transmitter capable of delivering less than 350 mW seems adequate. On the basis of a maximum of 3 calls of 100 seconds each per day per portable, with the assumption that the battery would require a recharge once every week, the battery capacity is estimated to be 250 Watt-sec (0.08 Amp-hr at 9V).

During the inactive periods the transceiver is required to listen passively to receive incoming calls. The transceiver can be designed to listen passively in a power saving mode. In this mode the receiver may follow a cycle of 10 milliseconds in ON state and 250 milliseconds in OFF state. In order that the transceiver does not miss an incoming call, the paging signal from the base station transmitter should therefore be repeated several times. This will, however, increase traffic on the signalling channel, but it may be an acceptable trade-off. White et.al. [11] estimate that a 1% duty cycle is achievable. If we postulate that a 64 bit message is necessary to set up a channel, it will require four milliseconds (at 16 kb/sec.) to receive the required information. If the receiver is off for 99% of the time and has no prior knowledge of when the information is to be transmitted, then we need to repeat information for a period of up to 400 millisecond. This is rather long, and might cause signalling channel congestion. Thus a 1% duty cycle is probably too low to be practical. A duty cycle between 5% to 10% may be more appropriate. The final choice, however, depends on traffic considerations.

9.6 Summary

This chapter introduced several mobile wireless communication systems and basic principles of their operation. Paging systems were introduced. Conventional, improved mobile telecommunications systems, and operating principles of cellular systems have been described in this chapter. Trunking systems using examples of APCO and TETRA systems have been discussed rather briefly. Battery life plays important role in cellular system. A technique to increase the battery life is also discussed. Several of the advanced cellular systems are discussed in Chapters 11 to 14.

Problems

Problem 9.1

The cellular concept was proposed in 1947. Discuss the reasons why it took nearly 35 years for the cellular system from concept stage to commercially operating systems.

Problem 9.2

(a) Discuss the modern trends in paging, and how integration of paging function with cellular systems will affect the paging industry.

(b) Discuss why there is no single world standard for paging system.

Problem 9.3

Discuss comparative advantages and disadvantages of MOBITEX, APCO-25, and TETRA.

Problem 9.4

Discuss reasons why the IMTS service was not very satisfactory.

Problem 9.5

Discuss what is the battery saving mode and show by diagram how it reduces the number of charging of a portable.

Problem 9.6

(a) A portable uses 350 mW during the call. A survey on the use of portables suggests that every users originates 3 calls of 100 seconds each day, and receives 5 calls of 90 seconds each. If the portable uses a battery of capacity 250 Watt-second (0.08 Amp-hr at 9V), estimate how many times in a week the user need to charge the battery?

Problem 9.7

(a) A technician is given a task of designing a battery saving system for the portable. The paging terminal has a duty cycle of 5%, that is it is the portable terminal is in off position 95% of the time. If a 64 bit message is typical and the transmission rate is 9.6 kbits/sec, then estimate how many times the base station should transmit the paging message so that it is received by the portable without excessive delay.

(b) Determine the probability that the paging message will be missed.

Problem 9.8

Repeat Problem 9.8 for duty cycles of 10 and 15%.

References

[1] N. Ehrlich, "Advanced mobile telephone service using cellular technology," *Microwave Journal*, vol. 26, pp. 119, August 1983.

[2] W.R. Young, "Advanced mobile telephone service: introduction, background, and objectives," *Bell System Technical Journal*, vol. 58, pp. 1-14, January 1979.

[3] K. Araki, "Fundamental problems of nationwide mobile radio telephone system," *Rev. of the Electrical Communication Laboratory*, vol. 16, pp. 357-373, May-June 1968.

[4] H.J. Schulter, Jr. and W.A. Cornell, "Multi-area mobile telephone service," *IRE Trans. on Comm.* Systems, vol. CS-9, pp. 49, 1960.

[5] R.H. Frenkiel, "A high capacity mobile radiotelephone system model using a coordinated small zone approach," *IEEE Trans. on Veh. Tech.* vol. VT-19, pp. 173-1977, May 1970.

[6] P.T. Porter, "Supervision and control features of a small zone radio telephone system," *IEEE Trans. on Veh. Tech.* vol. VT-20, pp. 75-79, August 1971.

[7] N. Yoshikawa and T. Nomura, "On the design of a small zone land mobile radio system in UHF band," *IEEE Trans. on Veh. Tech.* vol. VT-25, pp. 57-67, 1976.

[8] APCO-25

[9] Trans European Trunked Radio (TETRA) system, *Technical requirements specifications Part I (ETSI ETR 086-1), Part II (ETSI ETR 086-2), and Part III (ETSI ETR 086-3)* available from ETSI.

[10] K. Kobayashi, T. Nishiki, T. Taga and A. Sasaki, "Detachable mobile radio units for the 800 MHz land mobile radio systems," *IEEE conf. on Vehicular Technology*, Pp 6-11, 1984.

[11] P.D. White, R.G. Murray, and R.W. Gibson, "Cordless telephones - a system study," *IEE Colloquium on Short range Communication Systems and Techniques*, Digest No. 1983/5.

Chapter 10

WIRELESS DATA SYSTEMS

Data transmission over wired networks have been in widespread use for many years and it is having a major impact on modern telecommunications. During the past few years, the transmission rates have increased from ISDN rates to high speed ATM. In 2001, the data traffic volume surpassed the voice traffic over the fixed network. The trend is likely to continue. Similar growth in data applications is expected over wireless networks. The data is usually transmitted in the form of packets, the same technique is used over wireless channels.

Paging has been used over the wireless channels for many years. Though paging service is in some ways different from the conventional packet transmission, the paging messages before transmission are still bundled together in a package strikingly similar to packets. The packet transmission technology has been used for paging services long before the two way packet transmission started. There is also another major difference between paging and conventional packet data system. The former is always from base to mobile terminals, the latter is essentially a two way communications. In paging systems, the page recipient seeks another mode of communications, usually a telephone, to reach the source of the message.

Packet networks are being extended to the mobile wireless environment where it is finding many new applications. Indeed, packet radio systems forming a bulk of private land mobile radio systems (PLMRS) are rapidly proliferating. Packet radio networks provide data services to mobile users located over a wide geographical regions where either it is not possible or is very expensive to install wired connections between sources and destinations. In packet radio systems, network management functions such as packet formation, error control and channel access, are carried out without reference to any wired facility.

10.1 Packet Structure

In a reliable communication delivery of messages to the designated recipient with a minimum of corruption requires some kind of protocol. A protocol defines a procedure that must be followed so as to establish a reliable transfer of information between various entities forming the network. The protocol sets the rules to access the channel, transmit messages, ensure delivery of packets and arrange acknowledgments. For example, interrogation of a network node about the availability of a connection to the destination is a procedure known as random access. The interrogation can take place on one of the message channels or a separate channel may be designated for this purpose. It is important that the interrogation message uses a designated format otherwise the network nodes will fail to understand the message. Furthermore, the packet format is designed in a such a way to ensure error free delivery of messages to the designated destination.

The structure of the packet has a striking similarity to a parcel sent by a sender through the post office. The contents of the parcel must be packed properly; some post offices even specify a standard package e.g. size. The origination address written on the parcel identifies the sender. The parcel must also include destination address. In a similar manner, a communication packet includes a preamble, synchronization words to mark the boundaries of bits and symbols, addresses of the source and destination, error control bits (similar to redundancy in the address), and flags to mark the end of the message. A typical packet format is shown in Figure 10.1.

Dotting sequence:1010101010101010
Word Sync: Barker Codes

FEC: Forward Error Control
F: Flag

Figure 10.1 A typical packet format

Each packet starts with a preamble. This is usually a dotting sequence, which is a group of alternating ones and zeros. However, other patterns may also be used. A preamble is needed to establish bit timing of the received data. A longer preamble though helps the receiver demodulator to acquire bit synchronism more robustly, it increases the packet overhead and decreases the throughput. It also helps to preserve timing information during channel fades. Many designers consider that a longer preamble is worth the price to achieve greater reliability, since it has been observed that a larger number of errors are produced when the receiver fails to acquire correct timing. The preamble length is a function of the length of the packet and the channel characteristics. It is also obvious that for a short packet a long preamble is not acceptable because it would result in unacceptably low throughput. Dotting sequences as short as 16 bits and as long as 100 bits are generally used. In some applications, even longer preamble sequences have been used.

An equally important synchronization field follows the preamble; it is used to mark the beginning of message words. It is extremely important that the beginning and the end of the message words are correctly recognized. A wrongly marked word beginnings could result in total corruption of the message. A recognizable sequence with good correlation properties is usually used for word synchronization. Other desirable characteristics include low probability of false synchronization that is low cross correlation with preamble sequence or noise waveform. Although several such sequences are possible, Barker sequences are the most commonly used. Barker sequences of lengths up to 12 bits are listed below:

+ -	10
+ + -	110
+ + - +	1101
+ + + - +	11101
+ + + - - + -	1110010
+ + + - - - + - - + -	11100010010
+ + + + - - + + - + - +	111100110101

The Word Sync is followed by the Information Field, which consists of addresses of the source and destination, control signals if deemed necessary and the message text. Each user is allocated a unique sequence, which represents its address and upon recognition of this sequence the receiver opens a time aperture to receive packet(s) until the end-of-message flag is detected. The address sequences are chosen on the basis of good autocorrelation and low mutual cross correlation.

10.2 Error Control

A good number of packets may be lost due to collisions or excessive number of errors due to the presence of channel fading, noise and interference. Frequently a packet is not lost completely, but is received with a certain number of errors are detected and corrected to restore the packet to its original condition. Error detection is carried out by means of a block check bits that are appended to the information signal. Logical checking of the received message and the appended code bits determine whether the packet can be restored or not [1].

The performance of an error control method is measured by the probability of delivering the message successfully. Assuming that the errors occur independently in an n bit message length, the probability that the message is received error free is given by:

$$P_s = (1-p)^n \tag{10.73}$$

where p is the bit error probability and n is the message length. An important measure of the performance is the probability of producing a false call. False call arises when the address bits are decoded into other valid address. This happens when the number of errors is greater than the error correction capability of the code. The probability of falsing per message is given by

$$P_f = P(0, k)P(e > d, n)2^{-r} \tag{10.74}$$

where $P(0,k)$ is the probability that k bit synchronization word is received error free i.e. $P(0,k)=(1-p)^k$ and $P(e>d,n)2^{-r}$ is the false rate of cyclic code (or probability that n bit word contains d or more errors) with d as the minimum distance, n is the code word length, and r the check bits in a code word respectively. $P(e> d,n)$ is given by

$$P(e > d, n) = \sum_{i = d}^{n} \binom{n}{i} p^i (1-p)^{n-i} \tag{10.75}$$

A false message is produced when the decoder generates a valid frame sync and check bits out of noise. The probability P_{nf} is given by

$$P_{nf} = 2^{-(k+r)} \cdot \text{bitrate} \tag{10.76}$$

In the context of packet radio, recovery of a lost packet is known as data link control. Given an erroneous reception of the packet, one of three methods may be used to recover the lost packets.

(i) The most direct recovery technique is to inform the transmitter to resent the message. This method requires the receiver to send acknowledgments whenever one or more packets have been received correctly. This procedure is known as Automatic Retransmission reQuest (ARQ). In Chapter 7, ARQ has been discussed in some detail.

(ii) The second method of packet recovery requires that the receiver of the packet rebroadcast the received packet on a different frequency. Thus, each unit in the network receives its own packets and can determine whether the broadcast/rebroadcast sequence was free of a collision. If not, the packet is retransmitted. This technique is widely used in satellite packet networks.

(iii) Coding is extensively used in packet networks to recover packets with multiple error bursts. For this purpose error burst and cyclic codes have been proposed [2][3]. Pure burst correction codes may not be suitable for mobile channels, because over this channel burst errors are accompanied by random errors and burst correction codes such as the Fire code and burst trapping codes [4] fail to correct random errors. BCH codes and Reed-Solomon codes have been found to be most suitable for mobile data networks where short packets are used [1]. For long data streams, convolutional codes have been found to be useful. The selection of an optimum code is very difficult for mobile channels because of their extreme variability; in principle, some form of adaptive coding is found to be useful. However, a code adaptation procedure requires a link with sufficient reliability to transmit code changes. Similarly, the designer may decide whether to use a code which will provide require reliability for all possible channel conditions, or to rely on a less powerful code combined with an ARQ scheme. The latter is usually preferred. A relatively low complexity code which is able to correct a minimum number of errors, together with simple ARQ, form a powerful combination with reasonable complexity [1].

10.3 Paging Systems

Paging service is the most basic form of mobile radio communications. In the past two decades, paging systems have seen a tremendous growth in recent years and these services are used in most countries of the world. In the United States, there were more than 50 million users subscribers to paging services by the end of year 2000. A paging system essentially consists of a network controller, which may be connected to one or more base stations located within the paging zone. In more advanced paging systems, the controller is connected to the switched telephone network, which provides the facility to the originating party to dial the message directly into the controller.

Paging units with capacities ranging from 10,000 to 100,000 are commercially available. Recent advances in the radio and VLSI technology have made one-way "smart" pagers feasible. These, in addition to alerting the user, store and display messages. In 1984, pager technology went a step further when the New York and Tokyo stock exchanges started transmitting stock market reports to paging customers. The paging customer can select a particular segment of the market and receive market report on continuous basis. A similar service is now available in many parts of the world. Many more applications similar to these have emerged recently. Satellite based Global paging (or wide area paging) systems capable of delivering messages worldwide are now in operation.

The paging systems work on the principle of data transmission in packet mode in the forward direction (base to mobile). Several propriety systems designed by Motorola, NEC, Ericsson, and the British Post Office are now in operation in different parts of the world. It is rather unfortunate that regulatory authorities of different countries have not made efforts to agree on a single international standard. Several standards including the POCSAG system developed by British Post Office have been recommended by ITU-R.

10.3.1 Paging System Design Requirements

Low cost, small in size and light in weight requiring infrequent battery charging are preferred physical requirements. Because of relatively inexpensive VLSI technology used in implementation of advanced paging receivers, the cost of paging terminal has been falling continuously. VLSI technology has also helped in meeting the requirements low cost, small size and weight. More frequent charging is avoided with the use of battery saving techniques. This technique uses the concept of switching the terminal on periodically for short periods rather than leaving it on continuously. A duty cycle of 10 to 15% is recommended.

High paging success, low falsing rate, and large capacity are the technical requirements important to both the operator and the users. High paging success, desirable for operational reasons and high quality service primarily depends on the radio coverage design. The coverage design should provide acceptable signal strength over more than 90% of locations. The reliability of a paging system is measured by the rate of falsing. Falsing is regarded as a serious shortcoming of the system. Falsing occurs when a message is delivered to a user other than the one intended that is the message delivered to a wrong (false) address. Falsing can occur because of undetected error. However, most systems have adequate built-in safeguards like robust synchronization and error control.

Synchronization and error control procedures are designed keeping in view the desired paging success and failure rates. Obviously, any shortcomings in these procedures will result in poor service, which has a potential of customer base erosion and revenue. A paging system quality is defined in terms of radio coverage, system access strategy, paging terminals, services offered, and interface with other system services. The coverage design principles were detailed in Chapter 2. The access strategies were discussed in Chapter 8.

Paging terminals with capacities 10,000, 30,000, and 100,000 customers are commercially available. A paging system that meets the current demand and has a flexibility to expand capacity without disruption is usually preferred. Furthermore, to participate in nationwide or even worldwide paging systems, terminals that could handle all viable signaling formats have an advantage over the other systems. The modern paging terminals store and forward voice messages. The current generation of paging terminals has capability of eliminating pauses in voice messages so that system capacity could be increased. A good size voice storage and message forwarding capability is therefore desirable. Terminals with up to 800 seconds of storage capacity are now common. To the users, a system that offers many services is more attractive. In case the operator decides to offer several levels of service e.g. priority and emergency paging, group calls and direct dialling, it should select a terminal which is able to control the network as well as the radio coverage. For a smoother operation, the terminal should be equipped with a certain level of management functions like fault diagnostics, call count and invoicing.

The requirements on the page unit i.e. the user terminal are small size, rugged in construction, reliable in operation and low power consumption. It is of advantage if the terminals respond to multiple calling code formats; the desirable format designs avoid wastage of channel time and maximize loading. Several types of alert signals including tone, voice, and visual can be selected in a modern page unit. In voice pagers, reduction of transmission time and memory facility are provided. In modern paging units, ability to close the loop by interfacing them to answering machines, car and fixed telephone is becoming available. In this regard, merging between paging and cellular systems is taking place.

The paging message is assembled in the form of a data packet and to achieve high paging success and low falsing error rate (a reliable recovery of message), robust bit and word synchronization is a fundamental requirement. A typical packet structure is shown in Figure 10.2.

Preamble	Word Sync	Information	Check Bits

Figure 10.2 Structure of a Paging message

The bit synchronization is required to mark accurately the bit boundaries and for this purpose a dotting sequence (10101010...) or a preamble is often used. A dotting sequence of length from few tens to several hundred bits are used. The length of dotting sequence or preamble relative to the length of the message determines the paging message overhead, the system efficiency and the channel throughput. A short dotting sequence though results in an increases in the channel efficiency but it degrades paging reliability. The synchronization word marks the start of the first message word, therefore a considerable importance is attached to it because if the first word is marked erroneously, the whole message will be full of errors. Thus, synchronization word possessing high auto-correlation peak and low sidelobes is desirable. Low probability of being decoded spuriously in the presence of noise is another desirable attribute of the Synchronization Word. Barker or other pseudo-noise sequences are used for word synchronization. Synchronization takes place in two stages; first bit by bit search is done, which is followed by the lock mode.

Prior to transmission, several paging messages can be assembled into a frame. As shown in Figure 10.2, a frame consists of several fields e.g. preamble (a dotting sequence), a synchronization word (WORDSYNC), and information (INFO). The INFO field consists of a batch of address words. The last field in the frame is the frame-check field which adds redundancy for error detection and or correction. A higher paging success is obtained by repeating calls which consists of assemblage of messages from several users within the INFO field, instead of transmitting the synchronization word before every message.

10.3.1.1 Calculation of Paging Capacity

It is quite easy to calculate the paging capacity of a radio channel. A channel has a total channel capacity of 1 Erlang,[†] which is divided into a number of messages that can be accommodated in the busy hour. For example, if an average message length of 1000 bits is transmitted at a rate of say 1.2 kbits/sec, then a total of 4320 messages can be transmitted in one hour. Now if we consider that each user has one paging message per busy hour, then a paging channel will serve approximately 4320 users. Obviously, if the message length is shorter, more users can be accommodated. The above calculation does not take into account the repeat messages, frame overhead in the form of preamble and word synch, and the overhead bits needed for error control coding and flags. Thus, in practice the capacity[‡] will be lower than that indicated above. Note that this calculation is similar to the capacity calculation of forward paging channel described in Chapter 8.

The temporal traffic fluctuations may result in delays between the origination time of the message and the time of its delivery at the destination but it will not affect the system capacity. To meet the requirement of low mes-

---

†. 1 Erlang is equivalent to channel usage of 3600 seconds/hour.
‡. The paging capacity will be much lower if battery saving technique is used.

sage drops under heavy traffic conditions, the central control would require some message queueing strategy.

EXAMPLE E-10.3.1

A transmission rate of 2.4 kbits/sec is used in a custom designed paging system. Each frame consists of 8 messages, each message is repeated three times. The frame overhead consists of 32 bits for preamble, 11 bits for word sync, and 3 bits for flags. A standard message length of 180 characters is used. A (48,36) block code is used for protection against errors. Find the hourly capacity of the system.

Solution

Each frame consists of 34,560 characters of data which when encoded results in 46,080 bits of encoded data. Each frame has 46 overhead bits. Thus, the frame size is 46,126 bits or 19.219167 seconds. If the channel utilization is 100%, then the capacity of the system is found to be 1498 messages/hour.

10.4 Commercial Paging Systems

Several commercial paging systems have been deployed around the world. These systems use many propriety message formats and data rates. Unfortunately, a single universal standard does not exist. In addition to the POCSAG paging system developed by the British Post Office, ITU-R has recommended paging systems such as Golay Sequential Code Paging System developed by Motorola, NEC by NTT, and RDS by Radio Data System. In the following, the characteristics of the four paging systems mentioned above are described.

10.4.1 POCSAG Paging System

The POCSAG signalling format, shown in Figure 10.3, consists of a dotting sequence of at least 576 bits long is followed by a number of data batches. Each batch consists of a SYNCWORD and eight frames. Each frame consists of two blocks of either address or data. A unique word 01110011110100100001010111011000 represents the SYNCWORD. The transmission rate of 512 bits/sec uses direct FSK modulation. The address and data codewords are 32 bits long of which the most significant bit is transmitted first. The first bit of address is always 0 which distinguishes it from data whose first bit is always 1. Bits 2 to 19 are the most significant 18 bits of the 21 bits pager address while bits 20 and 21 are used to represent four functions. Bits 22 to 31 are checkbits and the bit 32 is chosen to give the address codeword an even parity. Thus, the system capacity is $2^{21} = 2,097,152$ four function addresses.

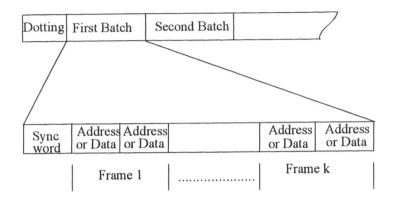

Figure 10.3 POCSAG Signalling Format

The bit 1 of the message codeword is always set to 1. Bits 2 to 21 are used for the message text. The framing rules do not apply to a long message, which is allowed to continue into the ensuing batches until the message terminates. In the absence of an address or message codeword, an idle codeword is sent to complete the batch. This code word is defined as: 01111010100010011100000110010111. Two message formats, numeric (4 bits per BCD character) or alphanumeric (7 bits per ASCII character) are used. For numeric only, the function bits in the address code word are set to 00 while 11 is used for alphanumeric. The 21 bits information message codeword is coded into (31,21) BCH codeword with the use of the generator polynomial $X^{10} + X^9 + X^8 + X^6 + X^5 + X^3 + X^1 + 1$. The generated code has a Hamming distance of 6. The code has the capability of detecting any 5 random errors. Both hard and soft decisions are used to detect and correct errors. In soft decision decoding the signal quality of each bit is preserved and a burst of up to 11 poor quality bits can be checked and corrected. Several other combinations of error detection and correction are possible.

The POCSAG system transmits at a page rate of 6 to 10 calls per second. The falsing rate is designed to be less than 10^{-8} for every call transmitted. Paging systems can be used for in-building as well as for other environments. In the system design, building penetration losses of 22, 18, 18, and 17 dB at 150 MHz, 250 MHz, 400 MHz, and 800 MHz respectively have been used.

Message success probability P_s, is given by $(1-p)^n$, where p is the probability of bit error, n is the message length. P_f the probability that the error caused a transmitted code word to be decoded as an other code word is given by,

$$P_f = P(0, s)P(>d, n) \times 2^{-r} \tag{10.1}$$

where r is the number of check bits in a code word, d is the code minimum distance, n is the code word length and $P(>d,n)$ is given by

$$P(>d, n) = \sum_{i=d}^{n} \binom{n}{i} p^i (1 \quad p)^{n-i} \tag{10.2}$$

$P(0,s)$ is the probability that the SYNC WORD is received free of errors. The falsing per unit time is given by

$$P_{nf} = 2^{-(s+r)} \times \text{bit rate} \tag{10.3}$$

where s is the length of the sync bit word.

EXAMPLE E-10.4.1

Find the falsing rate of POCSAG Paging system.

Solution

Here $s = 32$ bits, $r = 10$ and bit rate = 512 bits/sec. Substituting in (10.3), we get the falsing rate of 1.16×10^{-10}.

10.4.2 Golay Sequential Code (GSC) Paging System

The Golay Sequential Code (GSC) is based on (23,12) Golay code. Each of its transmission frame consists of a preamble packet, a sync code and a batch of up to 16 addresses and/or message blocks. The transmission frame format is shown in Figure 10.4. The data rate of 300 bits/sec is used for word sync and address. The data rate is doubled to 600 bits/sec for preamble and data. The decision to use faster rate for preamble is taken to reduce the time for synchronization. The slower rate for word sync and address are to ensure lower falsing rate. Direct FSK modulation with a deviation of ±5 kHz is used.

In each transmission frame, the preamble packet is followed by a SYNC Packet, address and an activation code packet to which display data is appended. The preamble packet consists of 28 bit long dotting sequence, 101010....10, which is transmitted at 600 bits/sec. It is followed by a selected preamble word repeated 18 times and transmitted at 300 bits/sec. The pream-

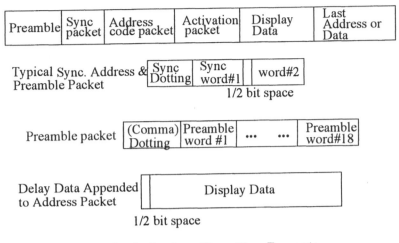

| Preamble | Sync packet | Address code packet | Activation packet | Display Data | Last Address or Data |

Typical Sync. Address & Preamble Packet

| Sync Dotting | Sync word#1 | | word#2 |

1/2 bit space

Preamble packet

| (Comma) Dotting | Preamble word #1 | ... ... | Preamble word#18 |

Delay Data Appended to Address Packet

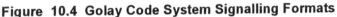

| | Display Data |

1/2 bit space

Figure 10.4 Golay Code System Signalling Formats

ble word is chosen from a table consisting of 10 specific (23,12) Golay code words each of which belongs to one of the ten battery saver groups. The SYNC packet starts with a 28 bit long dotting sequence again transmitted at 600 bits/sec. It is followed by two specific Golay SYNC WORDS which are transmitted at 300 bits/sec. Word 1 is 001011001001010000000011 and Word 2 is complement of Word 1. The last 11 bits of the Word 1 are parity bits.

A batch of 16 address packets may be sent after the sync packet. Each address packet consists of a comma (dotting sequence) plus two Golay address words. These are transmitted at 300 bits/sec. Only 50 of the 4096 (= 2^{12}) possible code words are used as the first address word. For the second, only 2000 code words are used. Four function addresses are obtained by using inverted/non-inverted combination of the two address words. Thus, there are $10 \times 50 \times 2000 = 1$ million four function addresses.

The information or data packet is identical in length to an address code word and may be freely substituted following a relevant address. Each data block has 56 information bits and 64 check bits. Each block of 7 data bits is encoded into a block of 15 bits using generator polynomial $X^8 + X^7 + X^6 + X^4 + 1$. The resulting 8 (15,7) BCH codewords are assembled by rows and transmitted by column at a rate of 600 bps. In this case a burst up to 16 bits can be handled.

The address is encoded using a generator polynomial, $X^{11} + X^9 + X^6 + X^5 + X + 1$. The generated code having a Hamming distance of 7 has a capability of correcting up to 3 random errors or an error burst of length up to 5 bits. The coded data has a Hamming distance of 5, which means it has a capability of correcting 2 random errors or an error burst of length up to 4 bits. Added protection against errors is obtained with the use of interleaving. The data blocks are assembled by rows and transmitted by columns.

10.4.3 NEC Paging System

The NEC Paging System, developed by Nippon Electric of Japan, transmits at a rate of 200 bits/sec using direct FSK modulation with a peak deviation of ± 5 kHz. The preamble is a dotting sequence. Two words repeated are used for word synchronization. The address and information use one of four groups of addresses and display information. Each group consisting of 20 bits (address or display data) and an additional parity bit is encoded using (31,21) BCH code. One of these 21 information bits is used to differentiate between the address and display data. The capacity of the system is $2^{20}=$ 1,048,570 single function addresses. The frame structure is shown in Figure 10.5.

Dotting (preamble)	Sync word	Sync word	Group 1 Message 20 BCH (32,21) words	Group 2 Message	Group 3 Message

Figure 10.5 NEC Format

10.4.4 RDS System

Radio Data System (RDS) is a paging/digital broadcast system which uses a 57 kHz subcarrier on FM broadcast stations. The deviation of ± 2 kHz and a data rate of 1187.5 bits/sec (subcarrier/48) is used. The coding structure uses 104 bit groups divided into 4 blocks of 26 bits each. Each block is encoded into (26,16) coded word, which is a shortened version of (31,16) BCH code. For encoding, the generator polynomial $X^{10}+X^8+X^7+X^5+X^4+X^3+1$ is used. The code has a capability of correcting 5 bit long bursts. The total system capacity is 350,000.

10.5 Packet Radio Systems

The Packet Radio Systems is a two way communications, therefore it uses channels in the forward and reverse direction i.e. from base to mobile and mobile to base. For communications in the reverse direction, the mobile terminal has to get an access to the system. For this purpose, the system sets aside one or more access channels, the actual number depends on the number of subscribers. The performance of packet radio network is generally measured in terms of network throughput, and the average packet service time or the time between the packet origination and its delivery to the destination. The presence of noise, interference and fading over wireless channels requires that a powerful error control technique be used to achieve acceptable access reliability and call falsing rate. The user preferred features of a packet radio network systems (PRNet) are transparency of use, wide area coverage, high throughput with minimum delays, rapid and convenient deployment, and provision of different modes of addressing and options on traffic routing.

In a packet radio network, the design of management and control functions to provide connectivity between users dispersed over a wide area is of major considerations. To provide reliable connectivity and to perform desirable network functions, the packet radio networks (PRNet) are in general organized into a hierarchical structure as defined in Open System Interconnection (OSI) Reference Model [5]. The OSI model consists of seven layers: Physical, Link, Network, Transport, Session, Presentation, and Application. Each OSI network layer performs functions assigned to it and establishes peer to peer communication, e.g. between same layers of the source and destination. Because each layer performs functions unique to it, the issues related to layer to layer communication differ. Thus, in a layered network, each layer can be designed independent of others. Inter-layer communication is established through so called layer interface. In this text, however, we shall discuss the issues pertaining to the radio portion of the network that is relating primarily to the physical layer. The link layer will be discussed in relation to error control issues. The design of the physical layer is important since any decision pertaining to this layer affects the performance and design of higher layers. For example, the issues of signalling, data rate, bandwidth, modulation encoding have substantial influences on overall network performance. For a discussion of other aspects of the OSI model, the reader is referred to several excellent publications [5][6][7].

As in the case of wired packet switched networks, the data to be transmitted is enveloped by additional headers comprising preamble, synchronization, coding and a set of flags. While the packet length might in principle vary between a few tens and several thousand symbols, it is usually restricted to a few hundred bits in mobile radio networks to enable them to operate in fast fading environments. When the wireless channel changes very slowly and the transmission rate is high, very long packet lengths can be used. Section 10.1 examines the formation and structure of packets for radio networks.

Channel sharing in some form is commonly used in packet radio networks. Time sharing of a wideband channel may be performed on a fixed time basis, polling is an example of this or alternatively random access in the form of either ALOHA and its variations or carrier sense multiple access protocol and its variations may be used. Still another multiaccess method called spread spectrum makes the access channel simultaneously available to all users. These methods have been discussed in Chapter 8. In this Chapter we describe systems that use data transmission in multiuser environment. In Section 10.8, several examples on wireless data communication systems are presented.

10.6 Packet Reservation Multiple Access

During the past few years, significant progress has been made towards the development of integrated networks for use in mobile environment. High speed transmission methods are being introduced to integrate voice, data, and image services. Straightforward integration is not possible since the services to be integrated have different performance requirements. For example, moderate error rate ($\approx 10^{-3}$) is acceptable for speech signals but transmission delays are not tolerated. Thus, the voice service is considered to be a real time service. On the other hand data transmissions require low error rates ($\approx 10^{-6}$) but transmission delays in transmission are tolerated. The image services are real time for moving images and these require low error rate. The data packets are characterized by short, high rate bursts that occur at random times. With this high peak to average ratios for the packet traffic traditional techniques like FDMA or TDMA are inherently inadequate. To circumvent these limitations, the Packet Reservation Multiple Access (PRMA) has been proposed. The PRMA may be considered as a hybrid multiple access scheme because it combines characteristics of TDMA and slotted-ALOHA. This combination is desirable because voice can not rely on purely contention based multiple access schemes like ALOHA however data can. Since the arrival of traffic types is random, flexibility in assigning priority to a particular service is a major requirement. PRMA provides such flexibility. PRMA is a derivative of Reservation ALOHA [8] but differs from it in several aspects.

(i) PRMA relies on very short round trip delays and is thus more suited to microcellular environment.

(ii) During congestion PRMA may drop those voice packets, which have been delayed beyond a specific time period. This is not so in R-ALOHA where packets keep accumulating.

PRMA is designed for a two way communication with star topology. While the reverse channel is controlled by PRMA which acts as a statistical multiplexer, the base station controls the forward channel. The forward channel contains information identifying which terminal did successfully broadcast in the previous time slot. In case of a collision on the reverse channel, the base station transmits a negative acknowledgment (NACK).

The messages on the reverse channel are organized into frames, which are divided into time slots. The speech terminal transmits one packet per frame, the rate of frames is equal to $1/T_p$, the rate at which packets are generated. When the data rate is higher than that of speech, PRMA allows a terminal to transmit data using more than one slot in a frame.

When the first packet is generated at the mobile terminal, it initiates reservation stage. At this stage, the terminal contends for an available slot and if the transmission fails, the terminal reschedules transmission in the next available slot with a probability p. This probability is called permission probability. Upon success, the base transmits a reservation confirmation message indicating that the requested slot has been reserved for that terminal use. No other terminal can use the slot until it is released by its owner. Once the slot is reserved the mobile is allowed to use the reserved slot without contention. The reservation lasts till the mobile exhausts data for transmission. The empty slot found by the base prompts it to transmit NACK and adds the slot to the pool of slots available for use by all terminals. The procedure of contending, reservation and transmission follows.

When transmission fails or terminal can not get a slot to transmit data, it keeps attempting until the packet is discarded. An unsuccessful packet is discarded after it has been delayed by D_{max}, the maximum holding time of the packet. D_{max} is determined by the quality of delivered speech. In case of random data traffic, D_{max} can be infinitely long. When the system is congested the packets (data as well as voice) experience longer delays thus the drop rate increases.

The performance of PRMA is measured by the maximum number of voice terminals that can share the channel without exceeding the packet drop probability threshold. The quality of voice transmission depends on the speed with which the speech detector can detect a talk spurt and the length of time a speech packet is held without being discarded by the terminal.

Some experimental results have been reported in [9] where channel having the transmission rate of 720 kbits/sec is shared by a number of users. The theoretical number of 32kbits/sec channels created from 720 kbits/sec channel is 22.5. When frame length of 16 *msec* is used, the channel throughput is estimated to be 0.48. For a $P_{drop} \leq 0.01$, the number of simultaneous users is found to be 37 which suggests that the channel capacity improved by 70%.

10.7 Packet Transmission Over Fading Mobile Wireless Channels

The environment in which mobile wireless systems have to operate is impaired by fading, shadowing, interference and intersymbol interference. The shadowing due to large scale obstacles in the propagation path is modeled as lognormal distribution, which is the distribution of the packet envelope. The flat fading is the result of localized multipath phenomenon and the signal envelope is known to follow Rayleigh distribution. Near/far interference is the most important type of interference in the case of packet transmission that use spread spectrum.

Two mechanisms are at work when channel impairment is present. First, all packet are subject to fading and shadowing and some of them fail to reach the base station in an acceptably healthy form. These include those which subsequently collide at the receiver. The channel impairments affecting different packets are usually considered to be independent and uncorrelated. The second mechanism is that of the receiver capture. In the presence of receiver capture, a desired packet may be recovered correctly if its power happens to be greater than the sum of powers of the interfering packets [10][11]. Surprisingly, under certain conditions, the overall impact of fading, near/far problem, and the capture effect is positive that is the performance in the presence of all these factors could actually improve.

In order to analyze the performance of packet radio system in the presence of channel impairments, we require propagation model, statistics of fading/shadowing, user distribution, statistics of desired power, the total interference power and its statistical distribution, and the capture properties of the receiver. We shall follow the analysis given in [10] and [11].

Consider the desired packet to have power P_d and I_k the total interference power of k interferers. Capture takes place when P_d exceed I_k by the capture ratio R_o. The desired packet is destroyed if and only if

$$\frac{P_d}{I_k} < R_o \tag{10.4}$$

where

$$I_k = \sum_{i=1}^{k} I_i \tag{10.5}$$

The probability that the desired packet collides with k packets conditioned on (10.4) being valid is given by

$$P\left(k \text{ collisions} \mid \frac{P_d}{I_k} < R_o\right) = \frac{G^k}{k!} \exp(-G) \qquad (10.6)$$

The unconditional probability of collision is

$$q = \sum_{k=1}^{\infty} P_k r_k \qquad (10.7)$$

where r_k is given by

$$r_k = \text{Prob}\left[\frac{P_d}{I_k} < R_o\right] \qquad (10.8)$$

The channel throughput is given by

$$S = Gp = G\left[1 - \sum_{k=1}^{\infty} P_k r_k\right] \qquad (10.9)$$

where $p = 1\text{-}q$ is the probability that capture will occur.

To evaluate (10.9), we need the probability distribution of the signal to interference ratio. Defining $P_d = R_I$ and $I = I_k$, the joint *pdf* $p(R,I)$ is obtained as

$$p(R, I) = Ip(P_d)p(I_k) \qquad (10.10)$$

and

$$p(R) = \int_0^{\infty} p(R, I)dI \qquad (10.11)$$

The distribution of R is then

$$P(R \le R_o) = \int_0^{R_o}\int_0^{\infty} p(R, I)dIpR \qquad (10.12)$$

10.7.1 System Performance

It is of considerable interest to compare throughput of various access schemes under fading conditions. This aspect has been studied by several researchers [1]-[3],[12],[13]. The performance of ALOHA, table driven CSMA and Polling access strategies are studied in [12] with regard to the application of TASI over mobile channels. Speech packets of length up to 2 seconds were considered and system performance was simulated. It was found that under fading conditions, polling remains the most efficient of the schemes and CSMA (Table Driven) provided the lowest throughput. Recently, studies on performance of ALOHA under fading conditions have received more attention. The emphasis has been on the reduction of throughput in a Rayleigh fading environment [2]. These studies have been extended to include the influence of shadowing on the throughput performance [8]. It has been shown that under heavy traffic conditions some throughput improvement can be realized.

The influence of fading on throughput has been examined through simulation for various access schemes. Studies show that, relative to ALOHA and slotted ALOHA, CSMA is most susceptible to fading; throughput reductions u to 60% of the steady channel performance were observed [1]. The ALOHA access has been found to be the most robust. However, because of already lower throughput in case of ALOHA systems, the resulting throughput of CSMA system is still higher than ALOHA. The polling access is found to be quite robust in presence of fading. Under fading conditions, all the access schemes, except CDMA, show that throughput curves is broader than one for steady channel but with a lower peak value. Figure 10.6 shows a performance comparison between several access schemes.

For indoor communication systems, the performance of direct sequence systems has been studied by Eschenbach [14] and by Khaverad and Mclane [15],[16]. In [17], a system was developed to study the feasibility of direct sequence spread spectrum system for office communication use. Khaverad and Mclane [16] have examined the use of spread spectrum systems for indoor communication. They have analyzed the system mathematically for the case of binary PSK modulation with or without diversity. Their model considers the indoor channel to have distinct paths, each with Rayleigh fading. Khaverad [15] has analyzed a system with several distinct paths, with or without Rayleigh fading. In his finding, the average probability of error is very sensitive to $E\{\beta^2_i\}$ $i=1$, where β_i is the gain of the selected path. If the selected path is not a fading one, then β_i is replaced with d_o^2. Figure 10.7 shows the average performance when the acquired and interference path envelope is Rayleigh distributed.

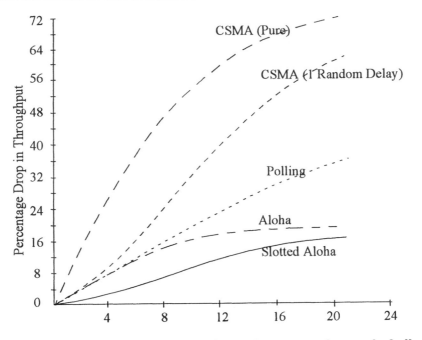

Figure 10.6 Throughput comparison of access schemes in fading

As seen from this figure, the performance of the system may not be acceptable with the implementation of certain diversity techniques and/or coding. Khaverad and McLane [16] have shown, however, that a combination of diversity and coding can restore the performance of DS-SS systems to an acceptable level. The presence of fading and interference can adversely affect the performance of DS-CDMA based transmission system. Figure 10.7 show the impact of fading and interference on the performance of DS-CDMA system.

10.7.2 Throughput Comparison of Access Schemes

The performance of an MFSK-FH system in a Rayleigh fading environment has been analyzed in [12]. It is shown that for a transmission rate of 32 kbits/sec, in the presence of mutual interference only, a 20 MHz wideband FH-MFSK system can support up to 209 users with an error probability of 10^{-3}, when the signal-to-noise ratio is 25 dB. However, under similar conditions but with frequency selective Rayleigh fading, the number of users that can be supported reduces to about 170. This is still three times that of PSK-SS as proposed by Yue [18]. The effects of lognormal fading on the performance have been studied by Muammer and Gupta [19]. The number of simultaneous users

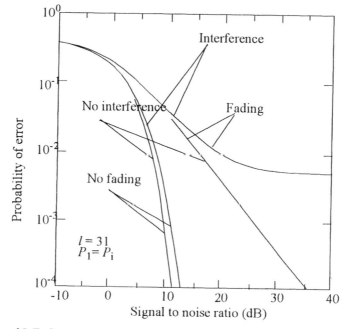

Figure 10.7 Average performance in the presence of fading

drops from 192 to 130 when Rayleigh fading is accompanied by lognormal shadowing.

10.8 Packet Transmission Systems

Private Land Mobile Data Systems (PLMDS) do not have a single universal standard. Several large manufacturers have developed systems based on their own propriety designs. Motorola's PCX system known as ARDIS is one of the largest mobile packet data system in North America. The network is used by IBM field service personnels. A variant of PCX network known as APCO-25B has been adopted in North America as a standard for police communications. MOBITEX, TETRA, CDPD and ATCS are the other examples of PLMDS.

10.8.1 ARDIS

ARDIS is a two way radio network, which is designed on the principle of distributed control architecture. The network consists of four network control centers (NCC) connected to 32 network controllers that provide intercon-

---

nection to 1250 base stations distributed over 400 cities in the continental U.S.A. The ARDIS network architecture is shown in Figure 10.8.

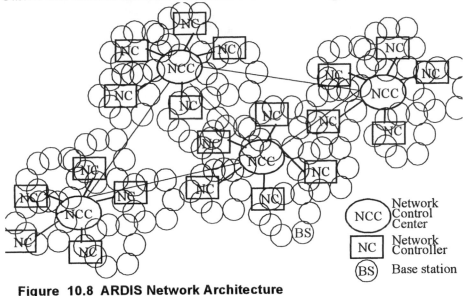

Figure 10.8 ARDIS Network Architecture

The system operates in frequency duplex mode in 800 MHz band with transmit and receive frequencies separated by 45 MHz. The initial transmission rate of 4.8 kbits/sec was upgraded to 9.6 kbits/sec in 25 KHz RF channels. FSK modulation is used. The base transmit power is 40 Watts which provides line of sight coverage up to 25 kilometers. The radio coverage is provided in the cellular form. A considerable radio coverage overlap is furnished to ensure reception of transmissions from the portable by at least one base station. The user data rate is about 8 kbits/sec. The transmission packet length is 250 bytes.

The data terminal is in the form of a portable radio computer which is linked to a dedicated base stations spread over United States. The portable transmits four Watts of power. The base stations use space diversity with maximal ratio combing. The channel controller sends the decoded messages over dedicated wired links to the network processor using HDLC protocol. The signal strength information is also sent to the network processor which eliminates duplicate copies. To avoid interference between two base stations, a scheduling protocol is used. Modified CSMA using periodic busy bits in the outbound data stream is used by the portable to access the system. The busy bit on the forward channel carries the information on status of the access channel.

10.8.2 MOBITEX

MOBITEX, is a wide area interconnected, trunked radio network designed by Swedish Telecom Radio Inc. and Ericsson. It has gathered an international stature as it is now used in many countries of the world. The first MOBITEX network became operation in Sweden in 1986, and since then other networks have been installed in Canada, Finland, the United Kingdom and the United States. In the U.S. the network is now covers over 100 metropolitan areas. It is basically a country wide digital dispatch system designed for text, data and voice. However, the networks in Canada and the United States are used for data only.

MOBITEX is an intelligent network with an open architecture that allows setting up of virtual networks. This feature facilitates mobility and expandability of the network. The system supports a number of standard network interfaces, including TCP/IP and X.25. The network is hierarchical as shown in Figure 10.9. It consists of a network control center, which is connected to national switches that form the top level of switching. These switches route the traffic between various service regions. At the intermediate level in the hierarchy the local switches each handling the local traffic are located. Multichannel base stations form the lowest level of hierarchy. The

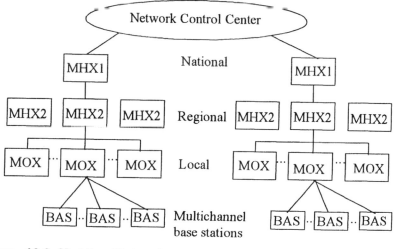

Figure 10.9 Mobitex Network Architecture

system works similar to a cellular system, except that handover between base stations is not permitted.

The mobile terminal locates the base station with the strongest signal and registers with it. As it leaves this area and enters an adjacent area, it registers with the new serving base station. The mobile units transmit in 896-901 MHz band and the base station at frequencies in 935-940 MHz band. The Gaussian Minimum Shift Keying (GMSK) modulation is used. The transmission rate of 8 kbits/sec, operates in half duplex mode in 12.5 kHz wide channels. The packet size is 512 bytes. Forward error correction as well as retransmissions protocol is used to ensure low BER. The portable terminals use Slotted ALOHA with dynamic frame length to randomly access the system.

The base transmits continuously and the mobile terminal transmits to the nearest base station only within designated time slot. The number of time slots in a frame and slot length are varied to optimize the channel throughput and mean waiting time.

10.8.3 APCO-25

The Association of Public-Safety Communications Officials (APCO) developed APCO-25 standard seeking a high level of performance, efficiency, capabilities and quality in two way radio communications. The APCO-25 standard uses CQPSK modulation in 6.25 kHz channel bandwidth and C4FM in 12.5 kHz bandwidth. A large number of individual and talk-group addresses are provided. The aggregate bit rate is 9600 bits/sec. In the case of data transmission, the data packets consist of header, overhead information and data. The modulation uses frequency deviations of 1.8, 0.64, -0.64, and -1.8 kHz for dibits 01, 00, 10 and 11 respectively.

The demodulator is designed to receive both C4FM and CQPSK signals. It consists of a frequency modulation detector, integrate and dump filter, and a stochastic gradient clock recover device. Improved Multiband Excitation (IMBE) vocoder operating at 4400 bits/sec is used. An additional 2800 bits/sec of forward error correction is added. The voice information is converted into digital data at a continuous average rate of 4.4 kbits/sec. Voice frames of length 88 bits for every 20 msec of speech are formed. The frame has several fields, which are collected into vectors, each frame consists of 8 information vectors. The frames are encoded into 144 bit words. The bits are protected using (23,12,7) Golay code, and (15,11,3) Hamming code words depending on the importance of the bits. The voice information is transmitted with additional encryption information, unit identifier, and low speed data to fully utilize the 9.6 kbps channel capacity.

APCO-25 supports several services, which include bearer, teleservices, and supplementary services. Data signals are processed either by direct modulation or by decomposing large data messages into small packets.

APCO-25 uses Open System Interfaces as shown in Figure 10.10. The air interface defines two types of interfaces - one between portable and base station and the other between mobiles and any station of RF subsystem. The latter interface is for talk around.

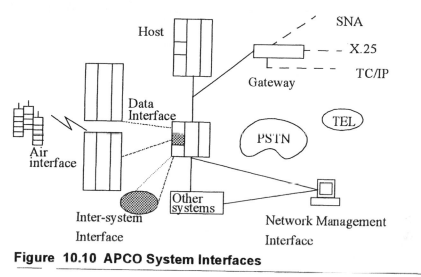

Figure 10.10 APCO System Interfaces

10.8.4 CDPD

Cellular digital packet data (CDPD) is now an emerging market. This system provides packet radio data services as an overlay to the existing analog cellular systems, which are still operating in some parts of the United States. It is also possible to overlay CDPD on North American TDMA. The applications of CDPD include computer aided dispatch and data base queries on a noninterfering basis using 30 kHz wide channels of analog cellular systems. Other applications include electronic mail, package delivery tracking, inventory control, credit card verification, security reporting, vehicle theft recovery, traffic and weather advisory services, and potentially a wide variety of information retrieval services [20]. The CDPD system integrates data with speech during the times when the analog channel is not used for voice communications. The system monitors activities on the available channels and when finds any channel idle, it transmits the data packets. Speech communications carries priority over data and the channel currently being used for data is allocated for voice, the data transmission is moves to another channel. The data transmission is aborted if another channel is not available.

The CDPD shares the transmission channels with the cellular system, therefore its structure is similar to that of cellular network. The air interface defines the protocol with which the terminals access the system through mobile data base station (MDBS) that are separate entities in the cellular network. The MDBS's are preferably co-located with the cell equipment that provides the cellular telephone service. Microwave or wired links connect the MDBS's with the mobile data intermediate systems (MD-IS). The MD-IS's also provide connections to a network management system, and support protocols for network management access to the MDBSs and M-ESs in the network. The service endpoints may be local to the MD-IS or connected through external networks when end points are remote [21].

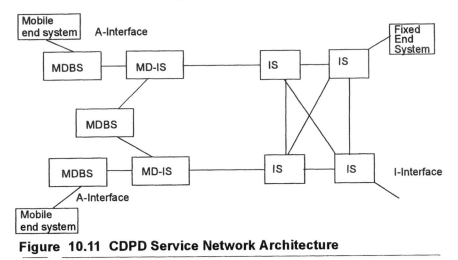

Figure 10.11 CDPD Service Network Architecture

The physical layer of CDPD radio link, on both forward and reverse links, uses GMSK modulation at the standard cellular carrier frequencies. The Gaussian shaping filter is specified to have bandwidth-time product of 0.5. The $B_b T$ product is chosen as a compromise to meet the adjacent channel interference requirements and the level of intersymbol interference so that a simpler signal detection circuit could be implemented. The data is transmitted at 19.2 kbit/sec. The CDPD system can also coexist with IS-54, the digital North American cellular system. The forward channel carries data packets transmitted by the MDBS, while the reverse channel carries packets transmitted by the M-ES's. In the forward channel, the MSBS form data frames by adding standard HDLC terminating flags and inserting zero bits, and then segments each frame into blocks of 274 bits. These 274 bits, together with an 8-bit color code for MDBS and MD-IS identification, are encoded into a 378 bit coded block using a (63,47) Reed-Solomon code over a 64-ary alphabet or six bits. A 6-bit synchronization and flag word is inserted after every 9 code symbols. The flag words are used to provide reverse channel status to the mobile terminals so that collisions with other terminals can be avoided.

In the reverse channel, when an M-ES has data frames to send, it formats the data with flags and inserted zeros in the same manner as in the case of forward link The M-ES may form up to 64 encoded blocks for transmission in a single reverse channel transmission burst. During the transmission, a 7-bit transmit continuity indicator is interleaved into each coded block, and is set to all ones to indicate that more blocks would follow, or all zeros to indicate that the current block is the last block of the burst. The forward and reverse channel block structure are shown in Figure 10.12.

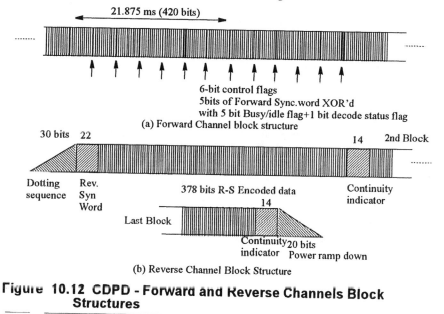

(a) Forward Channel block structure

(b) Reverse Channel Block Structure

Figure 10.12 CDPD - Forward and Reverse Channels Block Structures

10.8.5 TETRA

TETRA specifications place no constraint on the form of the radio network architecture. It consists of four sub-systems: base stations (BSs), switches, operation and management centre, and associated control and management facilities. The architecture is based on OSI reference model where lower three layers are associated with the network services and these are implemented in every node of the network i.e. infrastructure and mobile stations. The upper four layers provide services to the end users and are associated with the end users. The designated operating frequencies in VHF and UHF bands range 380, 410, 450, 800-900 MHz bands.Under the umbrella of the European Telecommunications Standardization Institute (ETSI) network, operators, administrators, equipment manufactu rs and users jointly developed a family of standards called Terrestrial Trunked Radio (TETRA). One of them is a set of radio and network standards for trunked voice (and data) services. The other is the air-interface standard, which is optimized for wide area packet data services for both fixed and mobile subscribers. TETRA specifications place no constraint on the form of the radio network architecture. It consists of four sub-systems: base stations (BSs), switches, operation and management centre (OMC), and associated control and management facilities. The architecture is based on OSI reference model where lower three layers are associated with the network services and these are implemented in every node of the network i.e. infrastructure and mobile stations. The upper four layers are associated with the end users and provide services to them. A simplified TETRA architecture is shown in Figure 10.13.

The figure shows a simplified model of TETRA architecture with three interfaces at which services are defined. The Um is the radio link interface between the base station and the mobile station. For the fixed users, a Fixed Network Access Point (FNAP) is provided. Fixed host computers and fixed data networks use typically standardized interfaces and protocols. The mobile users use the same standards. Mobile Network Access Point (MNAP) provides interface to the mobile data users. All the three interfaces support data services with packet mode protocols.

The designated operating frequencies in the VHF and UHF bands are in 380, 410, 450, 800-900 MHz bands. The transmission rate of 36 kbits/sec uses π/4-DQPSK modulation in a channel spacing of 25 kHz. The duplex spacing is 10 MHz, which is increased to 45 MHz in 800-900 MHz band. A TDMA frame structure with four 14.167 ms slots, each accommodates 510 modulation bits. Eighteen TDMA frames are assembled into one multiframe of 1.02 seconds and 60 multiframes constitute one hyperframe, which is 61.2 seconds long. The two versions of the standard will use a common physical layer.

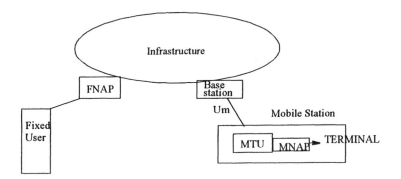

Figure 10.13 A simplified Tetra network architecture

TETRA provides a range of bearer services, teleservices, and supplementary services relevant to joint voice plus data (V+D) capability. The direct mode operation (DMO) provides direct mobile to mobile communications. It also supports packet data services under the banner of so called Packet Data Optimized (PDO). V+D and PDO specifications are based on the same physical radio platform defined by modulation and operating frequencies. It should be pointed out that though GSM can offer a host of data services, but the data is handled in the circuit switched mode.

The base station has 10 classes of transmission power; the lowest power being 0.6 W, while the highest is 40 W. The mobile unit has four classes; Class 1 transmits 30 Watts and Class 4 has a capability of transmitting 1 Watt. TETRA finds many applications in transmission of secure speech and data, road transport, fleet management, vehicle location, and railways.

10.8.6 ATCS

Advanced Train Control System (ATCS) is a series of standards related to railway information network developed as a part of major modernization effort undertaken by the Association of North American Railways. The main objectives of this initiative was to enhance safety of operations, reduce maintenance expenses, and improve efficiency. The ATCS offers many levels of sophistication. Level 10 system is the basic system and Level 40 provides the most sophistication.

The ATCS consist of a wireless part and a ground network part. Six radio channels with channel spacing of 12.5 kHz have been assigned to provide continental wide ATCS service. Gaussian Filtered FSK modulation with an index of 1.7 was chosen over GMSK because of the robustness of GFFSK. The wireless data link uses 4.8 kbits/sec transmission rate with (16,12;5) Reed-Solomon code to counter expected random and burst errors. This error control code is chosen after exhaustive testing of (152,112) Fire Code, rate ½ convolutional code, (48,36:6) and (16,12:5) Reed-Solomon codes on measured and simulated mobile channels. Although the chosen code was slightly inferior in performance to (48,36:6) RS code; the (16,12:5) code was selected on the basis of its relative low cost implementation of the encoder and the decoder and lower decoding delay [22]. Automatic Repeat reQuest is also provided in the access protocol in order to provide greater reliability.

The train locomotive exchanges information with the base station every one or two seconds; the information consists of current location (updated every two seconds), locomotive diagnostic parameters, trains axle parameters, condition of track etc.

The ground network consists of base stations connected to the central dispatch centre through cluster controller. Each cluster controller is connected to up to 8 base stations. The system handles up to 10 levels of priorities; the emergency situation is labelled the top priority.

10.8.7 Wireless LAN's

A wireless LAN as the name suggests, is a network that emulates a local area networks delivering all its benefits with no wires. WLANs can be connected to an existing wired LAN as an extension to it or it can act as a stand alone network. WLANs are particularly suited for indoors applications such as hospitals, universities and corporate offices. Most LANs use radio frequencies in the range of 2.4 GHz. The coverage radius could be up to 100 meters, and the data rates range from 1 to 11 Mbits/sec. Some LANs however work in Infrared frequencies.

The WLAN network when working as a extension relies on availability of access points with the wired network. These access points may be distributed in the area of application. The architecture is therefore similar to cellular system where access point has a function similar to a base station. The access points are interconnected via a distributed system or ethernet. The infrastructure etwork is shown in Figure 10.14. To standardize WLANs devices, IEEE 802.11 a series of standards has been formulated. The standard focuses on the media access control (MAC) and the physical (PHY) protocol levels. The frequencies of operation are in the Industrial, Scientific and Medical (ISM) bands - 902 to 920 MHz, 2400 to 2483.5 MHz and 5725 to 5850 MHz. The IEEE 802.11 defines application, TPC, IP, LLC, MAC and PHY layers at the mobile interface point as well as at the server. The interconnection between the mobile terminal and fixed part of the network is facilitated by LLC which defines interfacing between 802.11 and 802.3 MAC and PHY layers, which are defined for ethernet.

The physical layer of 802.11 define three versions, FHSS (Frequency hopping spread spectrum), DSSS (Direct sequence spread spectrum) and infrared. In FHSS defines minimum of 2.5 hops/sec. The data rate is up to 1 Mbit/sec that uses two level GFSK modulation. The DSSS defines rates of 1 and 2 Mbits/sec, using DBPSK and DQPSK respectively. Upper ceiling on the transmit power of 1 Watt in the USA and 100 mW in Europe is defined. The radiated power should not be less than 1 mW. The infrared LAN shall use 850-950 nm diffuse light type transmission with a range of 10m. Diffused IR provides moderate data rates and coverage, and are suited to moderate-size office offices, short distance battery oriented applications. Directed beam IR LAN's offer higher data rates with adequate coverage.

An enhancement to standard 802.11 was adopted in 1999. This standard known as 802.11b is now the leading standard for WLANs. It provides 3 separate channels in the unlicensed 2.4 GHz band with connectivity up to 11

Mbits/sec. It uses Complementary Code Keying (CCK) and works with 802.11 DSSS systems only. Further enhancements to 802.11 have been developed or are under development. For example, 802.11a operates up to 13 channels in the unlicensed 5 GHz band. It uses COFDM encoding schemes and can provide 54 Mbps of connectivity.

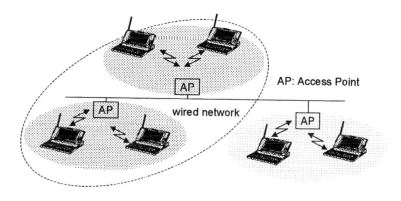

Figure 10.14 Infrastructure Network

Recently, architecture known as Wireless Adhoc Network is proposed. These networks do not have any infrastructure and are known as self organizing networks. Each terminal consists of a typical data terminal (a computer) and a router.

10.9 Summary

While cellular communications has definitely the most visible public presence of any form of mobile system, many other networks have been developed commercially. A feature frequently observed in these networks is a significant degree of independence from the switched telephone network. In addition, the use of such networks for data transmission is widespread. Drawing fro experience in wired data networks, designers have used the technique of packet transmission extensively. This development parallels the introduction of ISDN, where digital signalling allows the integration of voice and data.

The nature of the field and the extent of current activity has constrained the treatment of this chapter to brief consideration of some of the most important developments. A variety of access methods have been introduced. The packet structure in a radio environment has been considered. The topic of fading has been revisited in the context of mobile packet networks; it has been seen that the use of spread spectrum transmission, error control and related protocol considerations can substantially combat its effects.

The final chapters of this text treat a logical extension of the concepts of both Chapters 10 and 11. In it, we shall consider the evolution of the present cellular network to the more all-encompassing form of personal communications.

Problems

Problem 10.1

Evaluate autocorrelation function of Barker codes defined in Section 10.1.

Problem 10.2

(a) Compare the capacities of the four paging systems - POCSAG, GSC, NEC and RDS described in this chapter.

(b) What are the falsing probabilities of the systems in Part (a)?

Problem 10.3

(a) The packet arrival times in wireless data networks are modeled by Poisson process. In a pure ALOHA access system, the average packet arrival rate has been found to be 10^4 packets/sec. The transmission rate is 11 Mbits/sec. Each packet has 160 characters. Find the normalized throughput of the system.

(b) What packet length will maximize the throughput?

Problem 10.4

(a) A taxi operating company has to make a choice between analog and digital dispatch systems on the basis of typical message "taxi XYZ go to location JKLMN". The voice message typically takes five seconds to transmit. The digital transmission rate is 1200 bits/sec and every message is protected by 100% coding overhead. If the channel throughput over the reverse channel is limited to 18%, find the number of terminals which can answer the calls from the base station for both the digital and analog systems.

(b) Repeat part (a) when the access protocol is changed to slotted ALOHA.

(c) In your opinion state whether the traffic in taxi dispatch system describe above is symmetrical or asymmetrical.

Problem 10.5

(a) The average number of transmissions required to deliver a packet successfully is an important measure of performance. Evaluate the average number of transmission for the offered traffic, G, ranging from 0 to 4 Erlangs.

(b) In a design of Packet radio system, it is required that on the average not more than two tries should be needed to deliver a packet successfully. How far the channel be loaded to achieve this objective?

Problem 10.6

(a) Carrier Sense multiple access is an efficient random access method. Calculate the length of the collision window and the system throughput for cell radius of 20 km, remote attack time of 2 msec, decision time 15 μsec, set busy time of 50 μsec, and busy detect time of 1.3 msec. Assume the busy hang-time of 100 μsec. The average packet is 160 bits long and the transmission speed is 9.6 kbits/sec.

(b) Calculate the channel utilization.

Problem 10.7

The average number of trains within the coverage area of the base station of an ATCS system in a high density traffic route has been counted to be 85. Each train is updating its status twice every second. The information data is 180 characters long and is protected by (16,12:5) RS code. The transmission rate is 4.8 kbits/sec. If the trains use slotted ALOHA access protocol, estimate the number of channels required to handle traffic without causing congestion.

References

[1] A.U. Sheikh, "Design considerations for ATCS radio data links," *Proceedings of IEEE-ASM Joint Railroad Conference*, Toronto 1987.

[2] S.Lin, D. Costello, Jr., and M.J. Miller, "Automatic repeat request error control schemes," *IEEE Comm.Soc. Magazine*, vol. 22, pp. 5-17, December 1984.

[3] R.T. Chein, L. R. Bahl and T.D. Tang, "Correction of two erasure bursts," *IEEE Trans. on Information Theory*, vol. IT-15, pp. 186-187, 1969.

[4] E.N.Hum, A.U.H. Sheikh, and Hafez, R.H.M., "Performance evaluation of burst trapping codes for land mobile radio channels," *Proceedings, DMR IV*, Oslo, Norway, 1990.

[5] H. Zimmerman, "OSI reference model-the ISO model of architecture for open systems interconnection", *IEEE Trans. Communication*, vol. 28, pp. 425-432, 1980.

[6] F. Halsall, *Introduction to Data Communications and Computer Networks*, Addison-Wesley Publishing Company, 1985.

[7] U. Black, Data Networks - *Concept, Theory, and Practice*, Prentice Hall, 1989.

[8] D.J. Goodman, R.V. Valenzuela, K.T.Gayliard and B. Ramamurthi, "Packet reservation multiple access from local wireless communications", *IEEE Trans on Comm*, vol. 37, no. 8, pp 885-890, 1989.

[9 D.J. Goodman and S.X. Wei, "Efficiency of packet reservation multiple access," *IEEE Trans. Veh. Tech.*, vol. 40, no. 1, pp. 170-176, Feb. 1991.

[10] J.C. Arnbak and W. Van Blitterswijk, "Capacity of slotted aloha radio communications in rayleigh fading channels," *IEEE Selected Areas in Commun.*, vol. SAC-5, no. 2, pp. 261-269, February 1987.

[11] A.U.H. Sheikh, Y.-D. Yao, and X. Wu, "The ALOHA system in shadowed mobile radio channels with slow or fast fading," *IEEE Trans. on Veh. Tech*, vol. 39, no. 4, November 1990, pp. 289-298.

[12] A.U. Sheikh, and N. Nguyen, "Performance of access strategies for D-TASI over mobile radio fading channels," *Radio and Electronic Engineer (JIERE)*, vol. 57, no. 6 pp. S304-S310, November-December 1987

[13] N. Abramson," Cyclic Code Groups, "Problems of Information Transmission, *ACAD. Sci, USSR*, Moscow, vol. 6, No. 2, 1970.

[14] R. Eschenbach, "Application of spread spectrum radio to indoor data communications, *Proc. ICC*, 34.5.1-34.5-3, 1982.

[15] M. Khaverad, "Performance of non-diversity receivers for spread spectrum in indoor wireless communications," *AT&T Technical Journal*, vol. 64, pp. 1181-1210, July-August, 1985.

[16] M. Khaverad and P.J. McLane, "Performance of low-complexity channel coding and diversity for spread spectrum indoor, wireless communication," *AT&T Technical Journal*, vol.64, pp. 1927-1965, October 1985.

[17] Y.-D.Yao, A.U.H. Sheikh, and S.-X. Cheng, "Near/Far effects on packet radio networks with direct sequence spread spectrum signalling," *IEEE-Pacific Rim Conference on Communications, Computers and Signal Processing*, Victoria, 1989.

[18] O. Yue, "Frequency hopping, multiple-access, phase shift keying systems in a rayleigh fading environment," *Bell System Technical Journal*, pp. 861-879, July-August, 1980.

[19] R. Maummar and S.C. Gupta, "Performance of a frequency hopped multi-level fsk spread spectrum receiver in a rayleigh and lognormal shadow channel," *National Telecommunications Conference*, pp. 6.B.6.1-6.B.6.5, 1982.

[20] K. Pahlavan and A. H. Levesque, "Wireless data communications," *Proc. IEEE*, vol. 82, no. 9, pp. 1398-1430, September 1994.

[21] R.R. Quick, Jr. and Balachandran, "Overview of the cellular packet data (CDPD) system," *Proc. PIMRC'93*, pp. 338-343, 1993.

References

[22] R. G. Ayers, "Selection of a forward error correcting code for the data communication radio link of the advanced train control system," *IEEE Trans. Veh. Tech.*, vol. 38, no. 4, pp. 247-254, November 1989.

Chapter 11

WIRELESS CELLULAR SYSTEM DESIGN PRINCIPLES

The concept of cellular system emerged from the need to increase the mobile radio system capacity without outlaying a large portion of frequency spectrum. In the cellular system, the coverage area is divided into a number of elementary areas called cells. The available spectrum is also divided into a number of distinct channel sets dependent on the so called frequency reuse. Each set is then used simultaneously in designated areas or cells. With frequency reuse, the spectrum utilization increases several folds. These sets are created with a major consideration in view of not letting the ensuing interference to exceed a certain threshold. The threshold is selected in relation to a protection ratio relative to the minimum reception level of the desired signal. If the transmitted power is kept low and adequate physical separation between the reuse cells is allowed, co-channel interference can be minimized to the desired level. The division of spectrum into channel sets also implies that sufficient spectrum is available. The allocation of several megahertz of spectrum 1970's made installation of cellular systems possible[†].

In any area, the design of a cellular system begins with classification of the coverage area as quasi-smooth, irregular, rural, urban, or suburban. This is followed by division of the coverage area into hexagonal cells, which are arranged in so-called clusters. Each cell is assigned a set of frequencies, the membership of which is decided by a number of considerations of which co-channel and adjacent channel interference are of the foremost important. If the interference is totally eliminated, the theoretical capacity of the cellular system could be made infinite provided that all of the users are dispersed uniformly over the coverage area and are not concentrated in one cell. It is obvious that with extensive frequency reuse system complexity is bound to increase.

Each active mobile unit within a cell is assigned a pair of frequencies for two way communication with the serving base station. Since the mobile terminal is expected to roam in the coverage area, it crosses the cell boundaries from time to time. When the mobile unit crosses from one cell to

---

† *In the United States, for example, the FCC in an initial allocation of 40 MHz, resulted in 666 duplex channels of 30 kHz each. In 1986, an additional spectrum of 10 MHz was added allowing another 166 duplex channels. This brought the total number of 30 kHz wide duplex channels to 832. In the rest of the world similar allocation has been made.*

another, a new pair of channels is assigned so that the call could continue without any interruption. With extensive frequency reuse, a very large number of subscribers can be accommodated within the allocated spectrum.

The cellular mobile telephone service is designed to allow calls between two or more terminals, whether these terminals are mobile or fixed. A call from a mobile unit (or subscriber) proceeds as follows. The subscriber pre-dials the desired number and picks up the hand set. When the hand set is lifted, the transmitter sends a request message to the nearest base station[†] over one of several dedicated access channels, identifying both the calling and the called party. On receiving such a request from the mobile, the base station passes the request to a control centre. The latter determines the location of the called party and alerts the receiving station. At the same time an acknowledgment is sent to the calling party on a broadcast channel (also known as the paging channel) that such a request has been received; the acknowledgment message also includes designation of a pair of frequencies. The subscriber's equipment is then switched to the assigned transmit and receive frequencies. If the called party happens to be another mobile, the broadcast message also contains instructions for the called party to switch to a specific pair of frequencies. After switching to the allocated frequency pair, the mobile unit sends acknowledgment to the base station on the recently assigned reverse message channel. Upon the receipt of an acknowledgments from each party, the call is set up and conversation begins. At the end of the call, a supervisory signal indicating the end of the call is sent before the mobile transmitter is shut down.

Handover is an extremely important function of a mobile cellular system. The handover mechanism is illustrated in Figure 11.1. As long as the mobile terminal remains within Cell A, the Base station A will continue to serve it. When the terminal crosses the boundary between Cell A and Cell B, the mobile terminal receives much weaker signal from Base Station A while reception from Base Station B is relatively stronger. For the call to continue with acceptable quality, the mobile should switch to new receive and transmit channels, which would link the mobile to the base station B. This procedure is known as handover. If no such arrangement were made, the call would be prematurely disconnected. The handover sometimes fails because no free channels are available with the new base station. To accomplish handover, the location of the mobile must be known to the base station. This is done by monitoring the received signal strength from the mobile units (base station assisted handover) or from the base station (mobile assisted handover), as

---

†. *The nearest base station is the one which provides the strongest signal at the mobile terminal. This base station is identified at the time when the mobile terminal is first switched on and the search for the strongest is continued on a periodic basis afterwards.*

described in Section 11.6. For this purpose, each base station in the system is equipped with signal strength measuring receivers. If the signal level falls below the minimum required for an acceptable service, the control station instructs the base stations in the neighboring cells to measure and report back the signal strength from that particular mobile unit. The signal strength report is then used in estimation of the location of the mobile and the direction in which it is heading.

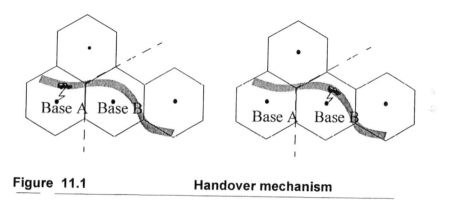

Figure 11.1   **Handover mechanism**

The mobile control station identifies the base station that receives the strongest signal from the mobile and instructs it to be ready for a handover. It also sends the mobile unit, interrupting the voice signal, a short digital handover order message, which consists of details on base station and the new pair of frequencies on which to establish link to the new base station to the mobile is being handed over. The mobile acknowledges the control station that the handover order is being implemented. The switch over to new frequencies by the mobile and the base station takes place after a short interruption of the call[†]. The call thus continues without any further interruption.

Wide area paging is another unique feature of cellular systems. When a cellular telephone number is dialled, the system automatically searches for the called party and if found makes the connection regardless of the recipient's location. The search proceeds on the basis that each user has a designated "home" base station (or operating area); mobile control station equipment continuously monitors the control channels and identifies users not in the home area. When a user is not in its home base station, it informs the

---

† *This interruption is called break before make handover. In TDMA and CDMA, breaking the connection before making it is not necessary. This kind of handover is known as make before break handover.*

nearest base station of its presence[†]. The nearest base station informs the home base station of the terminal of new location of the mobile terminal. This feature is similar to automatic call forwarding since all the cellular phone numbers uniquely identify the home base.

Some users consider this feature to be an infringement of privacy, since with this arrangement, every user can easily be tracked. On the other hand, the user can simply switch off his mobile telephone if he desires so, in that case the cellular system can not be used. The following sections provide a detailed description of the main features of a cellular system.

11.1 Cell Structure and Frequency Planning

A coverage area may consist of several types of terrain, different subscriber densities and traffic patterns, therefore in practice the cells dividing the coverage area are likely to be of different shapes and sizes. For the purpose of analysis, it is useful to assume a certain geometry for cells. Since the radiation pattern of an omni-directional antenna is roughly circular[‡], a circular cell appears to be more appropriate from the practical point of view. However, for theoretical analyses, circular cells pose analytical difficulties because an array of circular areas produces ambiguous areas, which are either provided with radio coverage by multiple cells or do not have any coverage at all. This difficulty is resolved by tessellating the area using one of the three shapes - triangular, rectangular or hexagonal. Figure 11.2 shows an area divided into cells of each of these three shapes. The use of triangular or rectangular shapes is not attractive because this would require a larger number of base stations compared with hexagonal cells. Also, larger number of frequency sets are also required.

Let R be the maximum distance from the cell center to its apex and D be the distance between the centers of closest cells that use the same frequency (co-channel cells). Then, the minimum number of channel sets, C, required to fully cover any area is given by [1]

$$C = \frac{1}{2}\left(\frac{D}{R}\right)^2 \tag{11.1}$$

for square cells. The allowable values of C are given by,

$$C = k^2 + m^2 \tag{11.2}$$

---

† *This phase of the call is known as initialization.*
‡ *The presence of environmental irregularities will make the cells of irregular shapes.*

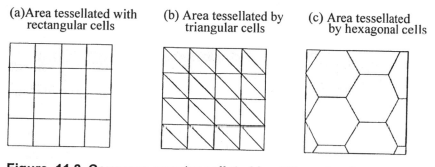

Figure 11.2 Coverage area tessellated by different cell geometries

where k and m are positive integers.
For hexagonal cells,

$$C = \frac{1}{3}\left(\frac{D}{R}\right)^2 \tag{11.3}$$

It is seen that the hexagon shaped cells have the largest area for the same D and R and its shape is closest to a circle, which happens to be the ideal shape from the point of view of the antenna radiation pattern. Also, for a given distance between reuse cells, hexagonal geometry is the least expensive in terms of the number of cells required, and therefore has the highest capacity

A given area may be furnished with radio coverage using a few large ones or many more small cells. Since base station sites are usually expensive, a single large cell will be the cheapest to install. A larger cell requires higher power transmitter, which have a higher maintenance and capital cost. Furthermore, a single cell will have a limited system capacity and the concept of increasing capacity by frequency reuse will be lost. On the other hand, a larger number of smaller cells could provide an increased capacity through higher degree of frequency reuse. The extent to which the size of the cell and transmit power could be reduced is limited by the practical problem of accurately locating the antennas relative to the cell center. It has been reported that dislocation of the antenna from the ideal location of cell centre up to about one quarter of the cell radius degrades the signal-to-interference ratio gradually; beyond this degree of dislocation, the degradation is rapid [2]. It may be also of an interest to note that the final choice of the cell size depends on the cell site elevation and the antenna gain required in a particular operating environment. A cell radius of about half a mile (or approximately 1 Km) appears to be the lower limit for vehicle based system. More frequent handovers for fast moving terminals is a strong reason for this limitation. For smaller cells, the low powered base stations do not require expensive real state, these can be installed at street intersections. For pedestrian or slow moving terminals cell radius as small as 50 meters are envisioned for the future high speed wireless systems.

In a practical system, a typical coverage area may divided into cells of several sizes. Three sizes of cells are used in Figure 11.3. The largest cells are meant to provide coverage to highly mobile terminals in low density areas. The medium size cells provide coverage to terminals in urban or moderately populated areas. The small cells are used to provide service to stationary or very slowly moving terminals in high density areas.

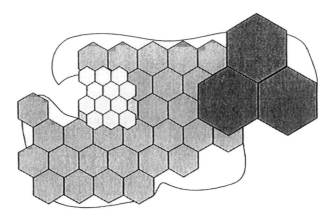

Figure 11.3 Area covered by cells

11.2 Frequency Assignments

In cellular system, the coverage area is divided into cells to facilitate frequency reuse and increase in the system capacity. The cells form distinct coverage areas each of which is served by a base station to which a set of frequencies is assigned. The main aim of such assignment is to increase the overall system capacity; to do so several performance criteria such as interference level are used. The interference level depends on the propagation loss factor, the type of terrain and the frequency reuse distance. In this regard, the concept of cell cluster and its size play an important role. The concept of frequency reuse can be explained as follows.

Consider that a given area, A, is tessellated by hexagonal cells of area A_c. The total number of cells is $M = A/A_c$. Suppose we are given a total of K carriers, which are to be assigned to the cells in the coverage area A in such a manner that these carriers can be efficiently used. Let us create a cluster consisting of N cells. We can divide K carriers into N sets, each set assigned to a cell consists of K/N carriers. The membership of each set is determined such that inter-channel, co-channel and intermodulation interference are minimized. Many frequency assignment strategies have been studied and reported in the literature [3][4][5][6][7][8]. We shall first consider the simple case where it is assumed that the signal from a given base station antenna is some how confined to its boundary, i.e. there is no inter-cell interference. It is easy to show that the normalized distance D_n between the centers of neighboring co-channel cells is then given by [2]:

$$D_n = \sqrt{k^2 + km + m^2} \tag{11.4}$$

where k and m are positive integers, and D_n is called normalized reuse distance.

The normalized co-channel reuse ratio D/R, and the number of cells per cluster, N, are related as,

$$\frac{D}{R} = Q = \sqrt{3N} \tag{11.5}$$

In practical systems, the choice of number of cells per cluster is governed by co-channel interference. For a given cell area and radiated power, an increase in the cluster size results in a reduced co-channel interference. However, increasing the cluster size also reduces the number of carriers in the set assigned to a cell. This consequently reduces the system capacity.

EXAMPLE E-11.2.1

Prove (11.4) and then confirm that the reuse distance is given by (11.5).

Solution

Equation (11.4) can be proven easily if we consider a set of coordinates that intersect at 60-degree geometry as shown in Figure E-11.2.1. The distance between two points d_{12} between two points is given by

$$d_{12} = \sqrt{(u_2 - u_1)^2 + (u_2 - u_1)(v_2 - v_1) + (v_2 - v_1)^2} \tag{E-11.2.1}$$

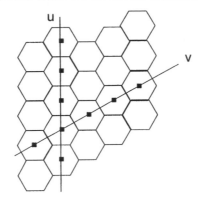

Figure E-11.2.1 A selection of convenient set of coordinates

It is observed that the distance between the centers of adjacent cells is unity and the length of cell radius is $R = 1/\sqrt{3}$. It can be seen that the distance between the center of the reference cell located at the origin and the cell with center at $(u,v) = (i,j)$, $i \geq j$. Using (E-11.2.1), we find

$$D = \sqrt{i^2 + ij + j^2} \qquad \text{(E-11.2.2)}$$

We visualize each cluster as a large hexagon which cover the plane with no gaps and overlaps similar to a cluster. The area of the large hexagon is considered to be equal to the area of the cluster. The distance between the two adjacent large hexagon is

$$\sqrt{i^2 + ij + j^2} \qquad \text{(E-11.2.3)}$$

The total number of cell areas, N contained in the area of the large hexagon is the square of the this factor, i.e.

$$N = i^2 + ij + j^2 \qquad \text{(E-11.2.4)}$$

Combining the above equations, we get the ratio D/R and the number of cells per cluster, N

$$D/R = \sqrt{3N} \qquad \text{(E-11.2.5)}$$

In the regular cellular geometry, the locations of co-channel cells are determined as follows. Refer to Figure 11.4, start in a given cell that uses frequency set S, move k cells along any chain of hexagons; turn counterclockwise (or clockwise) 60 degrees and move *m* cells along the chain that lies in this new heading to reach the co-channel cell. This cell is assigned the same frequency set *S*. Note that $k = m = 1$ in Figure 11.4.

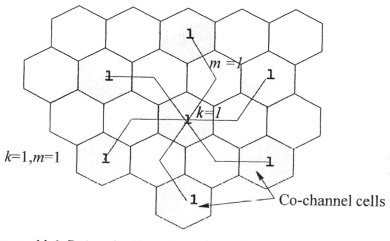

Figure 11.4 Determination of co-channel cells

To this point it has been assumed that cells contain centrally located antennas with omni-directional radiation patterns. These antennas can be also be used at the corners of hexagonal cells, however the total number of base stations required to provide radio coverage to the area remains unchanged. When the system is operating at or near capacity, a need arises to increase the capacity. This can be done by reducing the size of the cells. However, as previously noted, reduction of cell dimension will increase co-channel interference unless transmit power is reduced or antenna structure is changed to control interference. Inter-cell interference can be reduced with the use of sectorized antennas. For example, a cell may be divided into three sectors of 120°, or six sectors of 60° each. This will consequently allow smaller cluster cell. For a relatively small additional cost, it is thus possible to increase system capacity. For example, consider a system with 812 channels of bandwidth 30 kHz each. If for omni excited cells, a cluster size of 12 satisfies the interference criterion, then each cell will have 67 traffic (or message) channels plus 21 set up (access and paging) channels[†]. For 120° sectorized antennas at alternate corners, a cluster size of 7 will yield the same signal to co-channel

---

† *For details on traffic and set up channels see Sections 8.2, 8.3, 11.3 and 11.4.*

interference, and each cell will have 116 channels. For $60°$ sectorized antennas installed at each corner of the cell, a cluster of 4 cells will suffice and each cell will have 203 channels [9]. This simple example shows that a reduction of cluster size results in an increase in system capacity.

In the idealized case discussed above, it was assumed that propagation from an antenna located at the centre of the cell is confined within the cell and no interference between cells results. In practice the reuse of frequencies is limited by the extent of interference. As mentioned earlier, adjacent channel, co-channel and intermodulation interference are the main criteria used in design of frequency reuse. Because the D/R ratio affects the co-channel interference statistics, the co-channel reuse distance has a direct impact on transmission quality as measured by signal-to-interference ratio, and therefore it ultimately influences the capacity of the system. The amount of adjacent channel, co-channel and intermodulation interference determine the way frequencies are allocated.

The subject of frequency assignment has been studied and reported extensively in the literature [3]-[8],[10]-[12]. We shall use the approaches taken by French and Muammer. French has considered only a single interfering signal, whereas Muammer considers a more realistic situation of active users in the surrounding six co-channel cells.

Consider the cellular geometry having a cluster size of 3 shown in Figure 11.5. The co-channel cells are labelled A. The mobile unit in cell A receives a signal envelope s from its base station and composite envelope from unwanted signals s_j, $j = 1,2,....,N$ where N is the number of active co-channel cells. The mobile unit will suffer from co-channel interference when s does not exceed s_j by some protection ratio $\kappa^{\dagger}$.

The probability that the desired signal is less than the co-channel interference from n active co-channels is

$$P(CI|n) = P\left(s \leq \kappa \sum\nolimits_{j=1}^{N} s_j\right) \tag{11.6}$$

and

$$P(CI) = \sum\nolimits_{j=1}^{N} P(n)P(CI|n) \tag{11.7}$$

Three parameters determine co-channel interference power:

    (i) Fading due to the motion of the vehicle in the standing
        wave pattern created by multipath propagation.

---

†. *Protection ratio is defined as the ratio of the desirable signal level to the unwanted interference.*

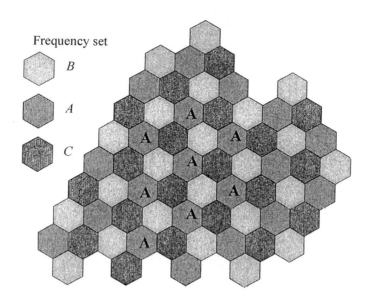

Frequency set

B

A

C

Figure 11.5 Location of co-channel cells

(ii) Radio shadow created by large buildings and natural terrain features. The mean value of the signal s_d when expressed in dB is Gaussian distributed, while s_{linear} is lognormally distributed.

(iii) The excess path loss due to the distance between the transmitter and the receiver, has a slope of approximately -40dB per decade of distance increase.

From [13] and equation (4.110),

$$p(s|\bar{s}) = \frac{\pi s}{2s^2} \exp\left[-\frac{\pi s^2}{4s^2}\right] \tag{11.8}$$

where σ = Standard deviation, and $m_d = \langle \overline{\sigma_d} \rangle$, both in dB and $\bar{s} = \langle s \rangle$.

$$p(\overline{s_d}) = \frac{1}{\sqrt{2\pi\sigma^2}} \exp\left(-\frac{(\overline{s_s} - m_d)^2}{2\sigma^2}\right) \tag{11.9}$$

Since $p(s|\overline{s_d}) = p(s|\bar{s})$, we can write

$$p(s) = \int_{-\infty}^{\infty} p(s|\overline{s}_d)p(\overline{s}_d)d\overline{s}_d$$

$$= \sqrt{\frac{\pi}{8\sigma^2}} \int_{-\infty}^{\infty} \frac{s}{s^2} \exp\left[-\frac{\pi s^2}{4s^2}\right]\exp\left(-\frac{(\overline{s}_s - m_d)^2}{2\sigma^2}\right)d\overline{s}_d \tag{11.10}$$

Also using $\overline{s}_d = 20\log_{10}s$

$$p(s) = \sqrt{\frac{\pi}{8\sigma^2}} \int_{-\infty}^{\infty} \frac{s}{\frac{s_d/10}{10}}\exp\left[-\frac{\pi s^2}{\frac{s_d/10}{4\times10}}\right]\exp\left(-\frac{(\overline{s}_s - m_d)^2}{2\sigma^2}\right)d\overline{s}_d \quad s \geq 0 \tag{11.11}$$

where,

$$C = \frac{10}{ln10} \tag{11.12}$$

M_s is the mean of s, given by

$$M_s = \langle s \rangle = \int_0^{\infty} sp(s)ds = \exp\left[\frac{m_d}{2C} + \frac{1}{8}\left(\frac{\sigma}{C}\right)^2\right] \tag{11.13}$$

and the variance Var(s) by

$$Var(s) = [\langle s^2 \rangle - \langle s \rangle^2] = \int_0^{\infty} s^2 p(s)ds - \left[\int_0^{\infty} sp(s)ds\right]^2$$

$$= \frac{4}{\pi}\exp\left[\frac{m_d}{2C} + \frac{1}{2}\left(\frac{\sigma}{C}\right)^2\right] - M_s^2 \tag{11.14}$$

The probability density function of co-channel signal levels, $Y = \Sigma s_j$ can be approximated, using the central limit theorem, as normal with mean and variance given by

$$M_y = nexp\left[\frac{m_{di}}{2C} + \frac{1}{8}\left(\frac{\sigma_i}{C}\right)^2\right] \tag{11.15}$$

$$VAR[y] = n\left\{\frac{4}{\pi}\exp\left(\frac{m_{di}}{C} + \frac{1}{2}\left(\frac{\sigma_i}{C}\right)^2\right) - \left(\frac{M_y}{n}\right)^2\right\} \tag{11.16}$$

where s_j are independent and identically distributed. The distribution $p(y)$ can be written following central limit theorem,

$$p(y) = \frac{1}{\sigma_y \sqrt{2\pi}} \exp\left(-\frac{(y-\sigma_y)^2}{2\sigma_y^2}\right) \tag{11.17}$$

$$P(CI|n) = P(s \leq \kappa Y) \tag{11.18}$$

$$P(CI|n) = \frac{1}{2\pi} \int_{-\infty}^{\infty} dx \int_{0}^{\infty} \left\{1 - \exp\left[-\frac{\pi u^2}{4\lambda^2 K(x)^2}\right]\right\} \exp\left[-\frac{(u-\lambda)^2 + x^2}{2}\right] du \tag{11.19}$$

where

$$\lambda = \sqrt{\frac{n}{\left(\frac{4}{\pi}\right)\exp\left(\frac{\sigma}{2C}\right) - 1}}$$

$$\tag{11.20}$$

$$20\log K(x) = z_d - 20\log n - \frac{\sigma^2}{C} + \sigma$$

$$z_d = m_d - m_{di} - R_R$$

and

$$R_R = 20\log\kappa \tag{11.21}$$

The difference between the lower bound ($n = 1$), and the upper bound ($n = 6$)[†] for constant probability of co-channel interference is shown in Figure 11.6. For example, for $\sigma = 0$ (fading only), 6 and 12 dB the difference is 15, 13, and 17 dB respectively.

Frequency assignments can be made on the basis of minimum total interference from the three sources: co-channel, adjacent channel and intermodulation interference. In [12] the frequency assignment problem is studied with an aim to achieve a minimum of aggregate interference; the aggregate interference includes co-channel, adjacent channel, and intermodulation products. Whatever be the criterion of frequency assignment, the assignment may be made in three possible ways: fixed, dynamic and hybrid.

In the fixed channel assignment strategy, channel sets are permanently assigned to cells and cannot be used again in any other cell of the same clus-

---

† *Here it is assumed that only six cochannel cells in tier 1 contribute to cochannel interference. The result can be easily extended for interferers in the other tiers.*

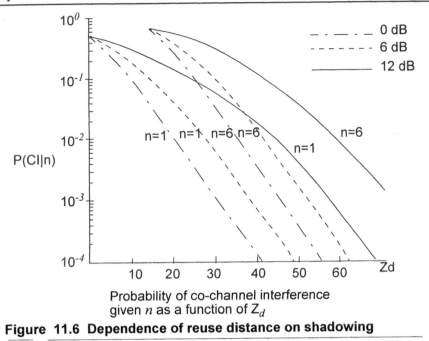

Probability of co-channel interference
given *n* as a function of Z_d

Figure 11.6 Dependence of reuse distance on shadowing

ter. In the dynamic channel assignment, all channels nominally belong to the control centre, which allocates the channels to the users in the system on per need basis. Naturally, following a request for a channel from a mobile, the control centre has to satisfy the co-channel, inter-channel, intermodulation interference or any other performance criterion prior to assigning a channel. This procedure takes some time to complete and a delay in making the assignment is naturally expected. Furthermore, an ensuing increase in the processing burden on the control center is also considered to be a disadvantage. The aim, however, remains of not allowing the resulting interference to exceed a certain threshold so that the reception quality is preserved.

The hybrid channel assignment is a combination of the fixed and the dynamic channel assignment. In this scheme a fraction of the total number of channels is permanently assigned to the cells and the remainder are left under the control of the central network control. The network controller assigns these channels in a particular cell on the basis of the need in that cell. Under certain circumstances, the need of a particular cell may become extremely high and many additional channels may be needed to meet a surge in demand[†]. Under these conditions, some channels in the neighboring co-chan-

---

†. *The example is traffic hot spot, which develops when some emergency occurs in a certain part of the coverage area and very high traffic is generated within a part of the coverage area. A hot spot may cover one or more cells.*

nel cells may have to be blocked from use so that the channel assignment criteria are not violated.

11.2.1 Fixed Channel Assignment (FCA)

In the fixed channel assignment, each cell is assigned a set of frequencies on a permanent basis, whenever the traffic demand exceeds the cell capacity, the calls within the cell will be blocked. In start-up (so-called infant) systems, FCA is the principal method of frequency assignment. The fixed channel assignment scheme has two major deficiencies:

(i) Spatial and temporal traffic surges could cause blocking of all message channels in a certain cell with the result that the service is lost in that cell, even though some message channels may remain unused in the neighboring cells.

(ii) Cell equipment failure could cause all users to lose service, although free channels may be available in the neighboring cells.

The number of channels to be assigned in a particular cell is determined by the expected traffic level. The traffic is estimated from the population size, percentage of the population likely to subscribe to the service (penetration rate), calling habits of the potential subscribers, average call duration and the grade of service desired by the subscribers. Erlang B (Erlang Loss Formula), described in Chapter 8, is used to determine the number of channels needed in each cell.

EXAMPLE E-11.2.2

It is desired to provide cellular services along a highway that has settlements of disparate population along it. These settlements, A, B, C, and D have population of 1000, 2,000, 3,000 and 1,500 respectively. In a survey, the average penetration of rates of cellular system in these settlements have been found to be 10, 30, 35, and 25 percent respectively. The average traffic generated by each subscriber in these settlements is estimated to be 1, 1.5, 1.75, and 1.3 calls per busy hour respectively and the average length of the call in all areas is estimated to be 125 seconds. If each settlement is covered by a single cell, calculate the number of traffic channel required. Assume the blocking probability of 2%.

Solution

From the penetration rates, the number of subscribers in settlement A, B, C, and D are calculated to be 100, 600, 1,050, and 375 respectively. The total traffic generated from each settlement is 3.5, 27.1, 63.8, 16.93 Erlangs.

From Figure 8.1, the number of trunks needed in each area is found to be 8, 35, 80 and 27 respectively.

11.2.2 Dynamic Channel Assignment

Some of the shortcomings associated with FCA are overcome with the use of dynamic channel assignment (DCA). In this scheme, control of all channels remains with the system switching center, which assigns channels to the cells on demand basis subject to rules governing the acceptable interference levels. The performance of dynamic frequency assignment schemes has been studied using simulations and is reported in [14].

Dynamic channel assignment performs very well under light traffic loads as in this case the blocking probability is found to be much lower than that for FCA. However, when the traffic increases, DCA suffers from excessive switching of channels at the switching center. The blocking probability also exceeds that in the case of FCA. Moreover, as the traffic within a certain cell increases, the frequency assignments pattern in the coverage area is modified which results in an increase in the frequency reuse distance. The higher average channel reuse factor degrades channel utilization and spectrum efficiency. If we try to maintain minimum channel reuse criteria in dynamic channel assignment, many channels could not be used with the obvious result of higher call blocking probability.

11.2.3 Hybrid Channel Assignment

A desirable frequency assignment scheme is the one which combines the best features of both FCA and DCA. Hybrid Channel Assignment (HCA) attempts this synthesis. With HCA, the total number of channels are divided into two categories - one category consists of cells that are permanently allocated to the cells while the second category consists of channels which are kept in reserve with the control center that assigns the channels on dynamic basis. A variation of HCA known as frequency borrowing has been shown to exhibit blocking probabilities lower than that achieved in FCA or DCA. With this technique specific channels are nominally allocated to each cell. Following a request from a user within a cell, a channel is allocated out of the pool of nominal channels if one is available; otherwise a channel is "borrowed" from a neighboring cell. In each cell, an ordered list of channels is kept and the channels at or near the top of this list are given preference to be used within the cell while those at or near the bottom of the list are given priority to be borrowed to serve temporarily in the neighboring cells.

When a request for a call occurs, the list is checked in an order of priority to find a free channel. If all of the nominal channels are busy, a channel is borrowed from an adjacent cell. Two criteria are used in making a decision on which channel is suitable for borrowing. First, the adjacent cell having the maximum number of free channels is sought. Second, the selected channel is the last ordered channel and its corresponding co-channel channels are also freed in the other two channel reuse cells. For example, Figure 11.7 shows that channel X is borrowed by cell 1 from cell 2. The use of the borrowed channel in Cell A, reduces the co-channel reuse distance, the new reuse distance is between cell A and cells B and C (in the figure). To meet the minimum signal to interference ratio, channel X is blocked in the other two closest co-channel cells.

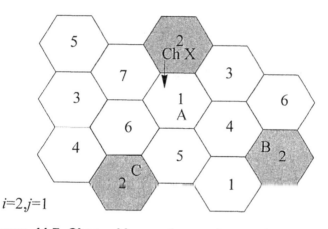

Figure 11.7 Channel borrowing assignment

This switching strategy ensures that the time for which a channel is borrowed is kept to a minimum. As soon as a nominal channel is freed in the borrower cell, switching to this nominal frequency takes place and the borrowed frequency as well as other blocked frequencies are released. In a similar manner calls are switched from the last ordered nominal channel to the other channel within the cell, as soon as it becomes free. The switching strategies are shown in Figure 11.8. It is obvious that the number of channel switchings in the case of channel borrowing strategy will be much higher than those for the fixed channel assignment and dynamic channel assignment.

The comparative performance of FCA, DCA, and Hybrid Channel Assignment schemes are shown in Figure 11.9. It is seen that FCA outperforms DCA when the traffic is increased by more than 50%. The channel borrowing scheme performs better than fixed channel assignment even for a 100% increase in traffic load. A penalty is of course paid in terms of more switching, both across the cell boundaries (for handoff purposes) and within

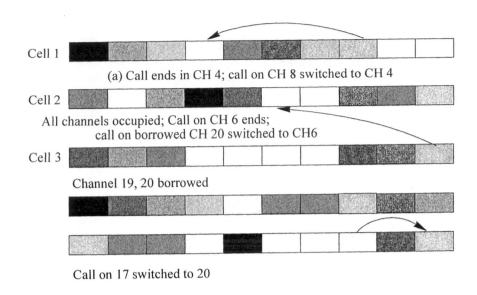

Call on 17 switched to 20

Figure 11.8 Channel switching mechanism

cells. The implications on channel borrowing on the system capacity has not been studied.

11.3 Cell Traffic and Channel Requirements

The number of traffic channels needed to meet the demand is determined by the number of potential users within the cell, call attempt rate, holding time, and the desired grade of service. The call attempt rate is the average number of calls that a user originates within a unit time. The unit time, in general, is taken to be the busiest hour of the day. The holding time is the average duration of a call. Usually the grade of service is expressed in terms of call blocking probability, although mean waiting time before a message channel is assigned may also be used to define the grade of service (GoS). Other parameters like call drop probability and outage probability may also be used as indicators of GoS.

The distribution of the call attempt rate has an important bearing on the number of channels required for a given grade of service. Generally it is assumed that each customer originates a call randomly, independent of all other users. The validity of independence of call origination is based on the

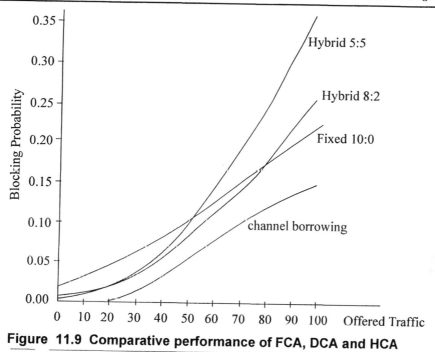

Figure 11.9 Comparative performance of FCA, DCA and HCA

assumption that the probability of a call originating and terminating within the same cell is negligible. This assumption is weakly valid when the number of mobile to mobile calls increases arbitrarily. The number of calls arriving in a time, t, is assumed to follow Poisson distribution, which is given by:

$$P_n(t) = \frac{(\lambda t)^n}{n!} exp(-\lambda t) \qquad (11.22)$$

Here $P_n(t)$ is the probability that n calls occur in $(0, t)$ and λ is the call arrival rate. The Poisson model describing the call arrival rate leads to an exponential distribution of call holding time [15][16]. The probability that any randomly selected holding time will equal or exceed length T is given by:

$$P(T) = 1 - exp(-\lambda T) \qquad (11.23)$$

where $1/\lambda$ is the average holding time of all customer calls.

The number of channels required to meet the user demand for a given service quality of GoS depends on the call handling discipline, which may be one of three types. The most commonly used method is Blocked Call Cleared, also known as Erlang B. In this discipline if a channel is not available at the time when request for channel is made, the call is dropped or cleared from the system. In other words, the user has to retry to get access to the system. In the second method, Blocked Call Delayed, a blocked call is held until a free channel becomes available. The third strategy, Blocked Call Held, is also known as Erlang C; a call that does not find a free channel is held for the average holding time, which is a design parameter. If a channel becomes available during the average holding time, the call is accepted for the time remained from the average holding time; otherwise it is cleared from the system. These models are discussed in Chapter 8.

11.3.1 Access Channel Requirements

In mobile radio systems channels are allocated on demand, and a request for assignment of a message channel is sent over an access channel. Since requests arrive randomly, collisions of request messages from more than one subscribers are expected to occur. The number of request channels needed in a system depends on the customer population, the rate at which requests for channel originate per customer, the length of request message and the maximum allowable collision probability. Several random access schemes are used in mobile radio systems, some of these are: ALOHA, Slotted ALOHA, CSMA and Polling. The random access protocols are described in Chapter 8.

11.4 Control Architecture

In contrast to the fixed telephone network, the link between a mobile and switching centre is somewhat unreliable, therefore cellular mobile radio systems require additional means of call supervision. As the subscriber moves, signal strength due to fading and shadowing fluctuates, which affects the reception quality. To maintain satisfactory call quality at all times, it is necessary to monitor the signal received at both ends of the link.

The control architecture may be defined in terms of interconnections between the system control elements, notably the mobile unit, base station and system control station. Two types of control architectures, centralized and distributed are used. In the centralized control, the base station acts merely as a relay point between the mobile control station (MCS) and the mobile terminal (MT). Decisions regarding handoff, frequency assignments, billing, and other call supervision functions are taken at the mobile control station or network's central control system. The latter is connected to several base stations through dedicated land lines. The mobile control station acquires the necessary information from the serving base station and takes necessary actions. The network control also issues directives to the other network modules. The main disadvantage of centralized control architecture is its susceptibility to processing overloads which usually results in access delays, call processing and handovers. Under heavy traffic conditions (busy hour or emergency), the system could result in a catastrophic failure. The advantage of centralized control lie in fault diagnosis, and network maintenance. For an equivalent system, the cost of system with centralized control may also be lower than that of system with a distributed control.

In a distributed architecture, system's control functions are divided among Regional Switching Group (RSG) stations and a Network Control Station (NCS). The regional switching control stations perform local switching functions, channel assignments tasks, and handle paging and access functions. These also provide relevant billing information to the central control *e.g.* call lengths, requests for trunk resources etc. The advantage of this technique lies in less likelihood of processor overloading. The cost of the system with distributed control is higher because high network intelligence requires more complex software. Schematics of the two systems are shown in Figure 11.10.

The distributed control schematic shown consists of cell sites, RSG (Regional Switching Group), NCS (Network Control Station), and Telco. Several base stations are connected to one RSG, which defines a certain region of a cellular system. Each RSG connects to local telephone exchange (Telco) and also forms a link to NCS.

In the centralized control, the base stations are directly connected via optical fibre, cable or microwave links to the mobile telecommunications switching office (MTSO). All decisions regarding the activities on the network are taken at MTSO. These functions include channel assignment, handovers, call supervision, billing, fault diagnosis and maintenance.

There is at present no formal technique for designing the overall control aspects of such a complex system. Computer simulations and 'walkthrough' methods have been used to validate the control system and, in particular, detailed flow charts are constructed for interface signalling.

11.4.1 System Control Elements

In a cellular system the three principal control elements (mobile unit, base station and mobile control station) are interconnected by interface elements whose functions are to provide compatibility between the control elements in terms of signalling waveforms, timing tolerances and many other fault conditions. The basic requirement is to avoid circuit lockouts and excessively long periods in releasing channels [17].

A mobile unit consists of three major parts; a transceiver, a control unit and an antenna. The transceiver consists of a duplexer†, a receiver, a frequency synthesizer and a logic unit. The transceiver and its logic unit are mostly realized with VLSI circuits to reduce size and weight. The frequency

---

† *A duplexer is not needed in terminals using TDM.*

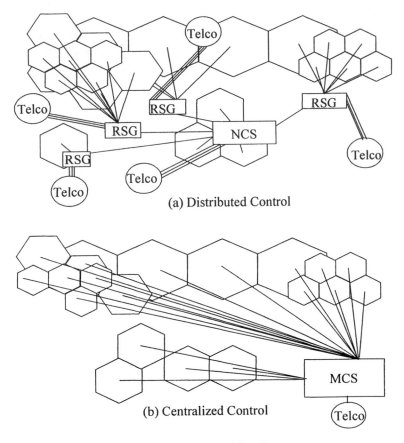

(a) Distributed Control

(b) Centralized Control

Figure 11.10 Network Control Architectures

synthesizer is designed to tune the transmitter and receiver and to minimize spurious emission and reception. The latter two are also controlled by proper designs of the transmit and receive filters. The logic unit, consisting of a microprocessors, memories (ROM and RAM) and peripheral circuits, performs several procedures related to link set up.

The control unit is the telephone set itself. It consists of a push button key circuit and other circuits necessary to store the dialled number, setup a radio speech path, and generate various signalling and alerting tones.

The base station consists of a group of transceivers and a switching system that links the mobile units to the land based network. Transceivers are used for both relaying voice/data signals and for the operation of control channels (paging and access). The base station is also equipped with a monitoring receiver, which periodically measures signal levels from mobiles in order to determine if a handover is needed.

The mobile control station (MCS) serves as a coordinating unit [18]. Its functions include administrating assignment of radio channels, coordinating the use of cell sites with moving subscribers, and maintaining system integrity.

11.4.2 Control Functions

The main control functions in a mobile radio system are those of paging and access, acquisition of network information, supervision, and handoff. As in wired networks, supervision is defined as detection of changes in the switch hook state caused by the customer e.g. lifting and replacing the hand set. In a cellular system monitoring of received signal strength is an additional responsibility. The techniques to implement these control functions differ from system to system. In the discussion that follows, we shall take AMPS as an example. The control procedures used in other systems, do not deviate much from those used in AMPS[†].

A combination of tone burst and continuous out-of-(voice) band modulation is used to supervise progress of calls[‡]. These are known respectively as the signalling tone (ST) and supervisory audio tone (SAT). In AMPS, three supervisory audio tones are used, each base station is allocated one of these three. As shown in Figure 11.11, such assignment increases the co-channel reuse distance of a particular SAT by a factor $\sqrt{3}$, thus enhancing the reliability of the supervisory functions. The mobile unit transponds the supervisory audio tone, the strength of the transponded SAT received at the base station is used as a measure of approximate distance of the mobile from the base as well as the signal strength of the voice channel. These measurements are used for mobile location and, if required, for handing off the mobile to another base station. The reception of transponded SAT by the base also used as an indicator of link continuity. The SAT frequencies are located outside the audio band (5970, 6000, and 6030 Hz in AMPS system) are FM modulated with ± 2 kHz deviation. The signalling tone is used for functions of alerting the user, to handoff, disconnection of the call and mid-call customer service. A 10 kHz frequency is chosen for this purpose.

11.4.3 Paging and Access

Paging channels are used to inform mobile units of system parameters and to alert the mobiles of incoming calls. The paging channels are always in

---

†. *In this chapter, we shall use AMPS as an example to describe design principles of a cellular systems.*

‡. *In G2 and G3 systems, control channels are used for this purpose. Details on these channels are provided in Chapters 12 and 13.*

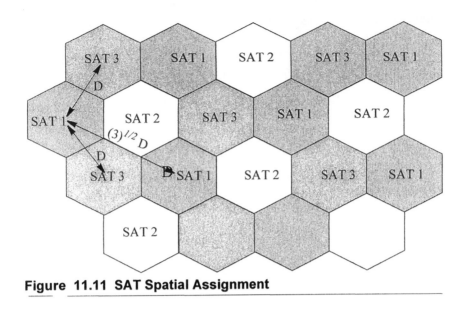

Figure 11.11 SAT Spatial Assignment

the forward (base to mobile) direction. An access channel is used by the mobile unit to get an access to initiate a call. On the access channel the mobile terminal supplies the network the mobile unit identification, authentication messages and the dialled subscriber number prior to the assignment of a channel. Access channel is used for mobile to base station communications.

In cellular systems, separate channels are assigned for paging and access functions. The number of access channels depends on the number of users, the need to avoid collisions between requests and the real time processing power available at the base station. The number of paging channels is determined by the potential number of paging messages to be transmitted to the users. The number of users in a start up system is usually small therefore a small number of channels are allocated. However, the number of channels should be sufficient to allow users with ten digit numbers to be paged within the cells without causing any delay in transmitting messages to the users. Some allowance is usually kept to account for future expansion. The raw transmission rate of a single paging channel is limited to approximately 10,000 bits/sec. However, since high redundancy is required, the net throughput of the channel is usually low. For a net transmission rate of 1200 bits/sec, and 300 bits per message, this is equivalent to 25 messages per second. Considering that one page/user/busy hour is required, one channel can support up to 90,000 customers. At present this limit has been exceeded in some major cities where substantial congestion exists.

In addition to the initiation of calls, the mobile unit can use the access channel to answer paging messages from the base station. The base station uses the forward access (paging) channel to send channel designation messages to the intended mobile unit, which then tunes to the assigned message channel pair (forward and reverse) so that the call can proceed. Since call initiation is a random event, call initiations from two or more mobile units may result in message collision. Part of the control function is to avoid collisions. To this end several measures are taken. First, the messages on the paging channel contain information on the current status (free or busy) of the access channel. For example, in AMPS, the base station uses every eleventh bit of transmission on the paging channel to indicate the status of the access channel. This information is scanned by mobiles prior to attempting access. Second, the mobile unit addresses the nearest base station by sending a digitally encoded precursor. Third, following the transmission of the precursor, the mobile unit monitors the status of the access channel on the forward channel to change from idle to busy. When such change occurs, the mobile terminal assumes that the change in the channel status occurred because of its initiation, and access channel is therefore available to it only. At this point, it can transmit the channel access message. Collision, however, is still possible if more than one mobile sends out the precursor at approximately the same time and assume that channel is available to them. However, since the precursor is a very short message, probability of collision is very small. In case a collision is detected, the mobile unit waits for a random time interval in order to avoid periodicity of repeated collisions while attempting to access the channel.

11.4.4 Data Requirements and Formats

Network control requires sizable transmission of data. The forward setup channel, the paging channel, the message channel and the land lines linking the various control elements are the main data paths. The first three of these use radio transmission over fading channels, so that error control is particularly important. Well defined protocols and some ways of providing data security are necessary. It is evident that for signals impaired by channel fading and interference considerable overhead is needed to maintain link reliability and system integrity. For example, in the AMPS system, over the forward setup (paging) channel 40 bits long coded data words are repeated five times. These words are interleaved with 10 sync bits, 11 word sync bits, and 42 busy-idle bits respectively, to form a packet of 463 bits as shown in Figure 11.12(a). The 40 bit long code word is formed by encoding 28 information bits to form a 40 bit long BCH code word. In a similar way the reverse (access) channel data is framed into a packet as shown in Figure 11.12(b).

The handoff orders are sent by blanking speech over the voice channel and sending a data burst. Since the handoff takes place under conditions of

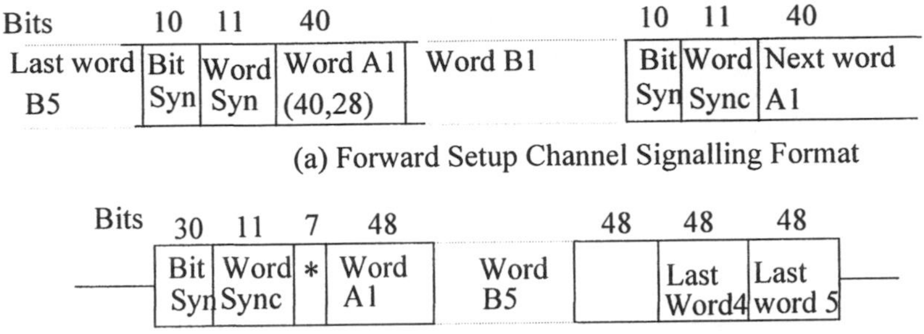

Bits	10	11	40			10	11	40
Last word B5	Bit Syn	Word Syn	Word A1 (40,28)	Word B1		Bit Syn	Word Sync	Next word A1

(a) Forward Setup Channel Signalling Format

Bits	30	11	7	48		48	48	48
	Bit Syn	Word Sync	*	Word A1	Word B5		Last Word4	Last word 5

(b) Reverse Setup Channel Signalling Format

Figure 11.12 Forward and Reverse Setup Channel Formats

adverse signal-to-noise ratio conditions, a greater data protection is needed at this stage. Thus, handoff messages from the base station to the mobile are repeated eleven times. After the mobile unit had switched to the newly assigned channel, the link signal-to-noise ratio improves and the message words are repeated five times on the reverse link. In both cases, after the data has been decoded, majority voting is taken. The voice channel data format is shown in Figure 11.13. In [19], it is shown that such heavy protection is not needed over an average mobile channel.

The information transfer rate between the base station and the base station controller must be sufficient so that large number of control functions are handled. It should also be flexible enough to allow system expansion. In the AMPS system a data rate of 2400 bps is used. The coding scheme chosen is a (32,26) BCH code, shortened from a (63,57) BCH code. Synchronization is effected by a preamble embedded at the beginning of each message. In subsequent systems link between the base and base station controller uses much higher capacity.

11.4.5 Operation of Cellular System

When the mobile telephone is energized, it scans the setup channels according to a program in its memory and selects the one with the highest signal strength and reads on this channel the system parameters. Thereafter the mobile unit continues to monitor the forward paging channel periodically and keeps an update on the strongest channel. In describing the cellular operation, two types, mobile completed calls and mobile originated, calls will be considered.

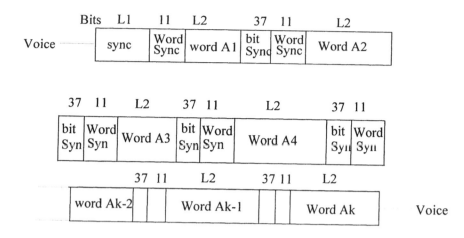

Figure 11.13 Data Format Voice Channel

Consider first the system responses in the case of a mobile completed call. The calling party dials the mobile telephone number and the information on the destination is directed by the wired network to the home base of the called unit. The mobile switching office sends the message to base stations in the coverage area, which transmits the message on a paging channel.

Since the mobile when in idle state keeps scanning the paging channels, it recognizes its paging call and replies to the serving base station; the latter directs the reply to the switching office. The switching office selects an idle channel pair and informs the base of the selection. This information is then paged to the mobile unit. The mobile unit after switching to the assigned (transmit and receive) channels, transponds the audio supervisory tone using the assigned transmit channel. The cell site recognizes its SAT and places the wired trunk in the off-hook condition. The mobile switching office interprets the action as the establishment of successful voice channel communications.

On receiving command from the switching office, the base station sends an alerting message on the newly assigned voice channel to signal the mobile terminal about the incoming call. A signalling tone is then sent by the mobile unit, which prompts the switching office to place the line connecting the calling party on-hook and provides a ringing tone to the calling party. When the mobile customer answers the call, the mobile originated signalling tone is removed, the switching office places the line off-hook, and conversation takes place. The call sequence is shown in Figure 11.14.

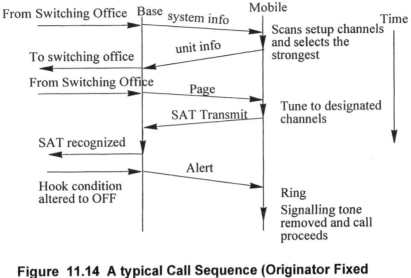

Figure 11.14 A typical Call Sequence (Originator Fixed Telephone)

When a mobile unit originates a call, the called party's number is entered. This places the mobile unit in the off-hook condition, a procedure similar to that described above for mobile completed calls is initiated. The stored digits, along with the mobile's identification, are transmitted over the access channel selected by the mobile unit. As in the previous case, the switching office designates a voice channel and establishes communication with the mobile unit. The switching office determines the routing and the charges information at this time by analyzing the dialled digits and sends the dialled digits over the wire line network to the called party. When the latter answers, conversation follows.

In theory, handoff[†] occurs when a cell boundary is crossed. In practice, the boundary is defined by signal strength rather than geographical position. When the signal strength is found to go below a nominal operating level, base stations neighboring the serving are directed to collect mobile location information by measuring the received signal strength. On the basis of this information, the switching office selects the new base station to which the mobile is to be handed off. A channel pair and wired channels are assigned in preparation of handover. The switching office sends a message containing new chan-

---

† *The term Handover is used in Europe while handoff is used in North America.*

nel and new SAT information to the currently serving base station. The latter relays this message to the mobile unit on the voice channel. The unit sends a brief burst of signalling tone and turns off its transmitter; it then retunes to the newly designated channel and transponds digitally encoded SAT. Upon receiving the SAT, the new base station sends an off-hook signal. The old base station receives the signalling tone and sends an on-hook signal. The switching office reconfigures its network, connecting the party with the appropriate wireline. The on-hook and off-hook messages are interpreted by the switching office as successful handoff.

Call disconnection can be initiated either by the mobile or the wired portion of the system. When the mobile unit initiates the disconnect procedure, it transmits a signalling tone to its base station; an on-hook signal is then placed on the appropriate wired trunk and the switching office places the line in the idle mode. It then directs the serving base station to shut down the transmitter associated with the call.

The first stage of a system initiated disconnect occurs when all of the switching office resources are taken off the call. The switching office transmits a release order to the serving base station. This order is relayed to the mobile unit on the voice channel. The mobile unit confirms the receipt by invoking the same release sequence associated with a mobile initiated disconnect procedure. Upon receipt of this sequence, the switching office orders the cell to shut down the radio transmitter.

11.5 Vehicle Location Techniques

Towards quality assurance of communications, it is necessary that during the call the signal strength be maintained above an absolute minimum. As mentioned in the previous section, the absolute location of the vehicle need not be known, only its location to the nearest base station is required[†]. Each time a cell boundary (defined in terms of signal strength) is crossed, a new set of channels served by a new base station are made available so that the call could continue without any perceptible degradation. The procedure used by the system to monitor vehicle location, its relocation and handoff is called the Location and Handoff Algorithm. Radio location techniques have been studied extensively even before the advent of cellular systems [20]-[28]. Trilateration, triangulation and pulse/phase ranging have been used in practice, the actual choice of the method depends on the application and the desired location accuracy.

---

†. *Recently, accurate location of mobile terminals in cellular systems has become a system requirement.*

Radio location techniques based on trilateration systems are common. An estimate of the vehicle location is made from a set of distance measurements from at least three base stations to the mobile unit. It is seen from Figure 11.15, that if we use only two base stations, location ambiguity results. The distance between the mobile and the base station may be measured by several methods, these include pulse transponding, phase comparison systems and signal strength. The pulse transponding is retransmission by the mobile terminal of the pulse received from the base station. The pulse arrival time of the transponded pulse is compared with the transmitted and the distance is calculated. The errors in the measurement distance is due to smearing of the leading edge of the pulse. Phase comparison involves measurement of phase difference between the transmitted pulse and the transponded pulse. The absolute signal strength can also be used for distance measurement. All these methods are subject to error, especially due to multipath propagation effects. Turin [20] showed that pulse ranging is relatively resistant to multipath effects. Location estimates for this technique could be in error by 200-400 feet for 90% of locations; errors for the remaining 10% may be as high as 700 feet. Engel [21] has estimated errors of up to 3000 feet for 90% of the locations when distance estimate is done by phase ranging.

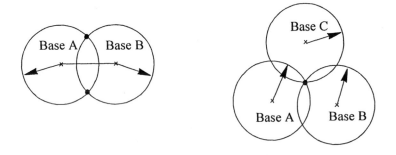

Figure 11.15 Trilateration with two and three base stations

Triangulation locates the vehicle through measurements of its bearing angle from several base stations, using a rapidly scanning narrow beam antenna or several fixed narrow beam antennas. The presence of radio wave scatterers is the major problem with this approach, since angle measurement defines the location of the scatterers rather than the mobile unit. It may be possible, however, to obtain a reasonable estimate of vehicle location if locations of the scatterers are known.

The most widely used method for vehicle location is based on direct measurement of signal strength. Errors are caused chiefly by signal strength fluctuation resulting from Rayleigh and lognormal fading. As described in Chapter 2, two types of signal fluctuations exist: instantaneous fluctuations due to fast fading of the signal and slow fluctuations of the median (or mean) signal level due to shadowing. The variation of the local median due to shadowing depends on terrain effects, and the long term median depends on attenuation due to the distance between the transmitter and the receiver. At each base station a special signal strength measuring receiver is operated to measure the signal strength. The receiver monitors a particular vehicle periodically for a short duration as shown in Figure 11.16. An error in the estimate arises because the received signal is not averaged for sufficiently long time. This results in a significant standard deviation in the signal strength measurement. A longer averaging time would require a larger number of number of measurement receivers at each base station. This section examines signal measurement with an aim to determine a suitable averaging time while keeping measurement error within reasonable limits. For this purpose, we follow the development in [24].

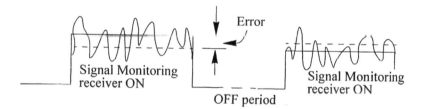

Figure 11.16 Received Signal Fluctuations in the Measured Mean

Consider first the variation in the measured short term mean. For a stationary random process $\zeta(t)$, let the received signal is averaged over T seconds and m_T be the T-second mean, i.e.,

$$m_T = \frac{1}{T}\int_{-T/2}^{T/2} \zeta(t)dt \qquad (11.24)$$

Then if the limit exists

$$m_{T \to \infty} = m_T = \langle \zeta(t) \rangle = m \qquad (11.25)$$

where m is the true mean.

For a fixed T, m_T is distributed about m and its variance is given by [22]

$$\sigma_T^2 = \frac{1}{T}\int_{-T/2}^{T/2}(1-\frac{|\tau|}{T})(R_\zeta(\tau)-m^2)d\tau \tag{11.26}$$

where $R_\zeta(\tau) = <\zeta(t+\tau)\zeta(t)>$ is the correlation function. We can determine the variance of m_T when $R_\zeta(\tau)$ and m are available.

For a Rayleigh distributed signal envelope $X(t)$, a logarithmic detector may be used to achieve a large dynamic range over which the signal could be measured accurately. As will be shown in the following, this type of detector has another advantage, which helps to reduce errors in the measurement of distance. If

$$y = f(X(t)) = 2alnX + b \tag{11.27}$$

where 'a' and 'b' are constants, and $R_y(\tau) = <y(t+\tau)\ y(t)>$

$$R_y(\tau) = \int_0^\infty\int_0^\infty f(x_1)f(x_2)p(x_1,x_2)dx_1dx_2 \tag{11.28}$$

where $X_1 = X_1(t_1)$, $X_2 = X_2(t_1+\tau)$, and $p(X_1,X_2)$ is the joint probability density function of X_1 and X_2. Assuming that the received signal has Rayleigh probability density function,

$$p(X) = \frac{X}{\sigma^2}exp\left(-\frac{X^2}{2\sigma^2}\right) \tag{11.29}$$

It can be shown that with the use of a logarithmic transformation the variance of a T-second mean is given by [24]:

$$\sigma_{MT}^2 = \frac{a^2}{T}\int_{-T/2}^{T/2}\left(1-\frac{|\tau|}{T}\right)(\rho^2(\tau)+0.2498\rho^4(\tau)+0.1108\rho^6(\tau)+...)d\tau \tag{11.30}$$

But $\rho(\tau) = J_0(2\pi f_D\tau)$ where $J_0(x)$ is the zero order Bessel function and f_D is the maximum Doppler frequency. Since $|\rho(\tau)| < 1$, (11.27) may be approximated as,

$$\sigma_{MT}^2 \approx \frac{a'^2}{T}\int_{-T/2}^{T/2}\left(1-\frac{|\tau|}{T}\right)J_0^2(2\pi f_D\tau)d\tau \tag{11.31}$$

where a'^2 includes the correction factor associated with the approximation. The importance of this result lies in non-dependence of σ_{MT}^2 on the envelope mean. In cellular systems, the mean received signal power depends on the distance of the mobile from the base and it is desirable that the variance in the mean should be consistent over the entire coverage area. The logarithmic transformation provides that desirable effect. For $\tau = 0$, the standard deviation $\sigma_{MT}(\tau = 0)$ can be obtained precisely. For the transformation $y = 20 \log_{10} X$, it becomes

$$\sigma_{MT}(\tau)\big|_{\tau=0} = \sqrt{\langle (y - \alpha_0)^2 \rangle} = 5.57 \qquad (11.32)$$

where $\alpha_o = -a\gamma + a \log_e (2a^2) + b$, and γ is Euler's constant.

If we set $\tau = 0$ in Equation (11.28) we get $\sigma_{MT}^2 = a'^2$. For $a' = 5.57$, Equation (11.28) agrees closely with (11.29) for $f_D T > 0.3$ [24].

The signal detection process in land mobile systems normally uses the median signal value taken over T seconds rather than the mean. Consider their standard deviations. If $T \to \infty$, they converge to zero. For $T \to 0$, we can calculate the standard deviation σ_{mo} of the median. We define X_o as the median when X is Rayleigh distributed. For the transformation $y = 20 \log_{10} X$, and $y_o = 20 \log_{10} X_o$, we get [23]

$$\sigma_{mo} = \sqrt{\langle (y - y_0)^2 \rangle} \qquad (11.33)$$

There is a difference of only 0.04 dB between the variance of the mean and the median. The T-sec median and T second mean agree when $T > 0.5$ seconds.

The above discussion did not include a lognormal distribution of the median signal strength. A logarithmically distributed long term (large area) median is considered to be modulated by a small area median. Let m_o (dB) be the mean of the received signal. If $X_1(t)$ is median superimposed by the modulation, then the composite signal is expressed as

$$X_2(t) = m_0 + X_1(t) \qquad (11.34)$$

where it is assumed that $X_1(t)$ is Gaussian noise with zero mean. i.e. $\langle X_1(t) \rangle = 0$. For a variance σ_s^2 the spectral density $S_{X1}(f)$ is given by:

$$S_{X1}(f) = \frac{\sigma_s^2}{2f_m}; \qquad |f| \le f_m \qquad (11.35)$$

$$= 0; \qquad elsewhere$$

Since the local area median varies much slower than the instantaneous variation i.e. $f_m < f_D$, we can write

$$\Re_{X2}(\tau) = \langle X_2(t) X_2(t+\tau) \rangle = \Re_{X1}(\tau) + m_o^2$$

$$\sigma_{ST}^2 = \frac{\sigma_s^2}{T} \int_{-T/2}^{T/2} \left[1 - \frac{|\tau|}{T}\right] \frac{\sin(2\pi f_m \tau)}{2\pi f_m \tau} d\tau \tag{11.36}$$

where we have used

$$\Re_{X1}(\tau) = \int_{-\infty}^{\infty} S_{X1}(f) e^{j2\pi f \tau} df = \sigma_s^2 \frac{\sin(2\pi f_m \tau)}{2\pi f_m \tau} \tag{11.37}$$

Thus, the measured deviation is a composite of the deviations of instantaneous variation and small area median. The T-second distributions of small area and large area are independent. If the distribution of the T-second median of the instantaneous variation is lognormal, the variance of the measured value σ_T^2 may be written as

$$\sigma_T^2 = \sigma_{MT}^2 + \sigma_{ST}^2 \tag{11.38}$$

Equation (11.38) indicates variance in the measurement of signal strength over a limited time ultimately depends on the sum of the variances due to Raleigh fading and lognormal shadowing. This is of course cumulative effect of fading and shadowing. It may be pointed out that shadowing effect can be virtually eliminated because variance of the area is in general known.

The aim of the vehicle location algorithm is to ensure that each mobile unit handled by a base station is provided with an adequate signal strength. Estimation of signal strength is followed by handover if the signal strength is found to be below reception threshold. Handovers require new channel assignments. Such assignments and number of channel switching by the mobile units are, in general, optimized. For example, if the signal-to-interference ratio at a mobile unit is 30 dB, then there is no point in increasing the S/I further by switching to another base station, which could provide higher signal level. However, a mobile unit whose S/I is less than 15 dB will greatly benefit from an improvement in signal to noise ratio. This should be given a high operational priority. Rapid switching of frequencies may improve the situation but could degrade the performance of the base station controller because in marginal situations it may give rise to a ping-pong effect, which will cause an unnecessary overload of the base station controller. On the other hand, too large a signal estimation (integration) time will require a larger number of estimators at each base station. It may allow the mobile unit to remain in a region of low signal strength for a longer period than necessary.

11.6 Handoff Considerations

The mobility of the user in the service area, divided into clusters and cells, necessitates that calls be handed-off from the serving base station to another base station that is likely to provide better signal strength. The need for a handoff is indicated whenever the averaged signal to noise ratio or signal to interference ratio at the mobile (or the base station) falls below a certain minimum value known as threshold value[†]. The threshold for signal to noise ratio or signal to interference is usually determined by the system's operating specifications, for example in North American analog system (AMPS), the signal threshold is -100 dBm for the noise limited systems and -95 dBm for the interference limited systems. The C/I threshold is chosen to be 18 dB for analog and around 14 dB for digital cellular system. The threshold is chosen so that the number of handoffs during an average call is not too high while the quality of reception is maintained. The average number of handoffs during a typical length of a call depends on the vehicular speed and the cell size. In a cell of fixed radius, the probability of handoff also increases with the call length.

The handoff initiation procedure starts when the parameter determining the handoff is found to be below the threshold level[‡] (e.g. signal strength slightly higher than the signal level that delivers an acceptable reception quality). We consider, network controlled handoff for illustrative purpose. When the signal level or signal to noise or interference ratio measured to be lower than the threshold value, the base station informs the mobile switching station of the situation, which orders the base stations adjoining the currently serving base station to monitor the signal strengths from the mobile. Several algorithms have been used to take a handoff decision. The main objective is to take a quick decision, avoiding unnecessary handoffs while maintaining the link quality. For example, if the decision for the handoff is taken as soon as the signal level drops below threshold, the number of handoff will not only be high but will include handoffs which may not be necessary. A high number of handoffs may also overload the mobile switching office. If the handoff decision is delayed the link quality may be affected and probability call drop may increase.

---

† *Several other handoff approaches are based on parameters such as signal to interference ratio, traffic volume, availability of spare channel etc. However, signal strength is the most widely used.*

‡ *The parameter can be measured by either the base station or the mobile terminal. When the base station measures the parameter, the handoff is known as network controlled handoff (NCHO). When the mobile measures the parameter and network takes the decision, it is known as Mobile Assisted Handoff (MAHO). When the mobile is the sole controller of the handoff process it is known as Mobile Controlled Handoff (MCHO).*

Several parameters may be included in the handoff decision process. These include: signal strength, bit error rate (data and digital speech), signal strength difference between the reports from two adjacent base stations, signal to interference ratio or even signal level crossing rates [36]. More recently, handoff control has been implemented by including intelligence in the form of neural networks or fuzzy logic [40]. The mobile switching station, depending up on the circumstances, may decide to advance, delay, force, or even prevent a handoff. For example, in the event that the current cell is too congested, the mobile switching office may force a handoff although it was not needed. For example, by increasing the difference between the thresholds A and B (see Figure 11.17), the MTSO forces the handoff earlier and by decreasing the difference the handoff is delayed. Several strategies could be used in the implementation of handoff. In [36], handoff mechanism is analyzed based on the level crossing theory. The level crossing at the base station is modeled as Poisson process, from which, an optimal handoff decisions are derived. This technique is shown to reduce the probability of unnecessary handoffs and handoff delays.

The quality of handoff technique is measured by one or more of several possible metrics; number of handoffs, call drop, call blocking probability, delay in executing handoff, and number of unnecessary handoff.

11.6.1 Handoff Based on Measured Signal Strength

The mobile cellular systems are interference limited system, therefore signal strength is a good indicator of the signal quality. Under the measured signal strength based handoff, a mobile is handed over from one base station to another when:

(i) the averaged signal strength from the currently serving base station falls below an absolute threshold, B

(ii) the averaged signal strength of the new base station becomes greater than that of the currently serving base station.

In Figure 11.17, a typical variation of the signal strength is shown. The average signal strength is also shown. At time L, the average signal strength drops below the threshold B, it recovers at M. However, at N, the average signal strength falls below the threshold A. It is seen that it is not necessary to execute handoff at L, whereas a handoff becomes necessary at E.

The signal strength in dB from the two base stations involved in the handoff, A and B, are given by:

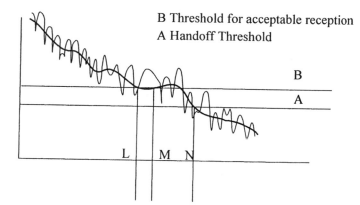

B Threshold for acceptable reception
A Handoff Threshold

Figure 11.17 Handoff Based on Measured Signal Strength

$$s_A(d) = k_1 - k_2 \log d + u(d) \qquad 0 \le d \le D$$
$$s_B(d) = k_1 - k_2 \log(D - d) + v(d) \qquad 0 \le d \le D \tag{11.39}$$

where k_2 is the propagation loss constant (usually 30 to 40) and k_1 is a constant. The terms $u(d)$ and $v(d)$ are zero mean independent Gaussian random processes representing shadowing. The autocorrelation function for shadowing fading is assumed to be exponential [38]

$$E\{u(d_1)u(d_2)\} = E\{u(d_1)u(d_2)\}$$
$$= \sigma_s^2 \exp(-|d_1 - d_2|/d_0) \tag{11.40}$$

where d_0 determines the rate at which the correlation function decays. If $P_{B|A}$ is the probability of handing off from B to A and $P_{A|B}$ from A to B, then the probability that a handoff occurs in the kth interval is given by

$$P_{ho}(k) = P_A(k-1)P_{B|A}(k) + P_B(k-1)P_{A|B}(k) \tag{11.41}$$

where $P_A(k)$ and $P_B(k)$ denote the probabilities of mobile being assigned at d_k to BS A, and to BS B respectively. These probabilities are given by [38]

$$P_A(k) = P_A(k-1)(1 - P_{B|A}(k)) + P_B(k-1)P_{A|B}$$
$$P_B(k) = P_B(k-1)(1 - P_{A|B}(k)) + P_A(k-1)P_{B|A} \tag{11.42}$$

Also,

$$P_{B|A}(k) = P\{B(k)|A(k-1)\}$$
$$P_{A|B}(k) = P\{A(k)|B(k-1)\}$$

(11.43)

A comment on determining the average signal strength is in order. As mentioned in Chapter 2, it is difficult to differentiate between the fast and slow fading. The probabilities given in the above equations are strongly dependent on the averaging interval since the rapidity with which the received signal envelope varies depends on the mobile speed or Doppler frequency. The averaging interval is primarily determined by the correlation distance of the slower (shadow) fading. For slower variations, the averaging interval should be increased and for higher Doppler it should be the decreased. In this regard, estimation of mobile velocity or maximum Doppler becomes necessary. The Rayleigh distributed envelope of the received signal, $x(t)$ is compressed logarithmically as $y_i = 20 \log_{10}(x_i)$. We define V as a measure of the squared deviation of y_i as

$$V = \frac{1}{N} \sum_{i=0}^{N-1} (y_{i+1} - y_i)^2$$

(11.44)

where N is the number of samples and samples are taken every τ seconds. For small $f_D \tau$, the estimate f_D is given by

$$\hat{f}_D = \frac{1}{\pi \tau} \sqrt{\frac{V}{124.1}}$$

(11.45)

In [37] an estimate of Doppler frequency f_D is obtained as

$$\hat{f}_D = \begin{cases} 15.25 \sqrt{V} & V < 3.8 \\ 4.033 V + 19.60 & V \geq 3.8 \end{cases}$$

(11.46)

where τ the time between measurements is equal to 1 msec. This approximation is valid up to $f_D = 100$ Hz. In the case when the received signal envelope is Ricean distributed, the estimate of Doppler is given by

$$\hat{f}_D = \frac{1}{2\pi \tau} \sqrt{\frac{V}{Var(x_i^2)}}$$

(11.47)

Thus, by estimating f_D, one can adaptively change the averaging interval.

11.6.2 Handoff Based on Signal Strength Difference

The problem of handoff is formulated as the one consisting of two base stations separated by a distance D meters, with the mobile terminal moving from BS A to BS B at a constant speed. The signal strengths at a distance d meters from A are given by (11.39), and d is equal to vt, where v is the velocity of the mobile terminal and t is the elapsed time. A handoff is affected when the average signal level from the new base station exceeds that from the currently serving base station by h decibels [36]. The difference between the averaged signal strengths from the two stations is given by

$$s(d) = \bar{s}_A(d) - \bar{s}_A(d) \tag{11.48}$$

where $\bar{s}_A(d)$ and $\bar{s}_A(d)$ are given by

$$\bar{s}_A(d) = \frac{1}{d_{av}} \int_0^d \exp(-x/d_{av}) s_A(d-x) dx$$

$$\bar{s}_B(d) = \frac{1}{d_{av}} \int_0^d \exp(-x/d_{av}) s_B(d-x) dx \tag{11.49}$$

where d_{av} determines the rate of decay of the averaging window.

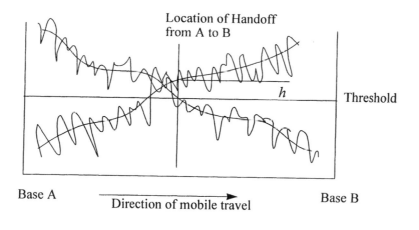

Figure 11.18 Handoff Based on Signal Strength Difference

The handoff decisions are based on the statistics

$$d_k = \sum_{j=1}^{m} (a_{k-j} - b_{b-k}) \tag{11.50}$$

It has been shown in [36] that expected number of handoffs is given by

$$E(nho) = \sum_{k=1}^{N} P(d_k d_{k-1} < 0) \tag{11.51}$$

The covariance, $Cov(d_k, d_{k-1})$ is given by [36]

$$Cov(d_k, d_{k-1}) = \frac{2\sigma_s^2}{(1-\rho)^2} [\rho^{m+2} + \rho^m - \rho^2(1+m) + m - 1] \tag{11.52}$$

where m is the length of averaging window length, ρ is the correlation coefficient, and σ_s is the variance of shadowing.

EXAMPLE E-11.6.1

Consider a situation where base stations are separated by 2km, and spatial sampling interval, d_s, is 1 m. The lognormal shadowing has a standard deviation of 6 dB. The correlation coefficient between the fading at two intervals separated by a distance x is exp(-x/20). Find the expected number of handoff for window lengths of 20 and 70.

<u>Solution</u>

We calculate $\rho = \exp(-1/20) = 0.95123$. The speed of the mobile is not given, so we assume it to be 10 m/sec or 36 km/hr. Thus, the averaging intervals equal to 2 and 7 seconds. The expected number of handoffs are calculated to be 5.02 and 13.64 respectively.

11.6.3 Handoff Decisions Based on Other Criteria

Several other algorithms are used to make decisions on handoffs. In [39] a direction biased handoff algorithm is proposed. The idea in this method is to bias the handoff decision in favor of the base station towards which the mobile is heading and the decision is disfavored towards the base from which the mobile is receding.

Fuzzy logic is also used to implement one of several algorithms [40]. In [41] an adaptive direction biased fuzzy handoff algorithm proposed. In this a new method of selecting best handoff candidate is selected under the constraints of keeping threshold and hysteresis values at nominal values. The goal of using these algorithm is to minimize the number of unnecessary handoffs while maintaining the signal quality prior to and after handoff.

11.6.4 Soft Handoff

Soft handoff is defined as "make before the break" handoff in contrast to the hard handoff which is characterized by "break before make". As in some handoff strategies similar to those used in TDMA systems, during the call, the MS continuously monitors the call quality and searches for cell sites which could be used to improve the call quality. To do so, the mobile listens to the broadcasts from the serving base station on the paging channel and compiles a list of neighboring base station that may be eligible to be the candidates for soft handoff. Most of the search time for neighbors is used up in determination of the PN offsets specified in the neighbor list found on the paging channel[†], i.e., offsets of BS1, BS2, BS3 and BS4 in Figure 11.19.

———	SHO Candidate Signals
– – –	Interfering Signals
———	BS-MTSO wired connections
———	PSTN-MTSO wired connections

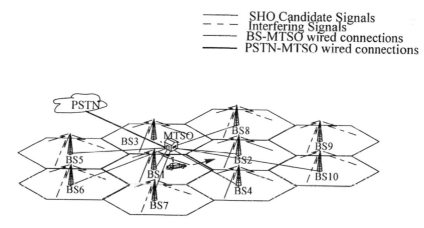

Figure 11.19 CDMA network of 10 Cells with MS moving in a straight line from BS1 towards BS2.

The mobile compares the strengths of the pilots from the neighboring base stations with the T_{add} threshold, which is a design parameter of the system. Whenever the MS identifies a pilot with strength exceeding the T_{add} threshold, it alerts the currently serving base station by sending a pilot strength message on the traffic channel by using either dim and burst or blank and burst signalling. This action initiates a soft handoff (SHO).

---

†. *In IS-95, the base stations are identified by offsets of PN long codes. For more details are found in Chapter 12.*

When the pilot strength message is received by the currently serving BS, it passes this request to the Mobile Switching Center (MSC). If the MSC has available channel cards, it passes the request to the second/third BS to see if a traffic channel is available for the SHO request. A second/third channel card is required at MSC for SHO to allow the MSC to be able to receive signals from 2 or 3 cell cites involved in the SHO. During SHO MS listens to the two/three cells on different codes simultaneously while BSs each listen to the same transmission from the MS. At the BSs the signals are transmitted via the network to the MSC, where a quality decision is made on a frame-by-frame basis. Either MSC selects the better frame or with some additional information could combine them. Thus two/three BSs act like a base antennas diversity system. As MS moves further away from its original BS, the MS could request that the link with the previously serving base station be terminated. This allows for smooth handoffs between cells.

We now consider an example of IS-95 CDMA system with regard to soft handoff. The mobile stations (MSs) are designed to support the SHO process by measuring and reporting to the serving BSs the strength of different received pilot signals, which are used to compute E_c/I_T for all the neighbors. Pilot here is referred to the E_c/I_T of Pilot Channel associated with the Forward CDMA Traffic Channel. The SHO procedure involves maintenance of an Active Set (AS) of BSs based on this received ratios E_c/I_T. The T_{ADD} is used as the threshold for deciding on the membership in the AS and T_{DROP} is the threshold for removing membership from the AS. Another parameter t_{DROP} is used to avoid ping pong effect in handoffs; it is the time that a member must remain below T_{DROP} for a time period at least equal to t_{DROP} before a handoff is executed.

If E_c/I_T of any BS not a member of the AS exceeds the T_{ADD}, the BS will be added to the AS as follows. The set of active base stations for MS i is given by $B_i = \{k|T_i^{(k)} \geq T_{ADD}\}$ where $T_i^{(k)}$, is the E_c/I_T of the received pilot signal from BS k at MS i. The MS i is said to be in SHO when B_i contains more than one BS. Similarly, if the E_c/I_T of a BS belonging to the AS drops below T_{DROP}, the MS starts a drop timer. The MS resets and disables the timer if the E_c/I_T of pilot goes above the T_{DROP} threshold before the timer expires. The BS is removed from the AS if the timer expires. T_{DROP} could be set relative to T_{ADD} or to an absolute value as is done in the case of IS-95.

An AS can have a maximum of 3 BSs at any time. The number of BSs in AS increases, subject to a maximum of 3, as the threshold T_{ADD} is lowered and/or threshold relative T_{DROP} is increased (i.e., absolute T_{DROP} is lowered) and/or t_{DROP} is increased. Consequently the voice quality improves because of the higher available diversity gain. On the other hand the NAUs will be larger thus higher control traffic load.

11.7 Power Control in CDMA

In Chapter 5, the problem of near-far affect was identified as of a major concern. In CDMA, if all mobiles, dispersed especially in a cell, transmit the same power then the signal from the mobile closest to the base is likely to dominate. The signal from a mobile located far from the base station may even be swamped despite processing gain available to it. Thus, the power from the mobile located closer to the base should be lowered while from the mobile located at a greater distance should be increased. The net effect of such power control is to equalize powers received at the base stations from all mobiles. The immediate effect of power control is to maximize the system capacity by minimizing the total interference. In controlling the power, some metric must be defined so that the transmit power is estimated accurately. For example, IS-95 requires power estimation with an accuracy of 1 dB in a dynamic range of 80 dB [47].

Because the error rate is a direct measure of voice quality, the system may be controlled by either measured error rates or signal to noise ratio, E_b/N_T. Thus, in order to assure satisfactory quality, it may not be sufficient to maintain a specific E_b/N_T, but it could be supplemented with bit or frame error rates. In order to provide a satisfactory voice quality for each individual mobile user, power control algorithms may be specified to maintain an "average FER" around the "desired FER".

11.7.1 Up Link Power Control Algorithm

In CDMA, MSs are power controlled to the minimum power that provides acceptable quality for the given conditions. As a result, each MS's signal arrives at approximately equal levels. In this way, the interference from one unit to another is held to a minimum. Two forms of power control are used for the reverse link: Open loop power control, and Closed loop power control.

11.7.1.1 Open Loop Power Control

Open loop power control is based on making the loss in the forward path similar to the loss in the reverse path. Open loop control sets the sum of transmit and receive power to a preset value. A reduction in signal level at the receive antenna will result in a command to increase the transmit. MS's open loop power control will change its transmitter power following a step change in the total received power, ΔP_R. The MS's mean output power will make transition to the final value to oppose the change ΔP_R.

11.7.1.2 Closed Loop Power Control

Closed loop power control is used to allow the power from the MS to deviate from the nominal as set by open loop control. The algorithm as specified by IS-95 standard is as follows:

  (i) Initialize E_b/N_T setpoint to a nominal power, *pnom*. *pnom* defines the nominal level of E_b/N_T at the beginning of the call.

  (ii) BS measures the average E_b on the first 1.25 msec long power control group received in a 20 msec frame, computes an estimation of E_b/N_T and compares this to the setpoint E_b/N_T.

  (iii) If the estimated E_b/N_T is greater than the setpoint, the BS commands the MS to reduce power by 1 dB for the next power group, if not the BS commands MS to increase power by 1 dB. This process, called the inner loop, is repeated for all the power control groups in the frame.

  (iv) At the end of the frame the BS computes a new setpoint E_b/N_T based on whether the last frame had errors or not. This calculation is done every 20 msec and is called the outer loop. The new setpoint is then used by the inner loop for the next frame.

11.7.2 Down Link Power Control Algorithm

The performance of the MS is limited by the level of interference received not only from other users in the same cell but also from the users from neighboring cells. Let MS2 receives a signal S2 from its BS2, and interference from BS1 (I2). The signal to impairment ratio for MS2 is S2/I2. Now, if mobile 1 (MS1) can operate at a lower signal from BS1 and still have acceptable voice quality, then the interference to mobile MS2 will be lower, which improves the voice quality to MS2. *Downlink power control attempts to reduce the BS power to the minimum required for each user served by that BS, and therefore reduces interference to all other users in the same cell as well as in the neighboring cells.* Downlink power control (DPC) adjusts transmit power at the BS for each traffic channel independently. MSs with favorable propagation and interference conditions may require on the average lower power, but mobiles with less favorable propagation and interference conditions may require on the average more power. The DPC algorithm attempts to achieve the desired frame error rate on the signal received at each MS. The algorithm as specified by the IS-95 standard is as follows:

  (i) Initialize MS with power control reporting frames.

  (ii) Initialize traffic channel to nominal power.

   (iii) MS detects downlink frame errors, and reports them to BS.

   (iv) BS adjusts traffic channel power to maintain preset FER objective. BS either increases traffic channel power when error reported is greater than the desired rate or decreases power when error reported is smaller than the desired rate.

   (v) If power after increase is greater than maximum, set power to maximum allowed. If power after decrease is less than minimum, set power to minimum allowed. BS does this step to restricts the power dynamic range so the transmitter power never exceeds a maximum that would cause excessive interference, or falls below the minimum required for adequate voice quality.

11.8 Base station and Subscriber's Equipment[†]

The base station functions as an interface between the wired telephone network and the mobile unit. In doing so, it provides a data communication path between the mobile unit and the mobile telephone exchange. Associated functions include vehicle location and channel control.

11.8.1 Base station Equipment

Base station equipment falls into four categories: radio channel units, interface modules, RF modules, and antenna. Consider first the radio channel unit; its sub-units comprise the control channel unit, the voice channel unit and the receiver location unit. The function of the control channel unit is to provide a method to establish contact with mobiles, to monitor their status and to receive system messages (for instance, call setup and acknowledgment). Data received from the mobiles is decoded and, after any necessary error correction, passed to base station controllers housed in the mobile telephone switching office. The busy-idle bit is also handled by the control channel unit.

The voice channel unit provides voice transmission links between the mobiles and another party. Also monitors the Supervisory Audio Tone (SAT), the Signalling Tone (ST) and received signal strength. As explained in Section 11.4.5, the SAT facilitates frequency reuse; it also indicates to the mobile that its is tuned to the correct channel and the base station. The ST is transmitted on the voice channel for acknowledgment and for other control functions,

---

† In this section, we use examples of AMPS. The specifications on frequency, power, and type of equipments for other systems are discussed in the following chapters.

especially on-hook and off-hook conditions. Upon reception of the ST, the exchange carries out the necessary action. The received signal strength unit determines whether or not a particular unit requires a handoff.

The interface modules provide the data transmission facilities between the base station and the mobile telephone exchange. Components typically include an audio processor, control I/O, a modem and a forward channel controller for data transmission. The audio processing unit provides pre-emphasis filtering in the transmitter; at the receiver it provides limiting, post limiter filtering and de-emphasis[†]. Thus, its functions are very similar to a standard land mobile radio.

The RF module consists of a transmitter module and a base station receiver. A frequency synthesizer and RF power amplifier are the two main components of the transmitter module. The synthesizer serves both as a receiver, local oscillator and as an input to the exciter mixer of the transmitter. The exciter signal is formed by filtering the mixer output. For example, in the North American AMPS system, the base station power amplifier usually delivers a maximum 45 watts over the frequency range 870-890 MHz. Output power is adjustable to allow for variations in cellular systems parameters such as antenna gain, length of antenna cables, and cell size. The base station receiver covers a frequency range of 825-845 MHz; over this range the mobile stations transmit. The base station is similar to a typical mobile radio but uses high stability and low spurious local oscillator radiations in order to reduce interchannel interference. In other systems, similar base equipment is used. Other systems like DAMPS, GSM, PDC are designed for different power outputs and frequencies. Low power versions of GSM are also available in the form of DCS1800 in Europe and rest of the world and DCS1900 in North America. These systems operate in 1800 and 1900 MHz respectively.

Several antenna configurations are used in cellular systems. These include tower top mounting omni-directional antennas, tower side mounting omni-directional antennas, and directional antennas. The actual choice depends on the type of coverage required. The tower top omni-directional antenna has several half wavelength dipoles installed on each of its sides. The signals from several transmitters are passed through four port hybrids which can feed several transmitting signals to one antenna without causing mutual interference. The tower side mounted omni-directional antenna makes use of half wave length antennas and vertical parabolic reflectors. These antennas are mounted on either side of the tower and are fed by the transmitter in a manner identical to that of the tower top omni-directional antennas. Modern cellular systems normally use antenna configurations which provide directional cover-

---

†. *Emphasis and de-emphasis are required in FM transmission and reception. In digital voice communications, these components are usually not needed.*

age. These may be either 60° or 120° radiation arcs in the horizontal plane. Dipoles with vertical reflectors are used to provide controlled coverage. Details on base station antennas are given in Chapter 6.

11.8.2 Mobile Unit Equipment

The mobile unit consists of a transceiver and control unit. Transceiver components include the receiver, duplexer, synthesizer, exciter and power amplifier. For example, in AMPS, the mobile terminal transmitter operating in 800 MHz band covers a frequency range 825-845 MHz. The power amplifier delivers between 7.5 and 12 watts to the antenna system in vehicle based equipment and up to 4 Watts for portable equipment. The power levels in the small cells higher frequency cellular terminal transmit even lower power. Usually an isolator is included in the path to reduce transmitter intermodulation products. The transmitter has a built-in capability of decreasing output power in several steps as necessary so that interference problems are reduced. The duplexer, designed to minimize transmitter noise at the receiver frequencies, allows both transmitter and receiver to be connected to the same antenna. The duplexer combiner is usually made of two quarter wave microstrip transmission lines printed directly on the main circuit board. The receiver is similar to that used in the base station, except the frequency range is 870-890 MHz. The duplexer is not required in the TDMA systems.

The mobile unit controller is based on a stored program microprocessor which coordinates a synthesizer, control head, audio processor, and transmitter power level. The frequency synthesizer, acting as a local oscillator for the transceiver, can be tuned to over 800 carriers. The synthesizer usually has a basic master clock and a phase locked loop to minimize spurious emissions.

Depending on the noise environment, the user can select an audio volume level which is controlled by the audio processing unit. The supervisory audio tone (SAT) circuits consist of a bandpass filter, phase locked loop (PLL), low pass filter and phase adjustment circuits. The PLL locks to the received SAT and passes the signals to the RF modem whose output contains the frequency information.

The mobile control unit consists of a handset. This is a simple terminal, consisting of a keypad for data entry, a display, a hook switch, a microphone and pre-amplifier, an ear piece, a clock reference and a microprocessor.

The mobile antenna is either a half wave dipole (sleeve type), or a whip type which is mounted on the roof or the side of the vehicle. It is designed to have an omni-directional radiation/reception pattern. As explained in chapter 6, however, the antenna patterns may be distorted due to their proximity to conducting surfaces.

11.9 Network Aspects of Cellular Systems

The cellular mobile telecommunications systems are designed to provides telecommunications services to mobile terminals. This aspect is one of the most glaring distinguishing feature from the fixed telecommunication systems. Mobile wireless communication systems have developed to a stage that there is not much difference between the two systems as far as services are concerned. The objective of interconnecting terminals, one of which is essentially mobile, is achieved by passing information bearing signals over radio part of the network to a base station, mobile switching system, and public switched telephone network (PSTN). The flow of information through PSTN is supervised by several protocols. Because of the added facility of mobility in the mobile network, additional network features such as handovers between switching centers, mobility management of terminals are required. The operation of wireless systems requires more complex network supervision and control. In modern wireless systems, many control modules operate in harmony for smooth operation of the network. Essentially, these modules must be interconnected and provisions should be made so that they could communicate between them. This requires a certain network architecture. The simplest version of such architecture is shown in Figure 11.20.

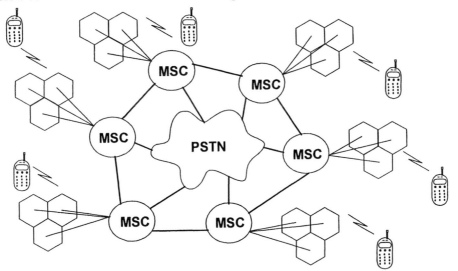

Figure 11.20 A simple example of signalling in cellular systems

In the figure, a number of mobile switching centers are connected to PSTN. Each MSC serves a number of cells which provide telecommunication services to mobile subscribers. Two types of supervisory data flows in the network. From mobile to base stations in the cells, access signalling is used, for which air interface is used. The supervisory data from MSC to PSTN is called network signalling. In some systems the link between base stations and MSC's either access or network signalling is used.

For network signalling a separate physical or logical channel is set aside. This channel carries signalling traffic for use by several bearer channels. This is why this channel is also called common channel and the protocol that this channel uses is called common channel signalling protocol. The use of separate signalling channel facilitates the use of enhanced features and supplementary services. It reduces the time to setup a call.

Signalling System #7 (SS7 in North America) is a large set of common channel signalling protocols defined by ITU-T. The basis of SS#7 is packet switching technology. It carries signalling information from an origination point to the destination through multiple switching nodes. The SS#7 messages contain addressing and control information which is used to perform many functions including but not limited to selection of routing of the signalling information through the network, performing management functions, establishing and maintaining calls, and invoking transactions initiated by a query or command and generating the response to that query or command. GSM (MAP) or (ANSI-41) signalling protocol provides operations of this type using the SS#7 (or SS7) signalling system. To define the functionality of signalling protocols and system, a network reference model is defined. A reference model consists of network entities that are connected by interfaces. For example, in a typical network, signalling points, signalling transfer points, and service control points are defined.

A wireless telecommunication network reference model has five basic network entities. These are radio systems, switching systems, location registers, processing centers, and representations of external networks.

The radio system consists of antenna systems, radio transceivers, and radio transceiver controllers. The switching systems transfer information from one circuit to another in the network. The switching systems consist of the transmission facilities and computing platforms that control the switch circuits to connect call between users. The location registers are data base systems that control subscriber services and contain records and stored information related to wireless subscribers. The network functional entities query these data bases to obtain current status, location, and other information to support calls to and from mobile users.

The processing centers are peripheral network computing platforms that provide services to enhance the capabilities of the network. These include authentication of subscribers. The representations of external networks are integral elements in wireless telecommunications network models. These represent interconnection between the wireless network and the PSTN or other networks. More detail on authentication, security, call processing and other functionalities is found in the following chapter where specific wireless communication systems are described.

11.10   Software Considerations

The mobile cellular systems and networks have two major components - hardware and software. The hardware part of cellular system has relied on miniaturization of electronic components using highly sophisticated VLSI implementations. The VLSI circuit layout and use of ASIC has become automated with the consequence that speed of these circuits is doubling every 18 months while the cost of producing these circuits is decreasing. The second component of software is now the major component of total system cost; it can cost up to 70% of total system development expenditure. These costs can be attributed to the design, coding, testing and documentation of software. System software is heart of the mobile cellular systems. The software programs are essential in providing system parameter monitoring, diagnostics, control and signal processing functions for maintaining network integrity in difficult signal environments. More specifically, software modules are created to handle the following functions:

(i)   setting up and clearing calls;

(ii)   routing with flow control;

(iii)   interconnection of base stations within the mobile
radio network;

(iv)   interconnection with the public switched network
(gateway functions);

(v)   call charging and recording for each subscriber;

(vi)   searching for a mobile subscriber (vehicle location)
over the entire network;

(vii)   call handoff when the mobile subscriber moves
from one cell to another;

(viii)   acquisition of diagnostic information concerning
the operation and maintenance of the exchange;

(ix)   acquisition of other information, including statistical
data, needed for administrative purposes.

The software modules are designed keeping in mind efficiency, clarity of task allocation, and high reliability as the foremost considerations. Because most of the functions above are carried out in real time, efficient algorithms must be accompanied by high speed processing. The software is generally distributed among the several control units, and each module is dedicated to a specific task. Management and maintenance of the software is facilitated by including automated debugging within the system structure. For example, in a digital cellular system the modular software structure has the following features [50]:

(i) task oriented subsystem which performs functions such as call processing, call charge recording, etc.;

(ii) program modules (the so called finite state machine) which perform special processing functions of the signalling subsystem for the CCITT R2 standard;

(iii) a standard message interface between program modules;

(iv) virtual machine concept; the system functions are programmed into hierarchical levels.

The advantage of this strategy is that if a function is programmed at a specific level, functions that have already been implemented in a lower level can be used as a basis.

11.11 Summary

The advent of cellular mobile communications represented a major step forward towards a class of telecommunications that fulfills the need for communications on the move. Indeed, it can be seen as the first step in the development of modern, widely available mobile and portable personal communication systems. Its inherent advantages include very efficient spectral utilization and a high degree of integration with the existing telephone network. Moreover, modern circuit techniques make it highly reliable and moderate in cost.

In this chapter emphasis has been on the design principles of wireless telecommunication systems. The chapter discussed principles of designing frequency reuse, channel assignments, vehicle location, handoff, power control, and network architectures. Topics on base station equipment and the architecture of mobile terminals and how these are interconnected through a network were discussed. Control signalling over the network and role of software in the wireless communication systems were also discussed.

Problems

Problem 11.1

To meet the expected demand of traffic, cell sizes differ from area to area; larger cells are used in less populated area while in urban areas these cells are split into smaller cells. What are the practical problems in splitting nominal size cells of 15 km? What is the practical limit to the cell size?

Problem 11.2

(a) Verify equations (11.11) to (11.19).

(b) To accommodate larger number of users in urban areas, the frequency reuse distance is decreased. However, to meet the cochannel interference criterion, the reuse distance must be increased. How would you reconcile these contradictory requirements?

Problem 11.3

In derivation of (11.19), we considered mobiles in only six co-channel cells of first tier. This model may be considered to be an approximation to a more realistic model where interference due to users in other tiers also contributes. Derive an expression for the probability of outage when interference accruing from outer tiers is also considered.

Problem 11.4

(a) Write a program using MATLAB to simulate fixed, dynamic and hybrid channel assignments and compare their results.

(b) Give intuitive reasoning why hybrid channel assignment performs better than the fixed channel assignment at lower offered traffic but not at higher offered traffic.

Problem 11.5

Compare advantages and disadvantages of centralized and distributed control architecture of cellular system. On the basis of this comparison which system would you recommend to your client?

Problem 11.6

(a) In vehicle location problem, a trade-off between accurately locating mobiles and the number of signal measurement receivers is used. In a single cell system, establish a criterion by which location accuracy and number of averaging receivers can be optimized.

(b) The variance of T-second averaging of the received signal is given by (11.38). If the averaging interval is selected to be 50 msec, find the accuracy in the measured distance to a mobile when the variance of shadowing in a given area is in error by 1.5 dB.

Problem 11.7

Why power control in DS-CDMA system requires such a high accuracy of 1 dB in 80 dB dynamic range? If the power adjustment error is less than 0.75 dB, what impact it will have on the system capacity?

Problem 11.8

In a certain cellular system, outage related to bit error is used to initiate handoff. The outage related to bit error is defined as the probability that the error rate exceeds a certain performance threshold. Consider a system that uses DPSK, and the maximum error that can be tolerated is 10^{-3}. If the tolerable outage rate is 10^{-2} relative to the performance threshold of 10^{-3}, find the average received signal to noise ratio.

Problem 11.9

Repeat Problem 11.8, when BPSK modulation is used.

Problem 11.10

Repeat Problem 11.8, when QPSK modulation is used.

References

[1] W.C. Jakes, Jr. *Microwave Mobile Communications*, John Wiley & Sons, 1975.

[2] V. H. MacDonald, "The cellular concept," *Bell System Technical Journal*, vol. 58, pp. 15-42, January 1979.

[3] L.G. Anderson, "A simulation study of some dynamic channel assignment algorithms in a high capacity mobile telecommunication systems," *IEEE Trans. on Comm.*, vol. COM-21, pp. 1294-1301, 1973.

[4] A. Gamst, "Some lower bounds for a class of frequency assignment problems," *IEEE Trans. on Veh. Tech.*, vol. VT-35, pp. 8-14, February 1986.

[5] D.C. Cox and D.O. Reudink, "Increasing channel occupancy in large scale mobile radio systems: dynamic channel reassignment," *IEEE Trans. on Comm.*, vol. COM-21, pp. 1302-1306.

[6] F. Box, "A Heuristic technique for assigning frequencies to mobile radio nets," *IEEE Trans. on Veh. Tech.*,vol. VT-27, pp. 57-64, May 1978.

[7] A Gamst, "Homogeneous distribution of frequencies in a regular hexagonal cell system," *IEEE Trans. on Veh. Tech.*, vol. VT-31, pp. 132-144, August 1982.

[8] T.J. Kahwa and N.D. Georganas, "A hybrid channel assignment scheme in large scale cellular structured mobile communication systems," *IEEE Trans. on Comm.*, vol. COM-26, pp. 432-438, April 1978.

[9] J. Williams, "Capacity dynamics in cellular mobile telephony systems," *Telecommunications*, vol. 17, pp.32, 1983

[10] J.S. Engel and M.M. Peritsky, "Statistically-optimum dynamic server assignment in systems with interfering servers," *IEEE Trans. on Comm.*, vol. COM-21, pp 1287-1293, November 1973.

[11] S.M. Elnoubi, R. Singh and S.C. Gupta, "A new frequency channel assignment algorithm in high capacity mobile communication systems," *IEEE Trans. on Veh. Tech.*, vol. VT-31, pp. 125-131, August 1982.

[12] G.K. Chan and S.A. Mahmoud, "A spectrum-efficient interference-free frequency allocation scheme for a cellular radio system," *IEEE Trans. on Veh. Tech.*, vol. VT-35, pp. 15-21, February 1986.

[13] R.C. French, "The Effect of fading and shadowing on channel reuse in mobile radio," *IEEE Trans. on Veh. Tech.*, VT-28, pp. 171-181, August 1979.

[14] G. Nehme and N.D. Georganas, "A simulation study of high capacity cellular land mobile radio communication systems," *Can. Elec. Eng. Journal*, vol. 7, No. 1, pp. 36-39, 1982

[15] E.C. Molina, "Application of the theory of probability to telephone trunking problems," *Bell System Technical Journal*, vol. 6, pp. 461-494, 1927.

[16] A.K. Erlang, "Solution of some problems in the theory of probabilities of significance in automatic telephone exchanges," *Post Office Electrical Engineers' Journal* vol. X, pp. 189-97, Jan. 1918.

[17] C.J. Hughes and M.S. Appleby, "Definition of a cellular mobile radio system," *IEE Proceedings*, vol. 132, Pt.F, No. 5, pp.416-424, 1985.

[18] Z.C. Fluhr and P.T. Porter, "Control architecture," *Bell System Technical Journal*, vol. 58, No. 1, pp. 43-70, 1979.

[19] H.M. Hafez, Sheikh, A.U.H. and B. McPhail, "Experimental evaluation of packet error rates for the 850 mhz mobile channel," *IEE Proceedings Pt.F, Special Issue on Mobile Radio*, August 1985, pp. 366-374.

[20] G.L. Turin, W.S. Jewell and T. T.L. Johnston, "Simulation of urban vehicle-monitoring system," *IEEE Trans. Veh. Tech.*, vol. VT-21, February, 1972.

[21] J.S. Engel, "The effects of Multipath transmission on the measured propagation delay of an FM signal," *IEEE trans. on Veh. Tech.*, vol. VT-18, pp. 44-52, May 1969.

[22] A. Papoulis, *Probability, Random Variables, and Stochastic Processes*, McGraw Hill, 1965.

[23] M. Hata and T. Nagastu, "Mobile location using signal strength measurements in a cellular system," *IEEE Trans. Veh. Tech.*, vol. VT-29, pp. 245-251, 1980.

[24] G.D. Ott, "Vehicle location in cellular mobile radio systems," *IEEE Trans. on Veh. Tech.*, vol. VT-26, pp. 43-46, February 1977.

[25] W.G. Figeel, N. H. Shepherd and W. Trammel, "Vehicle location by a signal attenuation method," *IEEE Trans. on Veh. Tech.*, vol. VT-18, pp. 105-109, November 1969.

[26] S. Rhee, "Vehicle location in angular sectors based on signal strength," *IEEE Trans. on Veh. Tech.*, vol. VT-27, pp. 244-258, November 1978.

[27] H. Staras and S. N. Honickman, "The accuracy of vehicle location by trilateration in a dense urban environment," *IEEE Trans. Veh. Tech.*, vol. VT-21, pp. 38-43, August 1973.

[28] S. Riter, "The effect of background noise on phase ranging measurements in urban vehicle monitoring systems," *IEEE Trans. on Veh. Tech.*, vol. VT-22, pp. 81-85, 1973.

[29] N Ehrlich, "Advanced mobile telephone service using cellular technology," *Microwave Journal*, vol. 26, pp. 119, August 1983.

[30] W.R. Young, "Advanced mobile telephone service: introduction, background, and objectives," *Bell System Technical Journal*, vol. 58, pp. 1-14, January 1979.

[31] K. Araki, "Fundamental problems of nationwide mobile radio telephone system," *Rev. of the Electrical Communication Laboratory*, vol. 16, pp. 357-373, May-June 1968.

[32] H.J. Schulter, Jr. and W.A. Cornell, "Multi-area mobile telephone service," *IRE Trans. on Comm. Systems*, vol. CS-9, pp. 49, 1960.

[33] R.H. Frenkiel, "A High capacity mobile radiotelephone system model using a coordinated small zone approach," *IEEE Trans. on Veh. Tech.*, vol. VT-19, pp. 173-1977, May 1970.

[34] P.T. Porter, "Supervision and control features of a small zone radio telephone system," *IEEE Trans. on Veh. Tech.*, vol. VT-20, pp. 75-79, August 1971.

[35] N. Yoshikawa and T. Nomura, "On the design of a small zone land mobile radio system in UHF band," *IEEE Trans. on Veh. Tech.*, vol. VT-25, pp. 57-67, 1976.

[36] R. Vijayan and J. M. Holtzman, "A model for analyzing handoff algorithms," *IEEE Trans. on Veh. Tech.*, vol. 42, no.3, Aug. 1993, pp. 351-356.

[37] J. M. Holtzman and A. Sampath, "Adaptive Averaging methodology for handoffs in cellular systems," *IEEE Trans. on Veh. Tech.*, vol. 44, no. 1, pp. 59-66, February 1995.

[38] N. Zhang and J. M. Holtzman, "Analysis of handoff algorithms using both absolute and relative measurements," *IEEE Trans. on Veh. Tech.*, vol. 45, no. 1, pp. 174-179, February 1996.

[39] M. D. Austin and G. L. Stüber, "Directed biased handoff algorithm for urban microcells," *IEEE-VTC Record*, pp. 101-105, 1994.

[40] A. U. Sheikh and C.H. Mlonja, "Performance of Fuzzy Algorithm based Handover Process for Personal Communication Systems," *Proc. IEEE International Conference on Personal Communications*, pp. 153-157, New Delhi, India, Feb. 1996.

[41] N. D. Tripathi, J. H. Reed, and H. F. VanLandingham, "Adaptive direction biased fuzzy handoff algorithm with unified handoff candidate selection criterion," *IEEE-VTC Record*, pp. 127-131, 1998.

[42] Ning Zhang and Jack M. Holzman, "Analysis of a CDMA soft handoff algorithm", *IEEE VTC Record*, pp. 819-823, 1996.

[43] P. Seite, "Soft handoff in a DS-CDMA microcellular network", *IEEE VTC Record*, pp. 530-534, 1994.

[44] S. L. Su and J. Y. Chen, "Performance analysis of soft handoff in cellular networks", *International Symposium on Personal, Indoor, and Mobile Radio Communications*, pp. 587-591, 1995.

[45] S. W. Wang and I. Wang, "Effects of soft handoff, frequency reuse and non-ideal antenna sectorization on CDMA system capacity", *Proc. IEEE VTC Record*, pp. 850-854, 1993.

[46] Suwon Park, Ho Shin and Dan Keun Sung, "Modelling and analysis of CDMA soft handoff," *IEEE VTC Record*, pp. 1525-1529, 1996.

[47] K. S. Gilhousen, I. M. Jacobs, R. Padvani, A. J. Viterbi, L. A. Weaver, Jr. and C.E. Wheatley III, "On the capacity of a cellular CDMA system," *IEEE Trans. on Veh. Tech.*, vol. VT-40, pp.303-312, 1991.

[48] Ching Yao Huang and Roy D. Yates, "Call Admission In Power Controlled CDMA systems", *IEEE VTC Record*, pp. 1665-1669, 1996.

[49] S. Grandhi, et. al., "Centralized power control in cellular radio systems", *IEEE Trans. on Vehicular Tech.*, vol. 42, no. 4, November 1993.

[50] T. Burian, "Switching philosophy of the digital cellular radio telephone system CD900," *Nordic Seminar on Digital Land Mobile Radio Communication*, pp. 271-278, Feb.1987.

Chapter 12

MOBILE WIRELESS CELLULAR SYSTEMS

The cellular concept for mobile communications is not new; the idea of frequency reuse and during the call handoff[†] when mobiles cross cell boundaries was proposed by D.H. Ring of Bell Laboratories in 1947. Between 1947 and 1974, the year when 40 MHz spectrum was allocated for trial systems, FCC received several dockets requesting designation of 40 to 75 MHz bandwidth for mobile cellular systems. In 1977, Illinois Bell was authorized to develop an advanced mobile system based on cellular concept. The trials of the prototype system took place in Chicago area in 1978. Simultaneously, American Radio Telephone Service (ARTS) was authorized to operate a cellular system in the Washington-Baltimore area on an evaluation basis. The success of field trials resulted in the development of first generation wireless systems known as AMPS in the United States and NMT in the Nordic countries. The elements of a typical cellular system and considerations to design cellular systems were described in detail in Chapter 11. In this chapter, details on several cellular systems including AMPS, NMT, and DS-CDMA (IS-95) are presented.

12.1 First Generation Cellular System

The analog cellular systems were actually designed in 1970s but were not implemented until 1981 because designation of spectrum for these systems was not confirmed till 1979 [1]. During 1981 and 1986, several cellular systems were designed and implemented in Europe and North America. These were: NMT in Nordic countries, AMPS and Aurora in North America, TACS in the UK, C450 in Germany, Radiocom in France, and NEC in Japan.

Though the first generation system consumed considerable investment, its outdated technology and insufficient system capacity could not meet the rapidly growing demand, therefore these systems were rather short lived. A replacement of the first generation systems was quickly sought and these systems were replaced by the second generation digital cellular systems at the

---

†. *Call handoff is used in North American Systems; other systems use call handover.*

beginning of 1990. The second generation digital systems offered higher capacity due to advancements in two key areas of technology.

The first is that of low rate speech encoding where it became possible to encode speech at rates lower than 8 kb/s, which reproduced the original speech with an excellent quality. The low rate speech encoding decreased the channel bandwidth requirement significantly. The other advancement was in the area of VLSI design. The reduction in size of the mobile terminal requiring complex signal processing was an important step towards revolutionizing mobile communications. The advancement in digital signal processing techniques made the implementation of elaborate slot time management possible where random delays created by dispersed mobile terminals over the cell area could be compensated. TDMA, multiple access also dramatically reduced the size and cost of the base station with the result that it became cheaper to install many more base stations in the coverage area. This helped in reducing the cell size, increasing frequency reuse, and increasing the system capacity. Furthermore, advancements in techniques to reduce interference were very helpful in reducing the cluster size.

Out of many second generation systems - CT-2 and CT-2 Plus, DCT-900/DECT/CT-3, GSM, DAMPS (IS-54), DS-CDMA (IS-95), PDC and PHS, only GSM, IS-54, PDC, PHS and IS-95 become the major second generation cellular systems of the world. The most recent survey indicates that GSM is far in front of other systems in terms of number of systems installed and the subscriber base.

Some systems like PCS, NAMPS, TDMA, ETDMA, CDMA, DCS-1800 may be classified as the Second Generation Plus systems. These systems are the result of some modifications to the second generation systems after these were admitted as standards.

In the following, we introduce the first generation systems. In this regard, we shall give details on Nordic and AMPS systems. The description of several second generation systems, included in these are D-AMPS (IS-54 and IS-136), GSM and DS-CDMA (IS-95) will follow.

12.2 The North American AMPS System

The AMPS system for mobile users requires complex control elements that are distributed throughout the coverage area. A mobile cellular system essentially requires three control elements: the mobile unit, the cell site, and a switching office as described in Chapter 11. The mobile terminal is connected to the base station via a standardized air interface which requires advanced techniques to overcome channel impairments, to supervise the link, and to execute handover when need arises. The base station forms an interface between the mobiles and the systems control unit (mobile switching station) and the network. Its functions include measurement of signal quality and submitting reports to the mobile switching station. We discussed the key functions of these elements in Section 11.3 through Section 11.6. In the following we describe in some detail, how these elements are implemented in AMPS.

12.2.1 System Control Elements

The AMPS cellular system consists of three principal control elements - mobile unit, base station and mobile control station. These are interconnected by well defined interface elements whose functions are to provide compatibility in terms of signalling waveforms, timing tolerances and many other functions that identify fault conditions. The basic requirement is to avoid circuit lockouts, call drops or failures, or excessively long periods in releasing channels.

A mobile unit consists of three major parts: a transceiver, a control unit and an antenna. The transceiver consists of a duplexer, a receiver, a frequency synthesizer and a logic unit. The transceiver with its logic unit is mostly realized with VLSI circuits to reduce size and weight. The frequency synthesizer is designed to tune the transmitter and receiver to the desired frequencies and to minimize spurious emission and reception. The logic unit, consisting of a microprocessor, memories (ROM and RAM) and peripheral circuits, performs various path set up procedures. The control unit is the telephone set itself. It consists of a push button keyboard and other necessary circuits to store the dialled number, setup a radio speech path, and generate various signalling tones.

The base station, consisting of a group of transceivers and a switching system, provides a link between mobile units with the land based network. The transceivers are used for both relaying voice/data signals and for the operation of control channels (paging and access). The base station also has a monitoring receiver dedicated to the measurement of received signal levels from active mobiles on specified radio channels for the purpose of call quality monitoring and handover.

The mobile telecommunication switching office (MTSO) or mobile switching centre (MSC) serves as a coordinating unit between the fixed and the mobile parts of the network. Its functions include administration of radio channels, coordination of links between cell sites and moving subscribers, and maintenance of system integrity.

12.2.2 Operation of Cellular System

When a mobile telephone is switched on (or energized), it starts scanning the setup channels predetermined by the service provider. The program to scan channels is stored in its memory and it selects the channel with the strongest signal. The mobile unit continues to monitor the forward paging channel periodically and keeps an update on the identity of the strongest setup channel. In describing the cellular operation, two types of call will be considered: mobile completed calls and mobile originated calls.

Let us consider first the operations in the system that take place in order to deal with a mobile completed call. The call from the calling party[†] is routed through the wireline network to the home base of the called unit. If the mobile is known to be in the home coverage area, the mobile switching office transmits the message over the paging channel covering the whole of the service area, otherwise the call is transferred to the base station in which the mobile was last known to be present.

Since the mobile is always in the process of scanning the strongest of the paging channels for messages, it recognizing its paging call, selects the strongest of the access channels (reverse control channel) and replies to the selected base station, which redirects the reply to the switching office. The switching office selects an idle voice channel pair (one forward and one reverse traffic channel) and informs the base of the selection. This information is then paged to the mobile unit. The mobile tunes to these channels and uses the allocated transmit channel to acknowledge the assignment of channels by transponding the audio supervisory tone received from the base station. The cell site recognizes the SAT and places the wire line trunk in the off hook condition. The mobile switching office interprets the action as the establishment of a successful communication link.

On command from the switching office, the base station sends an alerting message on the assigned voice channel, which signals the subscriber that there is an incoming call. This is the first time that the subscriber knows the presence of an incoming call. A signalling tone is then sent by the mobile unit which prompts the switching office to place the line off-hook and provides a ringing tone to the calling party. When the mobile customer answers the call, the mobile originated signalling tone is removed, and conversation takes place. The call sequence is shown in Figure 12.1.

In the second scenario, the mobile unit originates the call by entering the called party's number without lifting the handset. This is known as pre-dialing. This places the mobile unit in the off hook condition, and a procedure similar to that described above for the mobile completed call is initiated. The stored digits (called party's number) along with the mobile's identification are transmitted over the mobile selected strongest access channel. As in the previous case, the switching office designates a channel pair and establishes voice

---

†. *We consider here that the call is originated from a fixed telephone network.*

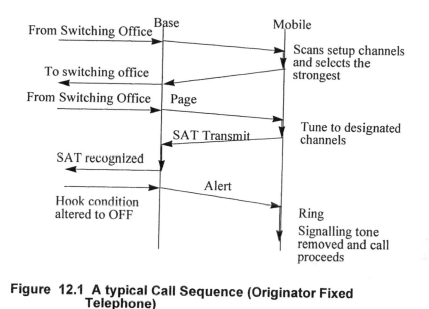

Figure 12.1 A typical Call Sequence (Originator Fixed Telephone)

communication with the mobile unit. In the meantime, the switching office determines the necessary routing and the charge information by analyzing the dialled digits. The switching office sends the dialled digits over the wire line network to the called party; when the latter answers, conversation follows.

From time to time, a mobile terminal may cross the cell boundary and from a serving cell enters in a new cell. Since each cell is allocated a different set of channels, the mobile must re-establish a link with the base station in the new cell. This phenomenon is called handover. In theory, handover occurs when a cell boundary is crossed. In practice, the boundary is defined by signal strength rather than geographical position. When the signal strength from the serving base station (or from the mobile at the serving base station) is found to go below a preset threshold or an acceptable level, the neighboring base stations are directed to measure the signal strength from the mobile in question and locate its position. The measured signal strength data is sent to the switching office, which uses this information and selects a base station[†] to which the mobile is to be handed over. A radio channel pair and wire line channels (from the new base station to the mobile station and mobile switching office respectively) are earmarked in order to set up a new link. The switching office sends a message containing new channel and new SAT information to the currently serving base station. The latter relays this message to

---

†. *The selected base station receives the mobile signal the loudest.*

the mobile unit over the voice channel currently in use. On receipt of this, the mobile unit sends a brief burst of signalling tone and turns off its transmitter before it tunes to the newly designated channel and transponds digitally encoded SAT on this channel. Upon receiving the SAT, the new base station sends an off hook signal to indicate completion of the link. The serving base station also receives the signalling tone and sends an on hook signal. The switching office reconfigures its network, connecting the party with the appropriate wire line. The on-hook and off-hook messages are interpreted by the switching office as indicators of a successful handoff.

Call disconnection can be initiated either by the mobile or the wire line portion of the system. When the mobile unit initiates the disconnect procedure, it transmits a signalling tone to its base station; an on hook signal is then placed on the appropriate wire line trunk and the switching office places the line in an idle mode. It then directs the serving base station to shut down the transmitter associated with the call.

The first stage of a system initiated disconnect occurs when all resources of the switching office committed to the call are forsaken. The switching office then transmits a release order to the serving base station. This order is relayed to the mobile unit on the voice channel. The mobile unit confirms receipt of these orders by invoking the release sequence associated with a mobile initiated disconnect procedure. Upon receipt of this sequence, the switching office orders the cell to shut down the radio transmitter.

12.2.3 Control Functions

The previous examples on the call initiation and termination show that elaborate control is needed in order to maintain reliable communication link. Thus, the network has to perform many control functions; of these, acquisition of network information, paging and access, call supervision, sending call control sequences and handoff are the main functions. As in the fixed networks, supervision is defined as detection of user initiated changes in the switch hook states e.g. lifting and replacing the hand set. Since wired connections between mobiles and base stations does not exists, monitoring of received signal provides a means of call supervision. Compared to the wireline networks, mobile systems perform an additional responsibility of measuring the signal strength from the mobiles.

12.2.4 Paging And Access Channels

In cellular systems, separate channels for paging and access are set aside. Paging channels (in the forward direction) are used by the base stations to transmit forward link control messages that includes informing the mobile unit of an incoming call and sending the mobile system parameters. Access channels (in the reverse direction) are used by the mobile unit to get an access to the network for placing a call through a base station. The access message consisting of the mobile unit identification and the dialed subscriber number is sent to the mobile switching office which assigns a channel pair so that the call can proceed.

The number of access channels depends on the number of users, the average number of calls generated per user per busy hour, the need to avoid interference, and the real time processing power available at the base station. The number of paging channels are designated on the basis of number of users in the system and the length of system paging message. The length of paging message is determined by messages consisting of ten digit telephone numbers. Allowing the users to move freely amongst cells without degrading access to base stations is also taken into account while deciding the number of channels needed. Some allowance is also made for future expansion. The transmission rate over a single paging channel is limited to approximately 10,000 bits/sec. High redundancy is required for the sake of reliability, therefore the net channel throughput is about 1200 bits/sec. At 300 bits per message, this is equivalent to 25 messages per second or 90,000 messages per hour. If on the average one message per busy hour/subscriber is needed, a paging channel could support up to 90,000 customers. At present this limit has been exceeded in several major cities in North America and Japan and has caused severe congestion.

In addition to the initiation of calls, mobile units can use access channels to answer a page from the base station. The base station uses the forward access (paging) channels to send channel designation messages to the mobile unit, which then tunes to the assigned voice channel pair (in forward and reverse direction) so that the call can proceed. Since call initiation is a random event, two or more mobile units may initiate a call at the same time, resulting in collisions. Part of the access control function is to avoid such collisions; to this end several measures have been taken in the mobile systems. For example, in AMPS, first the paging channel used by a base station contains information on the status of the access channel (busy or idle). To do this, the base station uses every eleventh bit in the forward frame to send the status of the reverse channel. This information is scanned by mobiles. Second, the mobile unit sends a digitally encoded precursor indicating which base station it is addressing. Third, following the transmission of the precursor, the mobile unit monitors the forward channel to see whether the channel status changes from idle to busy. When such a change occurs, the mobile terminal assumes that it has been asked to use the access channel and it can transmit message safely on the access channel. Despite these precautions against collisions, a collision with another mobile is still possible. To avoid periodicity of repeated attempts to obtain access, the mobile unit waits for a random time interval before attempting to access the system again.

12.2.5 Supervision

In AMPS system, a call is supervised by using a combination of tone burst and continuous out-of-(voice) band modulation. These are known as the signalling tone (ST) and supervisory audio tone (SAT) respectively. These tones, in addition to detecting changes in the switch hook state caused by the customer, ensure that subscribers receive adequate signal strength.

In AMPS, three SATs at 5930, 6000 and 6030 are set aside and only one of these three is used at any base station. The concept of SAT is to detect continuity in the radio link between the mobile and the base station (or MTSO). As shown in Figure 12.2, this increases the reuse distance of a particular SAT by a factor $\sqrt{3}$, thereby enhancing the reliability of the supervisory functions. A SAT transmitted by a base and received back after being transponded by a mobile indicates completion of the loop with that mobile. The supervisory audio tone transponded by the mobile unit is also used to measure the voice channel signal strength and approximate range of the mobile.

The SAT frequencies are selected close to each other so that only one phase locked loop is needed to keep the receiver locked to the strongest SAT. The SAT is FM modulated with a deviation of ±2 kHz. The SAT normally exists all the time except when wideband control signalling data at 10 kb/sec is being transmitted on the reverse channel. When present, it coexists with 10 kHz signalling tone (ST). When the mobile experiences fading or it goes out of coverage area, a valid SAT color code may not be received; when not detected for 5 seconds, the transmitter is shut down. This results in call drop.

The signalling tone (TS) is chosen to be 10 kHz. It is used in several ways. It is transmitted to alert a customer, to handoff a call from one base station to another, to disconnect the call, to invoke mid-call customer services (flashing), and to confirm orders. The transmission of ST ceases after ST time out of 60 seconds. The timer is not set during a legitimate sequence of operation which necessitates suspension of ST e.g. when wideband control signalling data at 10 kb/sec is transmitted during handoff orders.

12.3 Call Processing

Call processing is a procedure that is followed from initial call attempt to call termination. When a mobile is switched ON, the initialization process begins. Other functions related to call processing are paging channel selection, access (if call originates from the mobile), alerting, conversation and supervi-

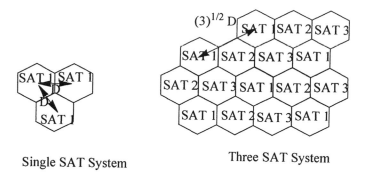

Single SAT System   Three SAT System

Figure 12.2 SAT Spatial Allocation

sion, and disconnection. Each procedure requires a number of steps and control functions. To illustrate this, some of these are described in the following.

12.3.1 Initialization

Each mobile terminal goes through the initialization process when the terminal is switched ON. The initialization procedure starts with the mobile scanning the dedicated control channels, which are set aside in the system for paging and access. In the North American AMPS and Digital AMPS (DAMPS), 21 top channels are designated as control channels. The mobile unit determines the strongest of the dedicated forward control (paging) channel and tunes to it within three seconds period. In case it fails to do so, it tunes to the second strongest within the next three seconds. If it is still unsuccessful, it restarts the scanning process. In any area of the United States, two competing systems are given operating licenses., These systems are known as Preferred System A and B. If the user subscribes to the preferred system is A, the system status is set to ENABLE, and for preferred system B it is set to DISABLE.

12.3.1.1 Paging Channel Selection

Following a successful tuning to the strongest or the second strongest forward (paging) control channel, the mobile terminal receives an overhead message and updates System Identification, Roam, and Local control status. In the idle state, when the mobile is not engaged in a call, it executes the following four tasks every 46.3 msec.

(i) Respond to overhead information: compares the stored SID with the received SID, if found not identical, the mobile reverts to initialization stage.

(ii) Page match: if roam status is disabled, the mobile station attempts to match one-word message or two-word message if the roam status is enabled.

(iii) Order: after matching MIN, it responds to the order.

(iv) Rescan access channels: since the mobile is usually in motion, it rescans the access channel after every 2 to 5 minutes and updates the status; the update interval is manufacturer dependent.

12.3.1.2 Access

An access to the system is gained in several steps. When the system access task is started, an access timer is set for a maximum of 12 seconds for Origination, a maximum of 6 seconds for Page response, Order Response and Registration.

12.3.1.3 Scan Access Channels

In this task one or at most two strongest signal channels are chosen. If the service request can not be completed on the strongest signal, the mobile can select the second strongest called the alternative access channel.

12.3.1.4 Seize Reverse Control Channel - (RECC)

BIS status = 1: The BIS status of 1 indicates the channel is ready for access data from the mobile terminal. The land station may ask the mobile station to check and wait for overhead message (WFOM) bit. If the received WFOM bit is 1, the mobile station waits to update overhead information and if it is 0, it delays for a random time between 0 and 92 ms and sends a service request. It is an access attempt.

BIS status 0: The BIS status of 0 indicates channel busy condition. Each time the mobile finds BIS status 0, it increments NBUSY by 1, and then waits a random time interval between 0 to 200 ms to check BIS status again whether changed from 0 to 1. When NBUSY exceeds MAXBUSY the call is terminated.

12.3.1.5 Access Attempt Parameters

Maximum number of access attempts is 10. There is a random delay interval of 0 to 92±1ms for each attempt at checking BIS status. If BIS is 1, the mobile station just waits for the transmitter power to come up and sends out the service request message. The random time delay is used to avoid two or more mobile stations attempting to access the network at the same time.

12.3.1.6 Updating overhead information

The Update Overhead Information (UOI) procedure consists of:

(i) Updating the Overhead Information (OHD): The Overhead Information is sent by the base station and is updated by the mobile;

(ii) Overhead Control Message (OCM): It contains information on the system whether overloaded or not.

Access Type Parameter Message sets the busy idle status bits in the BIS field. Access Attempt Parameter Message provides maximum number of seizures allowed and maximum number of busy occurrences. It is required that updating of the overhead information should be completed within 1.5 sec of initialization of the call. After the update overhead information is completely received, the mobile station waits for a random time interval between 0 to 750 ms and enters the **seize the reverse control** task.

It is conceivable that the seizure attempt fails. In this case the mobile must reexamine the access timer every 1.5 sec; if not expired, it reenters the access task after failure. Collision, failure of the land station to receiver the message bits due to fading, and a failure on the part of the land station to interpret the message because of message corruption are the main causes of failed attempts. After failure, the mobile must wait for a random time between 0 to 200 msec before making another attempt.

A whole package of service request messages must be continuously sent to the land station. The message consists of five words: A, B, C, D, E. After a complete message is sent by the mobile, an unmodulated carrier is sent for 25 msec to indicate the end of the message. If there is no response after 5 sec of sending the service request message, the call is terminated, and 120 pulses per minute fast tone is generated to the user. If uncoded MIN bits match within 5s, the mobile must respond with the following messages.

(i) If the access is because of origination or page response, the land station initiates voice channel designation message. For a directed-retry message the mobile station must examine the signal strength on each of the retry channels and choose up to two channels with the strongest signals. The mobile then must tune to the strongest retry access channel.

(ii) If the mobile encounters the start of a new message before it receives the directed-retry message, the call has to be terminated.

Within 100 msec of the receipt of the initial voice channel number designation, the mobile station must determine whether the channel number is within the set allocated to the home base station or from another source.

12.3.1.7 Alerting

(i) Wait for order

If an order can not be received in 5 seconds, terminate the call.

(ii) If an order is received within 100 msec, the action to be taken for each order is

If the order is Handoff, turn off the home land station.

This procedure has three steps.

(i) a 10 kHz signal tone remains ON for 50 msec after the SAT tone,

(ii) turn off the signalling tone,

(iii) turn off transmitter.

(iii) Turn on the new site. This procedure has four steps.

(i) tune to new channel,

(ii) adjust to new SAT,

(iii) set signalling color code (SCC),

(iv) turn on the new transmitter.

(iv) Alert - turn on the signalling tone and run it for 500 me, and enter the waiting for answer task.

(v) Release -

(i) send the signalling tone for 1.8 seconds,

(ii) stop sending signalling tone,

(iii) turn off the transmitter.

(vi) Audit - send order confirmation message to land station, remain in waiting for answer task, and reset the order time for 5 sec.

(vii) Maintenance - turn off signalling tone, run for 500 msec and enter waiting for answer task.

(viii) Change power -

a) adjust the transmitter to new ordered level,

b) send order confirmation to the land station,

c) local control.

Waiting for Answer

After requesting orders from the land station, the mobile station is in the Waiting for Answer status. An alert time must be set to 65 seconds. If no answer comes back within 65 seconds the call is terminated. Events occur in the same order as listed in the Waiting for Order section above.

12.3.1.8 Conversation

A release timer must be set to 500 ms during the conversation. The task can be used for the following conditions,

(i) if the user terminates the call,

(ii) if the user requests a flash,

(iii) within 100 msec of receipt of any orders, action will be taken by the mobile stations for each order.

12.3.1.9 Data Requirements and Formats

Network control requires sizable data transmission. The forward setup channel, the paging channel, the message channel and the land lines linking the various control elements are the main data paths. The first three of these use radio transmission over fading channels, so that error control is especially important. Well defined protocols and a means of data security are necessary. It is evident that considerable overhead is necessary to maintain data integrity. For example, in the AMPS system, the forward setup (paging) channel has five repeats of words (40 bits each) interleaved with 10, 11, 42 bits for bit sync, word sync, and busy-idle bits respectively, to form a packet of 463 bits as shown in Figure 12.3 (a). A BCH code is used to encode 28 information bits to form a 40 bit word. In a similar way, the reverse (access) channel data is framed into a packet as shown in Figure 12.3(b).

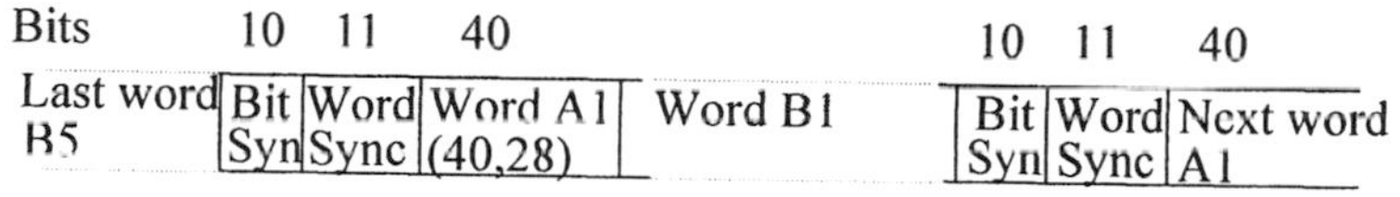

Bits	10	11	40			10	11	40
Last word B5	Bit Syn	Word Sync	Word A1 (40,28)	Word B1		Bit Syn	Word Sync	Next word A1

(a) Forward Setup Channel Signalling Format

Bits	30	11	7	48		48	48	48
	Bit Sync	word Sync	*	Word A1	word B5	Last Word3	Last Word4	Last word5

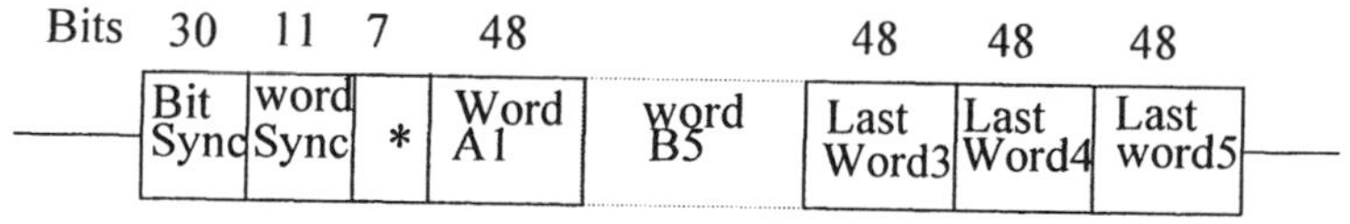

(b) Reverse Setup Channel Signalling Format

Figure 12.3 Forward and Reverse Setup Channel Formats

Function of each word:

Word A is an abbreviated address word. It is always sent to identify the mobile station. Word B is an extended address word. It will be sent on request from the land station or in a roam situation. In addition, the local control field and the other field are shown in this word. Word C is a serial number word. Every mobile unit has a unique serial number provide by the manufacturer. It is used to validate the eligible users. Word D is the first word of the called address. Word E is the second word of the called address.

When handoff occurs, speech is blanked and a burst of data is transmitted over the voice channel. Since the handoff takes place under conditions of low signal-to-noise ratio, the critical function of handoff messages are repeated eleven times in the forward direction. After the mobile unit switches to the newly assigned channel the signal-to-noise ratio improves and the messages are repeated five times on the forward channel. The voice channel data format is shown in Figure 12.4. The odd number of repeats are used to implement majority voting.

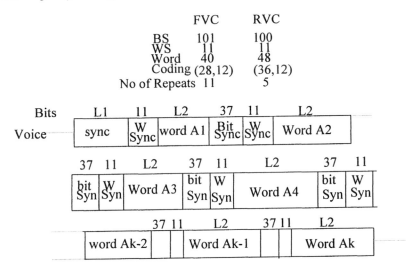

	FVC	RVC
BS	101	100
WS	11	11
Word	40	48
Coding	(28,12)	(36,12)
No of Repeats	11	5

Figure 12.4 Data Format Voice Channel

The rate of information transfer between base stations and the base station controllers must be sufficient to handle large number of control functions and flexible enough to accommodate system future expansion. In the AMPS system, a data rate of 2400 bps is used. The coding scheme chosen is a (32,26) BCH code, shortened from a (63,57) BCH code. A preamble embedded at the beginning of each message facilitates synchronization.

12.4 Nordic Mobile Telephone System

The Nordic mobile system was jointly developed by the PTTs of Nordic countries and was introduced in Sweden in 1981, as the first commercially operating mobile system in the world. The system architecture is shown in Figure 12.5. As is true for all cellular systems, the main control elements of the Nordic system are Mobile Terminals (MT), Base Stations (BS), Mobile Telephone Exchanges (MTX). Several base stations are connected to a MTX via either a high capacity cable or a fiber or microwave link. In a similar way, several MTXs are connected to a trunk exchange. Each base station is assigned a number of channels of which one is marked as Calling Channel. All mobiles not active in conversation are locked to this channel. The base station consists of transmitters, receivers, and a control unit. Channel selection, transmitter control, supervisory signal measurement, and general call supervision are the main functions of MTX. The signalling data generated at MTX transmitted over the lines and air is based on FFSK (or MSK) signalling at a rate of 1200 bauds, binary 1 represented by a full cycle of 1200 Hz and binary 0 in one and a half cycles of 1800 Hz. The signalling data containing the information is encoded using hexadecimal digits and is assembled into frames. The frames sent to the mobile contains five fields:

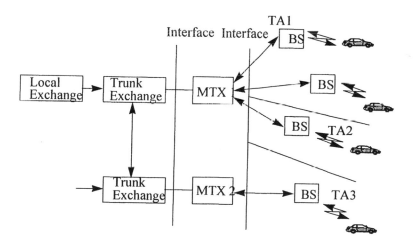

Figure 12.5 Nordic Mobile Telephone System Architecture

(i) Channel Number Field, $N1$, $N2$, $N3$. The channel numbers are also stored in the mobile. The mobile checks received channel numbers against its own channel numbers,

(ii) prefix P (type of frame identifier, there are 16 possible types),

(iii) traffic area information, *Y1, Y2,*

(iv) mobile station identity,

(v) information field.

The signal frames used for MTX to MS and MS to MTX transmissions are shown in Figure 12.6.

(a)

N1 N2 N3	P	Y1 Y2	Zx1 x2 x3 x4 x5 x6	j j j

channel number prefix   traffic area   mobile station ID   Information

(b)

N1 N2 N3	P	Zx1 x2 x3 x4 x5 x6	j j j j j j j

channel number   prefix   Mobile ID   Information

Figure 12.6 Signal Frame Structure, (a) MTX to MS,   (b) MS to MTX

In the reverse direction four fields, channel number, prefix, mobile identification number, and information, are used. The channel coding is rate ½ convolutional code capable of correcting 6 errors as long as there are at least 19 correct bits between the error bursts. A 15 bit long preamble is added for clock synchronization and an additional 11 bits word is used for frame synchronization. The basic functions of NMT system are shown in Figure 12.7.

The mobile terminal consists of a synthesized transceiver with transmitter and the receiver operating in frequency duplex mode. The signalling data is generated by a 1200 bits/sec modem. The supervisory tone from the base station is filtered from the speech signal before it is transmitted back to the base station. The transceiver also contains a microprocessor which handles control and signalling tasks.

12.4.1 Call Processing

When the mobile station is turned on, it scans for a channel labelled Calling Channel. The 180 channels are scanned with reduced sensitivity in order to find a channel with high field strength. This action also prevents the

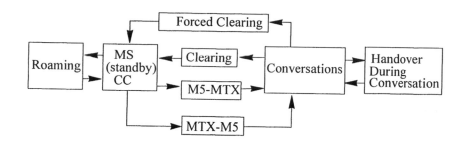

Figure 12.7 Basic Functions of NMT System

mobile from locking on to a channel from a distant base station. In case of unsuccessful first scan, the subsequent scans are done with increased sensitivity. A successful scan follows locking of the mobile to a calling channel whose frames are checked for legitimate channel number and traffic area codes. The traffic area code is stored in the mobile's memory even after the mobile is switched off. A match between the stored area code and detected area codes in the scan indicates that the mobile is still in the same area and can be accessed from a fixed telephone. Roaming[†] procedure is initiated if the recent area code does do not match with the stored code.

A call from the fixed telephone network to a mobile causes a paging message to be sent to all base stations in the area in which the mobile is previously known to be present. On receiving the message, the mobile sends acknowledgment to MTX, which checks the identification of the mobile. The positive identification triggers allocation of a free pair of message channels i.e. a forward and a reverse channel. For a call in the reverse direction, the mobile subscriber enters the desired subscriber number on the key board. When the handset is lifted, the mobile starts traffic channel scan over which a handshaking procedure is performed with the MTX informing it the mobile's identity and the number desired. The speech link is then established and the mobile subscriber hears the signals similar to those in the fixed network.

During conversation, the supervision and quality assurance of transmission is based on signal to noise ratio of the mobile transmitted 4 kHz supervisory signal and the field strength received from the mobile. If during the conversation the quality measure is degraded, a message is sent from the base to MTX. The MTX orders the neighboring base stations to monitor the

---

† *Roaming is defined as the ability to access a system other than that subscribed to by the user. It is important aspect of a system when international roaming is needed. In the future systems, worldwide roaming is envisaged.*

signal strength from the mobile on the channel being used by the mobile. The base station which receives the strongest signal is identified and the MTX sends a channel shift information to the mobile. This is in fact the channel designation information. The mobile tunes to the new channels, its identity is checked by the MTX, and if the identity is confirmed, the conversation is continued over the new channel. The supervisory signals and field strength are also used for forced clearing when the signal to noise ratio received is so bad that the link is presumed to be useless and handover could not be executed. The failure to handover the call to a new base station occurs under several conditions. Non-availability of a new channel for handover, mobile being at the end of coverage area, and very bad signal to noise ratio that commands can not be received correctly are main reasons of handover failure.

The supervisory signal is a tone of approximately 4 kHz generated in the base station and is used during conversation. The base transmitted supervisory tone is transponded by the mobile. This indicates that the link is alive. The supervisory tone from the base station is filtered from the speech signal before it is retransmitted by the mobile. The received signal by the base is also used to evaluate the signal quality. The base station also measures the signal strength which is used in conjunction with the tone measurement to decide about the handover. The signalling data is generated by a 1200 bits/sec modem.

Nordic systems, NMT-450 and NMT-900, have been designed to operate at 450 MHz and 900 MHz respectively. The channel spacing for both systems is 25 KHz, but the duplex frequency separation is 10 MHz for NMT-450 and 45 MHz for 900 MHz. Both systems transmit 1200 bauds using FFSK with a frequency deviation of ± 3.5 KHz. The mode of transmission is phase modulation. These systems have been now been replaced by GSM, the second generation European digital cellular system.

12.5 Other Analog Cellular Systems

Besides AMPS and NMT, several other analog cellular systems were put into commercial use. Only TACS, NEC, C450 are of any significance.

12.5.1 Total Access Mobile Communication System (TACS)

The TACS system, a modified form of AMPS, was developed for UK market. The channel spacing of 30 KHz in AMPS was reduced to 25 kHz to meet the European Mobile Radio Channelization Standard (EMRCS). The frequencies of operation was within 917-950 MHz for base transmit and 872-905 MHz for mobile transmit. The duplex frequencies were separated by 45 MHz.

The total number of duplex channels was 1320. Cell sizes between 2 and 20 km were used in the system.

For the signal spectrum to fit within the reduced channel spacing, other system parameters like voice frequency deviation, and signalling data rate were reduced to ±9.5 kHz, and 8 kb/sec respectively. The reduced deviation resulted in signal to noise ratio at the same distance from the base lower than that found in AMPS. With the same transmit power, the cell size and cell separation are smaller in TACS compared to those in AMPS.

12.5.2 NEC Japanese System

NEC Japanese system is similar to AMPS and TACS in most of system operational details. NEC system works in 870-855 MHz for base transmissions and 925-940 MHz for mobile transmissions. The mobile and base transmit frequencies are separated by 55 MHz. The channel spacings of 25 kHz and overlapped channels of 12.5 kHz have been chosen to increase frequency utilization. The total number of channels is 600. Cell sizes between 5 and 10 km radius are used. The speech signal is FM modulated that uses ±5kHz deviation. The control signals use FSK with a deviation of ±8 kHz. Data rate for signalling is relatively low at 0.3 kbits/sec.

12.5.3 The German Cellular System, C450

C450 is a German cellular system that operates in 450 MHz band. The base station transmit frequency is in 461-465 MHz band while the mobile stations transmit in 451-455 MHz. This gives the transmit and receive frequency difference of 10 MHz. The total number of channels 222 and each channel is 20 kHz wide. The cell size is between 5 and 30 km. The system uses FM modulation with 4 kHz deviation. The control signalling uses FSK with ±2.5 kHz. A data rate of 5.28 kbits/sec. The message protection is obtained by ARQ.

Table 12.1 summarizes the specifications of several analog cellular systems.

12.6 Digital Cellular Systems

Commercial operation of cellular systems began in 1981 in Europe and 1983 in the United States. Since their inception the growth of cellular mobile communications has exceeded predictions. By 1986, the cellular operators in the United States faced a severe shortage of capacity in Los Angles and in the New York-Newark corridor. They petitioned the FCC for additional spectrum. In July 1986, the FCC allocated an additional spectrum of 10 MHz, but the additional spectrum was considered to be still insufficient to

| Table 12.1 | | First Generation Cellular Systems Specifications | | | |

System	AMPS	C450	NMT	NEC	TACS
Transmission Frequency (MHz) Base station Mobile station	869-894 824-849	461-465 451-455	463-467 453-457	870-885 925-940	917-950 872-905
Separation between forward and reverse channel	45	10	10	55	45
Channel Spacing (kHz)	30	20	25	25, 12.5	25
Number of Channels	832	222	180	600	1320
Coverage radius (km)	2-20	5-30	1.8-40	5-10	2-20
Modulation: Audio Type Deviation (kHz)	FM ±12	FM ±4	FM ±5	FM ±5	FM ±12
Modulation: Control signal Type Deviation (kHz)	FSK ±8	FSK ±2.5	FSK ±3.5	FSK ±8	FSK ±6.4
Data Transmission Rate (kb/sec)	10	5.28	1.2	0.3	8
Message Protection	†	†	†	††	††

†. Principle of majority decision is employed
†. Message is resent when an error is detected
†. Receiving steps are predetermined according to the content of the message
††. Transmission rate is checked when it is sent back to the sender by the receiver
††. Principle of majority decision is employed

meet the growing demand with the installed analog cellular telephone system. However, despite injection of additional spectrum and capacity, the cellular systems in large cities remained in danger of facing saturation even if steps like extensive cell splitting and use of sectorized antenna to contain interference were taken to squeeze more capacity. The regulatory authority took the view that cellular operators should increase spectral efficiency to accommodate additional demand. A combination of subscriber base growth, the demand for new types of service and technological advances in communications led to the definition of new structures for mobile communications systems. This section presents design concepts associated with representative systems in both in the United States, Europe and to some extent Japan.

12.6.1 Developments in North America

Facing difficult future because of prevailing congestion and the requirements of CTIA (Cellular Telecommunications Industries Association) to increase the cellular capacity by at least ten folds, TIA (Telecommunication Industries Association) commissioned a study on a second generation cellular system with a primary aim to increase the capacity and also to ultimately make a transition towards personal communication systems. Telecommunication Industries Association (TIA) in North America created TR-45.3 study group to come up with solutions to increase the cellular system capacity without requiring additional spectrum. The mandate of TR-45.3 committee was to develop specifications for second generation cellular system which would provide approximately ten fold increase in the capacity of the system. Backward and forward compatibility of the new system was the other desirable feature. This meant that the user equipment for the proposed system must be usable with the first generation analog cellular system. The forward compatibility requires that the system design be flexible enough to accommodate newly emerging services using cellular systems.

Cellular system capacity can be increased by splitting cells and/or by reducing channel bandwidth. Cell splitting had already been implemented extensively in the analog cellular systems already in operation, and further splitting was not considered feasible due to impending interference problems arising out of high transmitted power. Digital cellular systems, however, offered substantial advantages but at technological complexities, which seemed to be difficult to surmount. The principles of lower power, increased frequency reuse, and efficient use of spectrum were the main starting points.

The transmission power can be lowered by making transition from analog transmission to digital and use of smaller cells. For example, 18 dB signal to noise ratio is required for acceptable speech quality with analog transmission; this can be dropped to around 14 dB for digital transmission. The digital system is also spectrally more efficient. For speech signals, the transmission bandwidth can be reduced from 30 KHz used in AMPS to about 10 kHz. This reduction in the transmission bandwidth is based on efficient voice encoding. Thus, an increased in the spectrum efficiency can be achieved by smaller cells, more extensive frequency reuse, and channel splitting. All these factors point to adoptation of digital transmission. In addition, new applications envisioned for mobile radio communications also required digital signal transmission. For example, provision of ISDN services over mobile radio channels or mobile Internet was seen as a distinct possibility, and it was projected that all those services currently offered by the switched telephone network will also be made available on the future cellular systems.

These include wide area paging, data transmission, integrated services, telefax, image transmission and other users defined services. Integration and migration to new personal communication systems was also a major consideration. The committee and its numerous subcommittees considered and evaluated number of technologies that could meet the requirements of additional capacity, migration and flexibility to offered yet to be identified services. These included considerations on access, network control, error control, modulation and data rates, channel bandwidth, and services. For access, three candidate technologies FDMA, TDMA and CDMA were considered, and TDMA was chosen because it allowed reasonable ease of transition from the first to second generation cellular systems. It also provided greater flexibility in data rates and services. Lower cost, high security and lower susceptibility to interference were the other reasons. The choice of TDMA had major consequences for other system parameters like protocols, modulation, interface with future ISDN network architectures. The choice of CDMA was thought to be interesting but the technology of the day was not fully muture to warrant its use. It needed few more years for Qualcomm to announce a cellular mobile communication system based on direct sequence code division multiple access. This development resulted in a lot of excitement in wireless communication circles [2]-[17].

12.6.2 Developments in Europe

In Europe, telecommunications has traditionally been under the control of national governments. To achieve a Pan European system, the emphasis of planners has been on the development of a system based on a commonly agreed standard. Their efforts resulted in the formation of Groupe Speciale Mobiles in 1982 under CEPT (Conference European des Poste et Telecommunications) which was given the mandate to study and design a Pan European Digital Cellular System. The group came to a decision that the new system will use TDMA, in which mobile units are allocated time slots in channels occupying a 200 kHz bandwidth. An important advantage of TDMA lies in its capability of rapid switching between time slots, as opposed to much slower frequency retuning necessary in single channel per carrier FDMA systems. Furthermore, absence of intermodulation products in TDMA made it more attractive in multichannel applications. The possibility of using smaller cells was seen another advantage of TDMA over FDMA [20].

The principle disadvantages of TDMA emanate from the harsh signalling environment presented by the mobile radio channel. The distances between the mobile units and the base station vary continuously, causing random time jitter to the slot boundaries. This necessitates a provision of guard times between time slots. Moreover, since the start of time slot of any user may be random, a slot alignment procedure must be is included in the design. Multipath propagation results in time varying intersymbol interference, in particular when the channel bandwidth is large. It becomes necessary to include adaptive channel equalization to mitigate the effects of ISI particularly in very irregular terrains.

12.7 North American Second Generation Systems

As mentioned earlier, in North America some degree of backward compatibility between the new digital system and the current analog was made mandatory. To some extent this need was met through dual mode equipment, whereby some system parameters and functions had to be identical to that used in AMPS, the first generation analog system. Another factor that added complexity is the use of a larger bandwidth, requiring channel equalization in highly irregular terrain or heavily built areas. While data rates in the GSM system are made compatible to those used in STN i.e. ISDN, this is not a prime consideration in the proposed North American system, where increased capacity has been the major consideration. Two distinct approaches were possible in order to achieve this end. The first is to decrease cell radius and cluster size with a proper choice of interference control techniques. The second is to decrease the effective channel bandwidth either through FDMA or TDMA. It was seen that capacity increase through cell splitting had been acheived to a maximum and there seems to be no room left for further cell splitting. Thus, the second approach of reduceing the bandwidth was considered to be more appropriate.

In NBFDMA (IS-54), the existing 30 kHz channel of AMPS was retained and this bandwidth was split into three TDMA channels giving a bandwidth reduction factor of 3; the effective throughput increased by more than 3 because of trunking gain predicted by the Erlang B formula. However, to provide toll quality speech within the effective bandwidth of 10 kHz, low bit rate good quality speech encoding is needed. An additional gain factor is provided by the antenna system. In present FDMA systems, high power signals are combined to feed the antenna giving rise to high intersymbol interference concerns. Antenna multicoupler technology becomes important in reducing the intermodulation products. The combiner becomes complex if the frequencies had to be tuned mechanically. Alternatively, signals can be combined at low power where combiner losses are not critical, but the combined signal then requires linear amplification. Since both of these approaches were costly, TDMA was seen as the preferred method. Furthermore, narrow band filter design becomes difficult and expensive; any compromise in the design will lead to an increase in the adjacent channel interference. IS-136, evolved from IS-54, includes PCS features such as enhanced battery life (low power transmission), 800 and 1900 MHz operation, private system, and digital control channel.

In the second generation North American Digital Cellular System, the speech signal is digitally encoded, as discussed in section Section 12.8.5. Each frame is divided into three voice slots resulting in three channels per 30 kHz. The data rate has been chosen to be 48 kbits/sec (1.6 bits/Hz). Full rate and half rate channels option is provided to accommodate future advances in speech encoding technology.

After the selection of TDMA multiple access method, attention was diverted towards choosing a suitable modulation scheme. A spectrally efficient and interference tolerant modulation are important characteristics, which could results in a significant increase in the system capacity. Interference tolerant modulation allows lower carrier-to-interference protection ratio, with the result that cell radii can be reduced considerably and capacity can be increased because of increased frequency reuse. It was also found in some research studies that with the adoption of digital transmission technology protection ratio can be lowered to 14 dB, a reduction of 4 dB from that required for analog systems. Further gains in capacity are possible if cell splitting is accompanied by a reduction in the channel bandwidth [21].

Thus, the chosen modulation had to be robust in the presence of channel impairments such as slow and fast fading, cochannel and adjacent channel interference, and channel frequency selectivity. It should have good spectral efficiency in bits/Hz. Its implementation should be simple and low cost. It should be transparent to different types of services. As previously discussed, most mobile data transmission systems use digital frequency modulation because of its ability to lock on to the strongest signal present (which should be the desired signal) and suppress interference. To minimize the effect of fading constant envelope continuous phase modulation schemes have advantages over other types. However, spectral efficiency of these modulation techniques is about one bit/sec/Hz.

Linear modulation schemes such as OQPSK (offset quadriphase shift keying) and ODQPK (offset differential quadriphase shift keying) were suggested as a possible modulation method for data transmission because of their nearly constant envelope and relatively high transmission efficiency. Offset is needed to make the signal envelope as constant as possible. However, envelope of the modulated signal still has envelope ripple of ±1 dB, which is fortunately within the definition of linearity for power amplifiers. The advantage of linear modulation lies in its ability to transmit two bits per symbol, or have a maximum efficiency of 2 bits/sec/Hz [6]-[12].

The ODQPSK modulation scheme has been adopted for the second generation North American Digital Cellular system. It requires linear amplification. The efficiency is more than 1 bit/sec/Hz, but relative to binary GMSK or TFM this method requires higher transmitted power for the same error performance. Quadrature differential phase shift keying (QDPSK) is preferred since phase recovery, in general a difficult operations in fading channels, is not needed. The basic assumption in differential detection is that the channel induced impairments do not introduce noticeable phase error over two consecutive symbols. Thus, phase comparison with the phase of the previous symbol provides reliable detection process [12].

It is well known that non-linearity anywhere in the transmitter chain is a serious impediment in the way of effective operation of linear modulation schemes. The non-linearities cause higher error rates. While class C amplifiers can be used at the transmitter for nonlinear systems, quasi A/B or class A amplifier are recommended for linear schemes. This restriction limits the power efficiency, i.e. RF Power/DC power conversion of the transmitter.

12.8 System Description

In this section we describe the North American second generation cellular system. In this regard, we focus on channel access and network control, modulation, channel bandwidth, data rates, and error control.

12.8.1 System Architecture

The system architecture of IS-54 is shown in Figure 12.8. The Public Switched Telephone Network (PSTN) is connected to a number of Mobile Switching Centers (MSC). With each switching centre, a Visiting Location Register (VLR), which keeps record of roaming by those subscribers who visit the area covered by the MSC. An area covered by several MSC has one Home Location Register (HLR), which contains database on the users registered with that particular cellular coverage area. The data base with HLR contains information on authentication and equipment identity number of the users registered with the system.

12.8.2 Frequency Parameters

In design, a provision of a total of 1023 channels is made. Table 12.2 lists the channel number designation:

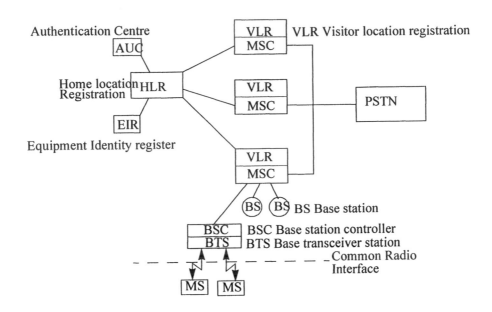

Figure 12.8 North American TDMA system Architecture

Table 12.2   *Channel Designation in DAMPS (IS-54)*

Transmitter	Channel number	Centre Frequency MHz
Mobile	$1 \leq N \leq 799$ $990 \leq N \leq 1023$	$0.030\ N + 825.000\ 0.030(N-1023)+825.000$
Base	$1 \leq N \leq 799$ $990 \leq N \leq 1023$	$0.030\ N+870.00$ $0.030(N-1023)+870.000$

These channels are equally divided among two competing systems System A and System B. The first generation analog cellular system channel bandwidth of 30 kHz is retained. This is because of the backwards compatibility requirement with the first generation analog system. The frequency tolerance is standardized to ± 2.5 ppm over temperature range of -30°C to +60°C for analog operations while for digital mode the frequency tracking capability is specified to be within ± 200 Hz. For each system, the control channels are divided into four groups: dedicated control channels, secondary control channels, primary paging channels, and secondary paging channels. A dedicated control channel is used for transmission of digital control information from

---

either a base station or a mobile station. A primary paging channel is a forward analog channel that is used to page mobile stations and send orders, and is supported by both EIA-553 and IS-54. The secondary control channels are developed specifically for the transmission of digital control information from either the base or mobiles stations. The secondary paging channels, a supplementary set of analog control channels is developed specifically for IS-54 compatible mobiles. These channels are used to page mobile and send orders.

12.8.3 Access and Control

12.8.3.1 Initialization

When power is applied to a mobile station, it enters initialization at the Retrieve System Parameters Task (RSPT) which is followed by the task called Scans the Primary Set of Dedicated Control Channels (SPSDCC). The mobile examines the signal strength on every access channel FIRSTCHDs (First channel) TO LASTCHDs (Last channel). The mobile then enters the UPdate Digital Overhead Information Task. The overhead messages are sent in a group called an overhead message train. The mobile then receives a system parameter overhead message and updates system identification, protocol capability indicator, number of paging channels, and serving system status. A time period of three seconds is allowed to complete these tasks. If the mobile fails to complete these tasks then it tunes to the second strongest dedicated control channel and repeats the procedure. The mobile then monitors mobile station control messages for orders.

12.8.3.2 Access

When a user initiates a call, the system access task indicating origination is entered. This sets the timer to 12 seconds and Last Try Code (LTs) to 0 before Scan Access is entered. The mobile examines signal strength of all the access channels available (FIRSTCHAs to LASTCHAs) and chooses up to two channels with strongest signals. The mobile tunes to the strongest access channel and retrieves the access attempt parameters task. The mobile station then sets the maximum number of seizure attempts and busy occurrences to 10. Followed by this, the mobile examines the Read Control Filler bit (RCFs) and reads the Digital Color Code (DCC). If it is unsuccessful in reading the DCC and the timer has not expired then the mobile tunes to the alternate access channel otherwise it goes into serving system determination task. On reception of DCC correctly, the mobile examines WFOMs for orders to proceed with orders to update overhead information or wait a random delay. After the overhead message train is received and processed, the mobile waits for random delay (between 0 and 750 msec) and enters the Seize Reverse Control Channel task.

The seizure procedure starts with reading of busy-idle bits of the channel. If the channel is busy then the mobile exits this task and execute a random delay and reads busy-idle bits. If the number of attempts to seize the channel exceeds $NBUSY_{max}$ then it enters serving system determination task. If the channel is idle, the mobile turns on the transmitter and then starts to send the message to the base station while continuously monitoring the busy-idle bits. On completion of the message the mobile sends an unmodulated carrier for 25 msec and then turns off the transmitter and waits for response from the base.

The base station sends information on initial voice channel designation message, initial traffic channel designation message, and directed retry message. Within 100 msec of the receipt of the initial traffic channel designation the mobile must determine whether the channel number is within the set allocated to the subscribed cellular system. If it is found to be within the allocated set then it must tune to the designated traffic channel. For digital channels the mobile tunes to the digital traffic channel and monitors fade timing status. If the fade timing status is enabled, a fade timer must be started; each time the fade timing status is disabled, the timer must be reset. If the timer counts to 5 seconds, the mobile must turn off its transmitter, and enters the serving system determination task. The mobile then transmits shortened bursts on the channel until a physical layer control order containing a time alignment setting is received.

Following the receipt of physical layer control order, the mobile enters Discontinuous Transmission mode on Digital Traffic Channel (DTDTC). In this mode the mobile sends Personal Identification Number (PIN), Random Challenge Memory (RANDs), and Call History Parameter (COUNTs) for authentication and waits for authentication response. The base compares PIN, RAND, and COUNT with the previously stored values and if any one of these mismatch, the mobile is denied service. If the mobile passes the authentication process, a traffic channel pair is issued. The base station may send a parameter update message on the FDTC. The mobile confirms the order on RVC or RDTC with a Parameter Update Confirmation Message, and tunes to designated channel and goes into conversation mode.

12.8.3.3 Supervision

The Coded Digital Verification Color Code (CDVCC) is used to supervise the link on a traffic channel. DVCC is used to distinguish the current traffic channel from traffic cochannels. DVCC is present in every slot and is transmitted by the base station. In the mobile station the received CDVCC is decoded and is compared to the DVCC that has been received in the traffic channel designation message. The mobile must in each burst set CDVCC to the coded value of DVCCs, irrespective of whether DVCC status is enabled or disabled.

Fast Associated Control Channel (FACCH) is a signalling channel for the transmission of control and supervision messages between the mobile and base station. The FACCH replaces the user information block whenever system considerations deem it appropriate to do so. Thus, speech signal is inhibited in order to use FACCH. The Slow Associated Control Channel (SACCH) is a signalling channel in parallel with the speech path. This is used for transmission of control and supervision messages between the base station and the mobile. Certain messages can be sent over either the SACCH or the FACCH. The SACCH is present in all slots transmitted over the channel whether these contain voice or FACCH information. The information structure for the messages is identical for both these channels, however, the forward error correcting methods differ.

Discontinuous transmission must be inhibited for 1.5 seconds after the mobile station enters conversation task. In the conversation state, the following may occur:

    (i) If the user terminates the call, the delay release must be examined and if the timer has not expired, the mobile station must wait until the timer has expired and then enters the release task,

    (ii) If the user requests flash, the mobile station must inhibit discontinuous transmission for 1.5 seconds if not capable of such transmission following which it turns on signalling tone for 400 mseconds,

    (iii) If the mobile is capable of discontinuous transmission and is in DTX low state, then it must wait for 200 msec following which it turns on signalling tone,

    (iv) If the mobile is in DTX low state and is capable of discontinuous transmission when the order arrives, the mobile station must enter DTX high and wait 200 ms and enters handoff procedure.

The handoff procedure begins with the mobile switching on signalling tone for 50 msec. Following this, it turns off signalling tone, turns off the transmitter, adjusts power level, tunes to the newly assigned channel, and sets the stored DVCCs to the DVCC field of the received message. Next, it sets the transmitter and receiver to digital mode, sets the transmit and receive rate and time slot to that indicated by the message type field. It configures codec rate indicated by the message type field, sets the time alignment offset to the value indicated by the TA field of the received message. Once synchronization is indicated, it turns on transmitter, resets fade timer, and enters the conversation task of the digital traffic channel.

12.8.3.4 Release

Discontinuous transmission is prohibited while the mobile station is in this task. The mobile station capable of discontinuous-transmission operation must remain in DTX high state. Any mobile in DTX low state must immediately enter the DTX high state, wait for 200 msec and take the following action.

If the user has terminated the call then mobile sends a release message. If the mobile station receives a valid message on FACCH containing base station release acknowledgment, then it must take action on release. It will send signalling tone for 1.8 second, stop signalling tone, turn off the transmitter and enter the serving system determination task.

12.8.4 Modulation and Coding

12.8.4.1 Analog Voice

The analog voice is processed in four stages and prepared for modulation. These are compression, pre-emphasis, deviation limiter and post deviation-limiter filter. The compressor and pre-emphasis circuit form a syllabic compandor. In 2:1 compressor, for every 2 dB change in input level within its operating range, the change in output level is a nominal 1 dB. The compressor has an attack time of 3 msec and nominal recovery time of 13.5 msec. The pre-emphasis circuit has a nominal + 6 dB /octave response between 300 and 3000 Hz. For audio (voice) inputs applied to the transmitter voice signal processing stages, the dual mode must limit the instantaneous frequency deviation to ± 12 kHz. This requirement excludes supervision and wideband signals. The deviation limiter is followed by a low pass filter whose characteristics are given in Table 12.3 [13].

Table 12.3   *Low Pass Filter Characteristics*

Frequency range	Attenuation relative to 1000 Hz
3000-5090 Hz	$\geq 40 \log(f/3000)$ dB
5900-6100 Hz	≥ 35 dB
6100-15000 Hz	$\geq 40 \log(f/3000)$ dB
above 15000 Hz	≥ 28 dB

Wideband Analog Data signals on the reverse voice control channel and reverse voice channel are coded such that each non-return-to-zero binary one is transformed to a zero to one transition and each non-return-to-zero binary zero is transformed to one to zero transition. The filtered wideband data stream is then used to modulate the transmitter carrier using direct binary frequency shift keying. Peak deviation of 8 kHz is used. This is similar to analog cellular system.

12.8.5 Voice Encoding and Digital Voice

The speech coding algorithm is a member of a class of speech coders known as Code Excited Linear Predictive coding (CELP), Stochastic Coding or Vector Excited Speech coding (VSELP). These techniques use code books to vector quantize the excitation (residual) signal. The speech coding algorithm is a variation on CELP called Vector Sum Excited Linear Predictive Coding (VSELP). This uses a code book which has a predefined structure such that the computation required for the code book search process can be significantly reduced. The VSELP coder uses two excitation codes each of 2^M code vectors which are constructed from two sets of M basis vectors, where M = 7. Each code vector in the code book is constructed as a linear combination of the M basis vectors [18][19].

The digital voice transmission proceeds through following stages: audio interface, bandpass filtering, preprocessing, and voice coding. The function of audio interface is to adjust transmit level of speech before passing through bandpass filter to avoid distortion due to aliasing of the input signal. The filtered voice is converted to PCM format with a minimum resolution of 13 bits. The speech is passed through a short term filter which serves as an LPC synthesis filter whose transfer function is given by:

$$A(z) = \frac{1}{1 - \sum_{i=1}^{N_p} \alpha_i z^{-i}} \qquad (12.53)$$

The short term predictor parameters are the coefficients, α_i's, of the short term predictor or synthesis filter. The short term predictor parameters are computed from the output speech. The order of predictor is chosen to be N_p = 10.

The speech encoder linearly interpolates α_i's for the first, second and third subframes of each frame. The fourth subframe uses the uninterpolated α_i's for that frame. The interpolated values of α_i's are checked to avoid any instability of predictive filter. In case such instability (value of any α_i is found to be less than 1.00) uninterpolated values from the previous frame are used in its place.

An energy value is computed and encoded once per frame. This energy value, $R(0)$, reflects the average signal power of the input speech over a 20 msec period which is centered with respect to the middle of the fourth subframe.

The 20 msec speech frame is subdivided into four 5 msec subframes. For each subframe the speech coder must determine and code the long term predictor lag, L, the two code words, I and H, and the gains, β, γ_1 and γ_2.

The speech coder uses a perceptual noise weighting filter of the form:

$$W(z) = \frac{1 - \sum_{i=1}^{N_p} \alpha_i z^{-i}}{1 - \sum_{i=1}^{N_p} \alpha_i \lambda^i z^{-i}} \qquad (12.54)$$

where α_i are the filter coefficient for the subframes. The synthesis filter used in the coder is different from that used in the speech decoder. Since in the encoder synthesis filter a noise weighting λ is used, therefore it is sometimes called weighted synthesis filter.

The speech coder uses the closed loop approach in choosing the long term predictor lag. In the closed loop case, the lag is determined from only past output of the long term filter and the current input speech. The long term filter response can be expressed as;

$$B_n(z) = \frac{1}{1 - \beta z^{-\left\lfloor \frac{n+L}{L} \right\rfloor L}} \qquad (12.55)$$

where $\lfloor x \rfloor$ is the floor function of x which evaluates to the largest integer $\leq x$ and n is the sample in the subframe. If $L<N$ then samples $n= L$ to $2L-1$, a delay of $2L$ is used. In this way the delay is always greater than n so that only long term filter states, which existed at the start of the subframe are used. This is shown in Figure 12.9. The lag, L, is determined and is encoded by using a search procedure choosing a code vector from the first code book followed by another from the second code book.

The speech coder is an analysis by synthesis coding system. Therefore, a version of the speech decoder is used in the speech encoder. The form of the synthesis filter used in the speech encoder is given by:

$$H(z) = \frac{1}{1 - \sum_{i=1}^{N_p} \alpha_i \lambda^i z^{-i}} \qquad (12.56)$$

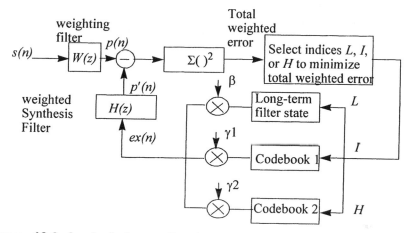

Figure 12.9 Analysis by synthesis procedure for long term predictor Lag and code search

Figure 12.10 shows the block diagram of the speech decoder, the parameters of which must be determined by the encoding process.

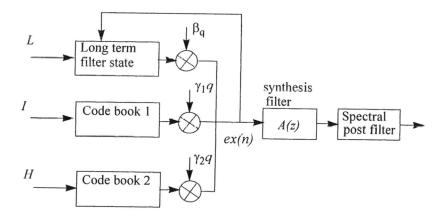

Figure 12.10 Block diagram of speech decoder

12.8.6 Channel Coding

The channel error control for the speech encoder data employs three mechanisms for the mitigation of channel induced errors. The first is to use rate one half convolutional code to protect more vulnerable bits of the speech

encoder data stream. The second technique interleaves the transmitted data for each speech coder frame over two time slots to mitigate the effect of Raleigh fading. The third technique is to use cyclic redundancy check over some of the most perceptually significant bits of the speech coder. After the error correction is applied at the receiver, these cyclic redundancy bits are checked to see if the most significantly perceptually significant bits were received properly. The channel coding procedure is given in Figure 12.11.

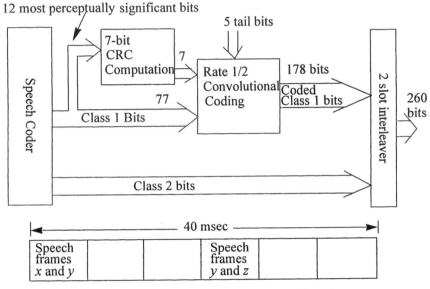

Figure 12.11 Channel Coding Procedure and interleaving

12.8.7 Receiver Functions

The π/4 shifted deferentially encoded quadrature phase shift keying can be demodulated by the use of several techniques. These include discriminator detection and baseboard differential detection. After the received signal is demodulated it is de-interleaved and data from an entire speech frame becomes available. This data, which was coded using convolutional code is decoded with use of any available technique as the standard does not specify this e.g. Viterbi Algorithm in conjunction with soft channel decoding. The CRC is used to check for errors in the decoded message. The frames containing large number of errors are masked. There is a possibility that a number of consecutive frames are corrupted. To avoid excessive degradation of speech as a result of masking, state machine strategy with six states is used. The State

0 indicates that CRC test was successful and the state six means that the past six frame were bad. If state six is reached the speech which was repeated in states 1 and 2, attenuated and repeated in states 3, 4 and 5, is muted in state 6.

The speech decoder takes the data at 7950 bps from the channel decoder and generates the received signal. The speech data is decoded for short term and long term lag parameters and synthesis filter is used to recover encoded speech.

12.8.8 Message Exchange on Digital Traffic Channel

12.8.8.1 Transmission of Messages

A mobile station may transmit a control message on either the FACCH or the SACCH. Some messages require acknowledgments from the base station. If the transmitted message requires an acknowledgment and time-out occurs before the acknowledgment is received, then the mobile must retransmit the message on the same associated channel. The time-out interval for an ACK response to a message requiring an ACK begins after the last bit had been transmitted. The time-out intervals are given in Table 12.4.

Table 12.4   *FAACH and SAACH Time out Intervals*

	FAACH	SAACH
Full Rate	200 msec	1200 msec
Half Rate	TBD	TBD

The mobile must observe the following rules when assembling messages into words:

(i) A message may span multiple consecutive words.

(ii) No part of another message or the filler code may be inserted between the segments of a message.

(iii) A filler message may be inserted as many times as desired in front of a message or following the end of a message.

The NA-TDMA (DAMPS or IS-54, or IS-136) shares analog set up channels with the analog AMPS system. On a digital channel (a 30 kHz TDMA) the frame rate is 25 frames per second. Each frame is 40 msec long and it contains 1944 bits (972) symbols, as shown in Figure 12.12. The frame has 6 time slots, each being 6.66 msec long, contains 324 bits (162 symbols). The bit duration is 20.57 μsec. Thus, the transmission rate over one radio carrier is 48.6 kbps but consists of only 24,000 symbols over the radio path. Each user is allowed two slots as shown in the diagram.

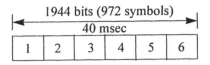

slot usage: 1, 4 for first user, 2, 5 for second user, and 3, 6 for the third

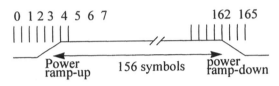

Figure 12.12 TDMA Frame Structure

There are two slot lengths, full rate and half rate. Each full rate traffic channel shall utilize two equally spaced time slots of the frame. It is necessary to control the TDMA time slot burst transmission (advancing or retarding) from the mobile unit so that it arrives at the base station in the proper time slot. An error in time alignment will cause signal in the two adjacent time slots to overlap at the head or tail of the time slot. The reverse and forward channel frame offset is given by

Forward frame = reverse frame + (1 time slot + 44 symbols)

= reverse slot + 206 symbols.

This means that the time slot 1 of the frame N in forward link occurs 206 symbol periods after the time slot 1 of the frame N in the reverse link.

The mobile station receives an initial traffic channel designation (ITCD) message (order code 01110) which is contained in word 2 (extended address word), and then moves to a traffic channel. The mobile station first synchronizes to the forward traffic channel. The time alignment is sent by a physical layer control message over a shortened burst transmission. The mobile station, while operating on a digital traffic channel, is transmitting over a slot interval 324 bits long at certain times. The mobile station continues to transmit a shortened burst at the standard offset reference position until a time alignment message is received from the base station. The mobile station adjusts it transmission time during the next available slot.

Time alignment during handoff becomes necessary. A mobile handoff message contains estimated time alignment information. Analog to digital and digital to analog handoff messages contains a shortened burst indicator (SBI) field:

SBI = 00 A handoff to a small diameter cell
SB1 = 01 A handoff from a sector to sector
SB1 = 10 A handoff to a large diameter cell.
The shortened burst format is *G1 RSDSDVSDWSDXSDYS G2*

G1 = 3 symbols, *R*=3 symbol length ramp, *S* = 14 symbol sync, *D* = 6 symbol length coded digital signal verification color code (CDVCC) on reverse channel, *G2* 22 symbol length guard time. The fields of *V, W, X, Y,* consists of

V = 4 zero bits, *W* = 8 zero bits, *X* = 12 zero bits, *Y*= 16 zero bits.

12.8.9 Reverse Digital Traffic Channel (RDTC)

A mobile station uses the reverse digital traffic channel to transmit to the base station. The digital channel transports user information and signalling messages. Two separate control channels, the fast associated control channel (FACCH) and the slow Associated Control Channel (SACCH) are provided. The FACCH is a blank and burst channel. The SACCH is a continuous channel.

The FACCH is a signalling channel for transmission of control and supervision messages between the system and mobiles. The FACCH block replaces the user information block (speech or data) whenever it deemed necessary. Each block is regarded as a signalling word. An FACCH message can consist of more than one such word. Messages over FACCH are interleaved over two consecutive bursts in a manner similar to that for speech.

The FACCH data is error protected by means of a rate 1/4 convolutional code. The coding uses the same start and end bit (tail biting), instead of extra tail bits, to reduce the overhead by avoiding explicit tail bits. The fields are presented to the encoder in the order starting from the signalling word header. Bits within the field are presented to the coder in the order most significant first. Each word is formatted as shown in Table 12.5.

Table 12.5   **FAACH Word Format**

Field	Description	Bits
Signalling word header	Continuation flag	1
Contents	Message	48
CRC	Cyclic Redundancy Check	16

The signalling word header consists of a continuation flag of one bit. The flag indicates to the receiver whether a word is the first or a subsequent in a message.

DVCC is the 8-bit color code distinguishing a designated traffic channel from its cochannels. It is used as a pointer to a traffic channel, in addition to identify both the originating address from a transmitter and destination address of the addressed receiver. The DVCC precedes the 49 data bits and is included in determination of the CRC. The 49 data bits in a FACCH word is appended with a 16 bit CRC to detect the presence of channel errors in the data as well as to provide a mechanism to distinguish the FACCH data from speech data.

The SACCH is a signalling channel for transmission of control and supervision messages between the digital mobile and the system. The SACCH uses twelve coded bits per TDMA burst.

Before transmission, the output of the convolutional coder is diagonally interleaved so that the twelve coded bits are transmitted over twelve time slots. The fields are presented to the SACCH convolutional coder in the order starting from the signalling word header. Bits within a field are presented to the coder in the order most significant bits first. Each word is formatted as shown in Table 12.6.

Table 12.6　　　　　　　　**SAACH Word Format**

Field	Description	Bits
Signalling Word Header	Continuation Flag	1
Reserved	Set to zero	1
Contents	Message	48
CRC	Cyclic Redundancy check	16

The word header consists of a continuation flag of 1 bit. A zero indicates first word of a message whereas a 1 indicates subsequent words of a multiple word message. The 50 data bits in a SACCH word is appended with a 16 bit CRC code to detect the presence of channel errors in the data. The CRC code is computed over the entire 50 bit data as well as the 8 bit DVCC.

12.8.9.1 Messages

Messages between the mobile and the base station on either the forward or the reverse digital traffic channel are formatted as shown Table 12.7. The message structure is the same on both the FACCH and SACCH. All messages contain an application message header and may contain mandatory fixed parameters, mandatory variable parameters and optional variable parameters. If the message is defined to have optional variable parameters, the remaining length field is required.

Table 12.7   **Message Format**

Message Header	Mandatory Fixed Parameters	Mandatory Variable Parameters	Remaining Length (6 bits)	Optional Variable Parameters
Message Format				

The message header consists of 8 bits for message type information and 2 bit protocol discriminator. The message type information identifies message purpose and its structure while the protocol discriminator field defines the specific protocol or instance of the protocol being used. The optional variable parameters defines the type and size of the parameter VAL-UEs. The parameter type is unique for every message type so every message can contain 16 different optional variable parameters. The number of values field indicate the number of values contained in the value field. The order in which the optional variable is sent is not important. Thus, it can be sent in any order. The remaining length field indicates the number of octets remaining following this field. In order to get an integer number of octets in a message, the end of the message is filled with zeros if necessary. For a message to be valid, all words in CRC block containing any part of a message must be received without any errors.

12.9 GSM (European CEPT) System

The GSM[†] (Global System for Mobile Communications) system was chosen on the basis of performance, lower complexity and data services flexibility. This system entered service at the end of 1991 and was available initially only in the principal cities of Europe. It is now a defacto standard of the world and is available in more than 150 countries. The Pan European service began when a number of national networks became available. The GSM system uses two bands of 25 MHz, 890-915 MHz for subscriber to base transmission (reverse channel) and 935-960 for Mobile to Base (forward channel). Versions of GSM, named as DCS1800 and DCS1900 systems, are also available; these operate in 1800 and 1900 band to provide low power microcellular

---

†. *GSM was previous known as Groupe Speciale Mobile. This group was established by CEPT to study second generation technology for mobile communications. The name was later changed to Global System for Mobile Communications.*

service. The available forward and reverse bandwidths are divided into RF channels of 200 kHz bandwidth. There are 124 possible RF carriers. The uplink and downlink RF channels or carriers are identified from their absolute radio frequency channel numbers (ARFCN), which are given as:

$$F_u(n) = 890.2 + 0.2(n-1) \quad \text{MHz}; \quad 1 \leq n \leq 124$$

$$F_d(n) = F_u(n) + 45 \quad \text{MHz}$$

(12.57)

where subscripts u and d identify uplink and downlink respectively.

The GSM system is defined in terms of functional entities and interfaces. Two interfaces *Um* (radio interface) and *A* (interface between the mobile services switching centre and the base stations) are mandatory. To facilitate roaming between European countries CCITT SS No. 7 has been specified. The functional architecture is shown in Figure 12.13.

The GSM system consists of three interconnected subsystems: Base Station Subsystem (BSS), Network and Switching Subsystem (NSS), and Operation Support Subsystem (OSS). The mobile is also a subsystem but it considered to be a part of BSS. The NSS handles the switching of the GSM calls between external networks and the BSC's. Between these subsystems, several interfaces are defined. These are shown in Figure 12.14. The *A* interface uses SS7 protocol called signalling correction control part (SCCP), which supports communications between MSC and the BSS. The BSS provides and manages radio transmission paths between the mobile stations and all other subsystems of GSM.

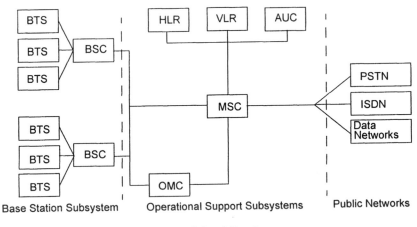

Figure 12.13 GSM Functional Architecture

In NSS, three different databases called the Home Location Register (HLR), Visitor Location register (VLR) and the Authentication Centre (AUC) reside. Networking and switching aspects are associated with the MSC, VLR and HLR. These entities manage the terrestrial channel, network management, call control and signalling, and are responsible for overseeing mobility of the users. The AUC handles the authentication and encryption keys for every subscriber in the HLR and VLR. This register contains a register called the Equipment Identification Register (EIR), which identifies stolen or fraudulently altered phones. In addition to its participation in the above functions, the BSS acts cooperatively to maintain system integrity. The total coverage area is divided into a number of location areas whose unique identity is conveyed via a broadcast common control channel (BCCH). The OSS supports one of several operational and maintenance centers (OMC), which are used to monitor and maintain the performance of each MS, BS, BSC and MSC within a GSM system.

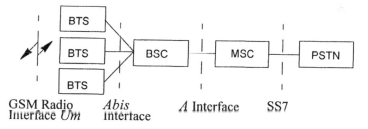

Figure 12.14 GSM Interfaces

The *Um* interface design is based on the need to support either network signalling information or teleservices (speech transmission, short message services, data message handling access, telex and facsimile) and bearer services that use circuit and packet modes. This interface is defined in a layered fashion, which has the following structure. Each RF channel is time shared between as many as eight subscribers using TDMA. The data using $BT = 0.3$ GMSK modulation has a transmission rate of 270.833 kbps. The effective rate per channel is 33.854 kbps and the signalling during is 3.692 μsec. The actual user data rate is 24.7 kbps. Each time slot has an equivalent time allocation of 156.25 channel bits out of which 8.25 bits of guard time and 6 total start and stop bits are provided to prevent overlap with the adjacent time slots. In order to improve performance in the presence of fading slow frequency hopping is also provided. The hop rate of 217.6 hops/sec is used over as many as 64 different channels [20][21].

12.9.1 GSM Channel Structure

The GSM system maintains a complex channel structure in which several channel types are used. The data bursts with time slots of frames have five specific formats, each with a specific purpose related to traffic and controls functions. A number of logical channels are provided and are divided into two categories: traffic channels and control channels. The traffic channels are designated to carry either speech or data. Full and half rate channels are defined.

The control channels are subdivided into BCCH (broadcast control channel), CCCH (common control channel), SDCCH (stand alone dedicated control channel), and ACCH (Associated control channel). BCCH is used to broadcast cell information (for a maximum of 16 cells as a group), and also carries synchronization and time information. CCCH is used on the uplink for random access and on the downlink for paging and for assignment of dedicated channels (TCH or SDCCH). SDCCH is used for transaction signalling and allows, for instance, handover preparation during the signalling exchange phase [20]. ACCH is a signalling channel which is always allocated in conjunction with a dedicated channel (TCH or SDCCH) and carries signalling information which is necessary during the main exchange on the dedicated channel (e.g. measurement data or billing information). When fast handover command is necessary, a fast ACCH (FACCH) channel using full capacity of dedicated channel is used. Normal bursts are used for Traffic Channel (TCH) and Dedicated Control Channel (DCCH). These are used on both the forward and reverse channels. The Frequency Control Channel (FCCH) and Synchronization Channel (SCH) bursts are used in Time Slot 0 (TS0) of specific frames to broadcast, on the forward link, the control messages for frequency and time synchronization. The Random Access Channel (RACH) burst is used by all mobiles to access services from any base station and the dummy burst is used as a filler information for unused time slots on the forward link. The structure of logical channels is shown in Figure 12.15.

The above description of the signals on different types of channels use a uniform convolutional code; all of these are adapted to a particular channel by a suitable puncturing scheme. For all channels two codes are concatenated: one block code for error detection, followed by convolutional code whose rate depends on the type of the channel.

In the TDMA system, it is advantageous to use mobile initiated handover, since mobile receiver/transmitters operate in a burst mode; when idle the mobile can monitor other channels. These measurements are then reported to the serving bases station regularly via ACCH channel as mentioned above.

Speech signals are transmitted at 13 kbits/sec, which necessitates that speech is encoded prior to transmission. Regular Pulse Excited LPC (RELP) with Long-Term Prediction is used for speech encoding and is described in the

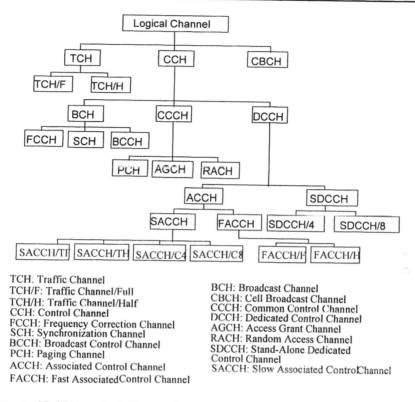

TCH: Traffic Channel
TCH/F: Traffic Channel/Full
TCH/H: Traffic Channel/Half
CCH: Control Channel
FCCH: Frequency Correction Channel
SCH: Synchronization Channel
BCCH: Broadcast Control Channel
PCH: Paging Channel
ACCH: Associated Control Channel
FACCH: Fast AssociatedControl Channel

BCH: Broadcast Channel
CBCH: Cell Broadcast Channel
CCCH: Common Control Channel
DCCH: Dedicated Control Channel
AGCH: Access Grant Channel
RACH: Random Access Channel
SDCCH: Stand-Alone Dedicated Control Channel
SACCH: Slow Associated ControlChannel

Figure 12.15 Logical Channels Structure in GSM

following [19]. The speech is sampled and encoded in frames of 20 msec duration, each consisting of 260 bits. Each frame is divided into 4 sub-blocks of 5 msec each.

The encoded bits are divided into four groups for error protection. Fifty most important bits, called Ia bits, have 3 parity check bits added to them. The next 132 bits along with the first 53 bits are reordered and appended by a 4 trailing zero bits. This constitutes a block of 189 bits. The block is encoded using a 1/2 rate convolutional code with constraint length $K = 5$, thus providing a sequence of 378 bits. The least important 73 bits are left uncoded. The final block of 456 bits in a 20 msec block is obtained. The 456 bits within each 20 msec frame are broken into eight 57 bit sub-blocks. These subblocks are spread over 8 consecutive TCH time slots. Out of the maximum end-to-end delay is approximately 75 msec, the codec is allowed 30 msec and more than 40 msec are allocated to the interleaving of the time slots and to channel coding.

12.9.1.1 Frame and Slot Structures in GSM

The transmission in GSM is organized in slots, frames, multiframes and super frames. The details of these frame structures are shown in Figure 12.16.

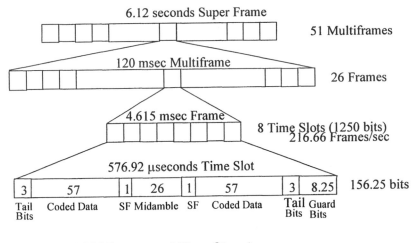

Figure 12.16 GSM Frame and Time Structures

A time slot consists of 156.25 bits and consists of 3 Tail Bits, 57 Coded data, 1 stealing Flag bit, 26 bits for Midamble, 57 Coded data, 3 Tail bits and 8.25 Guard bits. The slot is 576.92 μsec long. Eight slots constitute a Frame, which is 4.615 msec long and is transmitted at rate of 216.66 frames/ sec. A Multiframe, assemblance of 26 Frames, is 120 msec long and 51 Multiframes constitute a Super Frame. For every 26 frames, the thirteenth and the twenty sixth frame consists of slow associated control channel or the idle frame respectively. The Super Frame is 6.12 seconds long.

As mentioned earlier, the time slot bursts are of five types. Normal bursts carry 114 bits of message data. The forward control channel burst consists of 142 bits, all zeros. The synchronization channel burst consists of 78 bits of encrypted data and 64 bits for training. The random access channel burst consists of 41 bits for synchronization, 36 bits of encrypted data. The dummy burst consists of 116 mixed bits and 26 training bits. All bursts have 3 start, 3 stops bits, and 8.25 guard bits except in the RACH burst where the guard period is extended to 68.25 bits. These are shown in Figure 12.17.

There are three main types of control channels: Broadcast Channel (BCH), the Common Control Channel (CCCH) and the dedicated Control Channel (DCCH). Each control channel consists of several logical channels, which are distributed in time to provide the necessary GSM control functions.

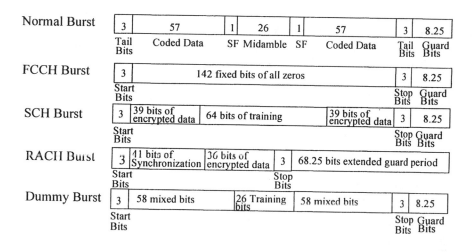

Figure 12.17 Time Slot Data Bursts

facilitate access to the system. The control multiframe in the forward and the reverse direction consists of 51 TDMA frames of length 235 millisecond. The forward control channel multiframe is organized into several types of bursts. However, the reverse channel is organized into 51 Reverse RACH bursts. The control multiframe is organized as shown in Figure 12.18.

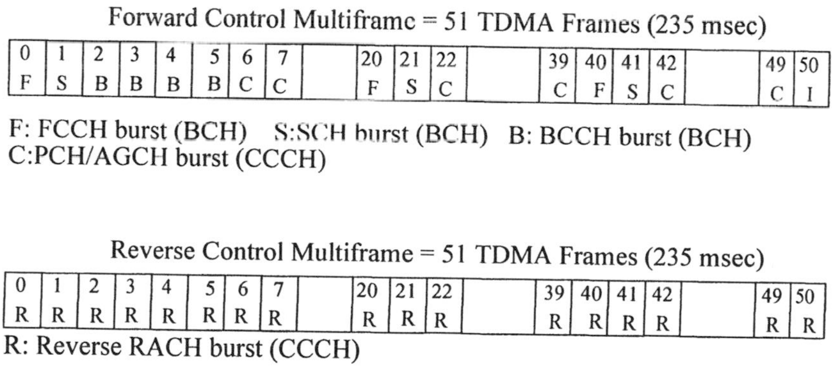

Forward Control Multiframe = 51 TDMA Frames (235 msec)

0	1	2	3	4	5	6	7		20	21	22		39	40	41	42		49	50
F	S	B	B	B	B	C	C		F	S	C		C	F	S	C		C	I

F: FCCH burst (BCH)   S:SCH burst (BCH)   B: BCCH burst (BCH)
C:PCH/AGCH burst (CCCH)

Reverse Control Multiframe = 51 TDMA Frames (235 msec)

0	1	2	3	4	5	6	7		20	21	22		39	40	41	42		49	50
R	R	R	R	R	R	R	R		R	R	R		R	R	R	R		R	R

R: Reverse RACH burst (CCCH)

Figure 12.18 Control Channel Multiframe Organization

The BCH operates on the forward link of a specific Absolute Radio Frequency Channel (ARFCN) within each cell and transmits data only in the first slot (TS0) of certain GSM frames. BCH serves as a beacon channel, which the mobiles identify and lock on to. BCH provides synchronization for all mobiles found within its cell; it is occasionally monitored by the mobiles in the neighboring cells so that received power from cell base stations could be measured and decisions on mobile assisted hand over (MAHO) can be made. The BCH is defined by three separate channels, which are given access to TS0 during various frames of the 51 frame sequence. There three types of BCH: Broadcast Control Channel (BCCH), Frequency Correction Control Channel (FCCH), and Synchronization Channel (SCH). The Common Control Channel is also of three types: Paging Channel (PCH), Access Grant Channel (AGCH) and Random Access Channel (RACH).

12.9.1.2 GSM Services

Several types of service are offered in the GSM system. These include bearer services, teleservices and supplementary services. Bearer services are meant for information transfer either between a user in the GSM and a user in a terminating network, or between a user in the GSM and an appropriate gateway that provides higher layer communication functions. Three types of bearer services are envisaged in GSM. These are data circuit duplex synchronous, pad access circuit synchronous, and data packet synchronous. The first group allows both transparent and non transparent support of data services at 300 bits/sec and 1200 bits/sec. The non transparent mode provides extra error protection at the expense of delay and throughput. The terminating equipment may be part of a PSTN, ISDN or GSM network. In the second group packet assembly and disassembly is provided with functionality otherwise it is similar to Group 1. The third group of data packet duplex synchronous transmission, provides access to an X.25 interface with transparent or non transparent support of data at 2400 bits/sec, 4800 or 9600 bits/sec.

In the teleservices, two groups - essential and additional teleservices, are supported. In the essential services group speech, short message, data message handling system (MHS) access and facsimile are provided, while in the additional services videotext access and teletex transmissions are provided. The implementation of these services is dependent on the specific type of the backbone network, which may different from one operator to another.

In the essential services, telephony is the major category, providing transmission of speech and audio signalling tones. Faster access for emergency calls is also provided under this category. Under essential services, the short message class includes only mobile terminated message services from a user of a telecommunication network. The correct reception of message is acknowledged. The maximum message length is recommended to be 180 characters.

The data message handling system, while not an essential element of GSM, is provided for the existing users in the fixed network. Speech and facsimile transmission is provided in both directions, i.e. both to and from mobile units.

Supplementary services are capable of modifying or supplementing other services offered in GSM. Call forwarding, call completion, charging, call restriction and multiparty call facilities are allowed. These are based on CCITT standards, but some of them may not be available in the fixed network. Call charging is based on the number of time units consumed by the call, and may include an additional fixed or minimum charge as well. The call restriction service allows certain classes of calls to be barred, e.g., local/international, incoming/outgoing calls.

12.9.2 GSM Call Flow Scenarios

The operation of GSM cellular system can be better understood by considering several call flow scenarios. These scenarios show the role that various components of GSM network play in linking mobile subscribers to the other users.

12.9.2.1 Call Initiated by a Mobile

We first consider a scenario of setting up a call from a mobile. The call originates from a mobile, which sends a request signal (SETUP_REQ) on a selected reverse control channel. The BSS relays this information to MSC, which accesses subscriber's data from HLR or VLR depending on whether the user is in its home cell or is visitor to another area. The HLR or VLR responds with SUB_DATA_RESP providing the particulars of initiator of the call. The MSC after authentication of the particulars of the mobile allows the call to proceed. At the same time it assigns a wireline network trunk and a radio channel. The BSS sends the assigned radio channel information to MS, which acknowledges the receipt of the assignment and acknowledges when tuning of the radio to the assignment is completed. The BSS sends a message Trunk and Radio Assignments complete to the MSC; following which the call proceeds. The sequence of events is shown in Figure 12.19.

12.9.2.2 Call Initiated by Land Network

The sequence followed by a land network based user to set up a call to a mobile is shown in Figure 12.20. Such call is also called mobile terminating call.

The call set up is initiated by the fixed network telephone by dialling the desired mobile telephone number. The PSTN sends the call setup request to the cellular network. The MSC provides the information NET_SETUP to

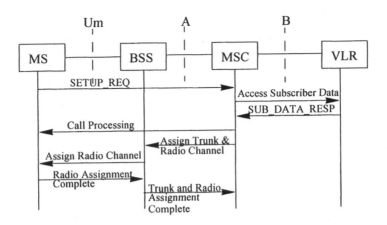

Figure 12.19 Call Initiated by Mobile

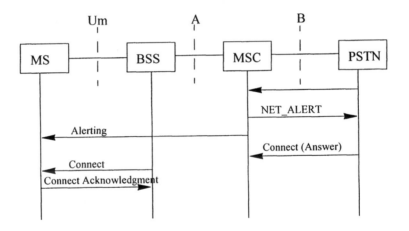

Figure 12.20 Call Initiated by Land Network

PSTN, which sends a NET_ALERT message to MSC. MSC then sends an alerting signal to MS along with the connect message. The alerting message also consists of information in the radio channels. The connection is acknowledged by the mobile with which the call setup is complete.

12.9.2.3 Mobile Initiated Call Release

Mobile to land call release initiated by mobile is shown in Figure 12.21. The release process takes place with the participation of MS, BSS, MSC and PSTN. The mobile sends CALL_DISC message to MSC via BSS, which sends a NET_REL request to PSTN. The MSC also sends CALL_REL message to the mobile confirming that the network is preparing to release the call. The mobile sends REL_COMP message indicating that mobile has completed the release procedure. On receiving this message the MSC sends out communication clearance message (CLR_COMM) to the mobile. This effectively orders the mobile to release the channel and shut down the transmitter.

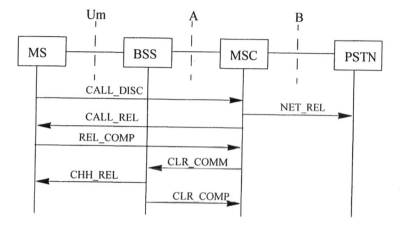

Figure 12.21 Mobile Initiated Call Release

12.9.2.4 Paging Land to Mobile Call

The sequence of events that take place when land initiated call requires paging to a mobile terminal. Such paging takes place when the mobile is interrogated by the MSC. The MSC delivers a PERM_PAGE message to BSS for broadcast to mobiles. The PAGE_MESS is broadcast on one of the forward control channels. On hearing paging message, the called mobile requests for a control channel. The BSS assigns a dedicated control channel by sending DSCH_ASS message. Mobile station sends a response to PAGE_RESP to BSS which is transferred to MSC and VLR to update the latest location of the mobile.

12.9.2.5 Land to Mobile Call - Routing

When a call is initiated by a land telephone, the incoming call inquiry is sent to MSC. The MSC requests HLR for route information by sending GET_ROUT. The HLR supplies the last known location of the mobile by sending ROUT_INF to MSC. This route information may point to a location outside the Home Register. MSC informs VLR about the incoming call. VLR

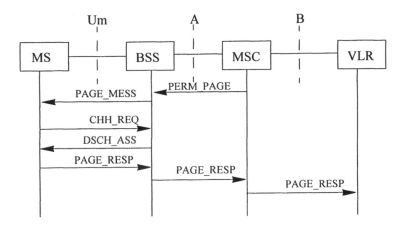

Figure 12.22 Paging - Land to Mobile Call

then sends PERM_PAGE to MSC, which is later broadcast to a particular cell. This procedure is shown in Figure 12.23.

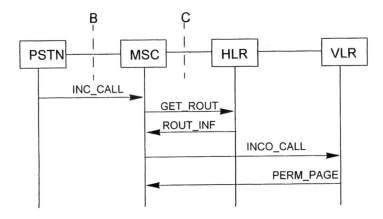

Figure 12.23 Land to Mobile Call - Routing

12.9.2.6 Handover in GSM

Handover is an integral part of any mobile cellular system, GSM is no exception. In GSM, mobile assisted handoff is used. The mobile assisted handover (MAHO) implies that the mobile provides assistance to the MSC to complete the handover procedure. In MAHO, the mobile measures the signal strength of the setup channels in the surrounding cells or candidate cells during the interburst interval and reports the measurements to MSC. MSC then takes appropriate steps.

Two types of handovers are possible, one is the inter-cell (or intra-MSC) handover and the other inter-MSC handover. In the inter-cell handover, the handover activity takes place between the base stations within the same coverage area and controlled by the same MSC. This means that the handovers are local to a certain MSC. The inter-MSC handover takes place between cells belonging to different MSCs. Obviously, the inter-MSC handover is more complex than intra-MSC. This is also similar to inter-system handover for wide area roaming.

The handover is initiated by mobile station which measures the signal strength and reports to the serving BSS. If the signal strength is found below the minimum required for acceptable service, then the BSS requests MSC for a handover. The handover request is then sent to the target BSS, which provides the best signal strength measurement. The target BSS acknowledges the request and indicating that it is ready to accept the call by sending HAND_REQ_ACK. The MSC sends HAND_COMM to the serving BSS, which relays it to the MS. The mobile on receiving this message sends a handover acceptance message. The target BSS receiving this acceptance message sends CHH_INFO to MSC and asks for handover detection message. The MS sends handover completion message to the target BSS, which relays the message to MSC. MSC then sends REL_RCH message to serving BSS, which returns the message with REL_RCH_COMP. The sequence of events that take place in Intra-MSC handover is shown in Figure 12.24.

Inter-MSC is more complex than the Intra-MSC handover because it involves the serving and target MSCs and the target VLR. In this case, the handover is also initiated by the mobile through the serving BSS, when it receives the signal strength below an acceptable level. The request is then sent to the serving MSC, which asks for handover permission from the Target MSC and the Target VLR. The Target VLR sends a message indicating the completion of necessary registration of the mobile identity in the Target VLR. The Target MSC sends a message HAND_REQ to Target BSS, which is acknowledged with a HAND_REQ_ACK message. On receipt of this message the Target MSC sends HAND_PER_ACK message to the serving MSC. The serving MSC sends NET_SETUP request to the Target MSC, which informs the serving MSC of completion of the network setup. This information is sent to serving BSS and the MS which accepts the handover decision and asks for channel information from the Target BSS. The target BSS supplies that information to the mobile, which tunes to the newly assigned frequencies and sends a message on the assigned channel. The Target BSS after

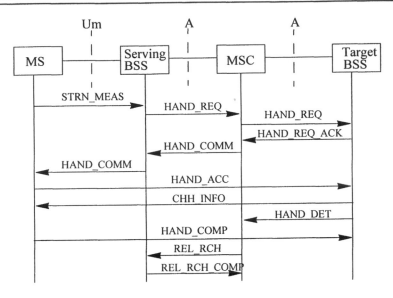

Figure 12.24 Intra-MSC Handover Signal Flows

detecting that the channel change over has been successful sends a message to the Target MSC to that effect. The MS also sends HAND_COMP message to the Target BSS, which it relays to the Target MSC. The Target MSC sends SEND_ENDSIG, which is followed by a reply from serving BSS and MSC confirming the release of the previous network. The details of Inter-MSC handover are given in Figure 12.25.

12.10 Cellular Spread Spectrum System

Prior to its use in public telecommunication system was considered, spread spectrum technology had been used for military communications for many years. The first application of spread spectrum in the civilian radio communications was for mobile satellite communications [24]. Direct sequence code division multiple access (DS-CDMA) technology was first proposed for use in terrestrial mobile environment in 1979 by Cooper and Nettleton [25]. The subject of spread spectrum communications for mobile radio systems became a center of considerable controversy whether CDMA is a right technology for mobile communications. The advantages and drawbacks of using this technology for terrestrial mobile radio communications have been discussed in some detail in Chapter 8. The successful operation of CDMA for mobile satellite communications led to application of this technology in ter-

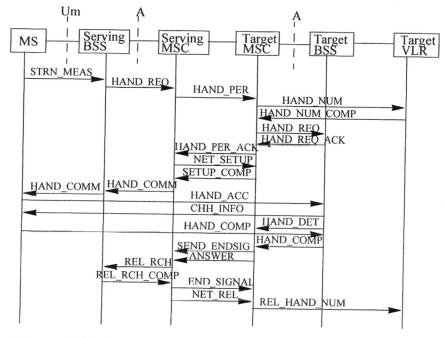

Figure 12.25 Inter-MSC Handover Signal Flows

restrial mobile radio communications. The DS-CDMA system proposed by Qualcomm was standardized by TIA as IS-95 [26]. The main aim of developing this technology was to exploit the inherent resistance of CDMA to interference and multipath propagation in achieving higher capacity in cellular systems. The spread spectrum systems are well suited to interference limited environment and the multiuser mobile radio systems operating in multipath environment is a good example of interference limited communications. This attribute of CDMA leads to the possibility of using universal frequency reuse (a frequency is reused in the adjacent cells); this being the main reason for several folds increase in the user capacity. Most of material in this section is taken from [27].

The first digital cellular standard, IS-95, is based on direct sequence code division spread spectrum technology. It is used along with the analog AMPS cellular standard with the use of dual standard portable handset. The maxim transmission rate is 9600 bits/sec, which is spread over 1.228 MHz wide bandwidth in a channel spacing of 1.25 MHz. This gives a spreading gain of 128. Intermediate data rates of 2400, 4800 are also used. The system takes advantage of voice inactivity, during which the transmission rate is reduced to 1200 bps, thus reducing the level of the multiuser interference. To maintain the interference level to the specified acceptable level, more users can be added to increase the system capacity. Figure 12.26 illustrates the basic principle of CDMA system.

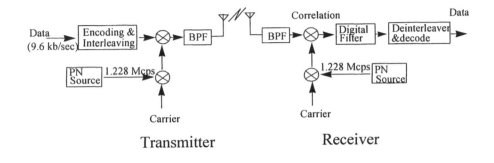

Figure 12.26 Simplified Schematic of CDMA System

In the figure, we show that the input data at 9.6 kb/sec is encoded and interleaved before it is spread by a carrier, which is BPSK modulated by a pseudorandom sequence running at 1.228 Mcps. Other modulation methods can also be used. These include QPSK or M-ary PSK or even QAM. The wideband signal thus produced is filtered by a bandpass filter/power amplifier before it is transmitted. The signal propagating through a radio channel is received by a front end filter and is simultaneously down converted and corre-lated by a pseudorandom sequence, which is identical to one used at the trans-mitter to spread the signal. The output of the correlator is low pass filtered to recover the information bearing spectrum. The output of the detector, recover-ing the data, is deinterleaved and decoded to recover the original message. The processing of data signal at the transmitter and the receiver is shown in Figure 12.27. It is seen that the relatively narrowband signal is spread over a much wider bandwidth. The transmitted spectrum is affected by the channel fading and multipath and the spectrum shape is no longer uniform over the band. Also, an interference signal, a composite of multiuser multipath signals, is also added to the transmitted signal during its propagation through the chan-nel. The correlation of the received signal with the locally generated sequence restores the desired signal spectrum whereas the interference is now spread over a wider bandwidth. With this process, the low signal to interference ratio of the received signal is thus transformed into high signal to noise ratio within the information bearing signal bandwidth. The improvement in signal to inter-ference ratio obtained through despreading of the received signal depends on the length and the correlation properties of the spreading sequence. The success in correlating the received and recovery of information depends on the accuracy of the synchronization of the locally generated sequence with that used at the transmitter.

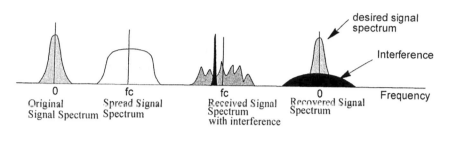

Figure 12.27 Processing of data signal at the transmitter and receiver

The system shown in Figure 12.26 is very simplistic but it conveys the essential idea of spread spectrum system. The practical systems require generation of suitable spreading codes, code timing alignment or code acquisition at the receiver, channel coding for reliability, interleaving against channel fading, power control to avoid near-far problem, and various other controls functions in order to achieve satisfactory system performance. To understand how well the above parameters contribute to working of the system, we start with the presentation of material on spreading codes including their generation and characteristics. Specifically, we shall provide details on orthogonal and pseudonoise sequences.

12.10.1 Orthogonal Functions

In IS-95, orthogonal functions known as Walsh functions are used for symbol covering in the forward channel. Two functions are orthogonal if the result of exclusive-or-ing them results in an equal number of 1's and 0's. Example, $1 1 1 1 \oplus 1 0 1 0 = 0 1 0 1$. Representing 1 by +1 and 0 -1, the result of exclusive- or-ing is zero. Thus, sequence 1111 is orthogonal to 1010. The orthogonal functions are created by repeating and inverting functions as shown in Figure 12.28. Starting from 1×1 matrix consisting of element 1, a 2×2 matrix with 1 repeated at (1,1), (1,2), and (2,1) and complemented at (2,2). This process is continued to obtain higher order Walsh functions.

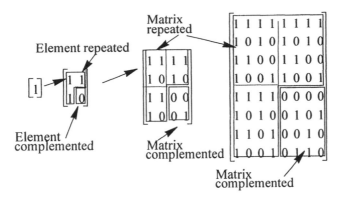

Figure 12.28 Creation of orthogonal functions

Examples of Walsh functions used in IS-95 are:

W0=
0000000000000000000000000000000000000000000000000000000000000000
PILOT

W16=
0000000000000000111111111111111100000000000000001111111111111111
TRAFFIC

W32=
0000000000000000000000000000000011111111111111111111111111111111
SYNC.

W48=
0000000000000000111111111111111111111111111111111000000000000000
TRAFFIC

W63=
0110100110010110100101100110100110010110011010010110100110010110
TRAFFIC

Orthogonal spreading is achieved by first exclusive ORing the data with the spreading sequence and taking complement of the result. For example, $1 \oplus 1100 = 0011$ and the transmitted sequence is 1100. Now the if the received data is 1100 and it is exclusive ORed with 1100, the result of the operation and complementing is 1111. This is decoded as 1.

12.10.1.1 Pseudorandom Noise (PN) Sequences

The pseudorandom noise sequences are generated using linear shift registers with a certain feedback connections similar to those shown in Figure 12.29, where a sequence of length seven chips is generated. In these full length sequences, also known as maximal length sequences, the all zero state is prohibited, with the result the number of states or sequence length is given by 2^n-1, where n is the number of stages in the shift register. In IS-95, long and short PN sequences are used. The long and the short sequences are generated using 42 and 15 stage shift register respectively. It is customary to use a seed to force the PN sequence to start in a particular state. In the figure a seed 100 is chosen.

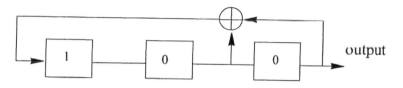

Seed = 100   Output before repeating = 0010110 (2^3-1)

Figure 12.29 PN Sequence Generation

A piece of information can also be imbedded in the PN sequence with the use of masking. For example, the electronic serial number (ESN) of the terminal can be imbedded into a sequence. This information can be used by the system for authentication. Masking can also be used to identify base stations. An example of masking is shown in Figure 12.30, where masking message is chosen to 101.

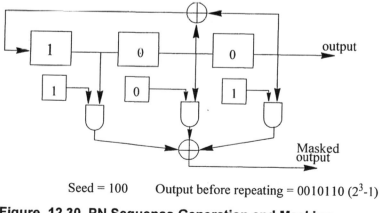

Seed = 100   Output before repeating = 0010110 (2^3-1)

Figure 12.30 PN Sequence Generation and Masking

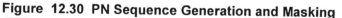

At the mobile terminal when the pilot is acquired, the mobile gets its code aligned with the short code but does not know the exact phase offset relative to 0 phase of the long code. SYNC channel provides PN offset with respect to 0 offset, which is used to determine the state of long code. Also, it can be used to derive the local time since the long and the short codes were aligned 00 GMT, on January 1, 1980.

12.10.2 IS-95 Channel Organization

IS-95 system uses many types of channels in the forward and reverse direction. Our discussion starts with architectures and generation of these channels. The composite forward channel consisting of a total of 64 channels, from W0 to W63 is organized as shown in Figure 12.31. The W0 channel is designated as the Pilot channel. Channels W1 to W7 are the Paging channels, channels W8 to W31 and W33 to W63 are the traffic channels and W32 is the Synchronization channel. Thus, a total of 56 traffic channels are available for transmission of user data.

12.10.2.1 Spreading and Scrambling Codes

Data on the forward channel is grouped into 20 msec long frames. The data is spread over the transmission bandwidth using a Walsh code and the short PN sequence clocked at a rate of 1.2288 Mcps. A 64 chip Walsh code is used for orthogonal covering of information symbol. A user assigned a Walsh code n uses the channel n (n=0 to 63). After covering by the orthogonal code, the symbols are spread in quadrature, I and Q channels with the use two short codes of length 32767 (12^{15}-1) chips, clocked at a rate of 1.2288 MHz. The two quadrature codes are generated by using different generator polynomials. This code is repeated every 26.6658 msec ($26.6658 \times 10^{-3} \times 1.2288 \times 10^{6} = 32767$). The short binary sequence is used to facilitate acquisition and synchronization by the mobile terminals and is also used to PSK modulate the signal. This short spreading sequence is called the Pilot PN sequence.

One long code (2^{42}-1 = 4.398×10^{12}) is used for scrambling and spreading the signal. The cycle time of this long codes in 41 days. Short and long codes were synchronized on Jan 01, 1980. The two codes will resynchronize or realign after 75 centuries. Their unique offsets serve as identifiers for a cell or a cell sector. The spread baseband signals in I and Q channels are filtered by I and Q pulse shaping filters. The I and Q channel signals are then up-converted using $\cos(2\pi f_c t)$ and $\sin(2\pi f_c t)$ sinusoids in the I and Q channels respectively. The I and Q signals from the other users are summed up in their respective channels before upconversion and transmission.

The long PN sequence is uniquely assigned to each user is a periodic long code with period 2^{42}-1 chips. Power control commands are sent on power control subchannel. A 0 sent on the subchannel to a mobile means a command to increase its power while a 1 is used as a command to decrease power. IS-95 specifies 16 possible power control group positions for the control bit. Power control bits are transmitted by puncturing the information bit and replacing them with a power control command. During a 1.25 msec, twenty four data symbols are transmitted. One of the first sixteen modulation symbols corresponds to a power control position. The 24 bits from the long code decimator are used to scramble data in a period of 1.25 msec. The contents of the last four bits of 24 bits are used to determine the position of the power control bit.

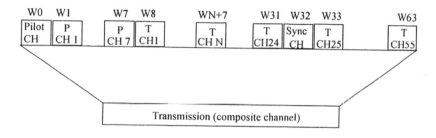

Figure 12.31 Organization of transmission composite channel

On the forward channel transmission rates of 9600, 4800, 2400 and 1200 bps can be used. The speech encoder exploits pauses in speech to reduce multiuser interference and during these pauses the transmission rate is reduced from 9600 bps to 1200 bps. This reduces the transmission power to increase the battery life and interference to other users. Indirectly it results in an increase in the system capacity. The make up of the forward CDMA channel is shown is Figure 12.32.

12.10.3 Forward Traffic Channel Generation

The forward channel frame structure is shown in Figure 12.33. There are four types of forward channel frame structure - 9600, 4800, 2400, and 1200 bits/sec. In each case, the frame is 20 millisecond long and depending on the data rates, the number of data bits per frame are different. Each frame has three fields, information field, frame quality indicator and encoder tail bits. The frame quality indicator bits are in fact CRC bits.

The frame quality indicator frame contains CRC bits. It is used for error detection and determination of transmission rate. A string consisting of 192 zeros is used as a traffic channel preamble frame. Frames of sixteen 1's followed by eight zeros are sent to represent null traffic frame. This is used as a "keep alive" indicator. Zero tailing is used to initialize the convolutional

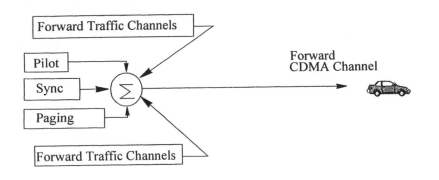

Figure 12.32 Making of a Forward CDMA Channel

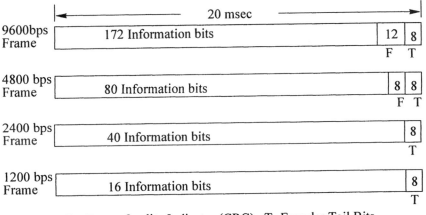

F = Frame Quality Indicator (CRC)   T- Encoder Tail Bits

Figure 12.33 Forward Channel Frame Structure

encoder. The frame staggering provision is made to allow the frames to start at offsets from the 20 msec boundaries of system time so that offsets between users created by congestion in the backhaul may be allowed at vocoders. The frame quality indicator (CRC) is used for error checking and transmission rate determination.

In the forward channel, the data is encoded using half rate convolutional coder with a constraint length $K = 9$. The encoded data is repeated once for 4800, three times for 2400 and seven times for 1200 bits/sec so that the output of the encoder delivers a constant rate of 19.2 kbps. This equalization of rates simplifies the block interleaver design. After convolutional coding and repetition, symbols are sent to a 20 ms block interleaving, which is 24 by 16 array, writing in columns beginning with the first of 16 columns. The interleaver transform this array into an output array with the use of two counters covering 9 bit array that uses a certain algorithm. The counter contents indicates the position in the interleaver matrix where the encoded data is to be written. The principle behind the interleaving procedure is shown in Figure 12.35. The output of the interleaver is at 19.2 kbps.

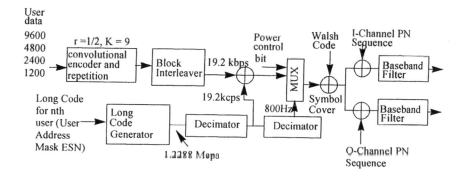

Figure 12.34 Forward CDMA Channel Modulation

	Second Counter			First Counter				
256	128	64	32	16	8	4	2	1

input array

1	25	·	·	·	·	·	· 360
2	26	·	·	·	·	·	· 361
3	27	·	·	·	·	·	· 362
·	·						·
·	·						·
·	·						·
24	48	·	·	·	·	·	· 384

First column
Output

First column Output	Counter contents
1	000000001
65	001000001
129	010000001
193	011000001
257	100000001
·	·
·	·
97	100101000

Figure 12.35 Block Interleaver Transformation

The long code is masked by the user address mask consisting of the Electronic Serial Number (ESN) of the portable. The masked long code is decimated by 64 to get 19.2 kcps, which is used to scramble data. The scrambled data is fed to a multiplexer so that the power control bit can be multiplexed at a rate of 800 bit/sec or every 1.25 msec. The multiplexed data and power control bit is symbol covered by one of 64 Walsh Codes. The spread signal is split in-phase and quadrature streams and are multiplied by I and Q channel PN sequences and filtered by wideband baseband filters. These streams are then upconvereted to the transmit RF frequency before transmission. The forward channel parameters are given in Table 12.8. The number of PN chips per modulated symbol is 64 and the number of PN chips/bit are 128, 256, 512, and 1024 for 9600, 4800, 2400 and 1200 bits/sec data rates respectively. The higher spreading factor gains for the lower data rates is compensated by reducing the symbol energy as much as to 1/16 of the energy for 1200 bits/sec. This procedure equalizes the bit energy for all data rates.

Table 12.8   **Forward Channel Parameters**

Parameter	Data Rate bps				Units
	9600	4800	2400	1200	
PN Chip Rate	1.2288	1.2288	1.2288	1.2288	Mcps
Code Rate	1/2	1/2	1/2	1/2	Bits/Code-Symbol
Code Repetition	1	2	4	8	ksps
Mod Symbol Rate	19.2	19.2	19.2	19.2	ksps
PN Chips/Mod Symbol	64	64	64	64	Chips
PN Chips/bit	128	256	512	1024	Chips
Frame Size	172	80	40	16	bits
Zero Tailing	8	8	8	8	bits
CRC	12	8	0	0	bits
Es	Eb/2	Eb/4	Eb/8	Eb/16	

12.10.3.1 Pilot Channel

The Pilot Channel plays an important role in IS-95 system. It is transmitted continuously to allow mobiles to acquire system parameters. It has a unique PN Offset code for each cell or a cell sector. Approximately 20% of the total radiated power lies in the Pilot. The pilot provides the mobile with the means by which it measures signal strength for comparison. Its functions include allowing the mobile to acquire the system (timing reference), providing mobile with signal strength for comparison and through unique offsets (phase reference), it allows for identification of a sector or a cell. The total number of possible offsets is 2^{15}. The pilot channel is generated as shown in Figure 12.36. The pilot channel is unmodulated so that its coherence time is limited by channel coherence time. It consists of all 0's, which are spread by W0 clocked at 1.2288 Mcps. The use of pilot establishes many to many relationships between the cell and the mobiles i.e. it allows for distinguishing cells. Since there is no need for phase locked loops for acquisition, therefore the system is robust. The spread pilot signal is scrambled by I and Q channel PN sequences. The Pilot channel provides to the users the phase and timing references and the means of measuring signal strength.

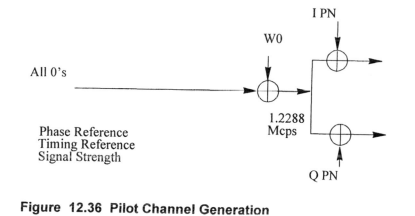

Figure 12.36 Pilot Channel Generation

12.10.3.2 Paging Channel

The paging channel provides mobiles with system parameter message, a list of base stations in the neighborhood of the currently serving base station, mobile location, message containing access parameters and a list of CDMA channels that the serving cell base station is using. The Paging Channel is used by the base station to page a particular mobile, to transmit overhead information and to assign traffic channels to mobiles. It uses transmission rate of either 9600 or 4800 bits/sec. Figure 12.37 shows generation of paging channel and long mask used for the paging channel.

The paging frame format is divided into 80 msec slots. The mobiles can operate in slotted or unslotted mode. The slotted mode is optional and mobiles only monitors 1 or 2 slots during a slot cycle. The slot cycle is in units of 1.28 seconds (Max length of slot cycle = 2048 slots). The main object is to preserve mobile terminal battery. In the non-slotted mode the mobile monitors all the time and the mobile must receive a valid message at least every three seconds.

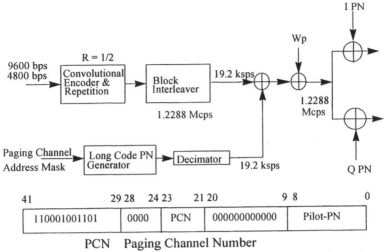

PCN   Paging Channel Number
Pilot-PN PN offset for the Forward CDMA Channel

Figure 12.37 Paging Channel Generation and Paging Channel Long Code Mask

The user data at 9600 or 4800 bits/sec is convolutional encoded at rate 1/2 and the coded data is block interleaved. The long PN code is masked by paging channel address mask. The masked code is decimated by 64, which outputs 19.2 ksps. This is used to scramble encoded data. The 19.2 kbits/sec data is then spread with the Walsh code, Wp, reserved for paging channel. The spread signal is then scrambled with I and Q PN sequence. The long code mask structure is also shown where bit 0 to 8 provides PN offset for the forward CDMA channel, bit 9-20 zero bits, 21-23 page channel number, and bits 29-41 a 12 bits sequence.

12.10.3.3 Sync Channel

The Synchronization Channel is used by mobiles to synchronize with the system clock. This channel transmits at 1200 b/sec rate. The sync message is accompanied by information on Pilot PN Offset, System Clock, Long PN Code, CAI Rev. Level, System ID, Network ID, and Paging channel data rate. Table 12.9 lists the sync channel parameters. In Figure 12.38, a schematic of sync channel generation is shown. The 1200 bits data is encoded using rate 1/2 convolutional encoder. The output is repeated and interleaved to get a rate of

4800 bps. The 4800 data rate is then spread by W32 Walsh code clocked at 1.2288 Mcps. The output is then scrambled by I and Q channel PN codes.

Table 12.9   ***Sync Channel Parameters***

Parameter	Data Rate (bps)	Units
	1200	
PN Chip Rate	1.2288	Mcps
Code Rate	1/2	Bits/Code Symbol
Code Repetition	8	ksps
Mod Symbol Rate	64	Chips
PN Chips/Mod Sym	256	chips
PN Chips/bit	1024	Chips

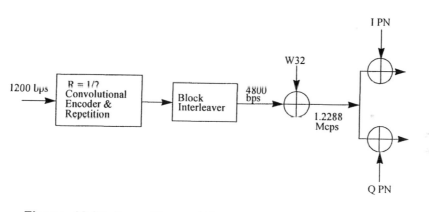

Figure 12.38 Sync Channel Generation

12.10.3.4 Convolutional Coding

The rationale for using convolutional codes is that it does not increase bandwidth. Though a higher constraint length yields better performance, it increases the decoding complexity. For example, when complexity and performance gain trade-off is considered, the $K = 11$ is not favored because only an additional gain of 0.5 dB is obtained with a substantially increased complexity. The other factors in choosing the code rate does not strongly influence complexity but they do influence performance. The encoder schematic is shown in Figure 12.39. The encoder is described by the generator polynomials:

$$G_0(x) = 1 + x + x^3 + x^5 + x^6 + x^7 + x^8$$

$$G_1(x) = 1 + x^4 + x^5 + x^6 + x^8$$

(12.58)

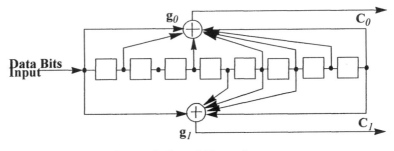

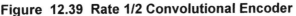

Figure 12.39 Rate 1/2 Convolutional Encoder

Whenever the user rate is less than 9600 bps, each symbol from convolutional encoder is repeated before it is interleaved using a block interleaver. When the information rate is 4800 bps, each code symbol is repeated once, the repetition rates for 2400 and 1200 each code symbol is repeated 3 or 7 respectively. Repetition maintains constant output rate of 19.2 kbit/sec, therefore the interleaver design is simplified. After the data streamed is interleaved, the repeated symbols are punctured according to some algorithm to remove repeated symbols. This increases the symbol energy. For example, symbols at 4800 bits/sec are twice as longer as the symbols at 9600 bits/sec.

The symbol repeating does not increase the energy of repeated symbols because corresponding power level is adjusted to maintain the energy needed to detect a symbol. For example, at rate 4800, twice as many bits are transmitted at half the energy, and for at 2400 bits/sec four times as many bits are transmitted at one quarter the energy.

12.10.3.5 Data Scrambling Decimator

The long sequence at 1.2288 Mcps is masked by a Long Code Mask. The forward channel public long mask is a sequence of bits 0 to 32, which defines a user specific permutated ESN. The permutated ESN is given by $\{E_0, E_{31}, E_{22}, E_{13}, E_4, E_{26}, E_{17}, E_8, E_{30}, E_{21}, E_{12}, E_3, E_{25}, E_{16}, E_7, E_{29}, E_{20}, E_{11}, E_2, E_{24}, E_{15}, E_6, E_{28}, E_{10}, E_1, E_{23}, E_{14}, E_5, E_{27}, E_{18}, E_9\}$. The bits 32 to 41 are 1100011000. The encoded and interleaved data at 19.2 ksps is modulo-2 added with the masked long sequence and then decimated by a factor 24. Each symbol of duration 52.0833 msec (19.2 kbps rate) is thus scrambled by 64 chips. The data scrambling is done to avoid probability of pilot reuse error, whereby the mobile might demodulate a distant cell that uses the same PN offset. This is of course unlikely if the code offsets are planned well. It also controls peak to average power ratio by randomizing the bits. For example, data stream consisting of a long sequence of 1's will result in high peak to average transmit ratio. Randomizing bits avoids this. It also reduces the chances of frame corruption. The decimated long code at 19.2 ksps is further decimated by 24 to generate sequence at 800 symbols per second, which is used to mark the instances at which power control symbols are inserted into the data stream. This is shown in Figure 12.40.

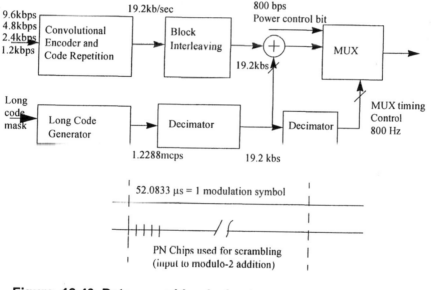

Figure 12.40 Data scrambler decimator

12.10.3.6 Demodulation of the Forward Channel

The spread spectrum signal received at the mobile is filtered and down converted to the baseband. It is passed to a set of four correlators, which includes a correlator known as searcher correlator. The searcher correlator searches for the pilot and provides the correct phase to other fingers. The receiver is capable of combining signals from three path separated by approximated 814 nanosec. The combining can be done by using one of several diversity combining algorithms. When the signal is combined using, for example, maximal ratio combining or equal gain combing, the signals on the three paths must be cophased. The phase aligning information is supplied by the search finger. The output of the combiner is deinterleaved and then decoded using Viterbi decoder. The decoded data is supplies to the vocoder, which delivers the speech signal. The forward channel demodulation process is shown in Figure 12.41. In Figure 12.42, the process of combining the outputs of the three finger is illustrated. The channel is assumed to have three paths and after the chip delays are compensated, the three paths may still have some residual phase difference. The three paths are co-phased and combined to get a larger signal output.

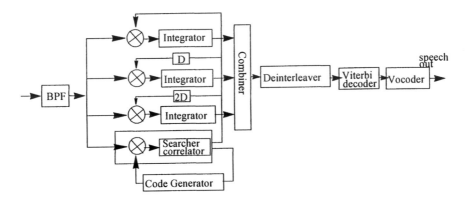

Figure 12.41 Demodulation of the Forward channel

The location of paths is separated by one chip interval (813.8 nanosecond), thus channel impulse response length of 2.4 µsecond can be handled. If the channel delay spread is less than 0.8 µsec, the Rake receiver combining will not provide any gain.

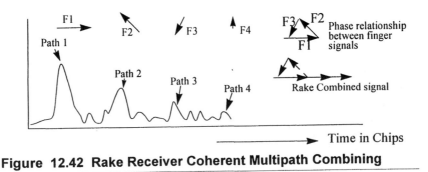

Figure 12.42 Rake Receiver Coherent Multipath Combining

12.10.3.7 Variable Rate Data Transmission

IS-95 provides variable rate data transmission. Data rate of 9600, 4800, 2400, and 1200 bits/sec are allowed. A variable data rate transmission over the forward channel example is shown in Figure 12.43.

A 20 milliseconds long data frame consists of 192 bits or 576 code symbols, or 96 Walsh symbols. The frame also consists of 16 power control groups. Each power control group is 1.25 second long. When the data rate is 9600 bits/sec all 16 power control group has data. When the data rate is 4,800 bits/sec, the data repeated once is punctured so that only 8 of the 16 power control groups has data. Puncturing of data takes place according to some algorithm. In the figure control group symbols 1, 3, 4, 7, 8 10, 13 and 14 are removed. Similarly when the data rate is 2,400, three repeats of data are removed according to some algorithm. In the figure, symbols 2, 5, 9, and 15 are retained while all others are removed. In a similar fashion, the data stream at 1200 is dealt.

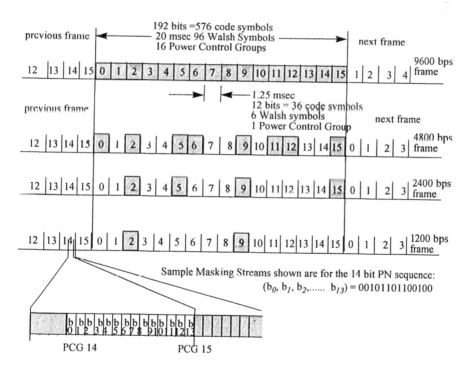

Figure 12.43 Variable Data Rate Transmission

12.10.4 Reverse CDMA Channel

Over the reverse channel which is used in the mobile to base direction, the transmission takes place over two types of channels - Access channel and traffic channel. This is shown in Figure 12.44.

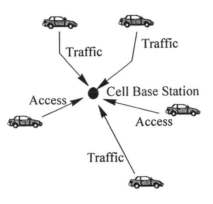

Figure 12.44 Making of a Reverse CDMA Channel

The reverse channels are addressed or identified by Long Code PN sequence. Out of a total of 63 reverse channels, some channels are reserved for system access. These channels are known as access channels. These channel share the same frequency assignment, but are identified by the user long code. The access channel is used by the mobile to gain access to the system and to respond to messages received over the paging channels. The data on the receivers channel are convolutionally encoded, block interleaved and modulated by a 64-ary orthogonal modulation and spread preceding transmission. The reverse channel parameters are listed in Table 12.10. The reverse traffic channel generation is shown in Figure 12.45.

The reverse channel is usually regarded as less reliable than the forward channel, therefore data on this channel given better protection compared to that on the forward channel. In the reverse channel rate 1/3 with a constraint length 9 convolutional encoding is used. As in the case of forward channel, coded bits are repeated before interleaving when the data rate is less than 9600 bps. After repetition, the coder delivers a fixed symbol rate of 28,800 bps. The block interleaver is an array with 32 rows and 18 columns and spans 20 msec. The code symbols are written into the matrix by columns and read out by rows. This provides an interleaving depth of 32. A 64-ary orthogonal modulation is used for the reverse CDMA channel. A group of six coded bits determines which Walsh function out of possible 64 is to be transmitted. Within each Walsh function a sequence of 64 Walsh Chips are transmitted.

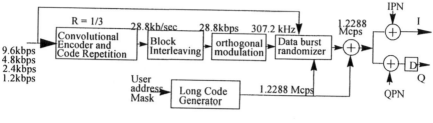

D = 1/2 PN Chip Delay

Figure 12.45 Reverse Traffic Channel

Table 12.10   *Reverse Channel Parameters*

Parameter	Data Rate (bps)			
User Data Rate	9600	4800	2400	1200
Code Rate	1/3	1/3	1/3	1/3
TX Duty Cycle(%)	100	50	25	12.5
Coded Data Rate(sps)	28,800	28,800	28,800	28,800
Bits/Walsh Symbol	6	6	6	6
Walsh symbol rate	4800	4800	4800	4800
Walsh Chip rate(kcps)	307.2	307.2	307.2	307.2
Walsh Symbol Duration μsec	208.33	208.33	208.33	208.33
PN Chips/Code Symbol	42.67	42.67	42.67	42.67
PN Chips/Walsh Symbol	256	256	256	256
PN Chips/Walsh Chip	4	4	4	4
PN Chip Rate (Mcps)	1.2288	1.2288	1.2288	1.2288

12.10.4.1 Rate 1/3 Convolutional Encoder

On the reverse channel rate 1/3 code is used instead of rate 1/2 rate used on the forward channel. This is probably due to a greater need of protecting data on the reverse channel. The encoder block diagram is shown in Figure 12.46. The output symbols are derived from the following three polynomials.

$$C_0 = 1 + X^2 + X^3 + X^5 + X^7 + X^8$$

$$C_1 = 1 + X + X^3 + X^4 + X^7 + X^8 \qquad (12.59)$$

$$C_2 = 1 + X + X^2 + X^5 + X^8$$

The repetition of symbols maintains constant 28.8 ksps output. The repetition rates for 9600, 4800, 2400, and 1200 are 0,1,3,7 times respectively. For example, each symbol of the input data at rate 2400 bits/sec is repeated 3 times, that is each symbol results in 4 identical symbols. The symbol repetition simplifies the implementation of the interleaver.

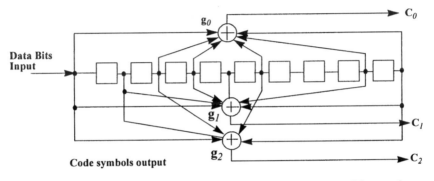

Figure 12.46 Convolutional Encoder for the Reverse Channel

Interleaving is done by first writing into the interleaver by columns and read out by rows. For this purpose a matrix 32×18 is used. The rate is 9600 bps. The orthogonal modulation block uses a Walsh look up table. The input data rate of 28.8 kbits/sec is divided into 6 bit words, resulting in an output rate of 4800 words per second. The output rate is thus 4800×64 = 307.2 kchips/sec. Each six bits word is used to select a 64 chip pattern. For example, binary 100011 is equal to decimal 35, so the Walsh code #35 is chosen, which is given by 01100110...0110100110011001...1001. The use of M-ary signalling set improves non-coherent demodulation process.

The output of the orthogonal modulator is sent to data burst randomizer. The use of randomizer is to reduce interference. It also makes the power control easier if symbols are sent at full power. The data is divided into 20 msec frames consisting of 192 bits or 576 code symbols. This is divided into 16 power control groups. Each power control group consists of six Walsh symbols. For 9600 bps frame, the code symbols given by the rows of the interleaver are transmitted in order. For lower data rates, since the code symbols are repeated therefore choice is made for which code symbol is to be transmitted. For 4800 bps, the code symbols are repeated once and therefore the randomizer discards one of each pair of code symbols. Thus, the transmission is rate is halved as compared to that at 9600 bps. The symbols have the same power but lower data rate means that the signal is off on the average half the time but the location of these symbols are randomized so that not all the mobiles transmit at the same time. This results in a reduction in interference and yields higher capacity.

Direct Sequence Spreading is accomplished by modulo 2 sum of 1.2288 mcps masked long code with 64 chip patterns of Walsh coded six bit symbols. The channel mask is derived from ESN. The reverse CDMA channels are addressed by long code PNs. The baseband 1.2288 Mcps chip sequence is filtered and then upconverted at the carrier frequency by using short codes in the *I* and *Q* channels. The 1/2 symbol offset in the OQPSK modulator is used. This is used to maintain a relatively constant envelope. The baseband filter is similar to the one used for forward channel.

Over the reverse traffic channels, frames can start at offsets from system time boundaries spaced 20 msec. The frames consist of Frame Quality Indicator and traffic channel preamble of 192 zeros. The null traffic data is sent at 1200 bps and is indicated by a frame of 16 ones's followed by 8 zero's.

Access Channels are used by mobiles to access system when a traffic channel is not assigned to it. Using this channel, mobiles register with the system, originate calls, and respond to page. Access channels are paired with Paging channels. There are 32 access channels corresponding to one paging channel. The access channel number is chosen by the mobile after scanning for the strongest. The transmission rate on these channels is 4800 bps. The structure of the access channel is similar to that of traffic channel except that the data burst randomization is not used in this case. The access channel frame consists of 88 information bits and 8 tail bits. There is no separate CRC for the frame as the message CRC is used for error detection. The access channel long code mask is given in Figure 12.47. The Base-ID is unique over entire system, but the Pilot_Pn number is periodically repeated.

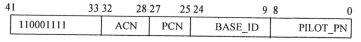

41	33	32	28	27	25	24	9	8	0
110001111		ACN		PCN		BASE_ID		PILOT_PN	

ACN = Access Channel Number
PCN = Paging Channel Number
BASE_ID = Base Station Identification
PILOT_PN = PN offset for the forward CDMA channel

Figure 12.47 Access Channel Long Code Mask

The parameters of the access channel are summarized in Table 12.11.

Table 12.11 **Parameters of the access channels**

Parameter	Data Rate (4800 bps)	Units
PN Chip rate	1.2288	mcps
Code Rate	1/3	Bits/Code Symbol
Transmit Duty Cycle	100%	%
Code Symbol rate	28.8	ksps
Modulation	6	code/sym/mod sym
Mod Symbol rate	4800	sps
Walsh Chip Rate	307.2	kcps
Mod Sym Duration	208.33	microseconds
PN Chips/mod Sym	256	PN chips/mod sym
PN Chips/Walsh Chip	4	Energy

12.10.4.2 Demodulation of Reverse Channel

The demodulation of the reverse CDMA channel is very similar to that for the forward channel. The demodulator structure is shown in Figure 12.48. The search correlator looks for preamble of zeros to search for the correct phase. Fast Hadamard Transform (FHT) is used in each finger to recover the signal. The energy for each Hadamard row is compared to make a decision on which row is received. The structure is shown in Figure 12.48.

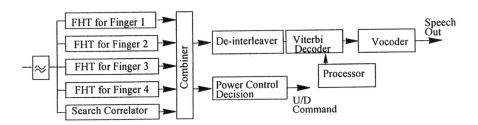

Figure 12.48 Demodulation of the Reverse CDMA Channel

The signal from each finger is combined to produced the demodulated signal. It is deinterleaved, decoded using Viterbi decoder to produce recovered speech signal. In Figure 12.49, the schematic of the finger demodulator is shown. The *I* and *Q* inputs are down converted and sampled at 8 samples/chip. The samples are digitized and despread using *I* and *Q* pseudnoise sequences. The sum of the *I* and *Q* outputs is converted into the data symbols using Fast Hadamard Transform. The outputs of four fingers are then combined.

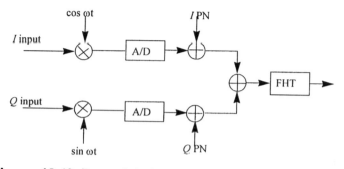

Figure 12.49 Demodulation of Finger signal

12.10.4.3 Power Control

In DS CDMA, the control of transmit power is necessary in order to counter the near-far effect. A mobile user near a cell base station can completely swamp out the signal from a user located at relatively longer distance if both users transmit same power. Depending on the relative distances from the base station, even with a processing gain of 128 (21 dB), it may not be possible to hear the desired user located at the edge of the cell. Power control is a viable solution to the problem of near-far effect. It has been shown that the optimum solution to this problem is to control the output powers of all users such that all signals reaching the base station from spatially dispersed users arrive at the base station having either the same absolute power or signal to noise ratio. IS-95 power control uses a two step process to eliminate near/far effect.

   (i) Open Loop Estimation

   (ii) Closed Loop Correction

In IS-95, the mobile transmit power, Txm, in dB uses the following algorithm.

Mobile Transmit Power (dB) = -73 - mobile receiver power (dB) + nominal power + initial power + access probe correction + closed loop correction.

The dynamic range of open loop power is 80 dB, whereas the closed loop corrections may fall in the range of ± 24 dB around the average open loop estimate power level. The open loop PC uses a coarse estimation technique and its purpose is to compensate for mean path loss variations and shadowing. Thus, it has a large dynamic range. It has also a relatively fast acting response, the closed loop can not keep up with the fast changes. If the response is too slow, it will not compensate for the shadowing fast enough. If the control is too fast, then it will compensate for fast fading on the forward link but will produce spikes on reverse link. After many tests a time constant of around 20 msec is chosen. The coarse estimation depends mildly on the initial *a priory* information or initial offset in access message. The basic open loop rule is,

TX Power (dBm) = -73 - Rx Power (dBm)

The mobile estimates transmitted power by measuring the total received power i.e. sum of powers from all users, in case of greater number of users, the received power will be higher. Note that for this measurement, demodulation of signal on the forward link is not necessary. All powers are measured in 1.23 MHz bandwidth. The figure of -73 dB reflects several system factors (e.g. Cell ERP, Cell Loading etc.). With -73 dB rule, closed loop correction usually averages close to zero dB. The general open loop rule is

Tx power (dBm) = -73 - Rx Power (dBm) + NOM_PWR + INIT_PWR + Access Probe Corrections

Open loop tolerance requirements recommends ± 6dB but ± 9 dB is actually required, because it is difficult to implement a tighter control. However, a loose tolerance is undesirable during access probing since closed loop is not available. Tolerance requirement is to be met over full range of NOM_PWR (-8 to +7 dB). Open loop should not be too fast, a reason for this requirement has been given above. The function of the closed loop power control is to keep the received power up to the correct estimate. Estimate is usually off the mark due to independence of forward and reverse link fading. The measurement time constant choice also results in different variance in the presence of fading. The measurement time constant of approximately 20 msec results in a loss of 0.2 dB.

12.10.4.4 Closed Loop Power Control

The open loop power control is susceptible to fading and shadowing and therefore is less accurate. Closed loop PC is faster than open loop PC; it can eliminate open loop errors. Also, since the forward and reverse links operate at frequencies that are separated by 45 MHz, closed loop correction can compensates for asymmetries in forward and reverse links. The variations due to mobile and cell path gains are also compensated. It has a dynamic range of

± 24 dB around open loop estimate, which is smaller than that for open loop. The basic idea of closed loop PC is to assist mobile in controlling the transmit power. The base station measures received SNR and compares it to a certain threshold. If the received power is greater than the threshold, then the cell sends the power down command, and if it is below the threshold send the power up command. Closed loop is integrated into modulation on the forward link by puncturing (removing) certain code symbols. The puncturing and replacement by power control commands reduces the coding gain slightly. This was mentioned earlier in the description of forward link. The process of puncturing data bit and replacement power control bit is shown in Figure 12.50.

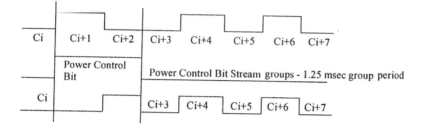

Figure 12.50 Insertion of Power control bit

The SNR is measured following a specific Procedure. In each power group there are six Walsh symbol, each Walsh symbol has 64 chips. During each Walsh symbol period, the maximum value of the 64 chips is computed. Six of these values are added together to get the measure of signal strength. This measure of signal strength is compared to a threshold. If the measure is greater than the threshold, a DOWN command is sent, if it is lower than the threshold, an UP command is sent.

The UP command relates to 1 dB increase in the transmit power and the DOWN command means a decrease in the transmit power by 1 dB. The power control channel has the effect of changing the voice activity factor since power control bits are sent at 800 bps with full rate power. Puncturing increases voice activity, because 1/12 (9600/800) of bits are always transmitted at full power. For example, if voice activity with no power control is 0.4, then voice activity with power control is 0.4×11/12+1×1/12 = 0.45. Thus, the voice activity on the forward link is slightly higher than that on the reverse link.

Power control bits are uncoded to reduce delay, this may result in an error rate, which is higher by 5-10%. If the power is measured on the reverse channel in (say) power group 5, then the power adjustment in the forward channel may be done in either power control group 6 or 7. Thus, power control adjustment suffers a delay d: $1 < d < 2$. This, however, has a minor impact on the overall performance of power control system but a delay lager than this may result in a degradation in the accuracy of power control adjustment, which in turn causes a loss of performance greater than in the case when control bits remained uncoded. In other words, delay is more critical than a slight increase in the error rate. Effectively, power control bits are transmitted only when the mobile is able to send data in the power control group. The lower the frame rate, the few power control commands are interpreted. For example, at full rate 16 Power Control Commands/Frame are available but at 1200 bps, only two Power Control Commands/frame are available.

A peculiar situation may arise during handover. Since the mobile may have communication links with more than one base station or sectors, it may receive conflicting power control commands from those cells or sectors. The general rule is biased downwards, that is the mobile is to follow the DOWN command if any sector or cell commandos that.

Forward power control is used to regulate forward traffic channel such that target forward FER of 1% is achieved. Target FER is usually a function of load i.e. calls in cell/sector.

Reverse open loop power control is based on the power received on the reverse channel (RX power). The reverse closed loop power control is based on SNR, while the forward power control (FPC) is based on FER. Forward Power control is slower process relative to reverse power control. In FPC mobile keeps track of number of received frames it determines to be in error, and the total number of frames received. Mobile sends a power measurement report message (PMRM) in two ways: periodically and after a number of frames in error cross a certain threshold. The base station takes action based on the reported FER: if FER > threshold, traffic channel is directed to increase gain, if FER < threshold, traffic channel is directed to decrease gain, and if no PMRMs are sent in some specified period decrease the gain.

12.10.4.5 Maximum Transmit Power

For each mobile class, maximum transmit power of CDMA mode is 5 dB less than the maximum of the analog mode, yet CDMA coverage is equal to or better than the coverage for analog systems. This backoff simplifies the linearity requirement of the power amplifier and decreases battery consumption at maximum power. The mean transmit power is usually 10 dBm. Field results show that mobiles seldom use maximum output power. Maximum output power is selected at the worst case scenario that is at the edge of coverage operation. For ease of production, the minimum transmit power is specified to be -50 dBm/1.23 MHz (-111 dBm/Hz), which is significantly higher than thermal noise floor of -174dBm/Hz. With the -50 dBm limit, mobiles are able to transmit to a cell over a short distance over which only 70dB transmission loss occurs. The -50dBm limit is about 80 dB less than the 31 dBm max ERP of class I mobiles. The station classes and their output power for analog and CDMA systems are shown in Table 12.12.

| Table 12.12 | Station Classes and Output Power Comparison with Analog System | |

Station Class	Analog Nominal ERP	CDMA Normal ERP
I	6 dBW	1 dBW
II	2 dBW	-3 dBW
III	-2 dBW	-7 dBW

The closed loop power control sub-channel use is shown in Figure 12.51. The 20 msec frame consisting of 96 Walsh symbols are divided into 16 power groups. Each power group consists of six data symbols. The signal to noise ratio is measured by the base station in one of the 16 power groups (power group 5 in the figure). The measured signal to noise ratio is converted into control bit and that control bit is transmitted in forward traffic channel.

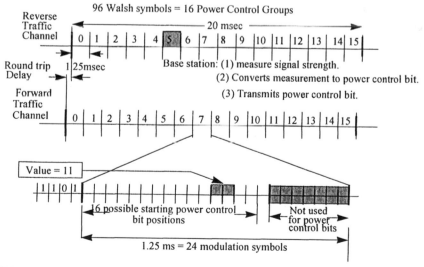

Figure 12.51 Closed Loop Power Control Subchannel

The round trip offset between the forward and the reverse channel does not allow insertion of the control bit in the same power group in which the measurement was made. Thus, it can only be inserted in the next two power groups (7 in figure). The power control bit can be inserted in the first 16 possible starting power control bit positions. This position has to be identified by the base station so that the mobile station can read it. This information is included in the power control group 6 in the last four modulation symbols of the group. For example, in the figure the last four symbols of the power groups 6 contains 1101, which indicates that the power control bit resides in the 11 symbol of the power control group.

12.10.4.6 CDMA Handoffs

The purpose of handoff is to retain continuity of the call whenever the mobile moves from one cell to another. Handoffs can be divided into two types: intra-system and inter-system. The intra-system handoff can be between multi-cells, multi-sector and multi-cell/multi-cell. The intersystem is a handoff between two CDMA systems or between CDMA and AMPS or any other system. The intra-system handoffs take place within a single system. These handoffs are mobile assisted and take place between channels on the same carrier frequencies. These handoffs use "make-before-break" concept. Multisector handoff is of softer kind. It can involve up to 2 sectors of the same cell. The voice data is combined at the cell and is passed as one frame to the telecommunication switching office. The mobile during this time receives the signal on the forward link from each cell that is signal spread by two Walsh codes.

In the multicell handoff up to 3 cells may be involved at a time. Each cell transmits a voice frame. In the multisector handoff, up to two sectors of the same cell take part. The cell transmits a combined voice frame. The number of channels required for multi-cell handoffs depends on the channel elements involved in the handoff. For example, a three way handoff will require 3 traffic channel elements. In CDMA, the multicell cell handoff increases the net system capacity. This increase is despite making a provision to receive signals at different base stations or sectors, which results in certain increase in the multiuser interference.

It has been estimated that for 60% of the time, no handoff is required, for 30% of the time two way handoff is required and for remaining 10% of the time 3-way handoff is required.

In handoff processing both mobile and base station take active part. The mobile detects new pilot signal with sufficient strength to be used for call handover, it sends the pilot strength measurements message to base station via existing reverse traffic channel. The base station on receipt of this message, allocates forward traffic channel associated with the reported pilot and sends handoff direction message to mobile to use new cell/sector. The mobile receives this message and begins decoding new traffic channel and sends handoff completion message on reverse traffic channels. The handoff thresholds are shown in Figure 12.52. Four sets of pilots are maintained by the mobile for handoff: Active Set, Candidate Set, Neighbor Set, and Remaining Set.

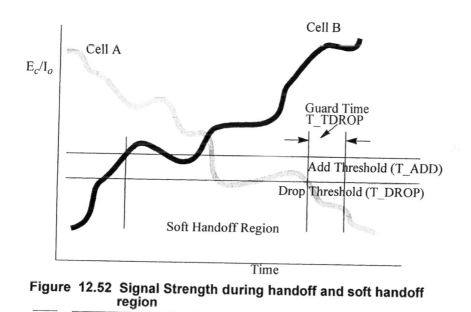

Figure 12.52 Signal Strength during handoff and soft handoff region

Search windows are specified in System Parameter Message in order to reduce the time to search for pilot signals. The search size is updated in both the system parameter message and the Handoff Direction Message. The Commands are SRCH_WIN_A, SRCH_WIN_N, and SRCH_WIN_R. Up to 16 sizes ranging from 4 chips to 452 chips long windows for the above mentioned sets are specified. The mobile maintains a certain maximum number of candidates in these sets (6 for active, 5 for neighborhood and 20 remaining). The pilots may move from one category of set to another depending on the Handoff direction message, and T_drop expiry. The membership of these sets keep changing depending on the signal conditions. A set transition takes place when the pilot strength exceeds T-ADD. The mobile station sends a pilot strength measurement message and transfers pilot to the candidate set. The base station in response sends a Handoff Direction Message. The mobile station transfers the pilot to the active set and sends Handoff Completion Message. When the Pilot strength drops below T_drop, the mobile station starts the handoff timer. After the expiry of the handoff timer, the mobile sends a pilot strength measurement message. The base station then sends a Handoff Direction Message. The mobile moves the pilot from the active set to the neighbor set and sends a handoff completion message. If the handoff direction message arrives before handoff drop timer expires then mobile station moves the pilot from the active to the candidate set.

12.10.4.7 Handoff Messages

In handoff procedure, Pilot Strength Measurement Message (PSMM), Handoff Direction Message, and handoff Completion Message are used. The pilot strength measurement message is sent whenever

(i) a new pilot detected with sufficient strength is to be used,
(ii) a neighbor set pilot or remaining set pilot is above T-ADD,
(iii) a candidate set pilot exceeds an active set pilot by a specified level (T_COMP),
(iv) strength of an active set pilot stays below T_Drop for T_TDrop seconds,
(v) when the base station has sent a handoff direction message for the sole purpose of forcing the mobile to send a PSMM.

The Pilot Strength Measurement Message Format is given in Table 12.13.

Table 12.13	Pilot Strength Measurement

Field	Length (bits)
MSG_TYPE	00000101
ACK_SEQ	000
MSG_SEQ	111
ACK_REQ	0
ENCRYPTION	00
REF_PN	000001010
PILOT_STRENGTH	010100
KEEP	1
PILOT_PN_PHASE†	000001010001100
PILOT_STRENGTH†	010100
KEEP†	1
RESERVED	0 -7 (as required)

†. One or more occurrence of the field
†. One or more occurrence of the field
†. One or more occurrence of the field

The Handoff Direction Message is used to direct the mobile station handoff. It specifies new set of forward traffic channels from base station to mobile station; it updates T_ADD, T_DROPs, T_COMP, Active Set; it contains search window sizes and also CDMA to CDMA handoff parameters. For each forward traffic channel assigned to the mobile, the Handoff Direction message contains, PN offset of each cell/sector assigned to the mobile active set, code channel with Walsh number being used on the channel, and one bit that specifies whether the associated channel is carrying identical power control symbols as the previous forward traffic channel listed in the message. The messages exchanged for handover are shown in Table 12.14.

Table 12.14	*Handover Messages*
Field	**Length (bits)**
MSG_TYPE	00000101
ACK_SEQ	000
MSG_SEQ	001
ACK_REQ	1
ENCRYPTION	00
USE_TIME	1
ACTION_TIME	000111
HDM_SEQ	01
SRCH_WIN_A	0111
T_ADD	10000
T_DROP	101000
T_COMP	0101
T_DROP	0011
FRAME_OFFSET	x1.25
PRIVATE_LCM	0
RESET_L2	0
ENCRYPT_MODE	00
FREQ_INCL	1
CDMA_FREQ	0 or 11
PILOT_PN[1]	000001010
PWR_COMB_IND[1]	1
CODE_CHA[1]	00111111
RESERVED	0 -7 as required

[1].One or more occurrence of the field

Handoff Completion Message contains PN offset of each pilot in the active set. The pilots are listed in same sequence as in the Handoff Direction Message. It is used in a hard handoff to tell new system the mobile's current active set. The messages related to this are listed in Table 12.15.

Table 12.15	Handover Offset Messages

Field	Length (bits)
MSG_TYPE	00001010
ACK_SEQ	000
MSG_SEQ	011
ACK_REQ	0
ENCRYPTION	00
LAST_HDM_SEQ	10
PILOT_PN†	000001010
RESERVED	0-7 (AS REQUIRED)

†. One or more occurrence of the field

12.10.4.8 Call Processing in IS-95

12.10.4.8.1 Registration

Registration is the process in which a mobile identifies itself to a cellular network by providing details on its location, identity, status, and its characteristics. In particular, the mobile provides its Electronic Serial Number (ESN), Mobile Identification Number (MIN), preferred slot cycle (SLOT_CYCLE), Station Class Mark (SCM) and Protocol Revision Number. The MIN points to a given HLR. The registration process begins with mobile sending a registration message, the base station relays the message, AuC verifies the mobile and MSC notifies the HLR. The mobile moves the HLR records to the new location and reroutes the mobile calls. The function of HLR is to check if the mobile is registered elsewhere and places the pointer to MSC. The MSC updates its VLR.

There are nine forms of registration; five forms are autonomous which are initiated by mobiles and conditioned upon roaming status. One form of registration is independent of roaming status and the other three types of registration are non-autonomous. The autonomous form of registration are power up registration, power down registration, timer based, distance based, and zone based registration. The power-up and power down registration starts when mobile powers on and powers down respectively.

12.10.4.8.2 Call Processing During Handoff (from cell A to Cell B)

Table 12.16 illustrates call processing procedure during handoff between Cell A and Cell B.Call Processing During Handoff

Table 12.16   *Call Processing Messages During Handovers*

Mobile Station		Base Station
(User Conversation using A)		User Conversation using A
Pilot B Strength exceeds T_ADD		
Sends Pilot Strength Measurement Message	>Reverse Traffic Channel>	A receives Pilot Strength Measurement Message
		B begins transmitting traffic on the Forward Traffic Channel and requires the reverse traffic channel
Receives Handoff Direction Message	<Forward traffic Channel<	A and B send Handoff Direction Message to use A and B
Acquires B: begin using active set {A,B}		
Sends Handoff Completion Message	>Reverse traffic Channel>	A and B receive Handoff Completion Message
Handoff drop timer of pilot A expires		
Sends Pilot Strength Measurement Message	>Reverse Traffic Channel>	A and B receive Pilot Strength Measurement Message
Receives Handoff Direction Message	<Forward Traffic Channel<	A and B send Handoff Direction Message to use B only
Stops diversity combining; begins using active set {B}		
Send Handoff Completion Message	>Reverse Traffic Channel>	A and B receive Handoff Completion Message
		A stops transmitting on the forward channel and receiving on the reverse traffic channel
(User Conversation using B)		(User Conversation using B)

It is also possible that during the time, the mobile is using cells A and B and at certain time the handoff drop timer of one base station expires and the pilot of another base station C exceeds T_ADD then the procedure of handoff is very similar to one given in Table 12.16.

12.10.4.8.3 Mobile Station Origination

The messages exchanged when a mobile station originates a call are given in Table 12.17.

Table 12.17	*Messages for Mobile Station Call Origination*

Mobile Station		Base Station
Detects User-initiated call		
Sends *Origination Message*	>Access Channel>	Sets up traffic channel
		Begins sending null traffic channel data
Sets up Traffic Channels	<Paging Channel<	Sends *Channel Assignment Message*
Receives N$_{5m}$ consecutive valid frames		
Begins sending the traffic Channel Preamble	>Reverse Traffic Channel>	Acquires the Reverse Traffic Channel
Begins transmitting null traffic channel data	<forward Traffic Channel>	Sends base station acknowledgment order
Begins necessary primary traffic in accordance with Service Option 1	<Forward Traffic Channel<	Sends *Service Option Response Order*
User Conversation		User Conversation

12.10.4.8.4 Mobile Station Initiated Call Disconnect

The exchange of messages when a mobile initiated call is disconnected are listed in Table 12.18.

Table 12.18	*Messages for Mobile Station Initiated Call Disconnect*

Mobile Station		Base Station
Detects User-initiated disconnect		
Sends Release Order	<Reverse Traffic Channel>	
	<Forward Traffic Channel<	Sends *Release Order*
Enters the *System Determination Substate* or the *Mobile Initialization State*		

12.10.4.8.5 Mobile Station Terminated Call

The messages exchanged between the mobile and base station for establishment of a mobile terminated call are listed in Table 12.19.

Table 12.19   ***Messages for Mobile Terminated Call***

Mobile Station		Base Station
	<Paging Channel<	Sends Page Message or Slotted Page Message
Sends Page Response Message	>Access Channel>	Sets Up Traffic Channel
		Begins sending null traffic Channel Data
Sets up Traffic Channel	<Paging Channel<	Sends Channel Assignment Message
Receives N_{5m} consecutive valid frames		
Begins sending the traffic channel preamble	>Reverse Traffic Channel>	Acquires the Reverse Traffic Channel
Begins transmitting null Traffic Channel data	<Forward Traffic Channel<	Sends Base Station Acknowledgment order
Begins processing primary traffic in accordance with Service Option 1	<Forward Traffic Channel<	Sends Service Option Response Order
Starts Ringing	<Forward Traffic Channel<	Sends Alert with Information Message (ring)
User answer call		
Stops ringing		
Sends connect order	>Reverse Traffic Channel>	
Begins sending primary traffic packets from the service option 1 application		
User conversation		user conversation

12.10.4.8.6 Fixed network initiated call procedure

The messages exchanged to complete a network initiated call are given in Table 12.20.

Table 12.20		*Messages for Network Initiated Call*
Mobile Station		**Base Station**
	<Paging Channel<	Sends Page Message or Slotted Page Message
Sends Page Response Message	>Access Channel>	Sets Up Traffic Channel
		Begins sending null traffic Channel Data
Sets up Traffic Channel	<Paging Channel<	Sends Channel Assignment Message
Receives N_{5m} consecutive valid frames		
Begins sending the traffic channel preamble	>Reverse Traffic Channel>	Acquires the Reverse Traffic Channel
Begins transmitting null Traffic Channel data	<Forward Traffic Channel<	Sends Base Station Acknowledgment order
Begins processing primary traffic in accordance with Service Option 1	<Forward Traffic Channel<	Sends Service Option Response Order
Starts Ringing	<Forward Traffic Channel<	Sends Alert with Information Message (ring)
User answer call		
Stops ringing		
Sends connect order	>Reverse Traffic Channel>	
Begins sending primary traffic packets from the service option 1 application		
User conversation		user conversation

12.10.4.8.7 Fixed User Call disconnect procedure

The messages exchanged when the fixed user disconnect the call are shown in Table 12.21.

Table 12.21   ***Fixed User Call Disconnect Messages***

Mobile Station		Base Station
		Detects Call disconnect
	\<Forward Traffic Channel\<	Sends release order
Sends Release Order	\>Reverse traffic Channel\>	
Enters the System Determination Substate of the Mobile Station Initialization State		

12.11 Proposals on Low Power Wireless Systems

The observations on the use of low power devices in confined spaces have led to several proposals on new personal communication systems. For example, proposals on CT2, CT2PLUS and CT3, originated from in the UK, Canada and Sweden respectively [28][29]. These proposals differ in details on available slots in frame organization. The other proposals for use in microcellular configurations are DECT [30], and WACS [31] Direct sequence spread spectrum systems have also been proposed for indoor systems [32]. Here we present some detail on DECT and WACS.

12.11.1 Digital European Cordless Telecommunications (DECT)

DECT is based on a microcellular mobile communication system which provides low power cordless access between portables and fixed points at ranges up to few hundred meters. The frequency band of operation is 1880-1900 MHz, with channel spacing of 1.728 Mhz. The total number of channels is ten. The transmit power is limited to 250mW, and each carrier consists of 24 slots per 10 msec long frame. Thus, the each slot is 0.4167 msecond. Time division duplex (TDD) has been chosen as the duplexing method with slots for forward and reverse directions on same RF carrier. The gross bit rate of 1.152 Mbits/sec is transmitted using GFSK (equivalent to GMSK with BT = 0.5). This modulation is chosen because it provides good C/I performance and better detection sensitivity when a low cost limiter discriminator detector is used. In North America, the DECT standard has been modified to allow for π/ 4-DQPSK modulation.

DECT uses dynamic channel selection (DCS) in order to take advantage of greater efficiency in three dimensional applications particularly in multi-storeyed buildings. It is three to four times more efficient in uncoordinated base station environment and six to eight times more efficient in coordinated system. Moreover, DCS does not need shadowing margin as it adjusts to compensate for shadows, increasing or decreasing the reuse distance which implies time variant cell dimensions. The highly unpredictable, irregular and fast changing shadowing effects are elegantly dealt with by the use of DCS and antenna diversity. The dynamic channel selection is made on the basis of signal to interference ratio in a slot.

12.11.2 Wireless Access Communications Systems (WACS)

WACS, proposed by Bellcore in 1992, is a method of accessing the local exchange (LEC) network for a Fixed Wireless Access (FWA) applications. The proposal though limited to a maximum of data rate of 128 kbits/sec for multibearer applications; it is nevertheless the first significant proposal to address the problem of replacing old and deteriorating local wireline loops by wireless access. The system is flexible in offering both connection (circuit) and packetized oriented modes. The systems also provides asymmetrical digital links to handle higher transmission rates in the forward direction.

The system uses Time Division Multiplex/Time Division Multiple Access architecture (TDM/TDMA). In the forward direction several fixed rate streams are multiplexed on to a carrier before being transmitted from a base station. In the reverse direction, the transmissions from mobile terminals are time sequenced and synchronized for TDMA. The total spectrum is divided into TDM channels or set of channels. These channels are reused at other radio ports (RP) which are sufficiently far apart to avoid cochannel interference.

The system uses Quasi-Static Autonomous Frequency Allocations (QSAFA). QSAFA achieves higher spectrum efficiency and eliminates the need for complex frequency planning. QSAFA algorithm assigns channels to each RP. Each RP scans all the transmitter frequencies channels and measures the received signal power. The frequency channel with the lowest received power is assigned to RP for transmission in the forward direction. The measurements is performed by each RP independently and asynchronously until the frequency reuse pattern remains the same as in the final iteration or until a specific number of iterations is reached. This follows assignment of a reverse channel to the portable terminal as is done in FDD method.

When an RP is measuring the power, it turns off its transmitter. To avoid large scale disruptions, the measurement of signal to interference ratio is measured during quiet periods. At call initiation, a portable terminal scans all frequency channels and selects an optimal frequency channel and corresponding RP. Following this the RP selects an idle time slot with the least interference to serve the requesting subscriber terminals.

The proposed frame structure of 2 mseconds is shown in Figure 12.53. The total number of time slots is 10 out of which 9 slots are assignable. The tenth slot is dedicated for alerting and broadcast messages in the downlink. In the reverse direction this slot is used for gaining access to the system. The individual time slot of 100 bits consists of 64 bearer (speech or data) bits corresponding to a multiplexed bit rate of 32 kbits/sec. There are 9 system control bits, 12 bits are used for error detection and synchronization. The remaining 15 bits in the forward direction are used for subscriber unit's frame initialization (14), and power control (1). In the reverse direction these are used as guard time (12), differential encoding (2) and 1 bit for other uses.

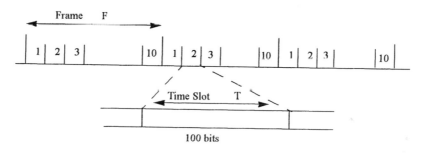

Figure 12.53 Details of Frame Structure of WAC Systems

12.12 Summary

In this chapter, salient features of several cellular systems were discussed. The philosophy behind using particular designs was presented. This chapter leans towards more descriptive material and rigorous mathematical theory behind designs has not been included. The chapter described AMPS, the first generation cellular system in some detail. Nordic, NEC, and TAC systems were described. The two main second generation systems, NAMPS and GSM were description in some detail. Several call scenarios using digital TDMA systems were included in the discussion. The call scenarios presented highlight the flow of signals between the mobile, base station and the network. The discussion then focussed on spread spectrum systems. The fundamentals behind the design of IS-95 were described in considerable detail. Most of the material presented in this chapter is taken from documents available in various forms, particular the documents describing the standards.

References

[1] The Technical Journal Bell System, ISSN0005-8580, vol. 58, January 1979.

[2] Users *performance requirements*, Issue 1, Approved by CTIA Board of Governors 9/8/88.

[3] TR 45.3 Working Group I, *"Requirements Task Group - feature services priorities,"* June 28, 1989.

[4] *Transmit modulation recommendation*, submission made by Northern Telecom Inc. Richardson, Tx. June 27, 1989

[5] *Modulation standards*, submission made by NEC corporation to 4.3 Digital Cellular Standards, March 1989.

[6] *Transmit modulation recommendation*, submission made by Northern Telecom Inc. Richardson, Tx. June 27, 1989

[7] *Modulation standards*, submission made by NEC corporation to 4.3 Digital Cellular Standards, March 1989.

[8] Description of Preferred modulation schemes, Submission to TIA Technical subcommittee made by Ericsson Radio Systems, March 14, 1989.

[9] TIA technical subcommittee TR45.3 Digital Cellular Standards, *Synchronization word/ training sequence length recommendation for digital voice channel structure*, Motorola Inc. September, 1989.

[10] TIA subcommittee TR 45.3, Working Group III Task force I, *Channel Structures*, submission by IMM, September 27, 1989.

[11] *Description of Preferred modulation schemes*, Submission to TIA Technical subcommittee made by Ericsson Radio Systems, March 14, 1989.

[12] *Discussion of π/4 Shift DQPSK*, submission to TR 45.3.3/89.03.14.05 working group III, modulation Task Group, by Motorola, Inc, Arlington Height, Il.

[13] *Design of transmit/ receive filters to meet requirements in time and frequency domain, submission* to TR 45.3.3.1/89.5.4.9 by AGT corporate Research and Development, Calgary.

[14] *Comments on verification and validation*, submission to TR 45.3.1.2/89.5.3 by AGT corporate Research and Development, Calgary.

[15] *A Possible Approach to control adjacent channel interference from a QPSK channel into an analog channel*, EIA TR 45.3 Digital Cellular Standards Doc. # EIA TR 45.3.3.1/89 (Novatel Communications Ltd).

[16] *Adjacent Channel Interference of π/4 DQPSK* submission to TIA technical subcommittee TR 45.3 WG III Modulation Task Group, Submission made by Motorola Inc. Arlington Heights, Il.

[17] π/4 DQPSK Modulation accuracy Specification, submission made by MCG/Spokane Division to TR45.3.1/89.05.24.02, Feb.1987.

[18] *Speech Processing for Digital Cellular Radio- Proceeding of US West AT Technical Symposium*, Denver, January 1989.

[19] B.S. Atal, "Predictive Coding of speech at low bit rates," *IEEE Trans. Communications*, vol. COM-30, pp. 600-614, April, 1982.

[20] B.J. T. Mallinder, "An Overview of GSM System," *The Third Nordic Seminar on Land Mobile Radio Communication*, Paper 3.1, Copenhagen, Denmark, 1988.

[21] Y. Akaiwa, *Introduction to Digital Mobile Communications*, J. Wiley & Sons, Inc., 1997.

[22] R. Vijayan and J. M. Holtzman, "A model for Analyzing Handoff Algorithms," IEEE Trans. on Veh. Tech., vol 42, no 3, Aug. 1993, pp. 351-356.

[23] T. Burian, "Switching Philosophy of the Digital Cellular Radio Telephone System CD900," *Nordic Seminar on Digital Land Mobile Radio Communication*, pp. 271-278, Feb. 1987.

[24] K.S. Gilhousen, I. M. Jacobs, R. Padovani, A. J.Viterbi, and L. A. Weaver, Jr., "Increased capacity using CDMA for mobile satellite communications," *IEEE Journal on Selected Areas in Communications*, vol. 8, no.5, pp. 503-514, May 1990.

[25] G. R. Cooper and R. W. Nettleton, "A spread spectrum technique for high capacity mobile communications," *IEEE Trans. on Vehicular Technology*, vol. 27, pp. 264-275, Nov. 1978.

[26] K.S. Gilhousen, I. M. Jacobs, R. Padovani, A. J.Viterbi, L. A. Weaver, Jr., and C. E. Wheatley III, "On the capacity of a cellular CDMA system," *IEEE Trans. on Vehicular Technology*, vol. 40, no.2, pp. 303-312, May 1991.

[27] Qualcomm, *Notes on IS-95, Code Division Multiple Access*, 1996.

[28] European Telecommunications Standards Institute, CT2 *Common Air Interface Version 1.1*, June 1991.

[29] NovaAtel and Ericsson Communications, *CT3-Common Radio Interface specifications for Canadian Cordless Communications*, December 1990.

[30] European Telecommunications Standards Institute, *DECT - Reference Document, Version 2.1 ETR 015*, March 1991.

[31] *Generic Framework Criteria for Version 1.0 Wireless Access Communication System (WACS)*, Bellcore Communications Research, June 1992.

[32] M. Kavehrad and B. Ramamurthi, "Direct-sequence spread spectrum with DPSK modulation and diversity for indoor wireless communications," IEEE Trans. on Comm., pp. 224-236, February 1987.

Chapter 13

3G WIRELESS COMMUNICATION SYSTEMS

The first generation of cellular systems had a relatively short life; these systems survived for nearly ten years. These systems had many limitations and shortcomings. These systems were phased out quickly because of their antiquated technology, expensive hardware, and inadequate capacity. For example, the first generation cellular systems were designed for users in vehicles, rather than those outside. The high powered bulky transmitters were not conducive to portability. The cellular mobile radio systems removed the tie of telephones to buildings and replaced that with a leash to vehicles. The change still did not provide a truly ubiquitous communication, which had been the ultimate desire of users. Being analog, these systems were very inadequate for data transmission, which was seen as the major growth area in telecommunications.

The first generation analog communications systems were replaced by the second generation digital cellular systems. Out of several second generation systems, four systems survived the test of time and at present are extensively employed in different parts of the world. These are GSM, which became defacto world standard; NTDMA, which reigned over North America; PDC whose application was limited to Japan, and Qualcomm CDMA (IS-95) systems, which are now installed in many countries of the world and has the distinction of introducing spread spectrum technology to public mobile communication systems. Two other systems, European DECT (Digital Enhanced Cellular Telecommunications) and Japan's PHS (Personal Handiphone System), were designed to provide cordless telephone services. PHS and DECT systems are typical of personal cordless systems that function as cordless phones inside buildings and become cellular terminals outside.

The second generation systems are also inadequate in many aspects. The most fundamental limitation remains to be low transmission rates of digital data. This came short of the desire of the users to see wireless communication systems providing services comparable to the level offered by the fixed telecommunication network or even surpassing them. The most important limitation is the inability of the network to provide the users fast and reliable access to Internet. The World Wide Web has become an important source of learning, education, current affairs and news on business and politics. Naturally, such facilities are possible when mobile communication systems are able to provide fast access and adequate transmission rates so that large data files can be downloaded in a short time.

Despite these limitations, the second generation cellular mobile radio systems are hugely popular and since their commencement in 1991, the number of users surpassed all projections. By March 2003, the number of subscribers had crossed 1.3 billion mark. The deficiencies of high power and large size portable equipment were solved by small low power terminals operating at higher frequencies in relatively smaller cells. However, these systems though capable of transmitting data, the rate of transmission is still not sufficiently high to allow for numerous newly emerging applications such as fast Internet access, fast file transfer and messaging service.

It was generally accepted that the next generation i.e. 3G of mobile systems shall provide a much wider range of services and will be truly portable in a personal sense. By personal sense, it is meant that the users and not the network operators have greater control over the choice of services, in a manner similar to or beyond the present day integrated services has over the switched telephone network. The next stage of development was envisaged to aim at providing users accessibility to a host of telecommunication services including voice, data, facsimile, and image in addition to providing freedom to the users to configure the network to suit their equipment and application without regard to their location whether in a vehicle, office or on foot in the street. In other words, a personal communication system that provides fast and variable rate services seemed to be the goal.

To provide additional services such as data communications and ISDN over wireless network, the new generation system was expected to be an integrand of the digital cellular radio for outdoor applications, indoor communications, and micro and macrocell communications. A wireless network is envisioned which is transparent to the users in the sense that they are not aware of how the service is provided but they still have some control over the network usage and could configure it to suit their applications. It was no doubt a very ambitious undertaking because the wireless channels characteristics are strongly dependent on the environment and so are the services that could be offered. It is hoped that generations of wireless communications beyond 3G will be able to meet these expectations.

Thus, we seek ultimate wireless communication systems with the following attributes [1]:

(i) The quality and reliability should be comparable to or even better than the services over the STN so that the users on the basis of quality and variety of services do not have any difficulty in deciding their preference to the mobile terminal over the fixed desk top telephone.

(ii) The users have the freedom to configure the network resources to suit their applications and the type of the equipment.

(iii) The system must be able to handle greatly increased subscriber density caused by an expanded consumer base. Since little additional spectral allocation is expected, the necessary higher spectral efficiency will be realized through frequency reuse, interference control, new modulation and multiple access methods.

(iv) Quick and easy call set-up.

(v) The handset should be small and light, preferably of "shirt pocket" size.

(vi) Long battery life - the battery should last for several days rather than hours before it needs recharging.

(vii) The mobile terminal should be convenient and user friendly.

(viii) Integrated (voice, data, and image) digital services should be provided.

In addition to these, there are other secondary requirements envisioned in the new systems. The reader is referred to CTIA report and other publications [1]-[9] for more detail.

The requirement (i) above is broad-ranging. It depends on the quality of the signal coverage and type of system architecture and available control functions. This requirement is difficult to meet unless powerful signal processing techniques are used in transmission and reception of signals. In (ii) the subscribers are given freedom to use any kind of equipment and demand a service compatible to that equipment. This requires a greater degree of flexibility in the network architecture. The requirement (iii) is perhaps the most important because it sets the minimum requirement on spectral efficiency that is the capacity in a given bandwidth over a defined service area. The requirements (v) and (vi) point to a very extensive frequency reuse and low power transmission which translates into a small low power personal terminal. These requirements also put a limit on the cell size, which may be reduced from several kilometers to few hundred meters.

All the above attributes could be satisfied in systems that are yet to appear. Systems of lesser capabilities known as 3G systems have been in operation in Japan since October 1, 2001. We devout this chapter to the descriptions of five 3G standards on air interfaces.

Before we embark upon describing the 3G systems, it is pertinent to discuss the situation vis-a'-vis the second generation systems in where a huge investment is tied. The question arises whether some transitory systems can be created out of existing 2G that are capable of meeting some if not all the requirements envisioned in the 3G. This question raising the issue of migration of 2G to 3G systems has been discussed at several levels and many migratory paths have been suggested. In the next section, we shall discuss these issues.

13.1 2G to 3G Migration Paths

Figure 13.1shows the development phases of cellular systems and related services offered by the three generations of systems. The first generation systems were focussed on analog voice with limited terminal mobility. Four first generation systems, NMT, AMPS, NEC and TAC, survived. These systems had not much commonality. In the second generation, transition from analog to digital transmission was made and the all digital second generation systems with limited data transmission capability were developed. Again among several, four systems, GSM, PCN, PDC and CDMA, are still in operation. The second generation systems had some improvements in both data transmission and terminal mobility rate.

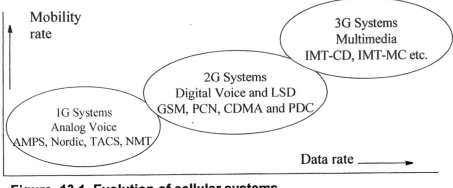

Figure 13.1 Evolution of cellular systems

It is clear that migration of G1 to G2 or G3 was quite difficult because of the analog nature of the former and the digital of the latter. However, migration from 2G to 3G systems was possible because both systems have digital platform. Such migration is envisioned because it is expected to be less costly than installing entirely new 3G systems. The migration path seems to be less painful and quicker in starting newer high speed services in different operating environments. Furthermore, in the light of enormous investment in the second generation systems, it was felt desirable to retain some commonality between IMT-2000 and 2G systems. Also, since IMT-2000 is likely to coexist with the earlier systems, terminal mobility between the second and the third generation systems need to be supported. In the evolutionary path towards IMT-2000, considerations have been given to maximize commonality with 3G systems in radio bearer adaptation and universal User Identity Module (UIM) functionalities. The migratory paths for the three 2G systems to transitory systems that meet some of the specifications of the 3G systems are shown in Figure 13.2.

The migration paths for GSM and NTDMA (IS-136) to 3G systems are suggested in two phases. In the phase one, the General Packet Radio Service

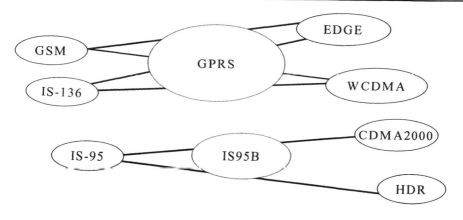

Figure 13.2 2G to 3G Evolutionary Paths

(GPRS) will increase the data rate to 115 kbits/sec. In the phase 2, the data rate will be further increased with the use of Enhanced Data Rates for GSM Evolution (EDGE) are shown in the above figure. The IS-95 system can migrate to CDMA2000 or high data rate (HDR) via IS-95B, a system with upgraded specifications.

In mapping these paths, considerations have been given to evolution and migration capabilities, coexistence and other inter-operation issues. For example, issues related to compatibility of services within IMT-2000 and with PSTN, coexistence and interworking with pre-IMT 2000 systems, and provision for continued expansion of network and services have been taken into account.

13.1.1 General Packet Radio Service (GPRS)

GPRS is a key step in migration of TDMA systems like GSM and IS-136, towards 3G systems. With regard to GSM, GPRS is the first enhancement, where a transition from circuit switched data to packet switched data is made. The theoretical maximum transmission speed of 172.2 kbits/sec is possible if all eight time slots in a GSM frame are used. However, the actual transmission speed is generally lower to around 115 kbits/sec because of the difficulty in allowing all eight slots to one user. Furthermore, it is expected that initially lower data rate services will be supported as only two or three slots may be made available for GPRS so that the system capacity is conserved. Since packet switching uses the channel resources more efficiently, GPRS is likely to be less expensive to the subscribers than short message service (SMS) and circuit switched data (CSD). The speed of circuit switched data on GSM is 9.6kbits/sec but high speed circuit switched data (HSCSD), a new implementation to transmit circuit switched data over GSM, can support rates up to 38.4 kbits/sec. This is accomplished by allocating all eight slots to a user. The main difference between HSCSD and GPRS lies in the transmission mode, which is circuit switched in the former and packet switched in the latter. Similar to GSM migration towards packet switched data services, GPRS service will also be deployed in North America over IS-136.

GPRS offers internet functionality by allowing inter-networking between the existing Internet and the newly created GPRS network. Practically all Internet services currently offered over the fixed network will be available over the GPRS network.

GPRS has several limitations and therefore it is likely to serve as a transitory step towards full implementation of 3G systems whether resulting from enhancements to GSM or by implementation of newer 3G systems. A limitation of GPRS relate to GMSK modulation, which will be replaced by 8-PSK when EDGE is implemented. Other limitations are transit delays, and limited cell capacity[†].

GPRS will offer a wide range of applications like CHAT, transfer of textual and visual information (still and moving images), web browsing, document sharing, corporate e-mail, and Internet e-mail. Other possible application is in-home automation or applications requiring machine to machine communications.

To enable GPRS on a GSM network, addition of two core network modules the Gateway GPRS Service Node (GGSN) and the serving GPRS Service Node (SGSN) are required. The GGSN acts as a gateway between the GPRS network and the public data network such as IP. GGSNs also connect to other GPRS networks to facilitate GPRS roaming. The Serving GPRS Support Node (SGSN) provides routing to and from the SGSN service area to all users in that service area. The Phase I implementation is illustrated in Figure 13.3.

In addition to multiple GPRS nodes and a GPRS backbone, other units like Packet Control Units, hosts in the Base Station Subsystem, mobility management to locate GPRS mobile stations, a new interface for packet traffic and new security features need to be added. The network will be further enhanced to include interface with UTRAN to provide 3G services.

13.1.2 Enhanced Data Rates for GSM Environment (EDGE)

Enhanced Data Rates for GSM Evolution renamed as Enhanced Data Rates for GSM Enhancement (EDGE) is a service platform to support high-speed data applications. EDGE is a new radio interface technology with M-ary modulation format that enhances both HSCSD and GPRS by three fold. With the use of 8-PSK modulation, the data rate can be increased to over 400 kbs/sec. The use of M-ary modulation, the coverage will be reduced in order to maintain the desired error performance. This is the consequence of using larger signal

---

† The cell capacity is limited due to allocations of multiple slot to a subscriber leaving lower number of slots for other users.

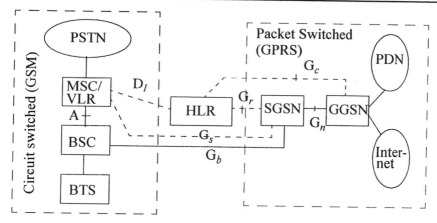

Figure 13.3 Phase I implementation of GPRS in GSM

constellation. Furthermore, the capacity of the system is not much degraded as number of bits/slot are increased. One of the features of EDGE is to offer multirate (multi-quality) voice services.

13.2 Third Generation Systems

Prior to the decision on air interfaces for 3G systems, the systems known as UMTS (Universal Mobile Telecommunication System) and FLMTS (Future Land Mobile Telecommunication System) have been under study for quite sometime by several ITU-R working groups [2]. At the beginning of studies, it was expected that the future systems will offer new services including real time multimedia over networks that use radio resources in a highly efficient manner. Flexibility of operation was a part of overall requirements. This requirement was set to allow the system operators to choose systems with their own specifications. For example, a system operator may opt for universal seamless roaming that requires interconnections with other networks while another operator may not need world wide roaming.

Seamless roaming at international level not only requires complex international agreements with many operators spread across the world but also a high degree of system compatibility particularly in the case of air interface is needed. At the outset of studies for new systems, it was expected that the third generation systems would meet the above requirements but the efforts ended up short of the ultimate goal.

It is now expected that 4th Generation or FLMTS will meet the objectives mentioned above as rapidly progressing technology will enable the developers to reach the goal. The FLMTS envisage the use of multiple size cell radii; cell radii as low as 50m and as large as several hundred kilometers in size have been cited[†]. The frequency of operation of these systems has not yet been determined but it is expected to be some where between 5 and 60 GHz.

13.3 W-CDMA System Design Concepts

The unexpected growth of wireless communications during the years 1992-96, and the projections that high speed data transmission will be the major application of telecommunications in the 21st century, compelled ITU-R to invite proposals in June 1996, for the next generation cellular systems, commonly called 3G systems. Note that the terms UMTS and IMT (or 3G) are synonymous. The ITU-R set the following as the main features supported by the third generation mobile communication system: [10]

(i) The 3G systems are likely to operate across the world. Therefore these should have high degree of compatibility in the access procedures. To do this, the new 3G systems should have a high degree of commonality of design between various systems around the Globe.

(ii) The services offered by the 3G systems and by the fixed network should be made compatible as far as possible.

(iii) The 3G systems should offer high quality services comparable if not better than those offered over the fixed telecommunication network.

(iv) These systems should support high speed data services including image services.

(v) The user terminal should be small, low cost and low power.

(vi) The system should support multimedia and voice services. The data rates are 144 kbits/sec and 384 kbits/sec for wide area applications and 2.048 Mbits/sec for local area applications. Wide area and local area applications are distinguished by the terminal speed i.e whether the terminal is on board a vehicle or not. Wide area implies that terminal move at higher speeds i.e. in vehicle users.

The Universal Mobile Telecommunication Systems (UMTS) are perhaps one generation earlier than FLMTS, the ultimate system envisioned by ITU-R. It was soon discovered that to a decision on the IMT-2000 specifications a consensus among representatives of several countries would be necessary. In June 1998, ITU-R received 15 competing proposals on IMT-2000, out of which 11 were for terrestrial and four for mobile satellites systems. These are listed in Table 13.1.

---

†. *This requirement assumes complete integration between terrestrial and satellite systems.*

Table 13.1   *Radio Transmission Technologies Proposals*

Proposal	Description	Environment				Source
		Indoor	Pedestrial	Vehicular	Satellite	
DECT	Digital Enhanced Cordless Telecommunications	x	x	-	-	ETSI Project(EP) DECT
UWC-136	Universal Wireless Communications	x	x	x		USA TIA TR45.3
WIMS W-CDMA	Wireless Multimedia and Messaging Service Wideband CDMA	x	x	x	-	USA TIA TR46.1
TD-SCDMA	Time Division Synchronous CDMA	x	x	x	-	China, CATT,
W-CDMA	Wideband CDMA	x	x	x	-	Japan ARIB
CDMA II	Asynchronous DS-CDMA	x	x	x	-	S. Korea TTA
UTRA	UMTS Terrestrial Radio Access	x	x	x	-	ETSI SMG2
NA:W-CDMA	North American Wideband CDMA	x	x	x	-	USA TIPI-ATIS
cdma 2000	Wideband CDMA (IS-95)	x	x	x	-	USA TIA TR45.5
CDMA I	Multiband Synchronous DS-CDMA	x	x	x	-	S. Korea TTA
SAT-CDMA	49 LEO Sats in 7 planes at 2000 km	-	-	-	x	S. Korea TTA
SW-CDMA	Satellite Wideband hybrid CDMA/TDMA	-	-	-	x	ESA
SW-CTDMA	Satellite wideband hybrid CDMA/TDMA	-	-	-	x	ESA
ICO RTT	10 MEO sats in 2 planes at 10390 km	-	-	-	x	ICO Global Communications
Horizons	Horizons satellite System	-	-	-	x	Inmarsat

Out of the 11 submitted proposals, five on the air interface for terrestrial cellular radio communications were approved in 2000. These proposals are named as IMT-DS, IMT-SC, IMT-MC, IMT-TC, and IMT-FT[†]. A sixth air interface standard known as UTRA TDD Low Chip Rate is also defined as a part of 3G air interfaces [10], [11]. With these decisions, the hopes of standardizing to a single universal air interface were again dashed. However, efforts towards harmonization between various standards resulted in a deci-

---

[†]. *IMT stands for International Mobile Telecommunication Systems. DS, SC, MC, TC, and FT stand for Direct Sequence, Single Carrier, Multi-Carrier, Time Code and Frequency Time respectively. IMT and UMTS are considered synonymous.*

sion that commonality between a group of core elements in the air interfaces be maximized.

To provide desired commonality a layered approach is considered. In this approach, radio interface functions are separated into transmission dependent and transmission independent functions [12] [13]. The transmission independent functions are: application protocol, call control, identity validation and confidentiality, registration and location control, acknowledgement control and error recovery procedure recovery in flow control, and logical channel structure and multiplexing. The RF functionalities, radio resource management, and error detection/correction have been identified as transmission dependent functions. The IMT systems have been commercially available in Japan since October 1, 2001. The installation of 3G systems in Europe has been delayed till 2003. The first 3G system in Europe is likely to be in Italy.

In this chapter, Sections 13.4 through 13.9 provide details on the five air-interfaces standards. A greater emphasis is given on W-CDMA (IMT-DS), the first system put in operation. In Section 13.11, a brief discussion on the emerging network architecture is provided. The role of intelligent networks in Personal Communications is discussed in Section 13.11 and some results of a study on performance are reported.

13.4 Wideband Direct Sequence System (IMT-DS)

The IMT-DS resulted from the joint efforts of ETSI and NTT-DOCOMO and it was known as wideband CDMA system or Universal Terrestrial Radio Access (UTRA) and when it is used in a network, it is called Universal Terrestrial Radio Access Network (UTRAN). The UTRA system has the following features.

High Efficiency Spectrum Utilization

The IMT-DS system achieves high spectrum efficiency by using voice activity in controlling transmission and the mobile terminal transmit power. The adaptation of these technologies with dynamic channel assignment achieves high capacity.

Frequency Management

The system allows the use of universal reuse of frequencies and therefore does not require any frequency managements. The management functions required in other systems using FDMA and TDMA - frequency in the former and both frequency and time slot management in the latter, are difficult to implement particularly in irregular terrain which gives rise to non-uniform propagation coverage patterns. The imperfection in assignment usually leads to lower capacity in other systems, however this is not of a concern in CDMA.

Low Mobile Transmit Power

CDMA delivers improved performance with the use of advanced receiver technologies such as Enhanced Rake receiver. With the use of the E-Rake, the gain due to path diversity lowers the mobile terminal transmit power. Because of resulting gains in Rake receiver, the CDMA mobile need to transmit power lower than that required in TDMA or FDMA systems. For example, in TDMA where intermittent power transmission is used, the peak power to send information symbols increases in direct relation to the number of time slots in use. The CDMA systems on the other hand uses continuous transmission thereby peak transmit power in CDMA can be lowered. The system also uses independent resources for uplink/downlink, as is needed in many applications e.g. Internet access. Asymmetrical resource allocation is required in these applications because of large difference between the transmission rates in the uplink and downlink channels.

In CDMA, it is simple to support an asymmetrical uplink/downlink structure. On the other hand, in other access methods such as TDMA assignment of different number of slots in uplink and downlink channels is not that easy. This requires complex slot management functions. This function is also difficult to realize in FDMA because in that case, the uplink/downlink carrier bandwidth needs to be different. On the contrary, the uplink and downlink rates can be set up independently in CDMA systems with the use of different spreading factors for uplink and downlink for each user. This allows better radio resource utilization.

Wide Variety of Data Rates

W-CDMA meets the IMT requirement of allowing multirate transmission. Since the proposed system uses wide bandwidth, therefore higher transmission rates are possible. It provides for efficient multirate services even in the situation where lower and higher rate services must coexist. This supports simultaneous coexistence of speech/fax services with high resolution video services in the same band. Furthermore, Internet can be accessed at the same rate as from the fixed communications network.

Improvement of Multipath resolution

The Rake receiver improves the reception performance by first separating the signals that arrive on different propagation paths and then diversity combining these signals to achieve significant signal to noise improvement. The use of wider transmission bandwidth improves the capability of resolving multiple paths with better resolution, thus greater number of paths can be combined. This improvement reduces the required transmit power. The lower transmit powers from mobiles improves the over all interference condition with the result that a further improvement in the spectrum utilization efficiency is realized.

Statistical Multiplexer Effect

The wider transmission bandwidth means that a larger number of traffic channels can be accommodated on one carrier. Further improvement in the use of frequency is thus achieved because of enhanced statistical multiplexer effect in the propagation environment. The efficiency of a narrowband CDMA system drops because of lower number of channels loaded on to a carrier. This effect is similar to higher trunking gain experienced in FDMA when greater number of channels are bundled together in a trunking set.

Reduction of Intermittent Ratio

The high transmission rate reduces the time taken to transmit system overhead information that is the mobile station when in a standby mode takes shorter time to receives this information. This results in lengthening of battery life and longer standby hours by the mobile station between charging.

13.4.1 Radio Interface Protocol Architecture

Figure 13.4 shows the network architecture from the physical to the network layers. The physical layer supports the actual transmission waveform over the air, which is supported by the air interface. The air interface is characterized by the radio frequency, multiple access protocol, system access method etc. [14].

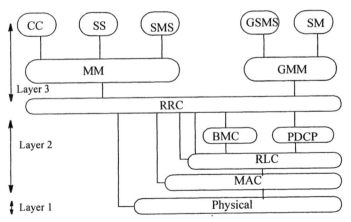

BMC Broadcast Multiaccess control
PDCP Packet Data Convergence Protocol
GMM GPRS Mobility Management
CC Call Control   MAC Media Access Control
SMS Short Message Service

RLC Radio Link Control
RRC Radio Resource Control
MM Mobility Management
SS Supplementary Service
SM Session Management

Figure 13.4 Air interface, physical layer to network layers

Media Access Control (MAC), Radio Link Control (RLC), Broadcast Multiple Access Control (BMC), Packet Data Convergence Protocol (PDCP) and Radio Resource Control are the functions of Layer 2. In the third layer,

Mobility Management and GPRS Mobility Management (GMM) sub-layers reside. The functions of these are to control the messages linked to call control (CC), Supplementary Services (SS) and Short Message Service (SMS), Sessions managements (SM) and support GPRS short message service (GSMS).

Figure 13.5 shows the radio interface protocol architecture for the IMT-2000, DS-CDMA. The ellipses between sub-layers indicate Service Access Points (SAPs). The physical layer (L1) interfaces the Medium Access Control (MAC) sub-layer of Layer 2 and the Radio Resource Control (RCC). It offers several Transport Channels to L2/MAC. A Transport Channel is characterized by the mode in which information is transferred over the radio interface. L2/MAC offers different logical channels to Radio Link Control sub-layer, L2/LAC. A logical channel is characterized by the type of information it transfers. Two duplex modes: Frequency Division Duplex (FDD) and Time Division Duplex are provided. Spreading code, frequency and in the uplink relative I/Q phase characterize the FDD mode. The TDD mode is characterized by the time slot. The RCC controls the physical layer.

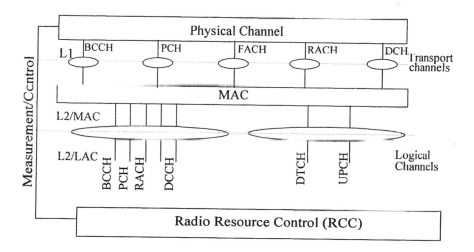

BCCH Broadcast Control Channel   PCH Paging Channel
RACH Random Access Channel   FACH Forward Access Channel
DCCH Dedicated Control Channel   DTCH Dedicated Traffic Channel
UPCH User Packet Traffic Channel   LAC Link Access Control
MAC Media Access Control

Figure 13.5 Radio interface protocol architecture

13.4.2 Inter-layer services

The physical layer is responsible for providing data transport services to higher layers. The physical layer performs the following functions:

Micro-diversity distribution/combining and soft handover execution, error detection on transport channels and their indications to the higher layers, forward error control (FEC) encoding/decoding are some of the functions of the physical layer. The physical layer also performs functions like multiplexing and de-multiplexing of coded transport channels, rate matching between the coded transport channels and physical channels, and mapping the composite coded channels on to the physical channels. In case the diversity is implemented the physical layer determines power weightings of the diversity branches prior to combining.

Other important functions of the physical layer include modulation, spreading/demodulation and de-spreading of physical channels, frequency and time synchronization, which includes chip, bit, slot and frame timings. The physical layer also measures radio characteristics of the physical channels. These measurements include frame eraser rate (FER), signal to interference ratio (SIR), and interference power. This layer reports the findings mentioned above to the higher layers. The inner loop power control is also the responsibility of the physical layer.

13.4.3 W-CDMA Radio Transmission Technologies

The parameters for the radio transmission over the air interface are shown in Table 13.2. Two modes, FDD and TDD are defined for operations with paired and unpaired spectrum respectively. The FDD mode is duplex method where uplink and downlink transmissions use two frequencies separated by 190 MHz. The TDD mode is specified for where uplink and downlink transmissions are carried out at the same frequency but at different time intervals. The time slots in the physical channel are divided into transmission and reception part. The information in the uplink and downlink are transmitted reciprocally. Thus, TDD is a combination of TDMA and CDMA.

In Universal Terrestrial Radio Access (UTRA), the normal bandwidth is 5 MHz, but in order to provide operational flexibility, a provision to expand the bandwidth to 10 or even 20 MHz is also made. At the same time, a lower bandwidth of 1.25 MHz is defined in order to make the system compatible with narrowband CDMA systems, e.g. Qualcomm IS-95 system, or IMT-MC.

The IMT systems are to operate in 1,920 to 1,980 MHz in the uplink and 2,110 to 2,170 MHz in downlink. The uplink and downlink operating bands are paired and these differ by 190 MHz. In Region 2 of ITU-R, the operating frequencies are 1,850 to 1910 MHz for uplink and 1,930 to 1,990 MHz for downlink. For TDD operation, the allocation is 1,900 to 1,920 MHz and 2,010 to 2,125 MHz. In Region 2, the frequencies are 1,850 to 1,910 MHz, 1,910 to 1,930 MHz, and 1,930 to 1,990 MHz. These allocations use unpaired bands.

Four classes of user terminal based on maximum output power are defined. These are Class 1 (+33 dBm), Class 2 (+27 dBm), Class 3 (+24 dBm) and Class 4 (+21 dBm). Both FDD (Frequency Division Duplex) and TDD (Time Divi-

Table 13.2   ***Parameters of W-CDMA Radio Transmission Technologies***

Item	Parameter
Radio Access Technology	FDD:DS-CDMA
	TDD:TDMA/CDMA
Operating Environment	Indoor/Outdoor to indoor/vehicular
Bandwidth	5 MHz
Chip rate	3.84 Mcps
Carrier Spacing	Flexible with 200 kHz carrier raster
Duplex scheme	FDD and TDD[a]
Channel bit rates	FDD (UL) 7.5/15/30/60/120/240/480/960/1920
	FDD (DL) 7.5/15/30/60/120/240/480/960/1920
	TDD (UL) variable from 240 kbps to 3.84 Mbps
	TDD (DL) 240 kbps to 3.84 Mbps
Frame Length (unit)	10 msec frame with 15 slots
Spreading factor	FDD (UL) variable 4 to 256
	FDD (DL) variable 4 to 512
	TDD (UL) variable 1 to 16
	TDD (DL) 1 to 16
Inter-cell Synchronization	FDD: Asynchronous
	TDD: Synchronous
Modulation scheme (DL/UL)	QPSK/BPSK (FDD)
	QPSK/QPSK (TDD)
Spreading (DL/UL)	QPSK/QPSK
Multi-rate	Variable spreading and/or multicode
Channel Coding	Convolutional (R=1/3 or 1/2, K =9), Turbo code
Spreading	Spreading code and scrambling code
Detection DL: Downlink	Coherent pilot symbol (time multiplexed pilot)
	Common pilot can also be used
UL: Uplink	Coherent pilot symbol (I/Q multiplexed pilot, time multiplexed pilot for TDD mode)
Packet Data Transmission	Adaptive Channel Switching between common and dedicated channel

a. For the TDD mode, in addition to the use of all slots within a frame, the TDMA structure where only part of the slots are used will be adopted.

sion Duplex) modes are duplex schemes. The UTRA TDD is also called IMT-FT; details on this variation are found in Section 13.6.

The chip rate is specified to provide maximum hardware flexibility so that gains through spreading the signal can be realized in the bands used. For instance, the chip rate is set at 3.84 mcps when 5MHz band is used. This gives a path resolution of 260 nanoseconds. The choice of 200 kHz carrier raster improves spectrum utilization efficiency by providing flexibility on carrier spacing that conforms to that of GSM. One 10 msec frame is divided into 15 slots. Each slot has 2,560 chips.

Most of the key parameters including the modulation/demodulation, chip rate, the frame length are made common to both modes. To provide greater flexibility in system deployment, precise intercell synchronization in both outdoor and indoor operations is not required. However, for added flexibility, intercell synchronization is also allowed. To keep the hardware complexity lower and allow multi-rate transmission, variable spreading factor is used for uplink and orthogonal variable spreading factor for downlink. It is possible to use multi-code transmission for high speed data.

For channel coding, several options are provided. In UTRA-FDD mode two options - convolutional coding and turbo coding are provided. In UTRA, the third option of no coding is also provided. Turbo codes that offer better error correction performance are also applied in the case of high speed data transmission. To deal with error bursts coding with interleaving is provided. The possible interleaving depths are 10, 20, 40 or 80 msec. The selection of codes are indicated at higher layers. Real time services use only FEC while non-real time services use a combination of FEC and ARQ. The convolutional code rates of 1/2 or 1/3 are used. For turbo codes, the rate 1/3 is selected.

The bandwidth spreading takes place in two operations, one consisting of channelization operation, which transforms every data symbol into a number of chips, thereby spreading the bandwidth. The number of chips per data symbol is the spreading factor (SF). The second operation is that of scrambling, which is applied to the spread signal. The two operations are shown in Figure 13.6. Over the downlink, the channelization codes the same that are used in the uplink. These codes preserve orthogonality between the channels. There are plenty of scrambling codes (a total of 262,144 (512 × 512) scrambling codes, numbered from 0 to 262,143), these codes are divided into 512 sets primary scrambling and 511 secondary scrambling codes. The scrambling codes are cell specific that is each cell is allocated one and only one primary scrambling code. In the uplink the scrambling codes are user specific. For separating channels (user signals) from the same source, channelization codes are derived from tree codes, which are discussed later.

For cell specific scrambling, in the FDD mode, Gold codes with 10 millisecond period with the code length of $2^{18}-1$ are used. For TDD mode, scrambling codes of length 16 are used. Spreading codes are orthogonal derived from

Walsh-Hadamard codes, and all codes are used commonly for all cells, thus minimizing interference between the users within a cell. For separating different User Equipments (UE), Gold codes with 10 msec periods or alternatively S(2) codes with 256 chips are used.

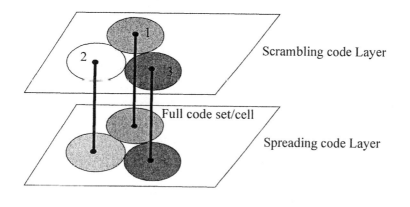

Figure 13.6 Two layered spreading code allocation

Pilot symbols are used for coherent detection and these are time multiplexed on the downlink so that the delay resulting from the transmitter power control is minimized. Data and the pilot symbols are spread with different codes and multiplexed on I and Q channels. This allows continuous transmission on the RF channel even when variable data rate is selected. This helps in minimizing the peak to average power factor of the transmit waveform, which reduces interference and improves the overall electromagnetic environment. Furthermore, with this technique, the requirements on the mobiles station amplifier are less stringent. However, in the TDD mode, similar to downlink, the pilots are time multiplexed on uplink. This is because there is no advantage in using IQ multiplexed pilots in a discontinuous TDD mode, and also because an asymmetric uplink/downlink structure is more desirable in the case of direct mobile to mobile communications.

A dedicated transmission scheme is adopted for pilot symbols transmission. The scheme allows for fast closed-loop transmitter power control on the downlink. It is also possible to apply a common pilot scheme, where the pilot symbols of common control channels are used by each traffic channel. This makes it possible to further improve the performance. The dedicated pilot scheme is an extremely effective solution in terms of securing expandability to adopt adaptive antennas and other technologies in the future.

The packet data transmission is performed to support asymmetric uplink/downlink transmission, and wide range of data transmission from low to high rates. For the packet transmission, adaptive channels depending on the traffic characteristics are used. For example, when the traffic is low both on the uplink and downlink, common physical channels (CPCH, RACH) are used. On the other hand, dedicated physical channels (DPCH) are used when the traffic is heavy on either uplink or downlink, or on both links.

13.5 System Description for IMT-DS

Three fundamental channels define the channel structure. First of these structures is that of the logical channels, which are offered by Layer 2/MAC to higher layers. The second structure is that of the transport channels, which are offered by Layer 1 to Layer 2/MAC. The third structure is that of the physical channels, which are handled within Layer 1. The mapping between logical and transport channels, and between transport channels and physical channels are also described [10], [14], [15].

13.5.1 Logical Channel Structure

The type of information transferred on the channel defines the type of logical channel. Two types of logical channels are defined: Control Channel (CCH) and Traffic Channel (TCH). The structures of these are shown in Figure 13.7.

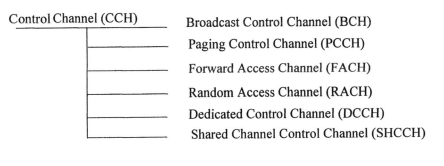

Control Channel (CCH)
- Broadcast Control Channel (BCH)
- Paging Control Channel (PCCH)
- Forward Access Channel (FACH)
- Random Access Channel (RACH)
- Dedicated Control Channel (DCCH)
- Shared Channel Control Channel (SHCCH)

Traffic Channel (TCH)
- Dedicated Traffic Channel (DTCH)
- Common Traffic Channel (CTCH)

Figure 13.7 Logical Channel Configuration

Broadcast Control Channel (BCCH) is a point to multipoint unidirectional downlink common channel. It broadcasts system and cell specific information. The information on this channel such as SFN (System Frame Number), uplink interference power etc. changes over time.

Paging Channel (PCH) is a unidirectional downlink channel that transfers paging information simultaneously with other notification data from a BS to a MS. Forward Access Channel (FACH) is a unidirectional channel for transmitting control information from BS to MS. This channel is used when the network knows the location of the MS or UE in a particular cell. Random Access Channel (RACH) is used for transmitting control information from MS (UE) to BS. Dedicated Control Channel (DCCH) is a point to point bidirectional channel that is used to exchange dedicated control information between MS to BS. The shared channel control channel (SHCCH) is bidirectional channel used to transfer information for uplink and downlink shared channels. This channel is used only in TDD mode.

Two types of Traffic channels (TCH) are used - Circuit switched mode Dedicated Traffic Channel (DTCH) and Common Traffic Channel (CTCII). DTCH is a point to point, bidirectional channel that transmits user information in circuit switched mode. The DTCH is point-to-multipoint channel that transfers dedicated user information to a group of users.

13.5.2 Transport Channel Structure

The physical layer offers information transfer services to MAC and higher layers. How and with what characteristics the data are transferred over the radio interface describe the physical layer transport services. In other words, the physical channels offer information services to MAC. These chan nels may be called Transport Channels. In general, these channels are classified into two groups: common and dedicated channels. Figure 13.8 shows the transport channel configuration.

Random Access Channels (RACH) is characterized by its existence in uplink only. This channel being a contention channel runs the risk of collisions with the information from other mobiles. It is used for initial access or non-real time dedicated control of traffic data. For example, it carries information on open-loop power control, and has a limited data field and provision for sending in-band identification of the MSs. The common packet channel is a contention based uplink channel used for bursty traffic. It is used only in the FDD mode.

Forward Access Channels (FACH) can be characterized by the existence in the downlink only. A FACH may carry a small amount of data. For example, it may carry information whether beam forming can be used or not. It is also used to indicate whether to use slow power control in case the fast power control is not available. This channel is also used for inband identification of MSs.

The downlink shared channel (DSCH) is shared by several user terminals for dedicated control or traffic data. It is associated with traffic data and does not exist alone. The uplink shared channel is shared by several user terminals in the uplink. It carries dedicated control or traffic information. It exists only in TDD mode. Broadcast Channel (BCH) exist in downlink only. A BCH is characterized by low fixed bit rate transmission and is used to broadcast system and cell specific information over the cell coverage area. Paging Channel (PCH) is characterized by its existence in downlink direction and it broadcasts in the whole coverage area. It can be used for sleep mode procedures.

Dedicated transport channel (DCH) is defined for use by one terminal only in either the uplink or down link. It is characterized by its ability to be used for beam forming, to fast (each 10 msec) change transmission rate, power control, and inherent addressing of MSs. The purpose of the fast signaling channel (FAUSCH) has not yet been defined.

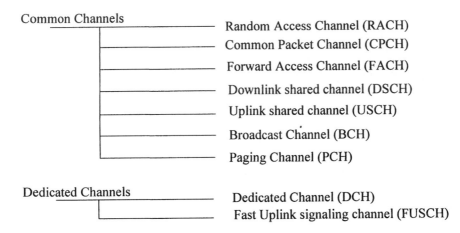

Figure 13.8 Transport channel configuration

13.5.3 Physical Channel Structure

In the FDD mode, a physical channel is defined by code and frequency, and in the uplink, also by relative phase whether in inphase or quadrature channel i.e. I or Q. In the TDD mode, code, frequency and time slot define a physical channel. Several physical channels are defined, their configuration is shown in Figure 13.9. Some physical channels are unidirectional, for use either in the downlink or uplink directions while a dedicated physical data channel is for use in both the uplink and down link transmissions [17].

The dedicated physical data channel (DPDCH) carries data generated at layer 2 and above. Several physical channels are defined for downlink use. On the common pilot channel (CPICH), the transmission is fixed at 30 kbps. It is of two types - primary and secondary. The primary CPICH provides phase reference to synchronization channel (SCH), primary CCPCH, AICH, and PICH and is the default reference for other down link physical channels. The secondary CPICH may be used as a reference for secondary CCPCH and down link

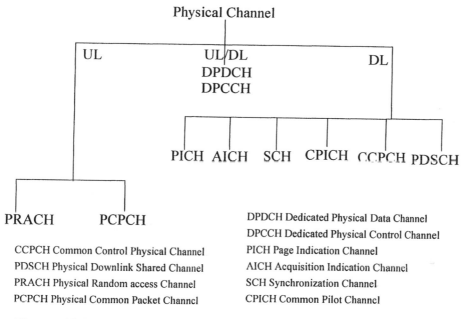

Figure 13.9 Physical Channel Configuration

DPCH. The primary common control physical channel (CCPCH) carries BCH and its transmission rate is also fixed at 30 khps. The secondary common control physical channel (CCPCH) is a variable rate transmission channel. It carries FACH and PCH. These can be mapped to the same or separate channels. Transmission on this channel takes place only when data is available.

The synchronization channel is used for cell search. It has two sub-channels - primary and secondary SCH. The physical down link shared channel (PDSCH) carries DSCH. It is always associated with a down link DPDCH. The acquisition indicator channel (AICH) carries acquisition indicators, the signatures for the random access procedure. The page indicator channel (PICH) carries page indicators to indicate the presence of a page message on the PCH. The purpose of AP-AICH is to carry AP acquisition indicator of the associated CPCH. CPICH carries CPCH status information. The CD/CH-ICH carries collision detection indicator only if channel assignment is not active, or both CD indicators and CA indicators at the same time if the CA is active.

In the uplink direction two channels are provided. The dedicated physical control channel (PCPCH) carries control information at layer 1. The physical random access channel (PRACH) carries random access channel. The physical common packet channel (PCPCH) carries the common packet channel. It uses DSMA-CD technique with fast acquisition indicator.

The system uses the concept of shared channels in order to increase spectrum utilization or capacity. The network does not assign a dedicated channel for every user but the channels are shared particularly when the traffic is bursty. On the shared channel, the resource is open to all users and each user can request a temporary allocation for a short time. For this purpose a special procedure is used.

13.5.3.1 Channel Mapping

Two types of channel mapping, Physical to Transport and Transport to Logical Channels are provided. The channel mapping for downlink and uplink are shown in Figure 13.10 and Figure 13.11 respectively. There are a number of physical channels but some of them are not mapped into transport channels because of their nature as indicator channels, which indicate some system parameters that are not of interest to the higher layers of the network [17].

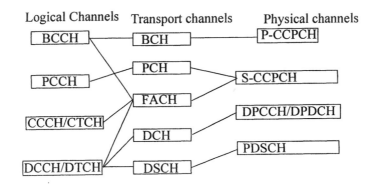

Figure 13.10 Mapping between downlink physical, transport and logical channels

13.5.4 Frame Formats

The system information on the physical channel is arranged in frames and super frames. Each frame is 10 msec long or consists of 2,560 chips. Each frame has 15 slots. Seventy two frames make a superframe and its length is 720 msec. The basic frame structure is shown Figure 13.15. The data in slots on the physical channel pertains to different functions associated with system operation. The slot contents differ from channel to channel. For example, the slot contents for the dedicated physical data channe (DPCCH) differ from that of paging channel.

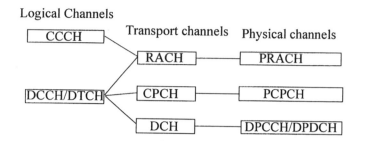

Logical Channels

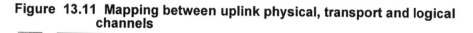

Figure 13.11 Mapping between uplink physical, transport and logical channels

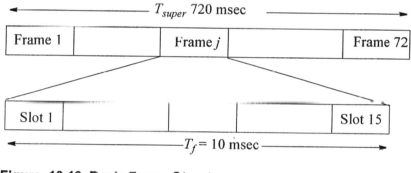

Figure 13.12 Basic Frame Structure

The slot contents used by DCH for the uplink and downlink are shown in Figure 13.13 and Figure 13.14 respectively. Over the uplink physical channel DPDCH and DPCCH are I/Q multiplexed while the downlink channel are time multiplexed. The user data is transmitted on DPDCH, which is associated with a DPCCH that carries Layer 1 control information. The transport format combination indicator (TFCI) field is used for indicating the demultiplexing scheme of the data stream. For static and blind format detection combinations, the TFCI field does not exist. The feedback information (FBI) field is used for transmit and site diversity function. The transmit power control bits (TPC) are used to control power.

For the uplink, the DPDCH bit rate can vary between 15 up to 960 kbits/sec using spreading factors 256 down to 4. If a subscriber desires to transmit higher data rate, several physical channels can be used. The bit rate of the DPCCH is fixed to 15 kbits/sec. The binary DPCCH and DPDCHs prior to spreading are represented by real valued sequences (binary 1 mapped to +1 while binary 0 mapped to -1). The DPCCH is spread to the chip rate by the channelization code C_c while the nth DPDCH called DPDCHn is spread to the chip rate by the channelization code $C_{d,n}$. One DPDCH and up to six parallel DPDCHs can be transmitted simultaneously. After channelization, the real valued signals are weighted by gain factors, β_c, for DPCCH and β_d for all DPDCHs.

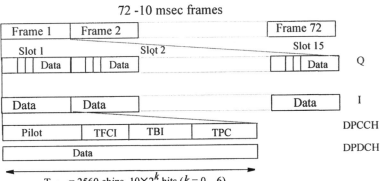

Figure 13.13 Frame structure for the uplink dedicated physical channel DPDCH/DPCCH

Over the downlink DPDCH, the bit rate can be varied between 15 to 1920 kbits/sec with a spreading factor (SF) ranging from 512 to 4. This implies that the bit rate is equal to the channel bit rate for the uplink and is half of the channel bit rate for the downlink.

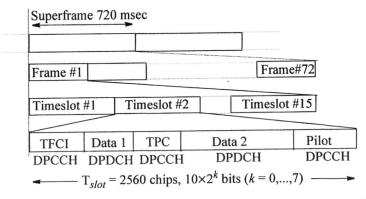

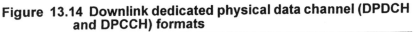

Figure 13.14 Downlink dedicated physical data channel (DPDCH and DPCCH) formats

The PCH is divided into several groups in one superframe of 720 msec long, and layer 3 information is transmitted in each group. Each group of PCH has information worth 4 slots, and consists of a total of 6 information parts: 2 paging indication (PI) parts - for indicating whether there are incoming calls or not, and 4 Mobile User Identifier (MUI) parts - for indicating the paged mobile user. In each group, PI parts are transmitted ahead of MUI parts. In all groups, 6 information parts are allocated with a certain pattern over 24 slots. By shifting each pattern by 4 slots, multiple 288 groups of PCH can be allocated on one Secondary Common Control Physical Channel. The PCH mapping method is shown in Figure 13.15.

In a similar way, FACH is mapped to Common Physical Channel. It has two modes - FACH-S and FACH-L. FACH-L is used for the information length longer than the prescribed value, whereas FACH S is used when it is shorter. FACH-S is transmitted with 4 FACH-Ss time multiplexed on one radio frame of FACH [15].

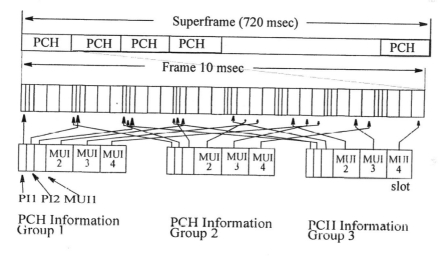

Figure 13.15 PCH Mapping Method

The Uplink Common Physical Channel carries PRACH. PRACH is transmitted by the Common Physical Channel on a frame by frame basis. There are four timings for Common Physical Channel that carries PRACH as shown in Figure 13.16. The Uplink Common Physical Channel Consists of one frame of length TRA or N_D data bits. Transmission starts in one out of four possible time instants (see Figure 13.16). The control part and the Data part are the two parts of the frame. The Data part is transmitted on the I-channel, while the control part is transmitted on the Q-channel. There are N_s identical signature field repeated on the Q-channel. Details on code allocations and spreading of PRACH can be found in [16].

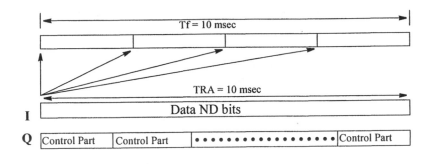

Figure 13.16 Frame structure for Uplink Common Physical Channel

13.5.5 Service Multiplexing and Channel Coding

The fundamental principle of multiplexing services is shown in Figure 13.17. One or more transport channels that require the same service quality are multiplexed using the same channel coding and interleaving unit. In order to obtain services of a particular quality, different channel coding and interleaving unit is used. If one transport channel data is too large to be handled by channel coding, it can be divided into plural small coding blocks. The data on similar quality transport channels, DCH*i*, are multiplexed, interleaved and encoded. The outputs from several channel coding and interleaving units are rate matched and multiplexed and resultant mapped into physical data channels.

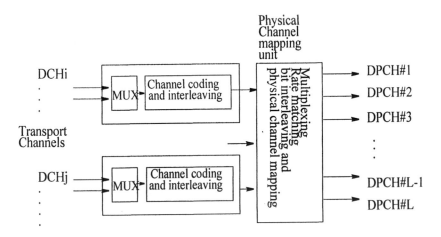

Figure 13.17 Service multiplexing structure

13.5.5.1 Channel Coding

Two types of forward error correction (FEC) coding schemes, convolutional coding and turbo coding are adopted to obtain efficient coding gains under different environmental conditions. The application of convolutional and turbo coding is shown in Table 13.3.

Table 13.3	*Inner coding parameters*	

Transport channel type (information rate)	Coding Scheme (Constraint length)	Code rate
BCCH		
PCH		
FACH	Convolutional code $K=9$	1/2
RACH		
DCH (Lower than 32kbps)		
DCH (equal or higher than 32kbps)	Turbocode $k=3$	1/3

In case of convolutional coding, the interleaving depth is chosen on the basis of delay requirement. The codes are designed for interleaving depths of 10 msec, 20 msec, 40 msec, and 80 msec.

The performance of turbo codes strongly depends on its interleaving block size and the number of iterations used in the decoding process to achieve certain degree of error control. Even if one transport channel is too large to be handled by channel coding, each interleaving depth is expanded to match the large data block. In turbo coding catalytic bit processing is used. In catalytic bit processing 12 bits are inserted as tail bits in certain unknown positions of the input data to the turbo encoder. The number of bits encoded by turbo codes of $R = 1/3$ has 12 more bits compared with convolutional code of $R = 1/3$ and $K = 9$. These 12 bits are punctured at the known positions of the coded data.

Rate matching is used to match the coded bit rate to a limited set of possible bit rates of a DPDCH. In the uplink, repetition and puncturing are employed to match the rates. In the downlink puncturing and repetition are used only for the highest rate, while discontinuous transmission is used for lower rates.

The W-CDMA offers two basic service classes with respect to FECs - standard service with convolutional codes and high quality service with turbo coding. In the former the BER of 10^{-3} is specified and for the latter the BER of 10^{-6} is set. Multiple service types belonging to the same connection are, in normal cases, time multiplexed. Time multiplexing may take place after inner or outer coding. After service multiplexing and channel coding, the multiple-service data stream is mapped to one DPDCH or, if the total rate exceeds the upper limit for single code transmission, to several DPDCHs.

We take two examples on how the procedure to multiplex services works [15]. In the first example, channel coding and service mapping for 8 kbps DCH is shown in Figure 13.18. Since the data rate is low, only convolutional coding is used. The information sequence is divided into 80 bits blocks. To these 80 information bits 16 CRC bits are appended. For convolutional encoding 8 tail bits are added. The 104 bits long sequence thus formed, is convolutionally encoded by rate 1/3 encoding. The 312 bits are passed to multiplexer and interleaving unit which maps the data on to a dedicated physical channel. The recovery of the message. information follows the same process but in reverse order. Figure 13.19 shows mapping to physical channel.

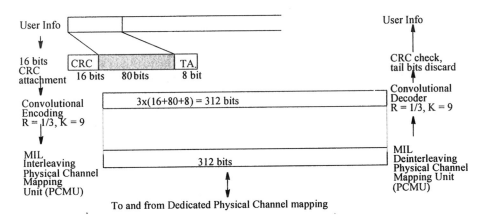

Figure 13.18 Channel coding and service mapping for DCH (8kbps)

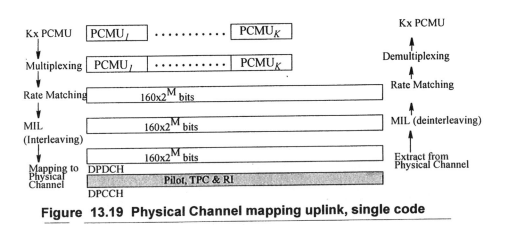

Figure 13.19 Physical Channel mapping uplink, single code

The data from the PCMU (Figure 13.18) in the form of units is then mapped into dedicated physical channels. The K units of data from PCMU are multiplexed and rate matched. The value of 2^M-1 gives the number by which the data bits are to be repeated. $M = 0$ implies no repetition. These are interleaved and mapped into DPDCH and DPCCH in I and Q channels as shown. The DPDCH carries user data and the DPCCH carries information on pilot, power control, and transmission rate.

The second example we consider is on 64 kbps unrestricted data transmission. The details on the process are shown in Figure 13.20.

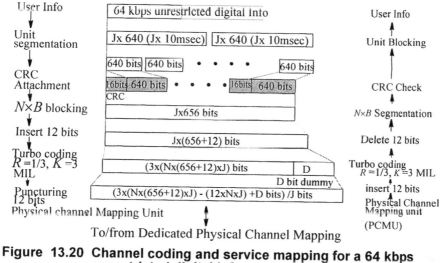

Figure 13.20 Channel coding and service mapping for a 64 kbps unrestricted digital information service

The 64 kbps unrestricted user information is segmented into 10 msec units, each unit consists of 640 bits of user information. To each 640 bits frame a 16 bits CRC is attached. Then Jx blocks of 656 bits are collected into a sequence. In each block, 12 bits are inserted prior to turbo encoding. The block is then encoding with rate 1/3 convolutional code. To the encoded bits, D dummy bits are added. Prior to mapping into a dedicated physical channel, the 12 inserted bits are punctured. The position of these bits are known from the encoding process. The data thus produced is mapped to a physical channel DPDCH after dividing the units into 15 time slots as shown in Figure 13.20. This technique can be extended to $N\times64$ kbps Unrestricted Data. In the highest rate service of 2 Mbps, unrestricted data service can be provided as shown in Figure 13.21. In this case, the user information data is mapped into several physical channel mapping units. Then these physical channel mapping units are mapped to physical channel, DPDCHs based on multicode transmission shown in the figure.

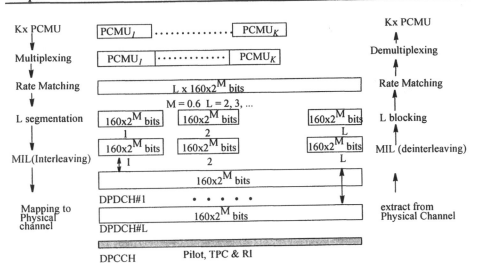

Figure 13.21 Physical Channel mapping uplink, single code

13.5.6 Spreading and Modulation

The Radio Transmission Technology (RTT) adopts a 2-layered code structure consisting of spreading codes (orthogonal codes) and scrambling codes (long codes). Orthogonal variable Spreading Factor (OVSF) codes are employed for the spreading codes in order to preserve the orthogonality between different rates and spreading factors in both uplink and downlink. The OVSF codes are generated from sets of orthogonal codes, e.g. Hadamard matrix, using the tree structure of orthogonal codes [18].

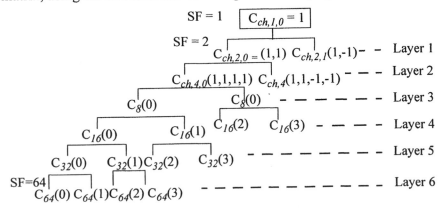

Figure 13.22 Orthogonal Code tree

In the uplink, two types of scrambling codes - long and short are used. The long scrambling codes are constructed from modulo 2 addition of 384,300 chip segments of two binary m-sequences generated by two generator polynomial of degree 25. These polynomials are $X^{25}+X^3+1$ and $X^{25}+X^3+X^2+X+1$. The short scrambling sequences are defined from a sequence from the family of periodically extended S(2). The nth quaternary sequence $z_n(i)$, $0 \leq n \leq 16777215$, is obtained by modulo-4 sum of the three sequences, a quarternary sequence $a(i)$ and two binary sequences $b(i)$ and $d(i)$, where the initial loading of the three sequences is determined from the number n. The sequence $z_n(i)$ of length 255 is generated according to the following relation

$$z_i(n) = a(i) + 2b(i) + 2d(i) \ \text{modulo } 4, \ i = 0, 1, ..., 254 \tag{13.1}$$

The code for scrambling uplink DPCCH/DPDCH may be of either long or short type. The nth long uplink scrambling code for DPCCH/DPDCH, denoted $S_{long,n}$ is defined as

$$S_{long,\,n}(i) = C_{long,\,n}(i), \quad i = 0, 1, ..., 38399 \tag{13.2}$$

where the lowest index corresponds to the chip transmitted first in time. The nth short uplink scrambling code for DPCCH/DPDCH, denoted by $S_{short,n}$ is given by

$$S_{short,\,n}(i) = C_{short,\,n}(i), \quad i = 0, 1, ..., 38399 \tag{13.3}$$

In the downlink, the scrambling code is a cell-specific 10 msec (38,400 chips) segment of a Gold code of length $2^{18}-1$ (262,143 chips). Scrambling is carried out after spreading with orthogonal codes.

The Walsh-Hadamard orthogonal codes of length 64 chips are used for scrambling. These codes are generated with the use of Hadamard matrices as discussed in Chapter 12. The 64 Walsh-Hadamard codes can be represented by a tree diagram as shown in Figure 13.22. There are six layers of codes. The codes in a lower layer have mother - daughter relationships with codes in the higher layer. For example, Codes $C_{64}(1)$ and $C_{64}(2)$ are daughters of mother code $C_{32}(1)$. Codes $C_{64}(1)$ and $C_{64}(2)$ do not form orthogonal sets with $C_{32}(1)$, therefore these can not be used for data channels on the same RF carrier. However, we can use $C_{64}(1)$ and $C_{32}(2)$ on the same carrier to provide two different rate data. For the highest data, higher layer codes are used.

13.5.6.1 Code allocations

The channelization codes are uniquely described as $C_{ch,SF,k}$, where SF is the spreading factor of the code, k is the code number, $0 \leq k \leq SF-1$. The DPDCH is always spread by code $C_c = C_{ch,256,0}$. When only one DPDCH is to be transmitted DPDCH1 is spread by code $c_{d,1} = C_{ch,SF,k}$, where $k = SF/4$.

When more than one DPDCH is to be transmitted, all DPDCHs have spreading factors equal to 4. DPDCHn spread by the code $c_{d,n} = C_{ch,4,k}$, where $k = 1$ if $n \in \{1,2\}$, $k = 3$ if $n \in \{3,4\}$, and $k = 2$ if $n \in \{5,6\}$.

The preamble signature s, $1 \leq s \leq 16$, points to one of the 16 nodes in the code-tree that corresponds to channelization. The sub-tree below these nodes is used for spreading of the message part. The lowest branch of the sub-tree, having a SF or 256 ($c_c = C_{ch,256,m}$, where $m = 16(s-1)+15$) is used to spread the control part. In other words, the SF for the data part is anywhere between 32 to 256.

In the case of PCPCH, the signature in the preamble points to one of the 16 nodes, which corresponds to channelization code of length 16. The sub-tree below the specified node is used for spreading the message part. The control part is always spread with a channelization code for spreading factor of 256. This code is chosen from the lowest branch of the sub-tree. The SF for the data part may be chosen from 256 to 4. A user equipment (UE) is allowed to increase its spreading factor during the message transmission. The channelization code for the PCPCH power control preamble is the same as that used for the control part of the message part.

13.5.7 Spreading and modulation

Figure 13.23 shows a possible schematic of downlink spreading/modulation unit. [15]

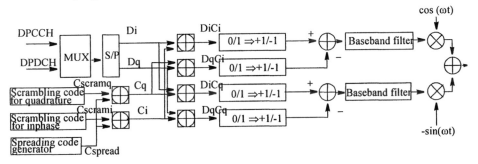

Figure 13.23 Schematic of downlink spreading/modulation unit

The data on DPCCH and DPDCH channels are multiplexed and converted into I and Q channels using serial to parallel convertor. The inphase component of downlink data, D_I, and quadrature-phase component, D_q, are spread with spreading code, $C_i + jC_q$, to generate spread signal $S_i + jS_q$. The code $C_i + jC_q$ is obtained from the scrambling and spreading codes. The spread I and Q signals are fed to a quadrature modulator and mapped onto QPSK constellation. The baseband filters having root raised cosine (RRC) spectrum with a roll-off, $\alpha = 0.22$, are used to shape the spectrum.

A schematic of uplink spreading and modulation unit is shown in Figure 13.24. The DPDCH and DPCCH are spread to the chip rate using the Orthogonal Variable Spreading Factor (OVSF). The DPDCH and DPCCH channels are spread using spreading codes, c and d respectively. The data, in the form logical 1 and 0, is converted into bipolar, ± 1, waveform. The signals are multiplied by long (scrambling) codes, C_i and C_q forming I and Q channels, which are added and passed through filters with a desired roll off. The power of the DPCCH is adjusted by a gain factor, G. The filtered base band signal is then up-converted in quadrature (I and Q) channels, which combined, amplified and transmitted.

If only one DPDCH is needed, DPDCH1 and DPCCH are transmitted. When multicode transmission is desired, several DPDCHs are combined. Multi-code transmission can be employed for the highest bit-rate, typically above 384 kbps in the vehicular based terminals, or several services of different rates may be transmitted in parallel while orthogonality between signals is maintained.

QPSK modulation is used for the carrier modulation on the uplink. The pulse shaping filters are root-raised cosine with a roll-off factor similar to that used for downlink [15].

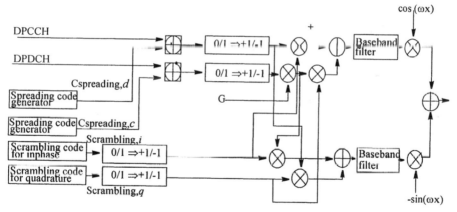

Figure 13.24 Configuration of uplink spreading/modulation unit

13.5.8 Radio Resource Function

The Universal Terrestrial Radio Access Network (UTRAN) supports both asynchronous and synchronous inter-cell operations. W-CDMA employs an intercell asynchronous network to make the network more robust and easy to install even in areas where Global Positioning System signal is not available. In order to realize smooth and quick cell acquisition even in an inter-cell asynchronous cell network, fast cell search function is required. This function is provided by the specific structure of the Synchronization channel.

Fast cell search is achieved by using a common spreading code to detect slot timing and identify the scrambling code group so that search range for scrambling codes can be narrowed down. Figure 13.25 shows the pilot channel structure.

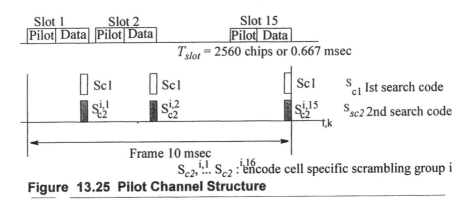

Figure 13.25 Pilot Channel Structure

The Ist search code is orthogonal Gold code of length 256 chips, transmitted once every slot. The Ist search code is the same for every BS in the system. The 2nd search codes are transmitted in parallel with the Ist search code. Each 2nd search code is chosen from a set of 16 different Orthogonal Gold codes of length 256. On the down link, each base station uses one of the 32 different scrambling code groups by which the base station is identified.

Search code symbols are spread only with orthogonal Gold codes, so that at the receiver these can be easily detected. With this pilot channel structure, the speed of cell acquisition by the MS is significantly accelerated. During the cell search process, the mobile station searches the base station to which is has the lowest path loss. The initial cell search process is carried out in three steps.

Step1: Slot/symbol synchronization step

Mobile station receives the Ist search code on the synchronization channel. Mobile station detects the symbol timing of search code symbol by scanning the Ist search code, which has a unique spreading code. The slot timing of the strongest base station is obtained by observing the output of the filter matched to the Ist search code.

Step 2: Frame synchronization and code group identification step

Mobile station receives the 2nd search code on the pilot channel with the slot/symbol timing obtained in Step 1. The group of the scrambling code (consisting of Nscrambling code) is then identified from the 2nd search code. The code group as well as the frame synchronization is determined by identifying the sequence that results in the maximum correlation value.

Step 3: Scrambling-code identification step

Mobile station then determines the exact scrambling code used by the selected base station. The scrambling code is identified through symbol-by-symbol correlation, with all the scrambling codes within the code group identified in the second step, over the pilot and data received on the pilot channel.

With the knowledge derived by reading the content of BCCH, on the designated spreading and scrambling codes, the received levels of the pilot channels in the neighboring cell/sectors can be the measured.

In Inter-cell synchronous operation, a specific code is used as the 2nd search code. In this case, since a common scrambling code is used for all BSs, and the identification of each cell/sector is based on the code phase offset, Step 3 of the intercell asynchronous operation becomes unnecessary. Thus, in this case the cell search is accelerated.

The physical channel operated in random access mode is called Random Access Channel (RACH). It is one of the constituent part of the uplink common physical Channel. In RACH access control for each message is handled independently.

Random access procedure is based on the slotted ALOHA algorithm. A mobile station transmits a radio frame with a random access message of RACH, and waits to receive an acknowledgment from the base station on FACH. The base station transmits the acknowledgment on Ack-mode FACHs after correctly receiving the radio frame from the mobile station on the RACH. The random message on RACH and the acknowledgment on FACH are related by the same PID (packet ID). The mobile station does not transmit a new radio frame on RACH until the previous radio frame is acknowledged or the time out period has elapsed. The mobile station retransmits the radio frame if it fails to receive the acknowledgment. Plural transmission timing offsets and/or plural spreading codes may be adopted to decrease the probability of collision.

Inclusion of the Transmitter Power Control (TPC) is essential to a CDMA architecture to overcome the near-far problem and to increase system capacity. W-CDMA uses an adaptive TPC method based upon desired signal level or signal to interference ratio (SIR). UTRAN uses two TPC forms; SIR-based fast closed loop TPC and open loop TPC. Open loop TPC is used on channels that can not apply closed loop TPC. For example, RACH uses open loop TPC. The receiver estimates the transmission channel's path loss. This is obtained by estimating and averaging the path loss over a sufficient number of fading periods. The power is then set.

The SIR based Fast Closed Loop TPC operates in the following way. The received SIR measurement is carried out every transmitter power control cycle (0.667ms). When the measured value is higher than the target SIR value, TPC bit 0 is sent. When the measured value is less than the target SIR-value, TPC bit 1 is sent. The receiving station transmits TPC bit to the station whose transmitting power is to be controlled. When TPC bit is received, soft decision on the TPC bits is performed. When the value is judged as 0, the transmitter power is reduced by 1 dB, whereas if it is judged as 1, the transmitter power is raised by 1 dB. When the TPC bit is not received, the transmitter power value is kept. When SIR measurement cannot be performed, the TPC bit has a default value of 1 [19].

13.6 IMT-TC (UTRA-TDD)

The IMT-Time Code (IMT-TC) system is designed using principles similar to those used in IMT-DS except that it operates in time division duplex (TDD) mode. For TDD operation, the unpaired frequency allocation is 1,900 to 1,920 MHz and 2,010 to 2,125 MHz. In Region 2, the frequencies are 1,850 to 1,910 MHz, 1,910 to 1,930 MHz, and 1,930 to 1,990 MHz. The TDD system is a flexible system as it can handle asymmetrical traffic by unsymmetrical slot assignment [19].

13.6.1 Physical Layer

The Physical Layer includes several building blocks having different functionalities. In the physical layer, the data on the transport layer is FEC encoded at the transmitter and decoded at the receiver. The error detected in the data on the physical layer is reported to the higher layers. The encoded data on the transport channels are multiplexed at the transmitter and the composite transport channel data is de-multiplexed in the receiver.

Rate matching of the data multiplexed on dedicated and shared channels is also done at the physical layer. After rate matching and multiplexing, the coded composite transport channels are mapped on to physical channels. As shown in Figure 13.24, power weighting of signals and combination of physical channels also take place at the physical layer. This also includes closed loop power control. The channel combining is followed by modulation and spreading of the signal and its reverse in the receiver. The physical layer is also used for frequency and time (chip, bit, slot, and frame) synchronization. The radio characteristic measurements including FER, S/I, interference power level and their indication to the higher layers is done at the physical layer. A support of timing advance on uplink channels is also provided at the physical layer.

13.6.2 Transport Channels to Physical Channel Mapping

Similar to the case of IMT-DS, the TDD has three types of channels: logical, transport and physical. The mapping between the three types for the downlink and uplink are shown in Figure 13.26 and Figure 13.27 respectively.

In the downlink, four physical channels, Primary Common Control Physical Channel (P-CCPCH), Secondary Common Control Physical Channel

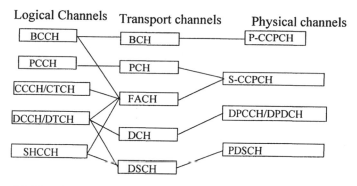

Figure 13.26 Channel mapping for TDD downlink

(S-CCPCH), Dedicated Common Control Physical Channel/Dedicated Physical Data Channel (DPCCH/DPDCH) and Physical Downlink Shared Channel (PDSCH) are created from five transport channels as shown in Figure 13.26. The Forward Access Channel (FACH) transports data from logical channels BCCH, CCCH/CTCH, DCCH/DTCH, and SHCCH to the physical channel S-CCPCH. The DSCH transport channel is used to map logical channels SHCCH and DCCH/DTCH to PDSCH. The other mapping is shown in the figure.

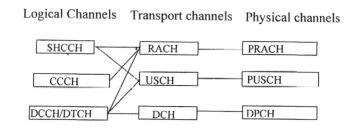

Figure 13.27 Channel mapping for TDD uplink

13.6.2.1 Physical Frame Structure

The transmission is in the form of superframes, which consists of 72 frames of 10 milliseconds. Thus, the superframe is 720 milliseconds long. Each frame has 15 slots, slot 0 to slot 14. Within each slot two types of bursts, Type 1 and Type 2 are quadrature multiplexed. Each burst has four fields, two data fields, a mid-amble, and GP field. The number of chips in each field for the two bursts are shown in Figure 13.28.

The data bits are QPSK modulated and the resulting symbols are spread with a channelization code giving spreading factors of 1 to 16. Due to variable spreading factor, each data part of one burst provides the number of symbols shown below.

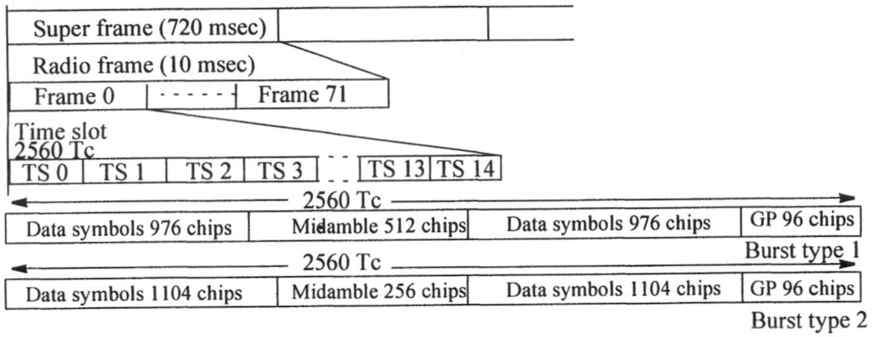

Super frame (720 msec)		
Radio frame (10 msec)		
Frame 0	- - - - - Frame 71	
Time slot 2560 Tc		
TS 0	TS 1	TS 2	TS 3	- - TS 13	TS 14

2560 Tc

Data symbols 976 chips	Midamble 512 chips	Data symbols 976 chips	GP 96 chips

Burst type 1

2560 Tc

Data symbols 1104 chips	Midamble 256 chips	Data symbols 1104 chips	GP 96 chips

Burst type 2

Figure 13.28 Basic Frame Structure TDD

Table 13.4   *Number of data symbols in TDD bursts*

Spreading factor, Q	Number of symbols, N, per data field in burst I	Number of symbols, N, per data field in burst 2
1	976	1104
2	488	552
4	244	276
8	122	138
16	61	69

13.7 Multicarrier CDMA (IMT-MC)

The IMT-Multicarrier (IMT-MC) is a system based on IS-95, which is a single carrier spread narrowband spectrum system. In this system, the physical layer supports RF channel spacing of $N \times 1.25$ MHz, where N is a spreading rate. In IMT-MC, the basic unit of channel spacing of 1.25 MHz is retained. The spreading rate numbers $N=1$ and 3 have been specified. N can be extended to 6, 9, and 12. This gives a maximum system bandwidth of 11.25 MHz.

The multiple access method is CDMA and the system operates in FDD mode. The effective chip rate in the systems in $N \times 1.2288$. In this system, interbase synchronization is required. In the uplink, a dedicated code division pilot is specified. However, in the downlink, a code division common or dedicated auxiliary pilot has been specified.

The channel interleaving depths from 5 to 80 msec have been specified. For channel coding convolutional codes of rates 1/2, 1/3, 1/4, 1/6 and Turbo codes of rates 1/2 to 1/4 are specified. The other significant parameters of IMT-MC are given in Table 13.5.

Table 13.5	*Summary of Major Technical Parameters*
Parameter	**Value**
Multiple access techniques and duplexing scheme	Multiple access scheme: CDMA; Duplexing Technique: FDD
Chip rate	$N \times 1.2288$ Mchips/s (currently $N = 1$ and 3 are specified) and N can be extended to $N = 6, 9, 12$
Inter-BS asynchronous/synchronous operation	Synchronous operation is required
Pilot structure	Code division dedicated pilot (UL); Code division common pilot(DL); and code division common or dedicated auxiliary pilot (DL)
Frame length and interleaving	5, 10, 20, 40, 80 ms frame and channel interleaving
Modulation and detection	Data modulation: BPSK; QPSK Spreading Modulation: HPSK (UL); QPSK (DL); Detection pilot aided coherent detection
Channelization code	Walsh codes and long codes (UL); Walsh codes or quasi-orthogonal codes (DL)
Scrambling (spreading code)	Long code and short PN code
Channel coding	convolutional code with $K = 9$, $R = 1/2, 1/3, 1/4$ or 1/6 Turbo codes with $K = 4$, $R = 1/2, 1/3$, or 1/4
Access Scheme (uplink)	Basic access: power controlled access; reservation access, or designated access
Power control	open loop; closed loop (800 Hz or 50 Hz update rate); power control steps: 1.0, 0.5, 0.25 dB.

13.7.1 Forward Physical Channels

The several forward physical channels are received at the mobile. These are shown in Figure 13.29. These channels fall into three classes - traffic channels, pilot channels, and paging and control channels. Of the traffic channels, eight are of type RC1 or 2 which are known as supplementary code channels. These are meant for rate 1 or 2. Three supplementary code channels are for use as RC3 to 9 rate channels. The traffic channels have also sub channels like power control sub channel, dedicated control channel and fundamental channel.

The pilot channel has four sub-channels - transmit diversity channel, auxiliary pilot channel, auxiliary diversity pilot channel and transmit diversity pilot channel. The function of the forward pilot channel is similar to that in the IS-95. The additional diversity channels are for the use of either receive or transmit diversity.

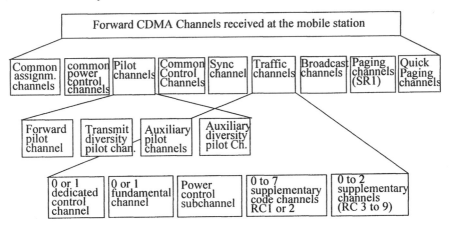

Figure 13.29 Forward CDMA channels received at the mobile station

In the class of paging and control channels, channels like common assignment channels, common power control channel, Synchronization channel, power control channels, and paging channels fall. Their functionalities are similar to those described in the previous sections.

Figure 13.30 shows the structure of the reverse traffic channel excluding the pilot channel. The pilot channel is multiplexed after the block interleaver. The reverse channel structure is very similar to that in IS-95.

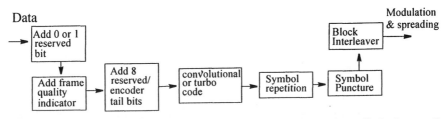

Figure 13.30 Reverse channel structure (excluding the pilot channel)

The reverse channel received at the base station for $R = 1$ to 3 is shown in Figure 13.31. These channels may be classified into three types - Reverse channel for RC1 to RC3 for user data, enhanced access channel operations, reverse channel control operation. The type consists of traffic channel RC1 and RC2 and 0 to 7 supplemental code channel. The supplemental code channels, eight in number, carry data spread by different codes. The enhanced access channel operation consists of reverse pilot channel and enhanced access channel, which is used for faster access to the network. The third type of channels are for common control operations on the reverse link. It consists of a reverse pilot channel and a reverse common control channel. The role of these channel are self explanatory.

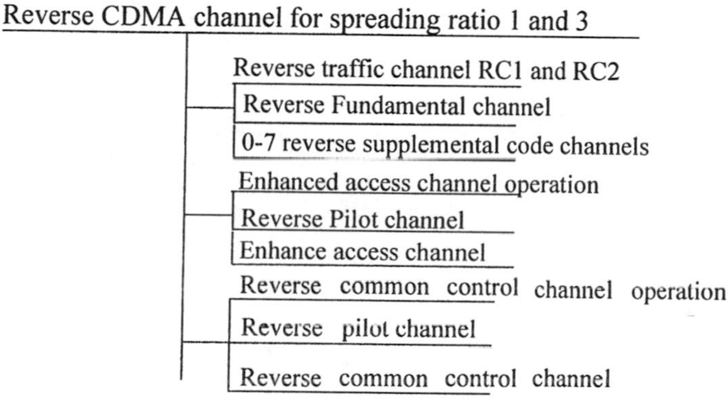

Reverse CDMA channel for spreading ratio 1 and 3

Reverse traffic channel RC1 and RC2
Reverse Fundamental channel
0-7 reverse supplemental code channels
Enhanced access channel operation
Reverse Pilot channel
Enhance access channel
Reverse common control channel operation
Reverse pilot channel
Reverse common control channel

Figure 13.31 Reverse CDMA channel received at the base station

Figure 13.32 shows an example of uplink spreading and modulation for radio configuration for three carriers and above.

The data on RSCH2, RDCCH channels after providing with Walsh covers are added to the Reverse pilot channel (RPCH) data. The output is fed to the spreading unit by masked long code and quadrature modulator. In a similar way, the data on RFCH, RSCH1, RCCH, or EACH if required, are Walsh covered and sent to spreading unit. The data on all these channels are gain adjusted to provide appropriate levels of power in the transmitted signal.

The output of the masked long code is fed to the offset quadrature spreading code. The spread signal is then modulated using QPSK modulator. The I and Q channels data are then up converted before transmission.

Figure 13.33 shows an example on scrambling, power control symbol insertion, puncturing and multiplexing. The masked long code is fed to an I & Q scrambling bit extractor, the output of which is used to scramble the encoded and interleaved data symbols. The scrambled symbols are then mapped into signal points. The mapped data are adjusted for channel gain. The long code is decimated and positions of power control bits at which the data is to be punctured and power control symbol inserted are extracted. The data symbols are then punctured and the forward power control bits are inserted. Sixteen power control bits in each 20 msecond frame or 4 bits per 5 msec

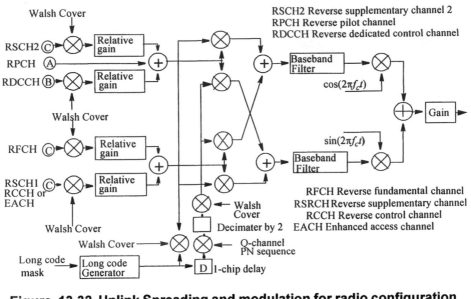

Figure 13.32 Uplink Spreading and modulation for radio configuration for three carriers and above

frame are inserted. Note that insertion of power control bits and scrambling process is similar to that of IS-95.

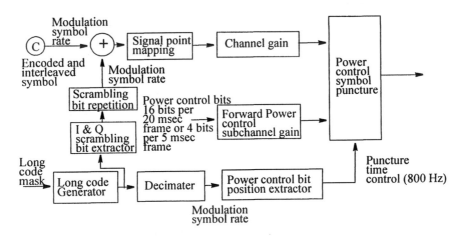

Figure 13.33 Example of scrambling, power control symbol puncturing and demultiplexing

13.8 IMT-SC (UWC-136)

The International Mobile Communication - single carrier system (IMT-SC) or Universal Wireless Communications-136 (UWC-136) was developed to maximize commonality between D-AMPS (TIA/EIA-136) and GSM GPRS and to meet the ITU-R's requirements for IMT2000. The strategy consists of enhancing voice and data capabilities of the 30 kHz channels through modulation enhancements to existing 30 kHz IS-136+ and defining a complementary wideband TDMA for services which are not possible with the 30 kHz system. 8-PSK modulation format and a new slot format is defined. RF carriers support both DQPSK and 8-PSK. By adding a 200 kHz carrier component for high speed data (384 kbits/sec) and high mobility component (136HS for Outdoor use), compatibility with GSM-GPRS or even EDGE is possible. For transmission of very high speed data (2 Mbits/sec), a carrier component at 1.6 MHz is added. These techniques define the migration path of IS-136 to a 3G system.

The system offers flexible spectrum allocation, high spectrum efficiency, and backward compatibility with the earlier systems, e.g. IS-136 and IS-136+. The flexibility of spectrum allocations is provided by allowing the system to work in the frequency range of 500 MHz to 2.5 GHz.

13.8.1 Services

The system is envisaged to provide voice, data and other ancillary services. The system is capable of supporting services with data rates up to 384 kbits/sec in pedestrian environment and up to 144 kbits/sec in high speed vehicular environment.

13.8.1.1 Voice Services

In the voice services, three voice coders are currently defined. These are Vector Sum Excited Linear Prediction (VSELP), Algebraic Code Excited Linear Prediction (ACELP), and US1. VSELP has two TIA standards, 7.95 kbps and 9.6 kbps for use in cellular mobile communications. The current GSM standard is based on 13 kbps RPE-LTP. The US-1 is a new low bit standard based on adaptive multirate speech coder.

13.8.1.2 Data Services

Scalable packet data service having rates from 11.2 kbits/sec to in excess of 2 Mbits/sec is supported. Packet data service is designated as GPRS-136. The users can access two forms of data network: *X*.25 and (IP)-based Internet protocol.

13.8.2 Ancillary services

The system supports many ancillary services. Examples of these are - short messaging service (SMS), message waiting, calling name indicator, extended standby time via sleep mode, wireless office, circuit switched data, over the air activation, over the air programming, encryption, broadcast Teleservices support, and authentication. The funtionalities of the networks layers 1 through 3 are described below.

13.8.3 Layer 1

IS-136+ supports both voice and data on a 30 kHz channel. Two modulation types are specified: mandatory π/4-DQPSK and an optional 8-PSK at a common channel symbol rate of 24.3 ksymbols/sec. The 136HS Outdoor bearer uses a 200 kHz carrier to enable the deployment of high speed indoor data services at greater than 2 Mbits/sec. The 136HS indoor bearer uses a 1.6 MHz RF carrier to enable the deployment of high speed indoors data services at greater than 2 Mbits/sec.

13.8.4 Layer 2

Layer 2 provides two different types of functionalities: circuit based operation and packet based operation.

In the circuit based operations, the modulated/encoded voice is carried over the digital traffic channel (DTC). The connection to the traffic channel is supervised using coded digital verification color code (CDVCC) on Layer 2. The Layer 2 of the digital control channel DCCH requests services from Layer 3 and provides indication to it. Various Layer 2 protocols can be used to carry RDCCH (Reverse Digital Control Channel) and Common Forward Digital Control Channel (CFDCCH) information in support of Layer 3 messages such as frame segmentation, their re-assembly, and ARQ mode of operation.

In packet based operations, Layer 2 is structured into logical link control (LLC), Radio Link Control (RLC), and Media access control (MAC). LLC is independent of the radio interface, it provides acknowledged or unacknowledged data transfers.

13.8.4.1 Layer 3

Layer 3 provides two types of mobility and resource management functionalities: the first type supports circuit switched based operation and the second supports packet based operation using defined GPRS-136 mobility and resource management unit.

The GPRS-136 mobility manager allows parallel operation of the GSM-based GPRS mobility management function and the 136 mobility management function. The GPRS-136 radio resource management (RRM) entity is similar to the DCCH procedures that control voice resources. It dynamically allocates the radio resource among users requiring bearer services.

13.8.4.2 Network

The defined network is a unique combination of TIA/EIA-136 TDMA radio interface and TIA/EIA-41 circuit switched and GPRS packet switched network. The network elements, their associated reference points, and interfaces are identified in Figure 13.34. For more details the reader is referred to [14].

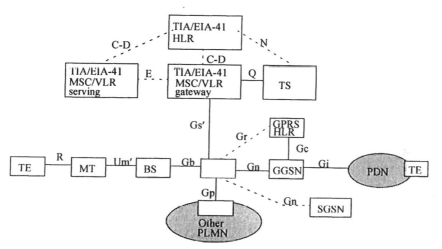

Figure 13.34 Network elements and the associated reference points

13.9 IMT-FT (DECT)

The international mobile telecommunication - frequency time is developed on the basis of (Digital Enhanced Cellular Telecommunication) DECT[†] technology. The reader should refer to Section 12.11 for details on the DECT.

13.9.1 General Access Technology

The radio interface covers only the specifications of air interface between the fixed part (FP) and portable part (PP). The profile specifications define IWU for various networks and are not a part of the Common Interface specifications. The Common Interface (CI) specifications contain general end-to-end compatibility requirements for relevant public network e.g. the Public Switched Telephone Network and Integrated Switched Data Network (PSTN)/ISDN. The common interface structure is shown in Figure 13.35.

---

†. *DECT stood for Digital European Cordless Telecommunication before its name was changed to Digital Enhanced Cellular Telecommunications.*

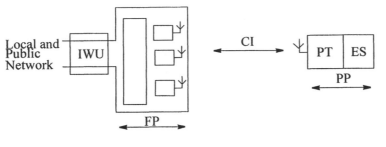

Figure 13.35 The CI structure

13.9.2 Summary of the Physical Layer

The tasks performed by the physical layer can be grouped into five categories:

(i) to modulate and demodulate radio carriers with a bit stream of a defined rate to create a radio frequency channel;

(ii) to acquire and maintain bit and slot synchronization between transmitters and receivers;

(iii) to transmit and receive a defined number of bits at a requested time and on a particular frequency;

(iv) to add and remove the synchronization filed and the Z-filed used for rear and collision detection;

(v) to observe the radio environment to report signal strength.

13.9.3 The Physical Channels

The system channel spacing has been chosen to be 1.728 MHz. For transmission, a TDMA structure with a frame length of 10 msecond is used. Each frame consists of 24 slots and each slot has two half slots. There is a provision of using a double slot which has a length of two full slots. The data is transmitted within the frequency, time and space dimensions using physical packets. The frame structure using full slots and physical packet types are shown in Figure 13.36.

The frame consisting of 24 slots is divided into two parts. The slots 0 through 11 are used for normal fixed part transmission and the slot 12 through 23 are for normal transmissions from the portable part. The frame structure is shown in (a). Four types of packets are handled. These are short physical packet P00, basic physical packet, low capacity physical packet P08, and high capacity physical packet P80.

Each packet consists of three fields, S-field, D-field and Z-field except the short packet which has two fields, S and D. The S-field is called the service field and it defines the type of packet service. The length of the S-field in each packet is 32 bits. The length of the 64 bits. Thus, the short packet is 96 bits long. The *D*-field in the basic packet is the data field is 388 bits, and *Z*-field is 4 bits long. The *D*-field in the low capacity packet is 148 bits, and the *Z*-field is 4 bits long. Thus, the low capacity packet is 184 bits long. The high capacity packet has a structure similar to the low capacity packet but the length of the data field is 868 bits. The total length of the packet is 904 bits.

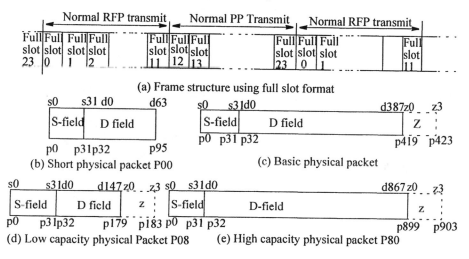

(a) Frame structure using full slot format

(b) Short physical packet P00

(c) Basic physical packet

(d) Low capacity physical Packet P08   (e) High capacity physical packet P80

Figure 13.36 Frame structure and physical packets types

13.9.4 RF carrier Modulation

Table 13.6	*Allowed combination of modulation schemes*		
Configuration	**S-field**	**A-field**	**B+Z-field**
1a	GFSK	GFSK	GFSK
1b	$\pi/2$-DBPSK	$\pi/2$-DBPSK	$\pi/2$-DBPSK
2	$\pi/2$-DBPSK	$\pi/2$-DBPSK	$\pi/4$-DQPSK
3	$\pi/2$-DBPSK	$\pi/2$-DBPSK	$\pi/8$-D8PSK
4a	$\pi/2$-DBPSK	$\pi/4$-DQPSK	$\pi/4$-DQPSK
4b	$\pi/2$-DBPSK	$\pi/8$-D8PSK	$\pi/8$-D8PSK

13.10 Networking Aspect of 3G Systems

The IMT-2000 defines the five air interface standards, out of which wideband CDMA is likely to be used most widely. The applications requiring packet data transmission are on the increase with circuit switched voice traffic suggest that the network should be a hybrid network - a combination of packet network and GPRS plus GSM network. The IMT-2000 network may be divided into two logical concepts, the generic radio access network e.g Universal terrestrial radio access network (UTRAN) and the core network, these are linked together through interfaces Uu and Iu as shown in Figure 13.37.

Figure 13.37 The IMT-2000 network architecture

The UTRAN is defined as a network that connects the user equipment to the core network. This is bounded by interface Uu towards the user equipment and by Iu towards the core network. The UTRAN has an architecture similar to the GSM Radio Access Network. It consists of Radio Network Controller (RNC) and Node Bs (base stations). These together form radio network subsystem. The components of UTRAN and interfaces are shown in Figure 13.38.

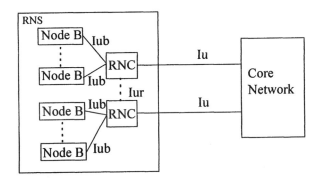

Figure 13.38 UTRAN components and Interfaces

The RNC controls one or more Node Bs via interface Iub. Node B is equivalent to base station transceiver and RNC to base station controller. The RNC is connected to an MSC (IuCS) or to an SGSN (IuPS). The interface between RNCs, Iur, is logical in nature. The Node B performs several functions, including mapping of logical resources to hardware resources, macrodiversity combining, uplink power control, measurement of interference and power, and transmission of system information messages according to scheduling parameters given by the RNC. In addition to these, RNC also contains the air interface physical layer. The functions of the physical layer are already discussed earlier. The UMTS network elements and interfaces are shown in Figure

13.39. Note that the core network portion is the same as in GSM-GPRS combination. The core network may serve both the UTRAN and GSM radio access network.

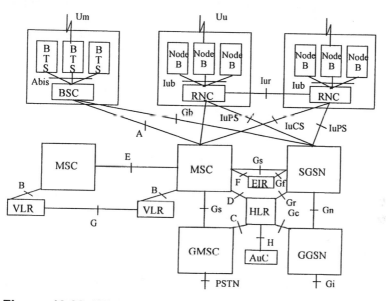

Figure 13.39 UMTS network architecture and interfaces

The core network consists of Mobile Switching Center (MSC), Visitor Location Register (VLR), Home Location Register (HLR), Equipment Identity Register (EIR), Authentication Center (AuC), Gateway MSC (GMSC), Service GPRS Support Node (SGSN), and Gateway GPRS Support Node (GGSN).

The functions of MSC include paging to the subscribers, dynamic allocation of the resources, coordination of call setup from other MSes, location registration, handover managements, billing, encryption parameter management, signaling exchanges between different interfaces, frequency allocation and managent and echo cancellation.

In the VLR resides the information about mobile stations that roam in MSC area including those belonging to this home network. The information in VLR is temporary. When a user becomes active in its own home network, the information is transferred temporarily in VLR. The data entry in VLR may contains the information on International Mobile Subscriber Identity (IMSI), mobile station's international ISDN number, location of home network of the user, and the last known location and the initial location of the MS.

The VLR may also contain supplementary service parameters like authentication procedures with the HLR and AuC. It also performs the task of tracking the state of all users in the area, and supporting the paging procedure.

The home location register (HLR) contains data on subscribers registered with the home network. These parameters do not change unless the subscriber deregisters or moves to other registration area. The permanent data with HLR includes international mobility subscriber number, mobile station ISDN, MS category number, roaming restrictions if any, supplementary parameter number, authentication key, and network access mode, which determines whether the user can access the GPRS networks, non-GPRS network or both.

The equipment identity register (EIR) stores the international mobile equipment identities in the form of lists like white list (equipment in good order), black list (stolen equipment), gray list (equipment known to have problems that are not fatal enough to be barred).

The authentication center is associated with the HLR. It stores the subscriber's authentication key and the corresponding IMSI. These parameters are entered in the center at the subscription time.

The gateway MSC is located between the PSTN and the other MSC's in the network. It routes the incoming calls to the appropriate MSC's.

The serving GPRS support node (SGSN) is the central element in the packet switched network. It contains subscription information, IMSI, temporary identities, PDP numbers, location information, the routing area where the mobile is registered, and VLR number. The GPRS Gateway Support Node (GGSN) corresponds to the GMSC in the circuit switched network. However, where the GMSC routes the incoming calls, GGSN must also routes the outgoing calls. The GGSN maintains information on subscriptions, IMSI, PDP addresses, location information, the SGSN address where the MS is registered.

13.11 Intelligent Networks

The users of wireless communications are expected to subscribe to a variety of services customized by them both in signalling and capabilities to match their needs at home. To allow for a slow or rapid change in the call routing and quality resulting from terminal mobility, certain specific functions like mobility management are needed in the network architecture. Ubiquity of service, from anywhere to everywhere, is a distinctive feature of future wireless communication systems. Several definitions of wireless communications for the future have also been proposed, these range from simple cordless telephones (portable voice service) to very complex all embracing terminals to provide voice, data and image services. The elegant definition of ubiquitous communication system where the users are given the command over communication resources is now widely accepted. This definition allows a greater flexibility in reconfiguring profile of the users when they change the definition of service to meet current application. Thus, to satisfy users' requirements, the system operator must offer ubiquity, mobility and control to the users over the services they subscribe The amalgamation of these characteristics requires considerable intelligence in the network [20]. The term intelligent network (IN), coined by the ITU-T which drafted recommendations for Intelligent Network Compatibility Set 1 (CS-1) specified in Q.1200 series [21], is used to describe an architectural concept intended for applications in advanced networks.

The IN concept attempts to standardize management of service logic which allows extensive use of information processing techniques in utilizing network resources more efficiently. This concept divides the network functions into modules which create integrated services by accepting subscriber's control over the services they desire and permit flexible allocation of network resources and functions and their portability to the physical entities of the network. Since, intelligent networks are expected to play an important role in the future wireless communication systems, we describe these networks and their functions in some detail.

13.11.1 PCS Service and Network Requirements

The IN functional requirements stem from the desire to provide the customer with tools to configure network capabilities in order to meet the requirements for the requested service. The functional set available to the users is of course dependent on the architecture of the network.

The user of available services is an entity external to the network. The network operator offers services that satisfy the telecommunication needs of the customers. The service currently used by a customer may be affected by the other users active on the network. In general, the requirements set by the users and the operator may be divided into two categories: service and its quality and network.

The task of the operator is offer a flexible network that offers IN services to the users. The users may need flexibility to access and keep control over the service that may span one or multitude of networks; define service on per call basis or over the duration of the call; to be able to call one or more parties simultaneously, to record service usage in the network for the purpose of supervision and carry out tests on performance and charges; and finally to control interaction between different invocations of the same service. These service requirements emphasize the network capabilities that the users seek to access intelligent network services.

13.11.2 Services Requirements

Intelligent Network concept defined by the ITU-T will allow service providers to increase the flexibility and speed of service. The concept lies in abstraction of network capabilities into Service Independent Building Blocks (SIBBs) used by the service logic in defining services. These building blocks enable service providers to deploy and modify IN based services rapidly.

13.11.3 Network Service Capabilities

One of the necessary network capabilities lies in providing basic and supplementary services using IN concept to the customer in a transparent manner i.e. the customer should not be aware whether or not IN is used in providing services. Therefore, no service processing requirement can be identified with a specific reference to the IN. Regardless of this, the IN should be capable of supporting a broad range of basic and supplementary services. With regard to processing a service, the following capabilities are identified.

Service Capabilities are necessary to support a broad range of basic and supplementary services. IN concepts can be applied to the support supplementary service for the following basic services [23]:

  (i) Bearer Services including circuit (unrestricted, speech, and audio) and switched mode services (packet switched data services, circuit switched data services and others);

  (ii) Teleservices includes Telephony, Telefax, Videotex;

  (iii) Broadband Interactive Services including conventional, messaging and retrieval services;

  (iv) Broadband Distribution Services including distribution service with or without user individual presentation control;

  (v) Mobility Services which includes location, handover, personal service profile, and paging.

13.11.4 IN Access Capabilities

The user needs some procedure to access the network and use a particular service. In order to allow freedom to access any of the services, it is necessary that there is sufficient degree of independence between the service capabilities on one hand and the access capability on the other. The following access capabilities are foreseen for IN:

  (i) Fixed Network: PSTN access, ISDN access, PSPDN access, CSPDN access;

  (ii) Private Network Access: Mobile Network, Cellular access, Microcellular access, Paging access, MSAT access;

  (iii) Broadband Networks: Asynchronous Transfer Mode (ATM) access, Synchronous Transfer Mode (STM) access.

13.11.5 Service Management

A proper operation of network needs tools for customers to manage services. The service management activities during deployment stage include limited deployment of services for testing purposes. Following successful testing, service logic, management logic etc. are downloaded into appropriate network elements in order to activate these logics. During the service provision phase, service management involves activities like creation of customer's service profile in the service logic and the activation of the service for the customer. Service activation is realized by initialization and activation of trigger conditions.

Service management activities following the service provisioning, i.e. during utilization stage, include:

(i) The service activation, service maintenance and service customizing activities give the customers abilities to modify service control parameters so as to match them with their immediate needs. They are also have ability to check whether a service is properly defined by the entered set of parameters, authorization and validity.

(ii) The activity, related to charges, collects data on service usage and generates report for billing either automatically or on demand. The network may provide services at a flat rate or charge the users on the basis of the call duration and the network resources used for the service.

(iii) The network management activities include service monitoring which collects and accumulate statistics, automatically or on demand, on a given service with a view to determine the quality of service. Typical manual operation are queries regarding status information and network configuration.

13.11.6 Network Requirements

The network requirements are set by the operator with a view of which IN services may be needed by the customers. The network uses communication protocols that allow flexibility in allocation of functions like allowing migration from existing network bases to target network bases in a practical, flexible and cost effective manner while reducing redundancies among network functions in physical entities. Another desirable feature of the network is its adaptability in providing new services in a cost effective and timely manner without compromising network stability whenever new services are introduced. It should be noted that the network must maintain service quality and network performance at all specified levels.

The service creation, service management, network management, service processing and network interworking are the capabilities identified to meet the service and network requirements. Figure 13.40 illustrates the network capabilities and their relation to the service and network requirements.

13.11.7 Network Management

The network management functions refer to the network capabilities that are necessary to support the proper operation of the intelligent network.

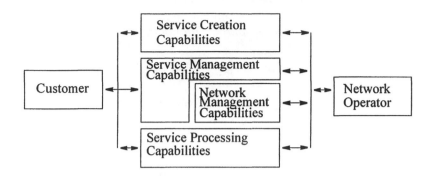

Figure 13.40 Network capabilities and their relation to the service and network requirements

They are organized into three management functions: performance management, fault management and configuration management.

The performance management function covers performance monitoring, traffic and network management, and quality of service observation. The fault management (maintenance) covers alarm surveillance, failure localization, and testing. The configuration management has the responsibility for provisioning, status and control, and initialization.

In addition to these management functions, two other functions may be included: accounts and security management.

13.11.8 IN Network Architecture

The Advanced Intelligent Network (AIN) architecture is dictated by the type of services that are to be provided by the wireless system. It is desired that the wireless service concept be based on mobility independence that is the quality of service be maintained over the duration of the call when the mobile terminal is in the state of continuous mobility. In contrast, the fixed network is geared towards discrete mobility only, which implies that over the duration of the call the position of the terminal remains fixed although the user may have altered its position between calls. The future system may be comprised of one system or several integrated systems. It may be thought to be an overlay of existing cellular radio network, an extension of STN or a completely independent network. From the operations and cost considerations, inclusion of STN as a backbone network is deemed necessary.

The functions of the intelligent network include creation of services, management of services and network, process of services and create interworking environment between the networks.

A three step approach is usually followed in service creation: service specification, service development and service verification. The service specification procedure follows steps of service description, generation of functional analysis, and transformation of high level structured design into detailed structured software design. The service verification is a stage where the developed service software is rigorously tested to ensure that service application meets the specification. The environment which allows creation of new services is known as Service Creation Environment.

Service portability across different environments; urban, suburban, rural, office, etc. is a fundamental requirement of the users. Such portability requires interworking of wired, terrestrial wireless and satellite networks. The flexibility gives the users the ability to configure the network to satisfy the needs of the application at hand coupled with the requirement of dynamic source allocation and routing. The latter two requirements imply dynamism in billing and other network management functions. All these requirements can be handled by intelligent networks. Considerable work has been done on intelligent networks during the past two decades, most of which though is related to switched telephone network [22][23][24][25].

13.11.9 Network Layers

A possible hierarchical network architecture, depicted in Figure 13.41, includes IN capabilities within the architecture which provides advanced services that may be partitioned into wireless and fixed portions. It consists of three layers: an access layer, a transport layer, and an intelligent layer [26].

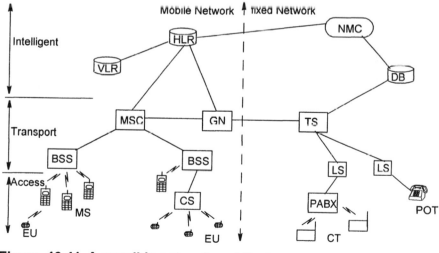

Figure 13.41 A possible network architecture

The access layer consists of equipment and functions to provide network access to all users irrespective of their location; this includes wireless access system. The access layer is used by the subscribers to reach for service.

The transport layer which includes equipment and functions related to the transmission and delivery of information across the network provides call control and switching functions. The functions of this layer are directly under the control of the intelligent layer.

The intelligent layer includes equipment and functions responsible for storage and retrieval of all kinds of information on all network users. The information is made available to the call control functions in the transport layer. It also contains functions responsible for the provision of customer defined services.

The key elements in the mobile side of the network are: the MS (mobile station), UE (user equipment), CS (Cell site) for UEs, BSS (Base station system), MSC (mobile switching centre), VLR (Visitor Location Register), and HLR (home location register). Other elements such as EIR (Equipment Identity Register), and AC (Authentication Check) may be added as independent elements but are generally combined into HLR.

On the fixed side of the network, POT (plain ordinary telephone), interfaced cordless telephone, local switch, toll switch, Public Branch exchange (PBX), and a Database (DB) site are the necessary elements. The MS and UE are the physical equipment used by the subscribers to access the network services. The MS may be vehicle mounted and is characterized by high rate of motion. The UE, usually a pocket size portable, can be used in several ways. For example, it could be used in a moving vehicle with MS acting as a tandem link to the network. UE can be used outdoors for rates of motion up to that of slow moving vehicle. Other access terminals e.g. MET (Mobile Earth Terminal) may be used for getting access to the users in remote and rural areas.

The BSS terminate the radio links to mobile stations and is composed of BST (Base station transceiver) and BSC (Base station controller). The distribution of functions within the BSS is not considered here and the term BS is used to denote the complete BSS configuration. The MSC sets up calls to mobile radio users via the BS and maintains connections to other users and the PSTN.

The HLR holds permanent parameters and features of users registers within its catchment area. Its pointer to the VLR assists in routing the incoming calls. The VLR is a dynamic database whose entries change with the movement of users coming into or going out of the defined service area. The VLR communicates with HLR to obtain user parameters and reports to it updates regarding the status of user location. The VLR also performs authentication of the visiting mobile and allocates temporary identity to the visitors.

13.11.10 Intelligent Network Configuration

The presence of AIN in the network architecture facilitates introduction of services and modifications to them in order to allow enhanced call control features, which can be used by the subscribers. Figure 13.42 shows a possible implementation of an intelligent integrated network architecture which contains Functional Entities (FE) as key elements. These FE's are mapped unto a physical IN architecture, that is functional relationship between control channels and traffic are established.

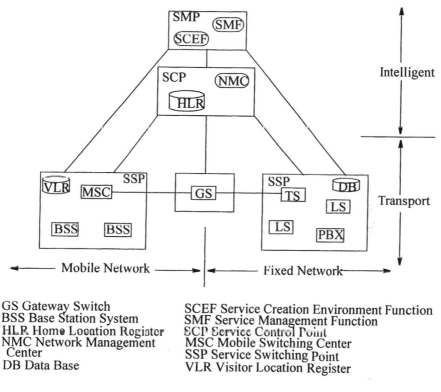

GS Gateway Switch
BSS Base Station System
HLR Home Location Register
NMC Network Management
 Center
DB Data Base

SCEF Service Creation Environment Function
SMF Service Management Function
SCP Service Control Point
MSC Mobile Switching Center
SSP Service Switching Point
VLR Visitor Location Register

Figure 13.42 A possible hierarchical network architecture

The Service Control Point (SCP) provides the functionalities of Home Location Register (HLR), the Authentication Check (AC), and the Equipment Identity Register (EIR). The AC is used to verify eligibility for a requested service and the EIR is used to determine whether or not a terminal is an authorized equipment for the requested service. The SCP also provides the functionality of the network Management Centre (NMC) which coordinates and monitors activities within the network. There are a number of different ways in which the Service Switching Points (SSP) can be implemented. The SSP contains the BSS, MSC and VLR. In the fixed network, the SSP contains the PBX, LS, TS and the fixed network database. The gateway switch (GS) linking the mobile network to the fixed network is also an SSP. The exchange of information between SCP and SSP takes place via SS7.

13.11.11 Service Configuration

Besides physical mapping of the functionalities, it is imperative to define IN services. These include subscribers's access to origination, call transfer, call termination, subscriber service call and personal mobility. These will need definition of access classes to subscribers even though the services may be provided by IN. In the following, we describe some features of services that can be provided in the Distributed Functional Plane (DFE).

13.11.11.1 Access, Identification and Authorization

This feature describes how a user gets access to services by dialling a special service code and identifying itself by his PTN and authorization code in a 'prompt and collect sequence'. The information flows (IF) needed to provide these features are: Initial Detection Point (DP), Request Status Report, Connect to Resource, Status Report, Request Report BCSM, Prompt and Collect User Info, Specialized Resource Report, and finally Query and Query results.

13.11.11.2 Procedure Identification

This feature shows how a PCS user can identify the required procedure for successful provision of a PCS service, e.g. procedure for location, registration, call setup, etc. This involves a query intended to check whether the PCS user is allowed to use the requested procedure. The IFs are: Prompt and Collect User Info, Request Status Report, Connect to Resource, Specialized Resource Report, Query and Query Results.

13.11.11.3 Network Access Registration

This feature registers the location of a user against a specific terminal whose physical address is known to the network. The IF's are: Analyze Info, Query, Prompt and Collect User Info, Update Data.

13.11.11.4 PCS Call Setup

This feature manages the call setup procedures for the network. The IF's are: Prompt and Collect User Info (called number), Query (the SCP's database is accessed for user authorization and for the location of the called user's home database, a second query to the home database requires the current location of the called user), furnish changes info (instructions to prepare changing record), disconnect forward (at the end of call setup), update data (charged info stored in user's home database).

13.11.11.5 Call Direction

This feature routes calls according to parameters such as called number and time of day previously set by the PCS user. The Information flows are: analyze info, query, query result, and route call.

13.11.11.6 Profile Status Interrogation

This feature provides PCS with the ability to query the settings and status of user profile. For example, the number of times a certain feature is used. The IF's are: analyze info, query, query result, prompt and collect user info.

13.11.11.7 Profile Modification

This feature provides the PCS users ability to modify the settings of his(er) profile. For example, the user may change the data associated with the call direction procedures for certain PTNs. The required IFs are: analyze info, query, query result, prompt and collect user info.

13.11.12 Call Processing

Figure 13.43 shows main phases of call origination while termination is

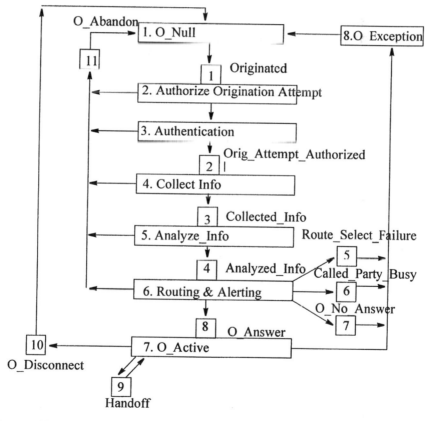

Figure 13.43 Main phases of call origination

shown in Figure 13.44. These figures depict the call states and interaction points where SSP and the SCP can exchange information and service control commands to design a customized call. The IN call has additional features of authentication, channel assignment, paging and handoff.

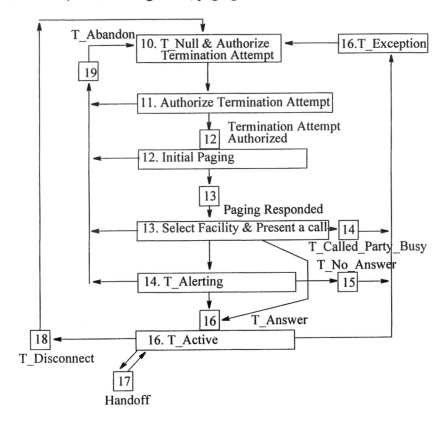

Figure 13.44 Main phases of call termination

Mobility and location registration are the main features of an integrated wireless communication system. Personal mobility refers to reachability to a subscriber using location registration and updating. Location registration can be retrieved by the network using access, identification and authentication features, procedure identification feature, and network access registration feature. The personal terminal can initiate the location registration periodically without the user intervention. Since frequent location registration can generate considerable traffic in the network, the frequency and type of location registration needs a careful study. The location registration begins when the terminal connects with a mobility management function residing in the SCP. The SSP after receiving the request triggers the Analyze_Info TCP. This initiates procedures for access identification and authentication at TCP. After the usual authentication checks out the procedure identification is initiated. This procedure is used to determine if the outgoing call is a registration request in which case a query to the SCP for further instructions is initiated. The SCP uses the Analyze Call Number FEA to invoke the mobility management program and responds with the instructions for the IP to prompt the users for a PTN and a password (PIN) using the Present Information FEA followed by Collect Information FEA. The IP passes the collected information to the SCP. The SCP again uses the Analyze Information FEA's to verify PTN and PIN format and uses the data Access FEA to query the database to verify their validity. Upon validation, the SCP then prompts for the new location. An area code and a subscriber number are used to provide the current location using the Analyze Calling Number FEA. The SCP updates the user's location in a mobility database and returns instruction to the SSP to terminate the call.

13.12 Summary

In this Chapter, the inadequacies of 2G systems are identified. The motivation behind the development of 3G systems are discussed. The user requirements of 3G systems as defined by ITU-R are introduced. Five new air interface standards are described in this Chapter. Details on uplink and down link channels and mapping between logical, transport and physical channels is discussed. W-CDMA is described in greater detail This chapter also discussed network aspect of 3G system. The role of intelligent networks in 3G network is brought into discussion.

References

[1] Sub-Committee for Advanced Radio Technologies, *Users Performance Requirements*, CTIA, June 1988.

[2] CCIR Report M/8 (Mod F), *Future Public Land Mobile Telecommunication System*, December 1989.

[3] H.L.Lester, S.S. Rappaport, C.M. Puckette, "PRCS - A consumer mobile radio telephone communication system with distributed control," *IEEE Global Telecommunication Conference*, pp. 1417-1424, 1983.

[4] C.M. Puckette, H.L.Lester, "900 MHz Private radio communication system - A Proposed solution for consumer communication needs," *IEEE Vehicular Technology Conference* pp. 389-392, 1983.

[5] D.C. Cox, H.W. Arnold, and P.T. Porter, "Universal Digital Portable Communications: A System Perspective," *IEEE Trans,. on Selected Areas in Communications*, vol. JSAC-5, 1987.

[6] D.C. Cox,"Universal Portable Radio Communications," *IEEE Trans. on Veh. Tech.*, vol. VT-34, 1985.

[7] A.J. Motley, "Advanced Cordless Telecommunication Service," *IEEE JSAC*, vol. SAC-5, no. 5, June 1987.

[8] S.W. Halpern, "Alternatives in Cellular System Design for Serving Portables," *IEEE Veh. Tech. Conf. Record*, pp. 162-167, 1984

[9] R. M. Singer and D.A. Irwin, "Personal Communication Services: Expanding the freedom to communicate," *IEEE Communication Magazine*, vol. 29, pp 62-66, February 1991.

[10] R. Prasad, W. Mohr, and W. Konhäuser, *Third Generation Mobile Communication Systems*, Artech House, Norwood, 2000.

[11] 3GPP TR25.834, v.4.0.0, *UTRA TDD Low Chip Rate Option: Radio Protocol Aspects*, 2000.

[12] ITU Recommendation, ITU-R M.1035, 1994.

[13] ITU Recommendation, ITU-R M.1308, 1997.

[14] J. Korhonen, *Introduction to 3G Mobile Communications*, Artech House Inc., 2001.

[15] ARIB IMT-2000 Study Committee, *Japan's Proposal to Candidate Radio Transmission Technology on IMT-2000: W-CDMA*, ARIB, 1998.

[16] *Third generation Partnership Project: Technical specification group radio access network; spreading and modulation*, (FDD) 3TS/TSGR-0125213U, 3GPP support office, Sophia Antipolis, Valbonne, France, 1999.

[17] *Physical channels and mapping of transport channels onto physical channel (FDD)*, 3GPP TS 25.211.v.3.4.0, 3GPP support office, Sophia Antipolis, Valbonne, France, 2000.

[18] F. Adachi, M. Sawahashi and K. Okawa, "Tree structured generation of orthogonal spreading codes with different length for forward link of DS-CDMA mobile radio," IEE-Electronics Letters, vol. 33, no.1, pp. 27-28, January 1997.

[19] ITU Recommendations, R-Series, March 2001.

[20] B. Jabbari, "Common channel signalling system number 7 for ISDN and intelligent networks," *Proc. IEEE*, vol. 79, no, 2, pp. 155-169, February 1991.

[21] CCIT Q.1200 Series Recommendations, *Intelligent Network Compatibility Set 1*, March 1992.

[22] G. M. Dean, "Network architecture and control: the defining element in IN evolution," *Proceedings IEEE ICC'91, vol. 1*, pp. 40.2.1-40.2.5, 1991.

[23] E. Fletcher-Haselton, "Service creation environment for intelligent networks," IEEE Communication Magazine, vol. 30, pp 70-76, February 1992.

[24] I. G. Ebert, P.S. Richards, and S. J. Harris, "PCS application for a service independent architecture," Proceedings Comforum (Rye Brook, N.Y.) June 10-11, 1991.

[25] M. Pierce, F. Fromm, and F. Fink, "Impact of the intelligent network on the capacity of network elements," *IEEE Communication Magazine*, vol. 26, pp 25-30, December 1988.

[26] D. N. Ashitey, An intelligent personal communication system, M.Eng. Thesis, Carleton University, Ottawa, Canada, 1993.

Chapter 14

WIRELESS SYSTEMS BEYOND 3G

With 3G revolution unfolding, the hottest question being asked at present is what lies beyond 3G. It is always difficult to make predictions on future events because these can turn out to be embarrassing wrong as happened many times in the past to several giants among the members of the scientific community. To forecast on the development of future events on scientific grounds, it is required that the history of the past events be examined. We can examine the past and the current status of wireless systems and try to establish the direction or the trend in which this rapidly changing technology will follow. If forecast on the trend is reasonably accurate then it is possible to open a window through which one can future events unfolding.

One dimension of the trend may be derived from the advances made in services offered in recent years over the fixed telecommunication networks. It is well established that developments over wireless networks have followed those over the wired network with a time lag of few years. Over the years, this lag time has been narrowing and with the introduction of 3G systems, the services over wireless networks will leapfrog to the level achieved over STN. Consider for example, the data traffic over the fixed telecommunication networks. The volume of the data traffic exceeded that of voice for the first time in year 2000. It is expected that the growth in data traffic will not only continue but will accelerate due to rapid increase in the rate hungry data services. These applications include proliferation of Internet usage. This trend is likely to catch on in the case of wireless communications.

The 3G system is relatively new and it has not been in extensive use except in Japan and so far its limitations have not been discovered because 3G systems with full potentials have not yet been implemented. From the subscribers' point of view, gaining access to information databases should be rapid and reliable. It is observed that the maximum rate of 2 Mbits/sec for stationary terminals and of 384 kbits/sec or even lower to 144 kbits/sec at vehicular speeds will become grossly inadequate within the next few years. It is likely that inadequacy of rates will create bottlenecks in networks; the primary reason of these will most likely be mismatch of rates between the lower speed over wireless network and higher speed over wired network. In order to make two parts of the network work harmoniously a higher degree of bandwidth compatibility between the two systems must be achieved. This may turn out to be a challenging task.

The limitation on sustainable data rates over mobile radio channel is due to several factors that relate to the character of wireless channels. Thus, to create harmonious working between the wired and the wireless networks, data rates over the two networks should be made compatible as far as possible. In other words the wireless mobile channels must be made suitable to carry transmission rates of several Mbits/sec[†]. This is seen as a major challenge to wireless communication system designers. How to overcome the limitation imposed by the channel is discussed in some detail in the following section.

The second peculiarity in which the network is used is the mode in which data is exchanged over the Internet. In this case, the network responds to infrequently occurring short commands from the user with large volumes of data towards users. It is desirable that the data be received in a shortest possible time. This requires high transmission speeds. This interactive data exchange on the Internet implies that the transactions are unsymmetrical. In future when the data volume exceeds that of voice by a significant amount, the links over wireless channels are likely to become asymmetrical. This requires techniques that use spectrum in an adaptive manner so as to maintain the spectrum efficiency high. Though all the five IMT standards are designed to use spectrum with high efficiency, the IMT-TC uses un-paired spectrum in a much flexible manner while taking into consideration the asymmetry in the traffic between the forward and the reverse channels. This flexibility lies in the adaptive use of frame slots.

The third desirable aspect of the future wireless networks lies in ubiquity of service, which is achieved by total integration of several networks like macro- and micro-cellular systems, indoors and short range wireless networks like Bluetooth and satellite systems. The total integration implies horizontal and vertical handovers that are within the peer systems and across systems at lower and higher hierarchical levels. The 3G systems resulting from migration from 2G systems i.e. IMT-SC, IMT-MC, and IMT-FT are likely to fit nicely into this scenario. This means that inter-system seemless handovers are easy to implement with the existing technology. However, because of distinctiveness of air interfaces of these systems, multi-standard terminals will have to be developed. It should be again emphasized here that the specifications of 3G systems have been developed with as much commonality as possible and minimum number of terminals would be required for universal roaming. In this regard, the use of software radio technology is likely to play a major role in

---

[†]. *Recently, NTT Docomo announced the beginning of 4G system trials. The data rates between 20 and 100 Mbits/sec are proposed.*

the development of multi-standard receivers, with a capability of adapting themselves to the available system.

It is hoped that total or global integration will eventually result in a new universal standard which will make the use of multi-standard terminals unnecessary. This could be the 4th generation systems.

One of the emerging application is Home Systems [1]. A home system embodies a communications network in home, which is interfaced with the outside world through the existing telecommunications networks. The In-home network will interconnect major components in home like refrigerator, VCR, computer, microwave ovens, cooker, grill, and other devices related to messaging and security. The idea is to create a possibility of controlling in-home devices from remote locations. For example, one would be able to remotely start cooking a meal while still at work and finds the food ready by the time one arrives home. Consumers and manufacturers of electronic equip-ment and appliances, and public utilities are likely to benefit the most from this technology.

The desire for global integration are likely to induce significant changes in the core network. The new network will most probably be based on packets but to achieve toll quality packetized voice over mobile channel where interpacket delay randomly vary because of continually changing loca-tion of mobile terminals is a major challenge. Thus, for a foreseeable future, packet and circuit switched networks are likely to coexist. It is therefore expected that the new architecture of the core network will be hybrid where both the circuit switched and packet switched services will be available. Also, the current GSM and IS-41 core network architectures will evolve into either purely packet or hybrid networks. We are observing that VoIP networks are getting exceedingly more transparent as the time passes.

In this chapter, we shall further expand on the challenges faced by sys-tem developers of 4G systems. We begin with a discussion on services which are likely to be offered by 4G systems.

14.1 Services in Future Wireless Systems

The 3G systems offer multimedia services, with a variable data rates and quality of service (QoS). The maximum transmission rate currently envi-sioned is either 1.92 Mbits/sec or 2.048 Mbits/sec. The maximum transmis-sion rate is meant for highest quality wireless links, which are found in-doors or in environments which are more or less remain unchanged over the link time. The choice of QoS is dependent on the customer requirement. The QoS requirement for data services is much tighter than that for voice services. For example, the data services require error rates batter than 10^{-6}, while an error rate of 10^{-3} produces good quality speech.

The envisioned services using either packet switched and or circuit switched services should be of acceptable quality for different radio environments. Service creation capabilities and tool kit for QoS measurements are to be given under the control of the user, who will be able to decide on the type of equipment to be used in order to achieve the desired quality of service. It is also desired that the service provided by the future wireless network be portable across other networks in a seamless manner. The new system should also facilitate the use of Internet services. To meet all the above mentioned requirements, advanced addressing mechanism, and multi-standard terminals may be needed unless a single terminal based on software radio technology becomes available.

The services mentioned above and related QoS are expected to remain a part of services offered by 4G systems but with an additional requirement of high transmission rates and service quality that does not depend on the channel vagaries. To meet the last requirement is a very challenging task. Not only sophisticated methods are required to improve the transmission reliability over the channels are needed, we need to create additional spectrum resources. In the light of the difficulty in acquiring new spectrum, the efficiency in using the currently assigned spectrum must be increased significantly.

14.2 Creating Spectrum Resources

The resource needed to introduce new services in telecommunications can be summed up in one word - spectrum. In fixed network, the signals are confined to the core of the cable or optical fiber and new spectrum resource can be created by simply laying new cables or fibers. Unfortunately this is not that simple to do in the case of radio communication, where radio signals scatter and disperse due to terrain irregularities.

The simplest method of adding to the spectrum resource is to seek new spectrum allocation. To some extent this is being done. In 1979, prior to the introduction of first cellular system in the United States, 40 MHz spectrum was allocated assigned to mobile cellular systems; this was increased to 50 MHz in 1986 and to 170 MHz in 1995. The 3G system has been assigned spectrum in the 2.2 GHz band. It is expected that new allocations will be made in 5 GHz band or even in 20-60 GHz band for short range high speed wireless systems. Moving into higher frequencies, several complications do accrue but these are not insurmountable. For example, the atmospheric effects drastically limit the propagation distance. Also, with the fluctuating weather conditions like rain, a large fade margins must be used in the system design. However, these may be considered to be blessings in disguise as interference between reuse cells will be reduced. Smaller cells, large in numbers are required. Because of smaller cells, more frequent handoffs are to be endured. These factors limits the use of 20-60 GHz spectral range to pedestrian environments.

Rather than finding new spectrum, making the spectrum usage much more efficient is another direction to conserve spectrum [2]. Several approaches can be used for efficient spectrum use. Appropriately designed multiple access methods can also improve the capacity of the wireless channels many fold particularly in the context of frequency reuse. This has been shown in the case of IS-95, which has higher capacity than DAMPS because CDMA reuses frequency in adjacent cells or even sectors. We could also increase the capacity by countering the impairments to the signal by the channel.

Multipath fading is the major obstacle in achieving reliable high speed transmissions over wireless channels, particularly when the channel fading is accompanied by ISI. Against flat fading, diversity is a major weapon. Diversity can be used in several domains including frequency, time, and space. To enhance channel utilization, simultaneous use of diversity in several domains can be used [3]. For example, diversity combining in time varying ISI and Doppler has been suggested [4]. The channel ISI can be effectively reduced by increasing the signal symbol length. An increase in the symbol length implies that the data rate is slowed down. A technique to do it is to convert high speed serial data into large number of parallel streams, each stream carry data that is several times lower than the original data. Since, the irreducible error rate is related to the ratio of channel delay spread to the symbol period, the lower the ratio, the better is the performance. Each slower data stream is then modulated to a separate carrier, which are multiplexed prior to transmission. This technique is called Multicarrier Communications (MC) [5][6]. OFDMA is an improvement over the multicarrier communications [7], where intercarrier spacing is kept to a minimum.

Taking a cue from the fixed network in laying more cables or fibers, we could create independent beams emulating new cables or fibers in the air. Antenna arrays can be used either at transmitter or receiver or both. In the latter case, many radio signal paths are created. If the paths are made to fade independently, phenomenal increase in capacity can be achieved. This is discussed later in this chapter. Antenna arrays can also be designed to create narrow beams which support different signals.

14.3 Multiple Access Methods

Wireless communications is essentially a multiuser communication system. It also reuses frequencies in a certain predetermined pattern. The frequency reuse results in interference that places a performance limit that is related to the transmission rate. To increase the channel capacity, several interference mitigation techniques have been suggested. These include serial, parallel and hybrid interference cancellation techniques. The other strategy could be to use certain multiple access technique which does not produce high levels of multiuser interference. For example, contrary to general belief, the wideband spread spectrum multiaccess could lead to higher spectrum efficiency. By increasing the complexity of receivers, the WCDMA efficiency can be increased nearly four fold [8]. Multiple access techniques like TDMA-CDMA hybrid, multicarrier CDMA, and orthogonal frequency division multiplexing (OFDMA), a form of CDMA multiple access that uses orthogonal carrier set to counter multitude of channel impairments. Another multiple access technique, though not legal at this time, is called impulse radio where information is modulated by the position of nanosecond pulses that results in infinitesimally small power spectral density of the transmitted signal [9].

14.3.1 Orthogonal Frequency Division Multiplexing

Some early development of OFDM can be traced back to 1950's [10]. The initial application of multicarrier system has been used in military communications. OFDM differs from MC system with respect to the transmission carriers which are always orthogonal in the case of OFDM but this condition is not maintained in MC. The initial versions of OFDM envisioned in using a very large number of sinusoidal generators and coherent demodulators. Obviously this was not found desirable from the implementation point of view. In 1971, Weinstein and Ebert applied the discrete Fourier transform to parallel data to modulation and demodulation process [10].

The modulated (PSK, QPSK, or QAM etc.) serial data is converted to parallel streams before modulated on to subcarriers. The subcarriers are then sampled at a rate N/T_s. The samples on each carrier are summed together to form OFDM sample. An OFDM symbol is generated by an N-subcarrier OFDM system consists of N samples and the mth sample of an OFDM symbol is

$$x_m = \sum_{n=1}^{N-1} X_n e^{j\frac{2\pi mn}{N}}, \qquad 0 \le m \le N-1 \tag{14.4}$$

where X_n is the transmitted data symbol on the nth carrier. (14.4) is equivalent to N point inverse discrete Fourier transform, which can be implemented using inverse fast Fourier transform. The baseband signal, thus, created is modulated by a carrier to become a bandpass signal, which is transmitted.

When the OFDM signal is transmitted over a multipath channel, the channel dispersion results in extension of the transmitted signal. This extension results in intercarrier interference (ICI). To reduce the effect of channel delay spread and preserve orthogonality between the carriers, the OFDM symbol is appended with a cyclic prefix which is derived from the last portion of the OFDM symbol.

An OFDM transmitter with N subscribers, a bandwidth of W Hz and symbol length of T seconds of which T_{cp} is the length of the cyclic prefix, has the waveform

$$\phi_k(t) = \begin{cases} \dfrac{1}{\sqrt{T - T_{cp}}} e^{j2\pi \frac{W}{N} k(t - T_{cp})} & \text{if } t \subset [0, T] \\ 0 \end{cases} \tag{14.5}$$

where $T = \dfrac{N}{W} + T_{cp}$. The transmitted baseband OFDM symbol, l, is

$$s_l(t) = \sum_{k=0}^{N-1} x_{k,l} \phi_k(t - lT) \tag{14.6}$$

where $x_{0,l}, x_{1,l}, ..., x_{N-1,l}$ are complex symbols that form a set of signal constellation points. The transmitter signal consisting of infinitely long sequence of OFDM signals,

$$s(t) = \sum_{l=\infty}^{\infty} \sum_{k=0}^{N-1} x_{k,l} \phi_k(t - lT)) \tag{14.7}$$

The OFDM receiver consists of a filter bank, matched to the last part $[T_{cp}, T]$ of the transmitter waveform $\phi_k(t)$, i.e.

$$\psi_k(t) = \begin{cases} \phi^*(T - t) & \text{if } t \in [0, T - T_{cp}] \\ 0 & \text{otherwise} \end{cases} \tag{14.8}$$

Considering that the channel does not change over the OFDM symbol interval, the sampled output is given by

$$y_k = \sum_{k'=0}^{N-1} x_{k'} \int_{T_{cp}}^{T} \int_{0}^{T_{cp}} g(\tau)\phi_{k'}(t-\tau)d\tau\phi_k^*(t)dt + \int_{T_{cp}}^{T} \tilde{n}(T-t)\phi_k^* dt \qquad (14.9)$$

The transmitter filters being orthogonal, the above equation is simplified as

$$y_k = \sum_{k'=0}^{N-1} \phi_{k'}(t)\phi_k^* dt + n_k = h_k x_k + n_k \qquad (14.10)$$

The above model can be discretized, the modulation and processing at the receiver may be replaced by an inverse DFT and a DFT respectively. We can write

$$y_l = DFT(IDFT(x_l) \otimes g_l + \tilde{n}_l)$$
$$= DFT(IDFT(x_l) \otimes g_l) + n_l \qquad (14.11)$$
$$= x_l DFT(g_l) + n_l = x_l h_l + n_l$$

OFDM signal is affected by channel delay spread, random phase and frequency fluctuations. The cyclic prefix effectively takes care of channel delay spread. However, to remove the effect of random phase and frequency (Doppler) fluctuations, channel estimation is deemed necessary. Channel estimation can be blind or symbol assisted. In the symbol assisted estimation, the placement of pilot becomes important [11]. Different arrangement results in different performance [12]. Block, comb, and other pilot locations have been proposed [11].

14.4 Antenna Arrays

Antenna arrays are expected to play an important role in the future wireless communications. The antenna arrays significantly lower the transmission power requirements, and could minimize interference by adapting their radiation patterns to the changing interference environment. By focussing the beam in the direction of signal arrival and nulls in the direction of interference, adaptive arrays can significantly reduce the channel delay spread. Furthermore, with proper control of the array radiation pattern, it is possible to create a number of independent beams pointing to different directions where each beam supports a distinct radio link. This is termed as space division multiplexing (SDM). Also, some of these beams created may be combined using appropriate algorithm to achieve the system design objectives.

The antenna array at the base station may be used in several configurations. For example, multiple fixed number of beams may be created. These fixed beams may be used as co-channel cells. Dynamically created beams, each of which could track a particular user will create an adaptive formation of cells. Also, interference can be minimized by creating nulls towards other mobile stations. For an M_B elements antenna array, M_B-1 nulls can be created.

14.5 Time-Space Processing

The basic idea behind space-time processing is to create a number of independent channels; a concept similar to laying parallel conduits or cables. The independence of channels depends on a number of factors such as temporal and spatial fading. In the case of temporal fading multipath intensity profile (MIP) or multipath power delay profile (MPDP) is important. Obviously, space-frequency, Doppler spectrum, space-time correlation determine capacity gains delivered by space-time processing [13]. Space-Time processing may be defined as the process of minimization of fading and Multiple Access Interferences (MAI) through the integrated use of multiple antennas, advanced signal processing, advanced receiver techniques and forward error correction.

The use of antenna arrays in future wireless communications is viewed to be a powerful technique in increasing channel capacity. When an array consisting of M_b elements at the base station and N_m elements at the mobile terminal is used, it is possible to create $M_b \times N_m$ channels. If the channels thus produced are independent, the channel capacity is then increased by $M_b \times N_m$ fold, similar to as if $M_b \times N_m$ new cables or fiber links have been installed. On each channel, data from many users may be multiplexed, the receiver, uses advanced signal processing techniques to recover individual signals. The use of multiple antennas at the transmitter and receiver is shown in Figure 14.1. The K streams of data are multiplied with a unique spreading code, and is modulated at a carrier ω_c. The modulated signals are then space-time processed to generate a number of signals which are transmitted over the wireless channels. The K streams of transmitted data is received by M receivers. The received signals are individually gain weighted and despread using the desired signal spreading code. The resulting MK branches of the signal are combined to deliver received signal. For each of the other data stream similar procedure is repeated.

14.5.1 Bell Labs Layered Space Time (BLAST) System

A specific development of space-time processing is BLAST. In BLAST (Bell Laboratories Layered Space-Time), the transmitting antennas

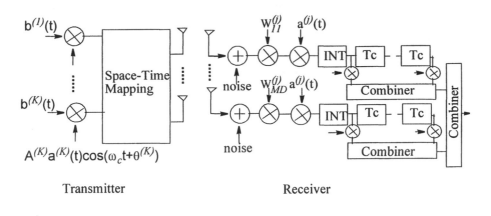

Figure 14.1 Multiple antenna CDMA transmitter and receiver

are fed independent data streams. On the receiver side the signals from the transmitting antennas are received by a number of receiver antenna. The receiver uses nonlinear multiuser detection methods to separate the incoming signals at the receiver. The BLAST system has demonstrated spectral efficiencies on the order of 40 b/s/Hz with 8 transmit and 8 receive antennas (8×8 system). This is forty times the achievable data rate with single element transmitting and receive antennas with the same bandwidth and power [14].

With M_b elements at the base station and N_m elements at the mobile terminal there are $M_b \times N_m$ number of paths. At the receiver, the noise $v(t)$ is complex N_m dimensional noise. The total power is considered constrained to $\widehat{P}$ regardless of N_m and the average signal to noise ratio at each receive antenna is $\rho = P/N$. The received signal is

$$r(t) = g(t) \times s(t) + v(t) \tag{14.12}$$

where $g(t)$ is the matrix channel impulse response. It is convenient to normalize matrix and write (14.12) as

$$r(t) = (\rho/n_T)^{1/2} \cdot h(t) \cdot s(t) + v(t) \tag{14.13}$$

It has been shown that the capacity is given by [14]

$$C = \log_2 det[I_{N_m} + (\rho/M_B)HH^\dagger] \quad b/s/Hz \tag{14.14}$$

where $H^\dagger$ is transpose conjugate of the normalized channel transfer characteristics.

Two types of BLAST architectures, Diagonal-BLAST and Vertical-BLAST, have been discussed in [13] [14]. The D-BLAST has a diagonal layered spaced-time coding architecture with sequential nulling and interference cancellation decoding. D-BLAST suffers from boundary wastage at the start and the end of the each packet which is significant for small packet sizes. The vertical-BLAST (V-BLAST) overcomes the limitations of D-BLAST by using independent horizontal layered space-time coding scheme. Unfortunately it does not utilize time diversity and therefore its capacity is lower.

14.6 Multiuser Detection

In multiuser CDMA systems, it is imperative to detect the desired signal with minimum of multiuser interference. Multiuser detection is gaining importance because the future high capacity systems are likely to be based on spread spectrum technology. When using correlator receiver to separate multiuser signals, the presence of MAI limits the performance of the correlator receiver. Verdú proposed a multiuser detector (MUD) consisting of a bank of matched filters followed by maximum likelihood sequence detector (MLSD) [15]. Multiuser detectors can be classified into two main categories - one-shot multiuser detectors and interference cancellation (IC) or iterative multiuser detectors. The one-shot multiuser detectors are also called joint multiuser detectors.

In the one-shot detection the basic receiver consists of a matched filter and a decision device. The received signal is matched to the spreading code of the desired user. For a single user case, the output signal to noise ratios is maximized and the resulting receiver is optimum. However, for the multiuser case, though the signal to noise ratio is still maximized but the receiver does not perform as a maximum likelihood receiver because of the presence of MAI and therefore is not optimum.

In the linear detection classes of multiuser detection, decorrelator detection and MMSE detector are most common type of receivers. Adaptive MMSE has also been proposed in [16][17][37].

For a single antenna receiver, the output of the user k is

$$s^{(k)}(t) = Re\left\{ A^{(k)} b^{(k)} a^{(k)} \exp\left[-j\left(\omega_c t + \theta^{(k)} \right) \right] \right\}$$
(14.1)

where $A^{(k)}$ denotes the received signal amplitude, $b^{(k)}(t)$ a binary data sequence with symbol period T_s seconds, $a^{(k)}(t)$ the binary spreading waveform with a gain N, i.e. $N = T_s/T_c$. The signal is transmitted over a L path wireless radio time variant channel with a low pass impulse response of

$$h_i^{(k)}(t) = \sum_{l=1}^{L} \beta_l^{(k)}(i) \exp(j\phi_l^{(k)}(t)) \delta[\tau - \tau_l^{(k)}] \tag{14.2}$$

Path l from user k at symbol interval i has a random strength $\beta_l^{(k)}(i)$. Each path has a phase, $\phi_l^{(k)}(i)$, which is uniformly distributed over 2π. $\tau_l(k)$ is propagation delay which is also uniformly distributed over 0 to T_s. For coherent detection, the receiver has the knowledge of carrier phase, multipath signal phases, and multipath time delays. The received signal vector and the noise vector are defined respectively as

$$\mathbf{r}(t) = (r_1(t), r_2(t), \dots, r_M(t))^T$$
$$\mathbf{n}(t) = (n_1(t), n_2(t), \dots, n_M(t))^T \tag{14.3}$$

For multiuser detection, vector of decision statistics over $P+1$ bits may be written as[13]

$$z = \mathbf{R}_z \mathbf{Cb} + \eta_z \tag{14.4}$$

where

$$\mathbf{b} = (b_0^{(1)}, b_0^{(2)}, \dots, b_1^{(1)}, \dots, b_P^{(K)})^T \tag{14.5}$$

$$\eta_z = (\eta_1^{(1)}(0), \eta_2^{(1)}(0), \dots, \eta_L^{(1)}, \dots, \eta_l^{(k)}(i), \dots, \eta_L^{(K)}(P))^T \tag{14.6}$$

$$\mathbf{C} = \text{diag}(\mathbf{C}_0, \mathbf{C}_1, \dots, \mathbf{C}_P) \tag{14.7}$$

$$R_z = \begin{bmatrix} R_0^{(0)} & R_0^{(1)} & & & & \\ R_1^{(-1)} & R_1^{(0)} & R_1^{(1)} & & & \\ & R_2^{(-1)} & R_2^{(0)} & R_2^{(1)} & & \\ & & & \ddots & & \\ & & & R_{P-1}^{(-1)} & R_{P-1}^{(0)} & R_{P-1}^{(1)} \\ & & & & R_P^{(-1)} & R_P^{(0)} \end{bmatrix} \tag{14.8}$$

Equation (14.4) contains all the necessary mobile radio parameters and represents the general algebraic structure of almost all the multiuser detection strategies. For example, if a multiuser detector is followed by a RAKE combiner, the corresponding detection statistic is

$$\mathbf{y} = (\mathbf{G}_R)^H \mathbf{R}_z \mathbf{C} \mathbf{b} + (\mathbf{G}_R)^H \mathbf{\eta}_z = \mathbf{R}\mathbf{b} + \mathbf{\eta} \tag{14.9}$$

We can define an optimal detector which maximizes $p(\mathbf{y}|\mathbf{b})$. When the all spreading codes are known to the receiver, $\mathbf{y}$ is Gaussian with $E\{\mathbf{y}\} = \mathbf{R}\mathbf{b}$ and $var\{\mathbf{y}\} = \sigma^2 \mathbf{R}$. The corresponding pdf is given by

$$p(\mathbf{y}|\mathbf{b}) = K_o \exp\left(-\frac{(\mathbf{y} - \mathbf{R}\mathbf{b})^H \mathbf{R}^{-1}(\mathbf{y} - \mathbf{R}\mathbf{b})}{2\sigma^2}\right) \tag{14.10}$$

Thus, estimation of the data vector is estimated as

$$\hat{\mathbf{b}} = min(\mathbf{y} - \mathbf{R}\mathbf{b})^H \mathbf{R}^{-1}(\mathbf{y} - \mathbf{R}\mathbf{b}) \tag{14.11}$$

14.7 Interference Cancellation

The mobile communication systems are interference limited. Thus, any reduction in interference directly results in capacity increase. This is particularly relevant to systems using wideband CDMA. The fundamental research carried out by Verdú showed that multiuser detectors with a property of cancellation of interference produce a large increase in capacity [8]. These detectors, however, are quite complex. The emphasis has been to reduce the complexity of the multiuser detectors. Besides using multiuser detectors to counter interference, adaptive equalizers have also been suggested to counter interference in systems using CDMA [18].

Two distinct scenarios are found in interference cancellation - uplink and down link. In most of the research related to uplink channel suffering from multiuser interference, it is assumed that the receiver at the base station uses the knowledge of the user spreading codes to estimate the interference in order to subtract from the received signal [19]-[20]. When the user spreading codes are known, the interference cancellation system lists the received signals in the order of received power and successively subtracts the signals beginning with the strongest and ending with the weakest. In some procedures, the strongest signal is detected, re-spread and subtracted from the received signal. This procedure is repeated successively - thus the name successive interference cancellation [23]. Successive interference cancellation

(SIC) results in a multistage detector. The other form of multistage detector is parallel interference canceller (PIC) [21]. To estimate interference linear as well as non linear decision functions have been used [22].

In the down link scenario, the processing takes place at the mobile which does not posses the knowledge of the spreading codes of the interfering signals. Some proposed systems first estimate the spreading code of one of the interferer and then use it to cancel it using one of several cancellation methods including equalization and subtraction [24],[25].

It is expected that the concept of using multicode transmission introduced in 3G systems will continue in the future generation of wireless communications. Inferences cancellation for multicode wideband CDMA will remain to be a topic of intense research [26]-[29]. Recently several papers appeared on this topic and different strategies have been suggested. For example, in [28], a concept of effective spreading code is introduced. The receiver estimates effective spreading code of the interference regardless of their spreading codes using Walsh transform correlators instead of regular correlators. Three interference cancellation techniques, subtraction, combined interference projection and interference signal subspace projection are discussed. In [29] a Rake receiver with successive interference cancellation is used. The receiver estimates the desired signal using channel estimate and the user spreading code. It is then subtracted from the received signal with a factor λ, and the desired signal estimate is derived again. This procedure is repeated successively. It is shown that the optimum cancellation factor in the kth stage is approximated by $\lambda_{opt}^{(k)} \leq 1 - 2P_e^{(k-1)}$, where $P_e^{(k-1)}$ is the error probability in the $(k\text{-}1)$th stage.

The SIC detector suffers from unacceptable delay. Considerable research into reducing the delay inherent in SIC detector. On the other hand, the parallel interference cancellers are quite complex. To reduce the overall complexity of PIC and inherent delay in SIC, hybrid detectors have been proposed [30][31][32][33][34][35].

Groupwise successive interference cancellation (G-SIC) architecture is a hybrid interference cancellation scheme where K transmitting users are partitioned into G groups yielding a bank of detectors; each detects the information symbols of users in each group. In a parallel group detection the group detectors operate independently, whereas in a sequential scheme each group detector uses the decisions of the previous stage of group detectors to successively cancel the interference. Figure 14.2 shows the architecture of a linear G-SIC.

The architecture shown consists of M-stages of successive interference cancellation; each stage has G parallel interference cancellers. The received signal, $\mathbf{r}$, enters the first stage of group 1, and delivers outputs $\mathbf{y}_{1,1}$. The error

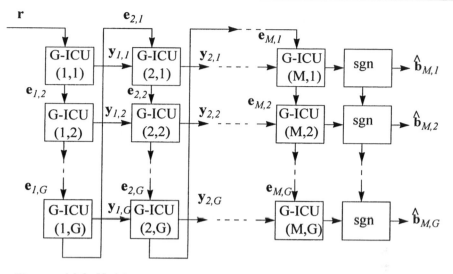

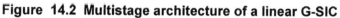

Figure 14.2 Multistage architecture of a linear G-SIC

between the received signal and $y_{1,1}$ is fed to the second group. The error between the user input of group (1,2) and output $y_{1,2}$ is fed to the next group. The output of the group G of stage 1 is fed to the group 1 of the second stage. This procedure is followed till the signal reaches the last stage, which outputs the estimates of signal vectors of G stages.

The block diagram of basic group interference cancellation unit is shown in Figure 14.3. It clearly shows how the error at each group interference cancellation unit is derived.

14.7.1 Mathematical Description of linear G-SIC

The vector of the decision variables of the first group interference unit (G-ICU) at the first stage is given by $y_{1,1} = F_1 S_1^T e_{1,1}$, where $e_{1,1} = r$ is the input to the first stage.

The input to the second G-ICU of the first stage is given by

$$e_{1,2} = e_{1,1} - S_1 F_1 S_1^T r = (I - S_1 F_1 S_1^T) r \qquad (14.12)$$

The corresponding vector of decision variable can be expressed as

$$y_{1,2} = F_2 S_2^T e_{1,2} = F_2 S_2^T (I - S_1 F_1 S_1^T) r \qquad (14.13)$$

The residual signal at the output of the gth G-ICU at the first stage may be written as

---

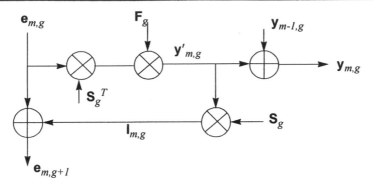

Figure 14.3 Group interference cancellation unit (G-ICU)

$$\mathbf{e}_{1,g} = \prod_{j=g-1}^{1} (\mathbf{I} - \mathbf{S}_1 \mathbf{F}_1 \mathbf{S}_1^T)\mathbf{r} = \Phi_{g-1}\mathbf{r} \qquad (14.14)$$

where $\displaystyle\prod_{j=g-1}^{1}$ defines the product of matrices with decreasing indices.

We can also write the corresponding decision variable as

$$\mathbf{y}_{1,g} = \mathbf{F}_g \mathbf{S}_g^T \mathbf{e}_{1,g} = \mathbf{F}_g \mathbf{S}_g^T \Phi_{g-1}\mathbf{r} \qquad (14.15)$$

Continuing in this manner, we can write for the decision variable of the gth G-ICU at the mth stage as

$$\mathbf{y}_{m,g} = \mathbf{F}_g \mathbf{S}_g^T \Phi_{g-1} \sum_{i=0}^{m-1} (\Phi_G)^i \mathbf{r} = \mathbf{G}_{m,g}^T \mathbf{r} \qquad (14.16)$$

Collecting the decision variable of all groups of in one matrix, we get:

$$\mathbf{y}_m = \mathbf{G}_m^T \mathbf{r} \qquad (14.17)$$

where

$$\mathbf{G}_m = \left[\mathbf{G}_{m,1} \ \mathbf{G}_{m,2} \ \cdots \ \mathbf{G}_{\dot{m},g} \ \cdots \ \mathbf{G}_{m,G} \right] \qquad (14.18)$$

and

$$G_{m,g} = \begin{bmatrix} g_{m,g,1} & g_{m,g,2} & \cdots & g_{m,g,m_g} \end{bmatrix} \qquad (14.19)$$

The linear G-SIC can be described as matrix filtering of the received chip-matched signal vector. If the spreading sequences and grouping of all users are known, then the decision variables of all users can be obtained without explicitly performing successive interference cancellation and group detection [31].

Several group detection schemes have been proposed. These include Group Matched Filter - Successive Interference Cancellation (GMF-SIC), Group Parallel Interference Cancellation - SIC (GPIC-SIC), Group Decorrelator - SIC (GDEC-SIC) and Group Minimum Mean Square -SIC (GMMSE-SIC).

14.8 Mobile Terminal or Personal Assistant

The future wireless infrastructure will be different from that in existence at present; it will be totally multidimensional in technologies, applications, and services. The wireless mobile terminal will have enough memory and is likely to serve as a personal assistant having multiple functions that will be based on broadband high speed access. In these terminals, broadband wireless transmission and fast access technologies will converge. The terminals will be reconfigurable and intelligent. As mentioned earlier, the wireless Internet will be the key application of this converged broadband wireless system.

In the light of developments mentioned above, the mobile terminal will be a hybrid terminal catering data at approximately ninety percent of the time and the voice for the remainder. The personal assistant will have enhanced security functions e.g. recognition of fingerprints to validate the user. The terminal without keyboard will recognize voice commands. It will be fully reconfigurable and adaptive [36]. The conflicting requirements of smaller size and large screen with good resolution is likely to be resolved with the use of lens of sufficient magnification so that the image of standard 13″ screen size appears at a virtual distance of 18 inches.

The terminals will resemble today's PDA's and will have all the functions required in a personal assistant. Already some terminals in that shape have started appearing.

14.9 Software Radios

Software radios are likely to play an important role in the future wireless communication systems. The concept of software radio is not new, it was first conceived in 1970's for use in military communications operating in HF band. The first operational software radio for use in digital HF communications was developed in 1980 [38]. The wireless communications of a modern armed forces rely heavily on reconfigurable systems.

Software radios provide enhanced fidelity and flexibility. The recent decision to standardize five air interfaces for 3G wireless communication systems requires that software radios be developed, which could change the air interface architecture from one system to another following determination of locally available system. Since most of the receiver functionalities are now provided by the software imbedded in it, a software radio could be developed around a complex and powerful signal processor. Software selectable options include station controller, international air interface standards, new service features such as fraud detection and identification of mobile terminal's geolocation.

The canonical software radio architecture is likely to consist of a power supply, an antenna (or antenna array), a multi-band RF converter, A/D and D/A converters, an on-chip general purpose processor and memory that could perform the radio functions and configure between air-interfaces. The digital receiver will use A/D and D/A converter placed as close as possible to the antenna. A possible functional architecture of a software radio for use at a base station is shown in Figure 14.4. The architecture consists of modules capable of working on real-time stream in near real time and On/Off line processing modes. The real time channel processing is characterized by discrete time point operations such as translation of a baseband signal to an intermediate frequency signal. A discrete time domain baseband waveform is multiplied by a discrete reference carrier to yield a sampled IF signal. This operation requires hundreds of MIPS or even Giga FLOPS [39]. The fundamental clock samples are decimated by different fractions at different stages of the receiver. Maintenance of isochronism through I/O interfaces and real-time embedded software in the real time data stream is seen as a real challenge. Multi-processing architecture seems to be the most suitable for multiple instruction multiple data streams (MIMD).

The blocks of the functional architecture shown in the above figure is divided into real time channel processing stream, the environment management stream and On-line and Off-line software tools [39]. The channel processing stream can be further partitioned into segments like antenna (if array

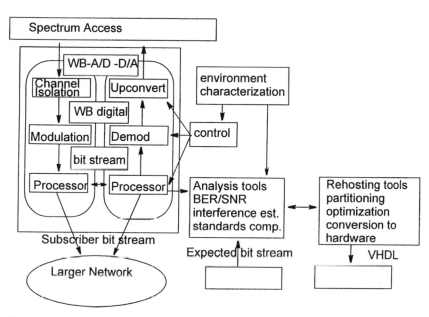

Figure 14.4 Software radio functional architecture

is used), RF conversion, IF processing, baseband processing, and the bit-stream segment.

14.10 Network Aspects of Future Systems

To support multidimensional wireless communications, the present day circuit switched network will have to evolve into hybrid or packet switched network. It is expected that once low cost high capacity systems become available, the user will shift its attention to services, flexibility to access them, and how easy to use them. The network operator has to attend to the user's requirements in order to run the network profitably. The current wireless network is based on the concept of circuit switching. Considering that the future applications will be mostly packet based, the GSM NSS (Network Switching subsystem) and ISDN/IN (Intelligent network) will evolve towards providing packet switching in the core network. The base transceiver system will act as an agent for clients to make adequate resources available for the application at hand. These will include CAI BIOS.

In [36] a possible architecture for the future wireless network is proposed. The access network will probably consist of gateways, which will act as proxies. The functional modules of the system are software definable and configurable. The core network will likely to be hybrid of circuit and IP back-bone and will consist of circuit and packet division multiplexers. The packet division multiplexer will have PSTN and Internet Gateways. It is likely that in the future this hybrid network will evolve into all packet switched core net-work.

The proposed core network consists of transaction server, billing server, management server and call_proc server. Their functionalities of these servers are self-explanatory. The evolved network will have new charging and accounting mechanisms, which may use the assigned bandwidth to appli-cation as well as the transaction time to calculate the charges. The network will support variable data rates and mixed traffic types and will have mobile-fixed convergence elements. The network will support service mobility across networks by using VHE (Virtual Home Network).

The network security has been and will continue to be a major issue. It will include protection against fraudulent use of the network, control over misuse and/or abuse of the network, and a provision of other security services to the user.

In the arena of operation and maintenance, the evolved network will allow automatic establishment of roaming relations in order to support multi-vendor networks.

14.11 Summary

In this chapter, an attempt is made to gaze into the future of wireless communications. The forecast is based on the degree of confidence in project-ing the current status of the wireless networks in to the future on the basis trends towards the evolution of new services and applications. The high speed services, expected in future, point towards wideband systems. High speed transmission is only possible if the spectrum is used efficiently. Several tech-niques of increasing available bandwidth have been discussed. We need some clever techniques to counter capacity limiting channel impairments without adding much to the system complexity. Small and multi-standard terminals based on software radios are expected to emerge. These radios will facilitate global seamless roaming. Some pointers towards new network architectures are also given.

References

[1] K. Wacks, "Home systems standards: achievements and challenges," *IEEE Commu. Mag.* vol. 40, no. 4, pp. 152-159, 2002.

[2] S. Verdu', "Wireless bandwidth in the making," *IEEE Communications Magazine*, vol. 38, no. 7, pp. 53-58, 2000.

[3] D.P Taylor et. al., "Wireless channel equalisation," *Euro. Trans. Telecommun.*, vol. 9, no. 2, Mar.-Apr., pp. 117-43, 1998.

[4] A. M. Sayeed and B. Aazhang, "Joint multipath-Doppler Diversity in mobile wireless communications," *IEEE Trans. Commu.*, vol. 47, no. 1, pp. 123-132, 1999.

[5] R. Chang, "Synthesis of bandlimited orthogonal signals for multichannel data transmission," *BSTJ* vol. 46, pp. 1775-1796, December 1966.

[6] K. Fazal and G. Fettwis, "Performance of an efficient parallel data transmission system," *IEEE Trans. Comm.* vol. 15, pp. 805-813, December 1967.

[7] R. Van Nee and R. Prasad, *OFDM for wireless multimedia communications*, Artech House Publishers, 2000.

[8] S. Verdu' and S. Shamai (Shitz), "Spectral efficiency of CDMA with random spreading," *IEEE Trans. Info. Theory*, vol. 45, no. 2, pp. 622-40, 1999.

[9] Pulson Communications, *Impulse radio communications*, Mclean, VA, June 1993.

[10] S. Weinstein and P. Ebert, "Data transmission by frequency division multiplexing using the discrete Fourier transform," *IEEE Trans. Commu. Tech.*, COMM-19, no. 10, pp. 628-634, 1971.

[11] S. Coleri, M. Ergen, A. Puri and A. Bahai, "Channel estimation techniques based on pilot arrangement in OFDM systems," *IEEE Trans. on Broadcasting*, September 2002.

[12] F. Tufvesson and T. Maseng, "Pilot assisted channel estimation for OFDM in mobile cellular systems," *IEEE-VTC Record*, pp. 1639-1643, 1997.

[13] P.v.Rooyen, M. Lötter, and D. v. Wyk, *Space-time processing for CDMA mobile communications*, Kluwer Academic Publishers, Boston, 2000.

[14] G. J. Foschini, "Layered space-time architecture for wireless communications in a fading environment when using multi-element antennas," *Bell labs Tech. Journal*, Autumn, pp. 41-59, 1996.

[15] S. Verdu', "Minimum probability of error for asynchronous Gaussian multiple access channels," *IEEE Trans. Info. Theory*, vol. 32, pp. 85-96, January 1986.

[16] T. J. Lim and S. Roy, "Adaptive filters in multiuser (MU) CDMA detection," *Wireless Networks*, vol. 4, no. 4, pp. 307-318, 1998.

[17] U. Madhow and M. Honig, "MMSE interference suppression for direct sequence spread spectrum CDMA," *IEEE Trans. Commu.*, vol. 42, no. 12, pp. 3178-3188, December 1994.

[18] M. Abdulrehman, A. U. H. Sheikh, and D. D. Falconer, "Decision feedback equalization for CDMA in indoor wireless communications," *IEEE-Journal of Selected Areas in Communications*, vol. 12, no. 5, pp. 698-706, May 1994.

[19] A. Johansson and A. Svensson, "Successive interference cancellation in multiple data rate DS/CDMA systems," *IEEE-VTC Record*, Chicago1995.

[20] P. Patel and J. Holtzman, "Analysis of a simple successive interference cancellation scheme in DS/CDMA system," *IEEE-Journal of Selected Areas in Communications*, vol. 12, pp. 796-807, June 1994.

[21] R.M. Buehrer, N. S. Correal-Mendoza and B.D. Woerner, "A simulation comparison of multi-user receivers for cellular CDMA," *IEEE Trans. Veh. tech.*, vol 49, no. 4, pp. 1065-1085, July 2000.

[22] H. Sugimoto, L. K. Rasmussen, T. J. Lim, T. Oyama, "Mapping function for successive interference cancellation in CDMA," *IEEE-VTC Record*, pp. 2301-2305, 1998.

[23] Y.C. Yoon, R. Kohno and H. Imai, "A spread-spectrum multipleaccess system with cochannel interference cancellation for multipath fading channels," *IEEE-Journal of Selected Areas in Communications*, vol. 11, no. 7, pp. 1067-1075, September 1994.

[24] X. Wang and H. V. Poor, "Blind multiuser detection: A subspace approach," *IEEE Trans. Inform. Theory*, vol. 44, pp. 677-690, March 1998.

[25] H. Yoshino, K. Fukawa and H. Suzuki, "Interference canceling equalizer (ICE) for mobile radio communications," *Proceeding IEEE-ICC*, pp. 1427-1432, 1994.

[26] M.C.-C. Chan, and J.C.-I. Chuang, "Multicode high speed transmission with interference cancellation for wireless personal communications," *IEEE-VTC Record*, pp. 661-665, May 1996.

[27] Y. Suzuki and K. Kobayashi, "Interference cancellation method for DS-CDMA multi-code-packet transmission," *Proceedings IEEE-Globcom '98*, pp. 3578-3583, 1998.

[28] M.F.Madkour and S.C.Gupta, "Successive interference cancellation algorithm for downlink W-CDMA communications," *IEEE Trans. on Wireless Commu.*, vol. 1, no. 1, pp. 169-177, January 2002.

[29] J. Chen, J. Wang, and M. Sawahashi, "MCI cancellation for multicode wideband CDMA systems," *IEEE-Journal of Selected Areas in Communications*, vol. 20, pp. 450-462, February 2002.

[30] M.K. Varanasi, "Group detection for synchronous Gaussian code-division multiple access channels," *IEEE Trans. on Information Theory*, vol. 41, pp. 1083-1096, 1995.

[31] A.L. Johanson, L.K. Rasmussen, "Linear group-wise successive interference cancellation in CDMA," *IEEEE-ISSSTA Proceedings*, pp. 121-126, 1998.

[32] S. Sun, L. K. Rasmussen, T.J. Lim and H. Sugimoto, "A hybrid interference canceller in CDMA," *IEEEE-ISSSTA Proceedings*, pp. 150-154, 1998.

[33] Z. D. Lei, T.J. Lim, "Simplified polynomial expansion linear detectors for DS-CDMA systems," Electronics Letters, vol. 34, pp. 1561-1563, August 1998.

[34] R. R. Muller, S. Verdu', "Design and analysis of low complexity interference mitigation on vector channels," IEEE Journal of Selected Areas in Communications, vol. 19, pp. 1429-1441, August 2001.

[35] M. M. A. Sessler, F. K. Jondral, "Rapidly converging polynomial expansion multiuser detector with low complexity for CDMA systems," Electronics Letters, vol. 38, pp. 997-998, August 2002.

[36] W.W. Lu, "Compact multidimensional broadband wireless: the convergence of wireless mobile and access," *IEEE Commu. Mag. vol. 38, no. 11,* pp. 119-123, November 2000.

[37] P. Rapajic and B.S. Vucetic, "Adaptive receiver structures for asynchronous CDMA systems," *IEEE Journ. Select. Areas in Commun.*, vol. 16, pp. 1437-1450, October 1998.

[38] J. Kennedy and M. C. Sullivan, "Direction finding and smart antennas using software radio architecture," *IEEE Commu. Mag. vol. 33, no. 5, pp. 62-68, 1995.*

[39] J. Mitola, "The software radio architecture," *IEEE Commu. Mag. vol. 33, no. 5,* pp. 26-38, May 1995.

APPENDIX A

SIGNAL ENVELOPE CHARACTERISTICS

An expression is desired for the autocorrelation function of the signal envelope due to E_z field. In general, the E_z field signal can be expressed as follows:

$$E_z(t) = X_c(t)\cos\omega_c t - X_s(t)\sin\omega_c t; with$$

$$X_c(t) = E_o \sum_{n=0}^{\infty} C_n \cos(\omega_n t + \theta_n)$$

$$\quad (A.1)$$

$$X_s(t) = E_o \sum_{n=0}^{\infty} C_n \sin(\omega_n t + \theta_n)$$

in which $\omega_n = (2\pi/\lambda)\, V \cos(\alpha_n)$.

For N sufficiently large (N > 10) both $X_c(t)$ and $X_s(t)$ are Gaussian random variables. Let

$$X_1 = X_c(t) \qquad X_2 = X_c(t+\tau)$$
$$Y_1 = X_s(t) \qquad Y_2 = X_s(t+\tau)$$

$$\quad (A.2)$$

To obtain an expression for envelope correlation function, consider the joint probability density function $p(X_1, X_2, Y_1, Y_2)$ or, in polar coordinate form: $p(r_1, r_2, q_1, q_2)$ in which

$$X_1 = r_1 \cos\theta_1 \qquad Y_1 = r_1 \sin\theta_1$$
$$X_2 = r_2 \cos\theta_2 \qquad Y_2 = r_2 \sin\theta_2$$

$$\quad (A.3)$$

Furthermore

$$p(r_1, r_2) = \int_0^{2\pi} \int_0^{2\pi} p(r_1, r_2, \theta_1, \theta_2)\, d\theta_1\, d\theta_2 \quad (A.4)$$

and finally,

$$\mathfrak{R}(r_1, r_2) = \int_0^\infty \int_0^\infty r_1 r_2 p(r_1, r_2) dr_1 dr_2 \qquad \text{(A.5)}$$

In general an *n*-dimensional Gaussian distribution is given by [A.1]:

$$p_{X_1, X_2, \ldots, X_n}(X_1, X_2, \ldots, X_n) = \frac{exp\left[-\frac{1}{|\Lambda|}\sum_{j=1}^{n}\sum_{k=1}^{n}|\Lambda|(X_j - m_j)(X_k - m_k)\right]}{(2\pi)^{\frac{N}{2}}|\Lambda|} \qquad \text{(A.6)}$$

in which Λ is the covariance matrix defined as:

$$\Lambda = \begin{bmatrix} E((X_1 - m_1)(X_1 - m_1)) & \ldots & E((X_1 - m_1)(X_n - m_n)) \\ \ldots & \ldots & \ldots \\ E((X_n - m_n)(X_1 - m_1)) & \ldots & E((X_n - m_n)(X_n - m_n)) \end{bmatrix} \qquad \text{(A.7)}$$

and, assuming that $E(X_i) = 0$, the covariance matrix reduces to:

$$\begin{bmatrix} E(X_1, X_1) & E(X_1, X_2) & \ldots & E(X_1, X_n) \\ \ldots & \ldots & \ldots & \ldots \\ E(X_i, X_1) & E(X_i, X_2) & \ldots & E(X_i, X_n) \\ \ldots & \ldots & \ldots & \ldots \\ E(X_n, X_1) & E(X_n, X_2) & \ldots & E(X_n, X_n) \end{bmatrix} \qquad \text{(A.8)}$$

For the case of four variables, X_1, X_2, Y_1, Y_2, assuming that $E[X_1] = E[X_2] = E[Y_1] = E[Y_2]$, the covariance matrix is given by

$$\begin{bmatrix} E(X_1, X_1) & E(X_2, Y_1) & E(X_1, X_2) & E(X_1, Y_2) \\ E(Y_1, X_1) & E(Y_1, Y_1) & E(Y_1, X_2) & E(Y_1, Y_2) \\ E(X_2, X_1) & E(X_2, Y_1) & E(X_2, X_2) & E(X_2, Y_2) \\ E(Y_2, X_1) & E(Y_2, Y_1) & E(Y_2, X_2) & E(Y_1, Y_2) \end{bmatrix} \qquad \text{(A.9)}$$

where

$$E(X_1, X_1) = \left\langle \left[E_o \sum_1^N C_n \cos(\omega_n t + \phi_n) \right]^2 \right\rangle$$

$$= \left(\frac{E_o^2}{2} \right) \sum_{n=1}^N \langle C_n^2 \rangle \tag{A.10}$$

or $E(X_1, X_2) = E_o^2/2 = \mu_{11}{}^2 = \sigma^2$. Similarly $E[Y_1, Y_1] = E[X_2, X_2] = E[Y_2, Y_2] = \sigma^2$, and

$$E(X_1, Y_1) = \left\langle E_o^2 \sum_{n=1}^N C_n \cos(\omega_n t + \phi_n) \sum_{k=1}^N C_k \cos(\omega_k t + \phi_k) \right\rangle \tag{A.11}$$

$$= 0$$

Similarly, $E[Y_1, X_1] = E[X_2, Y_2] = E[Y_2, X_2] = 0$, and

$$E(X_1, X_2) = \left\langle E_o^2 \sum_{n=1}^N C_n \cos(\omega_n t + \phi_n) \sum_{k=1}^N C_k \cos(\omega_k(t + \tau) + \phi_k) \right\rangle$$

$$= \left\langle \frac{E_o^2}{2} \sum_{n=1}^N C_n^2 \cos(\omega_n \tau) \right\rangle = \frac{E_o^2}{2} E[\cos(\omega_n \tau)] \tag{A.12}$$

$$= E[\cos(\beta v \tau \cos \alpha)]$$

$$= \upsilon = \sigma^2 J_o(\omega_n \tau)$$

Similarly, $E[X_2, X_1] = E[Y_1, Y_2] = E[Y_2, Y_1] = \upsilon$
i.e. since $\cos(x \cos \phi) = J_o(x) - \Sigma 2 \cos 2k\phi J_{2k}(x)$, $\langle \cos(x \cos \phi) \rangle = J_o(x)$,
$\langle [X_1, Y_2] \rangle = \langle E_o^2 \Sigma C_n \cos(\omega_m t + \phi_n) \Sigma C_k \sin(\omega_k(t + \phi_m) + \psi_k) \rangle$
$(E_o^2/2)\langle [\sin(\beta v \tau \cos \alpha)] \rangle = 0$

$$p(X_1, Y_1, X_2, Y_2) = \frac{1}{4\pi^2 \mu^2 (1 - \zeta^2)} \tag{A.13}$$

$$\exp \left[-\frac{\mu\left(X_1^2 + Y_1^2 + X_2^2 + Y_2^2\right) - 2(X_1 X_2 + Y_1 Y_2)}{2\mu^2\left(1 - \zeta^2\right)} \right]$$

where $\sigma^2 = \upsilon/\mu$

The above equations can now be used to obtain the joint probability distribution in polar coordinates.

Let $X_1 = r_1 \cos\theta_1, Y_1 = r_1 \sin\theta_1, X_2 = r_2 \cos\theta_2, Y_2 = r_2 \sin\theta_2,$

$$p(r_1, r_2, \theta_1, \theta_2) = \frac{r_1 r_2}{4\pi^2 \mu^2 (1 - \zeta^2)} exp\left[-\frac{r_1^2 + r_2^2 - 2\zeta r_1 r_2 \cos(\theta_2 - (\theta_1 + \phi))}{2\mu(1 - \zeta^2)} \right] \tag{A.14}$$

the Jacobian of transformation is $r_1 r_2$.

$$p(r_1, r_2) = \int_0^{2\pi} \int_0^{2\pi} p(r_1, r_2, \theta_1, \theta_2) d\theta_1 d\theta_2$$

$$p(r_1, r_2) = \frac{r_1 r_2}{\mu^2 (1 - \zeta^2)} exp\left(-\frac{r_1^2 + r_2^2}{2\mu(1 - \zeta^2)} \right) I_0\left(\frac{r_1 r_2}{\mu^2 (1 - \zeta^2)} \right) \tag{A.15}$$

where

$$I_0(x) = \frac{1}{2\pi} \int_0^{2\pi} exp(x \cos\theta) d\theta \tag{A.16}$$

Finally,

$$\Re_{r_1, r_2}(r_1, r_2) = \Re_{r_1, r_2}(\tau, \zeta) = \int_0^{\infty} \int_0^{\infty} r_1 r_2 p(r_1, r_2) dr_1 dr_2$$

$$\Re_{r_1, r_2}(\tau, \zeta) = \int_0^{\infty} \int_0^{\infty} \frac{r_1^2 r_2^2}{\mu^2 (1 - \zeta^2)} exp\left(-\frac{r_1^2 + r_2^2}{2\mu(1 - \zeta^2)} \right) I_0\left(\frac{r_1 r_2}{\mu^2 (1 - \zeta^2)} \right) dr_1 dr_2 \tag{A.17}$$

The above integrals can be transformed to elliptical integrals and solved as such. We can expand $I_0(x)$ and integrate term by term to obtain

$$I_0(x) = \sum_{n=0}^{\infty} \frac{x^{2n}}{2^{2n}(n!)^2} \tag{A.18}$$

and

$$\mathfrak{R}_{r_1, r_2}(\tau, \zeta) = \int_0^\infty \int_0^\infty \frac{r_1^2 r_2^2}{\mu^2 (1 - \zeta^2)} exp\left(-\frac{r_1^2 + r_2^2}{2\mu(1 - \zeta^2)}\right) \sum_{n=0}^\infty \frac{r_2^{2n}}{2^{2n}(n!)^2}\left(\frac{\zeta r_1}{\mu(1 - \zeta^2)}\right)^{2n} dr_1 dr_2 \quad (A.19)$$

Furthermore, using the fact that

$$\int_0^x x^m exp\left(-ax^2\right) dx = \frac{\Gamma\left(\dfrac{m+1}{2}\right)}{2a^{\frac{m+1}{2}}} \quad (A.20)$$

where $\Gamma[x]$ is the Gamma function defined as

$$\Gamma(x) = \int_0^\infty t^{x-1} exp(-t) dt \quad (A.21)$$

$$\mathfrak{R}_{r_1, r_2}(\tau, \zeta) = \int_0^\infty \int_0^\infty \frac{r_1^2 r_2^2}{\mu^2(1 - \zeta^2)} exp\left(-\frac{r_1^2 + r_2^2}{2\mu(1 - \zeta^2)}\right) \quad (A.22)$$

$$\sum_{\infty} \frac{\Gamma\left(\dfrac{2n+3}{2}\right)\left(2\mu(1 - \zeta^2)\right)^{\frac{2n+3}{2}}}{2 \quad {}_{-}2n, {}_{..}2}\left|\frac{\zeta r_1}{(\quad)}\right|^{2n} dr_1 dr_2$$

$$\mathfrak{R}_{r_1, r_2}(\tau, \zeta) = \mathfrak{R}(r_1, r_2) - 2\mu\left(1 - \zeta^2\right)^2 \sum_{n=0}^\infty \frac{\Gamma^2\left[\dfrac{2n+3}{2}\right]\zeta^{2n}}{(n!)^2} \quad (A.23)$$

Using the fact that

$$\Gamma\left[m + \frac{1}{2}\right] = \frac{1.35...(2n-1)\sqrt{\pi}}{2^m} \quad (A.24)$$

we obtain:

$$\Re(r_1, r_2) = 2\mu\left(1 - 2\zeta^2 - \zeta^4\right)$$

$$\left(\Gamma^2\left[\frac{3}{2}\right] + \Gamma^2\left[\frac{5}{2}\right]\zeta^2 + \frac{\Gamma^2\left[\frac{7}{2}\right]\zeta^4}{(2!)^2} + \ldots + \frac{\Gamma^2\left[\frac{2N+3}{2}\right]\zeta^{2N}}{(n!)^2}\right) \qquad \text{(A.25)}$$

$$\Re(r_1, r_2) = 2\mu\left(1 - 2\zeta^2 - \zeta^4\right)$$

$$\left[\left(\frac{\sqrt{\pi}}{2}\right)^2 + \left(\frac{3}{2}\frac{\sqrt{\pi}}{2}\right)^2\zeta^2 + \left(\frac{5}{2}\frac{3}{2}\frac{\sqrt{\pi}}{2}\right)^2\frac{\zeta^4}{(2!)^2} + \ldots\right]$$

$$\Re(r_1, r_2) = 2\mu\left(1 - 2\zeta^2 - \zeta^4\right)\left(\frac{\pi}{4} + \frac{9\pi}{16}\zeta^2 + \frac{225}{256}\zeta^4 + \ldots\right)$$

which reduces to

$$\Re(r_1, r_2) = \frac{\mu\pi}{2}F\left(-\frac{1}{2}, -\frac{1}{2}, 1, ;\zeta^2\right) \qquad \text{(A.26)}$$

in which $F(a, b, c; x)$ is the hypergeometric function defined as:

$$F(a, b, c, ;x) = 1 + \frac{a \cdot b}{c \cdot 1}x + \frac{a(a+1)b(b+1)}{c(c+1) \cdot 1 \cdot 2}x^2 + \ldots \qquad \text{(A.27)}$$

$$\Re(r_1, r_2) = \frac{\pi}{2}\mu\left(1 + \frac{\zeta^2}{4}\right)$$

$$\Re(r_1, r_2) = \frac{\pi}{2}\mu\left(1 + \frac{J_0^2(\omega_m\tau)}{\cdot \ 4}\right) \qquad \text{(A.28)}$$

APPENDIX B

SIGNAL PHASE CHARACTERISTICS

The predominant modulation system over mobile channels is FM. Signal phase characteristics are thus important since the derivative of the signal phase is instantaneous frequency, available at the output of the FM receiver. This appendix examines the probability density function of the phase derivative, its autocorrelation function, and subsequently the power spectrum of the random FM signal.

From Appendix A, we have:

$$p(r_1, r_2, \theta_1, \theta_2) = \frac{r_1 r_2}{4\pi^2 \mu^2 \left(1 - \zeta^2\right)} exp\left[-\frac{r_1^2 + r_2^2 - 2\zeta r_1 r_2 \cos(\theta_2 - (\theta_1 + \phi))}{2\mu\left(1 - \zeta^2\right)}\right]$$

(B.1)

$$p(\theta_1, \theta_2) = \int_0^\infty \int_0^\infty \frac{r_1 r_2}{4\pi^2 \mu^2 \left(1 - \zeta^2\right)} exp\left[-\frac{r_1^2}{2\mu\left(1 - \zeta^2\right)}\right] exp\left[-\frac{r_2^2 - 2r_1 r_2 C}{2\mu\left(1 - \zeta^2\right)}\right] dr_1\, dr_2$$

where $C - \zeta \cos(\theta_2 - \theta_1 - \phi)$ Using standard result for

$$\int_0^\infty r_2 exp\left(-\frac{r_2^2 - 2r_1 r_2 C}{2\mu 1 - \zeta^2}\right) dr_2$$

(B.2)

we have

$$\int_0^\infty r_2 exp\left(-a^2 r_2^2 + br_2\right) dr_2 = \frac{1}{2}\left[\frac{\sqrt{\pi}b}{2a} exp\left(-\frac{b^2}{4a^2}\right) erf\left(-\frac{b}{2a}\right) + 1\right]$$

(B.3)

For $a^2 = \dfrac{1}{2\mu\left(1 - \zeta^2\right)}; b = \dfrac{r_1 C}{\mu\left(1 - \zeta^2\right)}$, we have

$$\int_0^\infty r_2 \exp\left(-\frac{r_2^2 - 2r_1 r_2 C}{2\mu 1 - \zeta^2}\right) dr_2 \tag{B.4}$$

$$= \left(\mu\left(1-\zeta^2\right)\right)\left[\frac{\sqrt{2\pi}Cr_1}{2\sqrt{2\mu\left(1-\zeta^2\right)}}\exp\left(-\frac{r_1^2 C}{\mu\left(1-\zeta^2\right)}\right)erf\left(-\frac{r_1 C}{\sqrt{2\mu\left(1-\zeta^2\right)}}\right)+1\right]$$

$$p(\theta_1, \theta_2) = \frac{C}{4\pi^2\left(1-\zeta^2\right)}\left|\tan^{-1}\frac{\sqrt{1-C^2}}{-C} + C\sqrt{\left(1-C^2\right)} + \frac{\sqrt[3]{\left(1-C^2\right)}}{C}\right| \tag{B.5}$$

where $C = \zeta\cos(\theta_2 - \theta_1 - \phi)$ and $0 < \cos^{-1}(-C) < \pi$.

If two signals are transmitted at frequencies which differ by $\Delta\omega$ with a mutual delay of τ seconds, then Equation (B.5) can not be integrated directly. Jakes[B.1] arrived at a result

$$\Re_\theta(\Delta\omega, \tau) = \pi^2\left[1 + \Gamma(\zeta, \phi) + 2\Gamma^2(\zeta, \phi) - \frac{1}{24}\Omega(\zeta)\right] \tag{B.6}$$

where

$$\Gamma(\zeta, \phi) = \frac{1}{2\pi}\mathrm{asin}(\zeta\cos\phi)$$

$$\Omega(\zeta) = \frac{6}{\pi^2}\sum_{n=1}^\infty \frac{\zeta^{2n}}{n^2} \tag{B.7}$$

$$\Omega(1) = 1$$

phases θ_1 and θ_2 are random variables uniformly distributed from 0 to 2π; $p(\theta) = 1/2\pi$, thus $\langle\theta_1\rangle = \langle\theta_2\rangle = \pi$, and $\langle\theta_1^2\rangle = \langle\theta_2^2\rangle = 4\pi^2/3$ and δ is the mean delay.

$$\zeta = \frac{J_0(\omega_m \tau)}{\sqrt{\left(1 + (\Delta\omega)^2\delta^2\right)}} \tag{B.8}$$

$$\phi = \mathrm{atan}(-\Delta\omega\delta)$$

Reference

[B.1] W.C. Jakes, Microwave Mobile Communications, J. Wiley, 1974.

APPENDIX C

RANDOM FM CHARACTERISTICS

The probability density function of the random FM, $p(\theta')$, is obtained by integrating $p(r,\theta,r',\theta')$ over the range of r,r', and θ. $p(X,Y,X,Y)$, the joint probability density in rectangular coordinates can be transformed to circular coordinates.

$$p(x,x',y,y')dxdydx'dy' = q(r,r',\theta,\theta')drdr'd\theta d\theta' \qquad \text{(C.1)}$$

It has been shown (Appendix A.3) that

$$
\begin{aligned}
x &= r\cos\theta \\
y &= r\sin\theta \\
r' &= r'\cos\theta - r\theta'\sin\theta \\
y' &= r'\sin\theta + r\theta'\cos\theta
\end{aligned}
\qquad \text{(C.2)}
$$

$$\langle XY \rangle = \langle XY' \rangle = \langle YX \rangle = \langle YY' \rangle - \langle YX' \rangle - \langle YX' \rangle = 0 \qquad \text{(C.3)}$$

$$p(X,Y,X',Y') = \frac{1}{4\pi^2\sigma^2\upsilon^2}exp\left(-\frac{1}{2}\left(\frac{X^2+Y^2}{\sigma^2}+\frac{X'^2+Y'^2}{\upsilon^2}\right)\right) \qquad \text{(C.4)}$$

where $\upsilon^2 = E[Y'^2] = E[X'^2]$.
From (C.2), we have

$$p(r,\theta,r',\theta') = \frac{r^2}{4\pi^2\upsilon^2\sigma^2}exp\left(-\frac{1}{2}\left(\frac{r^2}{\sigma^2}+\frac{r^2\theta'^2+r'^2}{\upsilon^2}\right)\right) \qquad \text{(C.5)}$$

$$q(r,r') = \int_0^{2\pi}\int_{-\infty}^{\infty} p(r,\theta,r',\theta')d\theta d\theta' = \qquad \text{(C.6)}$$

$$\frac{r}{\sqrt{2\pi\sigma^2\upsilon^2}}exp\left[-\frac{1}{2}\left(\frac{r^2}{\sigma^2}+\frac{r^2\theta'^2+r'^2}{\upsilon^2}\right)\right]$$

$$p(\theta') = \int_0^{2\pi} \int_0^{\infty} \int_{-\infty}^{\infty} \frac{r^2}{4\pi^2 \upsilon^2 \sigma^2} exp\left(-\frac{1}{2}\left(\frac{r^2}{\sigma^2} + \frac{r^2\theta'^2 + r'^2}{\upsilon^2}\right)\right) d\theta\, dr\, dr' \qquad (C.7)$$

$$p(\theta') = \frac{\sigma}{2\upsilon}\left(1 + \frac{\sigma^2}{\upsilon^2}\theta'^2\right)^{-\frac{3}{2}} \qquad (C.8)$$

An expression for the autocorrelation function $\Re_{\theta'}(\tau)$ from Equation (B.6) when $\Delta\omega=0$ or by directly from $<\theta'(t)\theta'(t+\tau)>$ as:

$$\Re_{\theta'}(\tau) = \frac{\omega_m^2}{2J_0^2(u)}\left[\frac{J_0(u)J_1(u)}{u} - J_0^2(u) + J_1^2(u)\right]ln\left[1 - J_0^2(u)\right] \qquad (C.9)$$

where $u=\omega_m\tau$.

APPENDIX D

JTC POWER DELAY PROFILE MODELS

The wideband multipath propagation models were developed to enable effective performance comparison of various air interface technologies. Three tapped delay line models are specified for each of the expected environments. These models are based on complex Gaussian widesense stationary uncorrelated scattering model after Bello [1]. The time variant taps have delays that are multiples of 100 nanoseconds. Doppler spectra are specified to describe the time variation statistics at each delay. The physical environment is divided into indoor and outdoor; the latter is subdivided into pedestrian and vehicular. The classification is given in Table D.1.

Table D.1   *Environmental Classification*

Indoor		Residential
		Office
		Commercial
Outdoors	Pedestrian	Urban High-rise
		Urban/suburban low rise
		Residential
	Vehicular	Urban High-rise
		Urban/suburban low rise
		Residential

The indoor residential environment is typical of a single family home with one or two levels. Interior walls are made with the standard gypsum wall board; exterior walls usually consists of a layer of insulation covered by plywood and then siding or brick. The RMS delay spread is small and the terminal speed is in the range of 1.5m/sec.

The indoor office is a large open plan geometry; individual cubicles have partitions that can be reflective. Metal objectives such as bookcases, desks, and file cabinets are distributed throughout the environment. The delay spread can be moderate and the terminal speed in the range of 1.5m/sec.

The indoor commercial environment is considered to be the large open centers such as large retail stores, shopping malls, factories and airports. These have roof with a wide variety of obstructions. Terminal speeds are in the range of 1.5 m/sec.

The outdoor pedestrian high-rise is characterized by streets lines with tall buildings of several floors. Longer delay spreads are possible, the terminal speed is still in the range of 1.5m/sec.

The outdoor pedestrian urban/suburban low rise environment is made of wide streets with speed limits up to 22.4 m/sec. Building heights are less than three stories high. The terminal speed is still in the range of 1.5m/sec.

The outdoor pedestrian residential environment consists of single and double story houses. Roads are generally two lanes with cars parked on both sides. There can be light to heavy foliage.

The outdoor urban high-rise is the same as outdoor pedestrian urban high-rise with the exception that mobile terminal is much higher. Typical speeds are around 18 m/sec.

The outdoor vehicular urban/suburban low rise is the same as the outdoor pedestrian urban/suburban low-rise with the exception that the terminal speed are around 26.8m/sec.

The outdoor vehicular residential environment is the same as the pedestrian residential but the terminal speeds are around 11m/sec.

D.1 General Channel Model

The JTC recommends that standard discrete WSSUS channel models be used for simulation. The received signal is represented by the sum of delayed replicas of the input signal weighted by independent zero-mean complex Gaussian time variant processes. Specifically if $z(t)$, $w(t)$ denote the complex low pass input and output respectively, then

$$w(t) = \sum_{n=1}^{N} \sqrt{P_n} g_n(t) z(t - \tau_n) \qquad (D.1)$$

where p_n is the strength of the nth weight, and $g(t)$ is the complex Gaussian process weighting the nth replica. The power spectrum of $g_n(t)$, called the Doppler spectrum of the nth path, controls the rate of fading due to the nth path. To completely define this channel model requires only a specification of the Doppler spectra of the tap weights $\{G_n(v); n=1,...N\}$, and the tap weight strengths $\{p_n; n=1,...N\}$.

For outdoor channels, the Doppler spectrum is the classic Clarks model, i.e.

$$P_n(v) = P(v) = \frac{1}{\pi} \frac{1}{\sqrt{\left(\frac{V}{\lambda}\right)^2 - v^2}}; \quad |v| < \frac{V}{\lambda} \qquad (D.2)$$

For indoor channels, the Doppler spectrum is nearly flat and it is given as

$$P_n(v) = P(v) = \frac{\lambda}{2V}; \quad |v| < \frac{V}{\lambda} \tag{D.3}$$

D.2 Wide band Tapped Delay-Line Model

To provide a large variability of Doppler spread within a given environment three multipath channels are defined for each environment. Channel A is the low delay spread case, which occurs frequently, channel B is the medium delay spread case and channel C is the high delay spread case that occurs only very rarely. Table D.2 shows delay spread parameters for the channels A, B, and C.[3]

Table D.2   *Delay Spread Parameters*

Environment	S_A(nS)	S_B(nS)	S_C(nS)
Indoor Residential	20	70	150
Indoor Office	35	100	460
Indoor Commercial	55	150	500
Outdoor urban, high-Rise antenna	100	750	1800
Outdoor Urban/sub: low-rise-Low antenna	100	750	1500
Outdoor Urban-high Rise; high antenna	70	460	850
Outdoor Urban/Sub; high antenna	500	3250	8000
Outdoor urban/sub Low-rise: high antenna	400	4000	12000
Outdoor Residential High Antenna	350	2260	6450

For various categories of the channels defined above, the tap parameters are specified. These models are listed in the following tables [3].

Table D.3   *Indoor Residential tapped delay line parameters[a]*

Tap	Channel A		Channel B		Channel C		
	Rel Delay (nsec)	Avg Power (dB)	Rel Delay (nsec)	Avg Power (dB)	Rel Delay (nsec)	Avg Power (dB)	Doppler Spectrum
1	0	0	0	0	0	0	Flat
2	100	-13.8	100	-6.0	100	0.2	Flat

Table D.3 — *Indoor Residential tapped delay line parameters[a]*

Tap	Channel A Rel Delay (nsec)	Channel A Avg Power (dB)	Channel B Rel Delay (nsec)	Channel B Avg Power (dB)	Channel C Rel Delay (nsec)	Channel C Avg Power (dB)	Doppler Spectrum
3			200	-11.9	200	-5.4	Flat
4			300	-17.9	400	-6.9	Flat
5					500	-24.5	Flat
6					600	-29.7	Flat

a. For simulation modeling a ±3% variation about the relative delay is allowed.

Table D.4 — *Indoor Office tapped delay line parameters[a]*

Tap	Channel A Rel Delay (nsec)	Channel A Avg Power (dB)	Channel B Rel Delay (nsec)	Channel B Avg Power (dB)	Channel C Rel Delay (nsec)	Channel C Avg Power (dB)	Doppler Spectrum
1	0	0	0	0	0	0	Flat
2	100	-8.5	100	-3.6	200	-1.4	Flat
3			200	-7.2	500	-2.4	Flat
4			300	-10.8	700	-.4.8	Flat
5			500	-18.0	1100	-1.0	Flat
6			700	-25.2	2400	-16.3	Flat

a. For simulation modeling a ±3% variation about the relative delay is allowed.

Table D.5 — *Indoor Commercial tapped delay line parameters[a]*

Tap	Channel A Rel Delay (nsec)	Channel A Avg Power (dB)	Channel B Rel Delay (nsec)	Channel B Avg Power (dB)	Channel C Rel Delay (nsec)	Channel C Avg Power (dB)	Doppler Spectrum
1	0	0	0	0	0	0	Flat
2	100	-5.9	100	-0.2	100	-4.9	Flat
3	200	-14.6	200	-5.4	500	-3.8	Flat
4			400	-6.9	700	-1.8	Flat
5			500	-24.5	2100	-21.7	Flat
6			700	-29.7	2700	-11.5	Flat

a. For simulation modeling a ±3% variation about the relative delay is allowed.

Table D.6   *Outdoor Urban High-rise, Low Antenna tapped delay line parameters[a]*

Tap	Channel A		Channel B		Channel C		Doppler Spectrum
	Rel Delay (nsec)	Avg Power (dB)	Rel Delay (nsec)	Avg Power (dB)	Rel Delay (nsec)	Avg Power (dB)	
1	0	0	0	0	0	0	Classic
2	100	-3.6	200	-0.9	500	-2.1	Classic
3	200	-7.2	800	-4.9	800	-12.4	Classic
4	300	-10.8	1200	-8.4	2200	-4.1	Classic
5	500	-18.0	2300	-7.8	7000	-11.1	Classic
6	700	-25.2	3700	-23.9	10000	-19.7	Classic

a. For simulation modeling a ±3% variation about the relative delay is allowed.

Table D.7   *Outdoor Urban/Suburban Low-rise, Low Antenna tapped delay line parameters[a]*

Tap	Channel A		Channel B		Channel C		Doppler Spectrum
	Rel Delay (nsec)	Avg Power (dB)	Rel Delay (nsec)	Avg Power (dB)	Rel Delay (nsec)	Avg Power (dB)	
1	0	0	0	0	0	0	Classic
2	100	-3.6	200	-0.9	500	-0.2	Classic
3	200	-7.2	800	-4.9	800	-1.5	Classic
4	300	-10.8	1200	-8.4	2200	-2.6	Classic
5	500	-18.0	2300	-7.8	7000	-11.8	Classic
6	700	-25.2	3700	-23.9	10000	-6.9	Classic

a. For simulation modeling a ±3% variation about the relative delay is allowed.

Table D.8

Outdoor Residential, Low Antenna tapped delay line parameters[a]

Tap	Channel A Rel Delay (nsec)	Channel A Avg Power (dB)	Channel B Rel Delay (nsec)	Channel B Avg Power (dB)	Channel C Rel Delay (nsec)	Channel C Avg Power (dB)	Doppler Spectrum
1	0	0	0	0	0	0	Classic
2	100	-6.0	200	-1.4	200	-1.3	Classic
3	200	-11.9	500	-2.4	400	-2.7	Classic
4	300	-17.9	700	-4.8	2000	-13.4	Classic
5			1100	-1.0	2900	-6.1	Classic
6			2400	-16.3	4000	-16.8	Classic

a. For simulation modeling a ±3% variation about the relative delay is allowed.

Table D.9

Outdoor Urban High-rise, High Antenna tapped delay line parameters[a]

Tap	Channel A Rel Delay (nsec)	Channel A Avg Power (dB)	Channel B Rel Delay (nsec)	Channel B Avg Power (dB)	Channel C Rel Delay (nsec)	Channel C Avg Power (dB)	Doppler Spectrum
1	0	0	0	0	0	-1.2	Classic
2	200	-4.9	200	-0.3	700	0	Classic
3	500	-3.8	1300	-1.6	4000	-6.7	Classic
4	700	-1.8	6500	-6.1	10000	-3.6	Classic
5	2100	-21.7	13500	-9.3	22000	-4.3	Classic
6	2700	-11.5	19000	-23.6	50000	-19.5	Classic

a. For simulation modeling a ±3% variation about the relative delay is allowed.

Table D.10   *Outdoor Urban Low-rise, High Antenna, tapped delay line parameters^a*

Tap	Channel A		Channel B		Channel C		Doppler Spectrum
	Rel Delay (nsec)	Avg Power (dB)	Rel Delay (nsec)	Avg Power (dB)	Rel Delay (nsec)	Avg Power (dB)	
1	0	-1.6	0	-2.5	0	-4.8	Classic
2	100	-5.1	300	0	300	-0.3	Classic
3	200	0	8900	-12.8	800	-7.4	Classic
4	500	-7.6	12900	-10.0	8000	0	Classic
5	1200	-6.9	17100	-25.2	27000	-6.5	Classic
6	1600	-27.6	20000	-16.0	55000	-9.8	Classic

a. For simulation modeling a ±3% variation about the relative delay is allowed.

Table D.11   *Outdoor Residential, High Antenna, tapped delay line parameters^a*

Tap	Channel A		Channel B		Channel C		Doppler Spectrum
	Rel Delay (nsec)	Avg Power (dB)	Rel Delay (nsec)	Avg Power (dB)	Rel Delay (nsec)	Avg Power (dB)	
1	0	-3.8	0	0	0	-1.2	Classic
2	100	0	200	-0.3	700	-1.6	Classic
3	500	-6.6	1300	-2.3	4000	0	Classic
4	800	-1.2	5200	-0.2	10000	-12.5	Classic
5	1300	-18.4	12000	-20.8	22000	-21.4	Classic
6	1700	-23.7					Classic

a. For simulation modeling a ±3% variation about the relative delay is allowed.

References

[1] P.A. Bello, "Characterization of Randomly Time-Variant Linear Channels," IEEE Trans. on Commu. Systems, vol. CS-11, December 1963, pp. 360-393.

[2] R. H. Clarke, "A statistical Theory of Mobile Reception," BSTJ, vol. 49, pp. 957-1000, 1968.

[3] Technical Report on RF Channel Characterization and System Deployment Modeling, JTC(AIR)/94.09.23-065R6, 1994.

Index